Lecture Notes in Computer Science 3062

Commenced Publication in 1973
Founding and Former Series Editors:
Gerhard Goos, Juris Hartmanis, and Jan van Leeuwen

Springer
Berlin
Heidelberg
New York
Hong Kong
London
Milan
Paris
Tokyo

John L. Pfaltz   Manfred Nagl
Boris Böhlen (Eds.)

Applications of Graph Transformations with Industrial Relevance

Second International Workshop, AGTIVE 2003
Charlottesville, VA, USA, September 27 - October 1, 2003
Revised Selected and Invited Papers

 Springer

Volume Editors

John L. Pfaltz
University of Virginia
School of Engineering and Applied Science, Department of Computer Science
151 Engineer's Way, P.O. Box 400740, Charlottesville, VA 22904-4740, USA
E-mail: jlp@virginia.edu

Manfred Nagl
Boris Böhlen
RWTH Aachen University, Department of Computer Science III
Ahornstr. 55, 52074 Aachen, Germany
E-mail: nagl@i3.informatik.rwth-aachen.de; boehlen@cs.rwth-aachen.de

Library of Congress Control Number: 2004106200

CR Subject Classification (1998): D.2, F.3, F.4.2, E.1, F.2.1, I.2, G.2.2

ISSN 0302-9743
ISBN 3-540-22120-4 Springer-Verlag Berlin Heidelberg New York

Springer-Verlag is a part of Springer Science+Business Media

springeronline.com

© Springer-Verlag Berlin Heidelberg 2004
Printed in Germany

Typesetting: Camera-ready by author, data conversion by DA-TeX Gerd Blumenstein
Printed on acid-free paper   SPIN: 11011156   06/3142   5 4 3 2 1 0

Preface

This volume consists of papers selected from the presentations given at the International Workshop and Symposium on "Applications of Graph Transformation with Industrial Relevance" (AGTIVE 2003). The papers underwent up to two additional reviews. This volume contains the revised versions of these papers.

AGTIVE 2003 was the second event of the Graph Transformation community. The aim of AGTIVE is to unite people from research and industry interested in the application of Graph Transformation to practical problems. The first workshop took place at Kerkrade, The Netherlands. The proceedings appeared as vol. 1779 of Springer-Verlags's Lecture Notes in Computer Science series. This second workshop, AGTIVE 2003, was held in historic Charlottesville, Virginia, USA.

Graphs constitute well-known, well-understood, and frequently used means to depict networks of related items in different application domains. Various types of graph transformation approaches – also called graph grammars or graph rewriting systems – have been proposed to specify, recognize, inspect, modify, and display certain classes of graphs representing structures of different domains.

Research activities based on Graph Transformations (GT for short) constitute a well-established scientific discipline within Computer Science. The international GT research community is quite active and has organized international workshops and the conference ICGT 2002. The proceedings of these events, a three volume handbook on GT, and books on specific approaches as well as big application projects give a good documentation about research in the GT field (see the list at the end of the proceedings).

The intention of all these activities has been (1) to bring together the international community in a viable scientific discussion, (2) to integrate different approaches, and (3) to build a bridge between theory and practice.

More specifically, the International Workshop and Symposium AGTIVE aims at demonstrating that GT approaches are mature enough to influence practice, even in industry. This ambitious goal is encouraged by the fact that the focus of GT research has changed within the last 15 years. Practical topics have gained considerable attention and usable GT implementations are available now. Furthermore, AGTIVE is intended to deliver an actual state-of-the-art report of the applications of GT and, therefore, also of GT implementations and their use for solving practical problems.

The program committee of the International AGTIVE 2003 Workshop and Symposium consisted of the following persons:

Jules Desharnais, Laval University, Quebec, Canada
Hans-Joerg Kreowski, University of Bremen, Germany
Fred (Buck) McMorris, Illinois Institute of Technology, Chicago, USA

Ugo Montanari, University of Pisa, Italy
Manfred Nagl, RWTH Aachen University, Germany (Co-chair)
Francesco Parisi-Presicce, Univ. of Rome, Italy
and George Mason Univ., USA
John L. Pfaltz, University of Virginia, Charlottesville, USA (Co-chair)
Andy Schuerr, Technical University of Darmstadt, Germany
Gabriele Taentzer, Technical University of Berlin, Germany.

The program of the workshop started with a tutorial on GT given by L. Baresi and R. Heckel (not given in the proceedings). The workshop contained 12 sessions of presentations, two of them starting with the invited talks of H. Rising and G. Karsai, respectively. Two demo sessions gave a good survey on different practical GT systems on the one hand and the broad range of GT applications on the other.

At the end of the workshop five participants (G. Taentzer, H. Vangheluwe, B. Westfechtel, M. Minas, A. Rensink) gave a personal summary of their impressions, each of them from a different perspective. In order to enliven the workshop there were two competitions, namely for the best paper and for the best demo presentation, which were won by C. Smith and aequo loco by M. Minas and A. Rensink, respectively. The proceedings contain most of these items.

The workshop was attended by 47 participants from 12 countries, namely Belgium, Brazil, Canada, France, Germany, Italy, Poland, Spain, Sweden, The Netherlands, the UK, and the USA. The success of the workshop is based on the activeness of all participants contributing to presentations and discussions. Furthermore, it is due to the work done by referees and, especially, by the members of the program committee.

A considerable part of the workshop's success was also due to the familiar Southern State atmosphere we witnessed at Charlottesville. Omni Hotel, the workshop conference site, gave us complete support from excellent meals to any kind of technical equipment. On Wednesday afternoon, the main social event was a visit to the homes of Thomas Jefferson (Monticello) and James Monroe (Ash Lawn), followed by the workshop dinner. Jefferson was the 3rd, Monroe the 5th president of the United States. Especially, Thomas Jefferson, also being the founder of the University of Virginia and the author of the Declaration of Independence, had a strong influence on the Charlottesville area.

A more comprehensive report about AGTIVE 2003, written by Dirk Janssens, was published in the "Bulletin of the European Association for Theoretical Computer Science" and in the "Softwaretechnik-Trends" of the German Association of Computer Science.

The workshop was made possible by grants given by the following organizations: Deutsche Forschungsgemeinschaft (the German Research Foundation), the European Union Research Training Network SEGRAVIS, the United States

National Science Foundation, and the Society for Industrial and Applied Mathematics. In particular, the donations have allowed researchers from abroad as well as young scientists to come to Charlottesville by partially financing their travel expenses. Furthermore, the grants covered part of the organizational costs of the workshop.

Last but not least, the editors would like to thank Peggy Reed, Scott Ruffner, and Bodo Kraft for their help in the organization of the workshop.

March 2004   John L. Pfaltz
                                                              Manfred Nagl
                                                              Boris Boehlen

List of Referees

Our thanks go to all those who helped in reviewing the papers:

U. Assmann
R. Banach
S. Becker
M. Bellia
B. Böhlen
P. Bottoni
A. Cisternino
G. Cugola
J. Desharnais
H. Ehrig
C. Ermel
F. Gadducci
F. Gatzemeier
T. Haase
M. Heller
D. Hirsch
K. Hoelscher
B. Hoffmann

P. Inverardi
B. Kraft
R. Klempien-Hinrichs
P. Knirsch
H.-J. Kreowski
S. Kuske
J. de Lara
A. Marburger
O. Meyer
U. Montanari
C. Montangero
M. Nagl
F. Parisi-Presicce
J. Pfaltz
A. Schürr
G. Taentzer
I. Weinhold
B. Westfechtel

Table of Contents

Web Applications

Data Structures and Data Bases

Engineering Applications

Agent-Oriented and Functional Programs, Distribution

Object and Aspect-Oriented Systems

Natural Languages: Processing and Structuring

Re-engineering

Reuse and Integration

Modelling Languages

Bioinformatics

Management of Development and Processes

Multimedia, Picture, and Visual Languages

Demos

Summaries of the Workshop

Graph Transformation
for Merging User Navigation Histories

Mario Michele Gala, Elisa Quintarelli, and Letizia Tanca

Dipartimento di Elettronica e Informazione — Politecnico di Milano
Piazza Leonardo da Vinci, 32 — 20133 Milano, Italy
galam@tiscali.it
{quintare,tanca}@elet.polimi.it

Abstract. Web Mining is a promising research area which mainly studies how to personalize the Web experience for users. In order to achieve this goal it is fundamental to analyze the user navigations to get relevant informations about their behavior. In this work we consider a database approach based on a graphical representation of both Web sites and user interactions. In particular, we will see how to obtain the graph summarizing a set of user interactions from the graphs of single interactions by adopting the graph transformation technique.

Keywords: Semistructured Data, User Navigation History, Web Mining, Graph Transformation.

1 Introduction

In recent years the database research community has concentrated on the introduction of methods for representing and querying sem istructured data. Roughly speaking, this term is used for data that have no absolute schema fixed in advance, and whose structure may be irregular or incomplete [1]. A common example in which semistructured data arise is when data are stored in sources that do not impose a rigid structure, such as the World Wide Web, or when they are extracted from multiple heterogeneous sources. It is evident that an increasing amount of semistructured data is becoming available to users, and thus the need of Web-enabled applications to access, query and process heterogeneous or semistructured information, flexibly dealing with variations in their structure, becomes evident. More recently, interest on semistructured data has been further increased by the success of X M L (eX tensible M arkup Language) as an ubiquitous standard for data representation and exchange[22].

Most available models for semistructured data are based on labeled graphs (see, for example, OEM [20], UnQL [5], and GraphLog [8]), because the formalism of graph supports in a flexible way data structure variability. These models organize data in graphs where nodes denote either objects or values, and edges represent relationships between them.

In the context of semistructured data, proposals presented in the literature for representing temporal information also use labeled graphs [6, 12, 19]. Recently,

J.L. Pfaltz, M. Nagl, and B. Böhlen (Eds.): AGTIVE 2003, LNCS 3062, pp. 1–14, 2004.

it has been recognized and emphasized that time is an important aspect to consider in designing and modeling Web sites [3]: semistructured temporal data models can provide the suitable infrastructure for an effective management of time-varying documents on the Web.

When considering semistructured data, and more in particular Web sites, it is interesting to apply the classical notion of valid time (studied in the past years in the context of relational databases) to the representation of user browsing. With Jensen et al. [15], we regard valid time (VT) of a fact as the time when the fact is true in the modeled reality.

In this work we represent a Web site by means of a semistructured, graph based temporal model called TGM [19]. By browsing through a document (for example a hypermedia representation) each user chooses a particular path in the graph representing the document itself and in this way defines a personalized order between the visited objects. In this context, temporal queries, applied to an appropriate representation of the site, can be used to create site views depending on each user's choices to the end of personalizing data presentation [11].

Monitoring and analyzing how the Web is used is nowadays an active area of research in both the academic and commercial worlds. The Web Usage Mining research field studies patterns of behavior for Web users [21]. Personalizing the Web experience for users is a crucial task of many Web-based applications, for example related to e-commerce or to e-services: in fact, providing dynamic and personalized recommendations based on their profile and not only on general usage behavior is very attractive in many situations.

Some existing systems, such as WebWatcher [16], Letizia [17], WebPersonalizer [18], concentrate on providing Web Site personalization based on usage information. WebWatcher [16] "observes" users as they browse the Web and identifies links that are potentially interesting to them. Letizia [17] is a client side agent that searches for pages similar to the ones already visited. The WebPersonalizer [18] page recommendations are based on clusters of pages found by the server log for a site: the system recommends pages from clusters that most closely match the current session. Basically, these systems analyze the user navigations and propose some kind of related (useful) information. Instead, we have a different approach, based on gathering the navigation information into a (semistructured, graph-based) database that can be queried, for example, with a SQL-like query language.

Mining mainly consists in the analysis of usage patterns through pattern recognition algorithms that run on user navigation information. This information can be represented in different ways, e.g. by means of log files. Here we consider for users' log a graph-based representation using the same temporal model we adopt to represent Web sites.

The novel idea of our proposal is to use a generic graph based data model called TGM [19] to store uniformly represented Web sites and user interactions, and to apply graph transformation as an algorithm to collect in a unique graph the information about the navigation of a group of users. From this new data structure we can directly extract relevant information by using a SQL-like query language commonly adopted to query static or dynamic information stored in databases. Indeed, queries applied on our graph-based structures can be used to find more frequently visited pages and traversed links. These information can be used to optimize usability of a site by rearranging links.

The structure of the paper is as follows: in Section 2 we present the graphical model we use to represent information about sites and user interactions with them. In Section 3 we recall the basic notion of graph transformation, and in Section 4 we apply the graph transformation technique for deriving global information about user navigation activities. Some conclusions and possible lines for future work are sketched in Section 5.

2   A Semistructured Temporal Model for Representing Web Sites and User Navigation Activities

The TGM temporal data model [19] we will use in this work is based on labeled graphs and makes use of temporal elements to store the validity of objects and their relationships. Here we represent a Web site as a sem istructured tem poral graph $G = \langle N, E, \ell \rangle$, where N is the set of nodes, E is the set of edges, and ℓ is the labeling function which assigns to each node or edge its set of labels. In particular, each node has a unique identifier, a name, a type (complex or atomic) and a temporal element (i.e. the union of one or more time intervals). C om plex nodes (depicted as rectangles) represent Web pages, atom ic nodes (depicted as ovals) represent the elementary information contained in the pages. The identifier is depicted as a number in the upper-right corner (it is reported only for complex nodes for better readability). Edges have a name and a temporal element: edges between complex nodes represent navigational links and are labeled "Link", whereas edges between complex nodes and atomic nodes represent containm ent relationships and are labeled "HasProperty". For readability reasons we will omit edge labels in the examples. The tem poral elem ent of each node or edge states its validity time, that is, the validity period of that piece of information in the Web site.

For example, in Figure 1 we show the representation of the Web site of a university. Note that although the Professor nodes are quite similar to each other in the structure, there are some differences: the Name and Office number can be simple strings or can be contained in an object with subobjects listing the Professor Name and the Office number.

The analysis of how users are visiting a site is crucial for optimizing the structure of the Web site or for creating appropriate site views. Thus, before running the mining algorithms one should appropriately preprocess the data (see for example [9]): sequences of page references must be grouped into logical units representing user sessions. A user session is the set of page references made by a user during a single visit to a site and is the smallest "work unit" to consider for grouping interesting page references as proposed in [7, 10].

It is important to note that, given a Web site, modeled by a semistructured temporal graph G, in a specific time instant t a user U can interact with the

4   Mario Michele Gala et al.

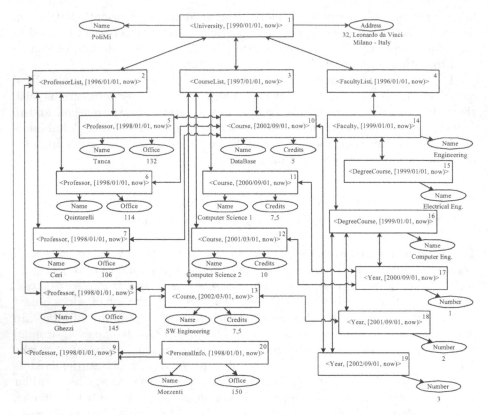

Fig. 1. A Semistructured Temporal Graph representing a Web Site

current view (i.e. the currently valid portion) of the site itself, which is represented by the so-called snapshot of G at time t. For simplicity we assume that a site does not change during user navigation sessions, thus a user interacts only with the current snapshot in a given session.

By means of an interaction, a user defines a path through the snapshot.

A **path** in a semistructured temporal graph $G = \langle N, E, \ell \rangle$ is a non-empty sequence $p = \langle n_0, n_1, \ldots, n_m \rangle$ of nodes s.t.: $\forall i : 0 \leq i \leq m, n_i \in N$ and $\forall i : 0 \leq i < m, \langle n_i, \mathrm{Link}, n_{i+1} \rangle \in E$.

The **interaction** of a user U in a session S with a Web site represented by a semistructured temporal graph G is a pair $\langle p, \mathrm{TT} \rangle$ where:

1. $p = \langle n_1, n_2, \ldots, n_m \rangle$ is a path in the snapshot of G at time t;
2. $\mathrm{TT} : N_p \to \mathcal{V}$ is a labeling function such that:
   N_p is the multiset of the nodes that compose the path p, defined as $N_p = \{\!| n_1, n_2, \ldots, n_m |\!\}$, $\mathcal{V}$ is the set of all possible temporal elements, and $\forall i : 0 \leq i \leq m$, $\mathrm{TT}(n_i) = t_i$, where $t_i \in \mathcal{V}$ is a temporal element representing the "thinking

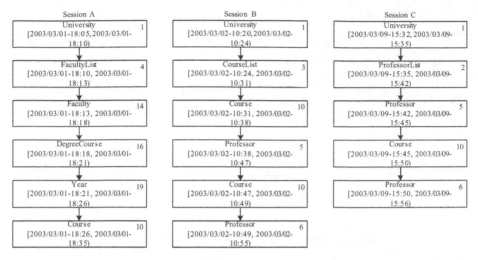

Fig. 2. The Global Interaction of a user

time" of user U on node n_i (that is, the time spent for visiting, or the valid time related to the user browsing activity).

The set of interactions of a user U with a Web site, represented by a semistructured temporal graph G, is called the global interaction of U with G. In Figure 2 we can see an example of global interaction. For readability reasons, in the graphs representing user interactions we will draw only complex nodes, that correspond to visited pages, and not atomic nodes, that are just attributes related to the complex parent object and can be inferred from the Web site graph.

At this point we could examine directly the global interaction, but we may find some problems by analyzing this structure:

- its size grows linearly with the users' page visits (the number of nodes is equal to the total number of visited pages), hence a visiting algorithm may require much computing resources;
- we need to examine it thoroughly to get some information like the visit time of a specific page, because we have to sum the visit times of all the instances of that page in the global interaction;
- our graph is actually a set of lists, thus it doesn't exploit the opportunity of the graphs to have multiple and complex connections between nodes;
- it represents raw data, and it is not intuitively readable for humans.

For these reasons we consider to merge a set of sessions in a unique graph, called merge graph.

The **merge graph** G_A of a global interaction $A = \{I_1, I_2, \ldots, I_k\}$, composed of single interactions $I_j = \langle\langle n_{j1}, n_{j2}, \ldots, n_{jm_j}\rangle, \mathrm{T T}_j\rangle$, for $j \in \{1, 2, \ldots, k\}$,

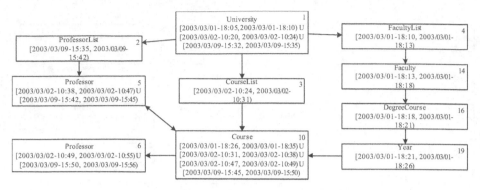

Fig. 3. A Merge graph

on a semistructured temporal graph $G = \langle N, E, \ell \rangle$, is a graph $G_A = \langle N_A, E_A, \text{TotalTT}_A \rangle$, where:

1. $N_A = N_A^{complex} \cup N_A^{atomic}$ where: $N_A^{complex} = \{n_{11}, n_{12}, \ldots, n_{1m_1}\} \cup \{n_{21}, n_{22}, \ldots, n_{2m_2}\} \cup \ldots \cup \{n_{k1}, n_{k2}, \ldots, n_{km_k}\}$ and $N_A^{atomic} = \{m | \ell_T(m) = atomic \wedge \exists n \in N_A^{complex}(\langle n, \text{HasProperty}, m \rangle \in E)\}$;
2. $\forall a, b \in N_A$, $\langle a, \text{Link}, b \rangle \in E_A$ if and only if there is a path in A containing the edge $\langle a, \text{Link}, b \rangle$;
3. $\forall a, b \in N_A$, $\langle a, \text{HasProperty}, b \rangle \in E_A$ if and only if $\langle a, \text{HasProperty}, b \rangle \in E$;
4. $\text{TotalTT}_A : N_A \to \mathcal{V}$ is a labeling function that gives the temporal element related to each node $n \in N_A$, i.e. $\text{TotalTT}_A(n) = \bigcup_{\{n_{jq} \in I_j | n_{jq} = n\}} TT_j(n_{jq})$ (the union of the thinking time intervals of node n in the global interaction A).

Figure 3 contains the merge graph of the global interaction of Figure 2.

The merge graph represents the activities that a user takes in a set of sessions (i.e. visited pages and clicked links). From this graph we can extract some relevant information useful to define the behavior of the user, e.g. the time spent by the user on each page, or the page that is, the last visited one.

The merge graph represents the user navigations in a more compact way than the global interaction, with some advantages:

- even with a large user interaction, the size of the merge graph (in terms of the number of nodes) will be limited by a constant upper bound that corresponds to the size of the Web site. Note that the temporal element associated to nodes may grow hugely, thus to solve this problem we can consider to shrink temporal elements periodically by deleting older information;
- we can get useful information just by a local analysis, for example to get the visit time of one node we just need to examine that node and no others;
- here we have a "real graph", we have a more complex linked structure that may bring more information than a set of lists;

  – it is more intuitively readable for humans, it can be seen as the visited subset
  of the Web site.

The merge graph can be considered a sort of "personal view" of the Web site:
in fact, it represents the portion of the site visited by the user; moreover, each
node in the merge graph has as temporal element (i.e. interaction time) a subset
of the temporal element of the corresponding node in the Web site graph (where
the time notion represents the valid time).

However, reducing our information to such a small data structure has some
drawbacks. The process of merging interactions does not keep all information:
the graph can contain cycles, so it is not possible, in general, to derive neither the
exact sequence of visited pages nor the information about the sessions. Anyway,
the information content of the merge graph is enough for many applications
to get the desired knowledge about user navigations; for the mining activities
that require the missing information, we can think to directly query the original
global interaction.

So far we have considered the interaction of a single particular user, but we
can extend our approach on a more general scenario and consider the interaction
of groups of different users to analyze their behaviors. This extension requires
some further considerations on simultaneous visits. When building the merge
graph for a group of users, we have to take into account that different users could
access the same page at the same time, therefore thinking time intervals may
overlap. Indeed, if we merge them, we would lose the information that the page
is visited more times instead of one: hence, we need to keep the time intervals
separated in order to have the real time spent on a page, and the union of time
intervals will be, as in the case of a single user, just their concatenation (i.e. the
union without merging overlapping intervals). We will mark the concatenation
of time intervals with the symbol $\uplus$.

Note that these considerations hold also in the case of a single user working
in parallel sessions: in fact, if a user could make different simultaneous sessions,
we should keep the information that in a specific instant t a node can be visited
contemporarily through different sessions instead of being visited in just one
session, thus this case can be led to the one related to the group of users.

If we keep depicted the union of time intervals that overlap partially or totally,
we can get a non-tidy notation where some time instants belong to more than
one interval. To avoid this problem we can consider a more powerful structure in
which any time instant can appear at most in one interval: we can rearrange the
temporal elements in a set of non-overlapping simple intervals (that may span
contiguous time periods), each with an associated integer number called weight
representing the number of times all the instants of that interval appear in the
temporal element.

A **weighted temporal element** $\mathrm{WTE} = [s_1, e_1]^{w_1} \uplus [s_2, e_2]^{w_2} \uplus \ldots \uplus$
$[s_n, e_n]^{w_n}$ is the union of weighted intervals $[s_i, e_i]^{w_i}, i \in \{1, 2, \ldots, n\}$ where s_i
is the start time, e_i the end time, and w_i the weight of the interval.

The utility of the merge graph structure (and in general of graph-based structures reporting information about user navigation) arises if we think about the

queries we can apply to it, in order to obtain information useful for Web analysis purpose. As an example, we propose some intuitive SQL-like queries without entering into the details of the query language, defined in [19, 13].

- Find the most visited Degree Course page:
  SELECT DegreeCourse.*HasProperty*.Name
  FROM Merge Graph
  WHERE DegreeCourse->**COUNT** >= **ALL** (
  SELECT DegreeCourse->**COUNT**
FROM Merge Graph)

- Find the courses of Prof. Ghezzi which have been visited from his page:
  SELECT Course.*HasProperty*.Name
  FROM Merge Graph
  WHERE EXISTS Professor.*Link*.Course
  AND Professor.*HasProperty*.Name = "Ghezzi"

- Find the office number of the professors whose pages have been visited in the week of Christmas 2002:
  SELECT Professor.*HasProperty*.Office
  FROM Merge Graph
  WHEN Professor **OVERLAP** [2002/12/23-00:00, 2002/12/30-00:00)

The possibility to directly derive this kind of information from a graph based structure, by means of a simple query language, motivates our interest in applying a powerful and elegant technique, such as graph transformation, in a database context.

3 Basic Notions on Graph Transformation

Graphs are well-acknowledged means to formally represent many kinds of models, e.g. complex nodes, databases, system states, diagrams, architectures. Rules are very useful to describe computations by local transformations, e.g. arithmetic, syntactic and deduction rules. Graph transformations combine the advantages of both graphs and rules [4, 14]; here we will apply them for formalizing the algorithm to manipulate user interaction graphs.

Actually, our use will exploit only a minor potential of graph transformations, and we refer to [2] for some more enhanced applications.

To define the concept of graph transformation, we consider the algebraic, more general, notion of graph (also called multigraph). The TGM model considers edges as in the relational notion of graph (i.e. the edge set is a binary relation on the set of nodes), with the (label) extension which allows one to insert multiple edges between the same two nodes, if these edges have different labels (i.e. the edge set is a ternary relation between the set of nodes and the set of edge labels). Note that in our application of semistructured temporal graph the edge label is determined by the type of the nodes it connects: it is a "Link" for complex→complex relationships and "HasProperty" for complex→atomic relationships. It follows that we do not have parallel edges, but we use the multigraph data structure anyway for convenience of notation: we can refer to the functions $source : E \rightarrow N$ and $target : E \rightarrow N$, which give the source and target node of a given edge, being understood that there cannot be parallel edges.

A graph transformation basically consists of deriving a graph from another one by applying a rule that transforms it by replacing a part with another subgraph.

Very briefly, a graph transformation rule $r = \langle L, R, K, glue, emb, appl \rangle$ consists of: two graphs L and R, respectively called the left-hand side and the right-hand side of r; a subgraph K of L called the interface graph; an occurrence glue of K in R, relating the interface graph with the right-hand side; an embedding relation emb, relating nodes of L to nodes of R; a set $appl$ specifying the application conditions for the rule.

An application of a rule $r = \langle L, R, K, glue, emb, appl \rangle$ to a given graph G produces a resulting graph H, if H can be obtained from G in the following steps: choose an occurrence of L in G; check the application conditions according to $appl$; remove the occurrence of L with the exclusion of the subgraph K from G, including all dangling edges, yielding the context graph D of L (that still contains an occurrence of K); glue D and R according to the occurrences of K in them (that is, construct the disjoint union of D and R and, for every item in K, identify the corresponding item in D with the corresponding item in R), yielding the gluing graph E; embed R into E according to emb: for each removed dangling edge that was connecting a node $v \in G$ and the image of a node $v_1 \in L$ in G, $\forall v_2 \in R$ such that $(v_1, v_2) \in emb$, add a new edge (with the same label) between v and the glued image of v_2 in G.

4 Applying Graph Transformations

In the previous section we have seen the basics of graph transformation, and now we apply this technique to build the merge graph from the global interaction, and also to combine temporal elements.

4.1 Merging User Navigation Histories

For the purpose of transforming the global interaction to get the merge graph, we just need one transformation rule, called R_{merge}, shown in Figure 4. Intuitively, each application of the rule merges two nodes that have the same id and label, unifying their temporal elements and collapsing all their ingoing and outcoming edges to the same node.

In Figure 5 we show the application of the rule in a general case. Figure 6 briefly shows the steps to build the merge graph from the global interaction depicted in Figure 2. For simplicity, a node just contains its identifier originally reported in the Web site graph. The nodes involved in every step are marked with dashed circles and arrows.

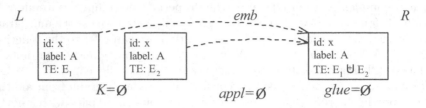

Fig. 4. The transformation rule R_{merge} for creating the merge graph

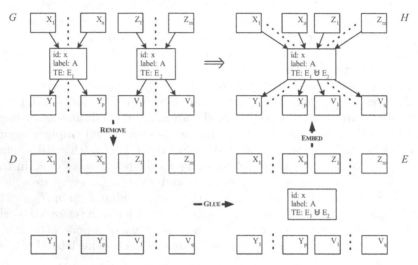

Fig. 5. Rule application in a general case

In graph transformation, non-determinism may occur, and this can happen at two levels:

1. if we have a set of transformation rules, we have to choose one among the applicable rules: in our particular context, this issue will not cause concern because we have only one transformation rule;
2. given the chosen rule, there could be several occurrences of its left-hand side in the graph. This question requires some deep considerations: given a graph, it is possible that we have to take a decision about which occurrence we should first apply the rule to. But this is a trivial problem: in fact, it can be proved that the iterated application of our rule will always converge to the same solution, unambiguously given by the definition of merge graph, independently of the taken decisions.

Proposition 1. Given a global interaction A over a semistructured temporal graph G, applying R_{merge} to A yields a unique merge graph G_A.

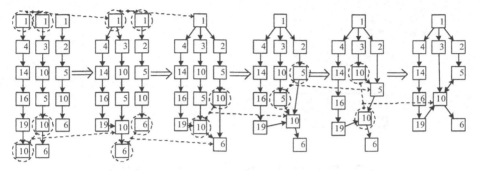

Fig. 6. Examples of rule application

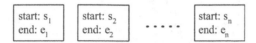

Fig. 7. Graph representation of Temporal Elements

Basically, the proof uses the facts that the merge graph contains all the nodes of the global interaction but without any repetition, and on the commutativity of the union (between time intervals in our context) operator.

We can easily compute how many times we have to apply the rule: given a Web site whose graph representation is composed of nodes from the set $N = \{n_1, n_2, \ldots, n_m\}$, the global interaction defines a function $num\_occ : N \to \mathbb{N}$ that returns the number of occurrences of each node in the interaction itself. Hence the number of rule applications is $\sum_{i=1}^{m} \text{MAX} \left(num\_occ(n_i) - 1, 0 \right)$.

4.2   Union of Temporal Elements

We can apply graph transformation for the purpose of merging temporal elements too. As we previously said, if we unify these time-based structures we lose the information that a node is visited more than once: indeed, we consider the following rules just as a base step to define the ones for weighted temporal elements, as we will see in the next paragraph.

To be able to apply graph transformation, we represent a temporal element with an unconnected graph where each node corresponds to a single time interval and has two attributes, start and end. In Figure 7 we represent a generic temporal element $[s_1, e_1) \uplus [s_2, e_2) \uplus \ldots \uplus [s_n, e_n)$.

In Figure 8 we show the two graph transformation rules that merge temporal elements. The first rule is used to merge two simple intervals such that one is contained in the other (total overlapping); the second rule is applied when there is just a partial overlapping between two intervals. In both cases we replace the nodes with one new node having an interval that covers the original union interval. For this graph transformation the interface K of the left-hand side L is empty and its occurrence *glue* in the right-hand side R is empty as well. The

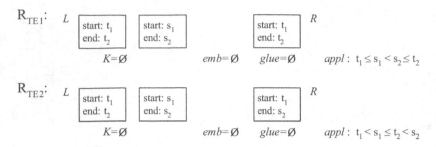

Fig. 8. Transformation rules for Temporal Elements

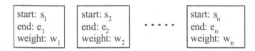

Fig. 9. Graph representation of Weighted Temporal Elements

embedding function is empty, as a consequence that there are no edges, in fact the embedding function has just the purpose of restoring the removed dangling edges.

4.3   Union of Weighted Temporal Elements

We can extend the application of graph transformation to unify weighted temporal elements. To represent them with a graph, we need to add the weight attribute to nodes. In Figure 9 we represent a generic weighted temporal element $W\ TE = [s_1, e_1)^{w_1} \uplus [s_2, e_2)^{w_2} \uplus \ldots \uplus [s_n, e_n)^{w_n}$.

The transformation rules (see Figure 10) have some similarities with the ones of Figure 8, in the sense that we still have two rules that merge total and partial overlapping intervals, respectively. The difference is that here, in the left-hand side, we have two nodes with weights w_1 and w_2 respectively. Thus, the result of the rule application will transform them in nodes, corresponding to time intervals, with three possible different weights: w_1, w_2 and $w_1 + w_2$ for intervals containing respectively time instants in only the first, only the second and both the first and the second time interval of the original graph. In general, the result of a rule application will produce three nodes, but some of them may have the same start and end time: in this case a redundant node can be removed, and the third rule applies to this purpose.

5   Conclusion and Future Work

In this work we used a graphical temporal model for representing time-varying Web sites and user interaction activities while navigating the Web. More particularly we discussed in some detail the possibility to apply graph transformation

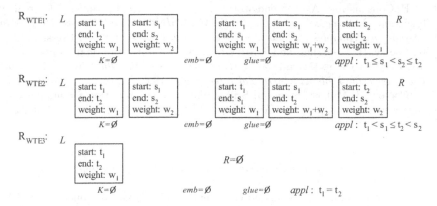

Fig. 10. Transformation rules for Weighted Temporal Elements

in order to obtain a graph-based structure containing a summary about the navigation activities of a user or a group of users.

As a future work, we plan to implement a system, based on this work, to customize the Web experience by using a graph transformation tool.

References

[1] S. Abiteboul. Querying semi-structured data. In *Proceedings of the International Conference on Database Theory*, volume 1186 of *Lecture Notes in Computer Science*, pages 262–275, 1997. 1

[2] M. Andries, G. Engels, A. Habel, B. Hoffmann, H. J. Kreowski, S. Kuske, D. Plump, A. Schurr, and G. Taentzer. Graph transformation for specification and programming. Technical Report 7/96, University of Bremen, 1996. 8

[3] P. Atzeni. Time: A coordinate for web site modelling. In *Advances in Databases and Information Systems, 6th East European Conference, ADBIS 2002*, volume 2435 of *Lecture Notes in Computer Science*, pages 1–7. Springer-Verlag, Berlin, 2002. 2

[4] L. Baresi and R. Heckel. Tutorial introduction to graph transformation: a software engineering perspective. In *Graph Transformation, First International Conference, ICGT 2002*, volume 2505 of *Lecture Notes in Computer Science*, pages 402–429. Springer, 2002. 8

[5] P. Buneman, S. B. Davidson, G. G. Hillebrand, and D. Suciu. A query language and optimization techniques for unstructured data. In *Proceedings of the 1996 ACM SIGMOD International Conference on Management of Data*, pages 505–516. ACM Press, 1996. 1

[6] S. S. Chawathe, S. Abiteboul, and J. Widom. Managing historical semistructured data. *Theory and Practice of Object Systems*, 5(3):143–162, 1999. 1

[7] M. S. Chen, J. S. Park, and P. S. Yu. Data mining for path traversal patterns in a web environment. In *Proceedings of the 16th International Conference on Distributed Computing Systems*, pages 385–392, 1996. 3

[8] M. P. Consens and A. O. Mendelzon. Graphlog: a visual formalism for real life recursion. In *Proceedings of the Ninth ACM SIGACT-SIGMOD-SIGART Symposium on Principles of Database Systems*, pages 404–416. ACM Press, 1990. 1

[9] R. Cooley, B. Mobasher, and J. Srivastava. Web mining: information and pattern discovery on the world wide web. In *Proceedings of the 9th IEEE International Conference on Tools with Artificial Intelligence*, 1997. 3

[10] R. Cooley, B. Mobasher, and J. Srivastava. Data preparation for mining world wide web browsing patterns. *Knowledge and Information System*, 1(1):5–32, 1999. 3

[11] E. Damiani, B. Oliboni, E. Quintarelli, and L. Tanca. Modeling users' navigation history. In *Workshop on Intelligent Techniques for Web Personalisation. In Seventeenth International Joint Conference on Artificial Intelligence*, pages 7–13, 2001. 2

[12] C. E. Dyreson, M. H. Böhlen, and C. S. Jensen. Capturing and querying multiple aspects of semistructured data. In *VLDB'99, Proceedings of 25th International Conference on Very Large Data Bases*, pages 290–301. Morgan Kaufmann, 1999. 1

[13] M. M. Gala. Web log analysis by applying graph transformation on semistructured temporal data. B.Sc. thesis, Politecnico di Milano, 2003. 8

[14] R. Heckel and G. Engels. Graph transformation and visual modeling techniques. *Bulletin of EATCS*, 71:186–202, June 2000. 8

[15] C. S. Jensen, C. E. Dyreson, and M. H. Bohlen et al. The consensus glossary of temporal database concepts - february 1998 version. In *Temporal Databases: Research and Practice. (the book grow out of a Dagstuhl Seminar, June 23-27, 1997)*, volume 1399 of *Lecture Notes in Computer Science*, pages 367–405. Springer, 1998. 2

[16] T. Joachims, D. Freitag, and T. M. Mitchell. Web Watcher: a tour guide for the world wide web. In *Proceedings of the Fifteenth International Joint Conference on Artificial Intelligence, IJCAI 97*, volume 1, pages 770–777. Morgan Kaufmann, 1997. 2

[17] H. Lieberman. Letizia: an agent that assists web browsing. In *Proceedings of the Fourteenth International Joint Conference on Artificial Intelligence, IJCAI 95*, volume 1, pages 924–929. Morgan Kaufmann, 1995. 2

[18] B. Mobasher, R. Cooley, and J. Srivastava. Creating adaptive web sites through usage-based clustering of urls. In *Proceedings of Knowledge and Data Engineering Workshop*, 1999. 2

[19] B. Oliboni, E. Quintarelli, and L. Tanca. Temporal aspects of semistructured data. In *Proceedings of The Eighth International Symposium on Temporal Representation and Reasoning (TIME-01)*, pages 119–127. IEEE Computer Society, 2001. 1, 2, 3, 8

[20] Y. Papakonstantinou, H. Garcia-Molina, and J. Widom. Object exchange across heterogeneous information sources. In *Proceedings of the Eleventh International Conference on Data Engineering*, pages 251–260. IEEE Computer Society, 1995. 1

[21] J. Srivastava, R. Cooley, M. Deshpande, and P. N. Tan. Web usage mining: discovery and applications of usage patterns from web data. *SIGKDD Explorations*, 1(2):12–23, 2000. 2

[22] World Wide Web Consortium. Extensible Markup Language (XML) 1.0, 1998. http://www.w3C.org/TR/REC-xml/. 1

Towards Validation of Session Management in Web Applications based on Graph Transformation

Anilda Qemali and Gabriele Taentzer

Faculty of Electrical Engineering and Computer Science
Technical University of Berlin
Berlin, Germany
{aqemali,gabi}@cs.tu-berlin.de

Abstract. One of the challenges faced by Web developers is how to create a coherent application out of a series of independent Web pages. This problem is a particular concern in Web development because HTTP as underlying protocol is stateless. Each browser request to a Web server is independent, and the server retains no memory of a browser's past requests. To overcome this limitation, application developers require a technique to provide consistent user sessions on the Web. Before implementing a Web application, developers have to decide which session data is to store. In this paper, we provide a modelling approach for powerful and flexible Web session management, based on UML. We propose the definition of a session model which contains version management issues. The validation of a session model concerning consistency issues is possible, due to the formal basis of our approach using graph transformation.

1 Introduction

State management is the process by which you maintain state and page information over multiple requests for the same or different pages. As is true for any HTTP-based technology, Web form pages are stateless, which means that they do not automatically indicate whether the requests in a sequence are all from the same client or even whether a single browser instance is still actively viewing a page or site. Furthermore, pages are destroyed and recreated with each round trip to the server; therefore page information will not exist beyond the life cycle of a single page. There are various client-side and server-side options for state management.

Storing page information using client-side options doesn't use server resources. However, because you must send information to the client for it to be stored, there is a practical limit on how much information you can store this way. Client-side options are URL extensions, hidden fields, and cookies. State information can be added to a URL (Uniform Resouce Locator) as additional path information. By hidden fields, state information may be stored inside the fields of an HTML document. Hidden fields are useful in Web forms, since the

J.L. Pfaltz, M. Nagl, and B. Böhlen (Eds.): AGTIVE 2003, LNCS 3062, pp. 15–29, 2004.
© Springer-Verlag Berlin Heidelberg 2004

value associated with the field is send to the server when the form is submitted. Another very popular approach to do session management is to use cookies.

To design session handling in Web applications adequately, several questions have to be answered. First of all, what is the important session-specific data to be held? Then, which changes have to be recorded? Further questions occur concerning the security level of different kinds of information.

In general, a Web application is well designed following the model-view-controller approach (compare e.g. Struts in the Apache Jakarta Project [1]). The Web pages (client and server pages) build the view on the business model which contains persistent business data and business logic. It is e.g. capsuled in an Enterprise Java Bean on the server site. A dispatcher in form of a servlet controls the input and leads it to the business model. For a first design of the session management, we propose a session model which contains structural as well as behavioural aspects of session handling. It contains a representation of the session-specific data together with version management aspects to keep track of all important state changes in sessions.

Several approaches to UML-based ([14]) Web engineering exist which are sketched and compared in [8]. To describe the structural aspects of Web pages we use the Web Application Extension (WAE) [3] to UML. Besides the structural aspects, also the workflow of the guided input processing has to be modelled. This is done by state diagrams containing different kinds of actions, e.g. the user can follow a link, submit a form, etc. A comprehensive approach is the UML Web Engineering approach (UWE) [9]. Here, UML is used to design a Web application on different layers. Especially the runtime layer deals with sessions and provides a history functionality for all activities performed by the user. The user can browse through instances, modifiy them or is just inactive. Our approach can be seen in close relation with UWE where sessions are modelled in a way that the revision structure of session objects is explicitly handled. It is possible to navigate within an hierarchical revision structure and to compose complete previous session states from there.

Having an executable session model at hand, the consistency of session management can be tested already at design time. But the testing of a session model is confronted with some principal problems: Sessions are needed for complex transactions in Web applications which can incorporate a number of dependent Web pages. The user has the possibility to arbitrarily jump forward and backward on such dependent Web pages by navigation facilities. This behaviour leads to an explosion of test cases. Furthermore, it is hard to attempt all possibilities of entries in a given form submission which might lead to different follow-up pages.

Due to this testing dilemma, we propose a session model validation concerning certain consistency constraints. Such a semantic consistency checking is possible if we translate the session model (given as UML model) into the semantic domain of graph transformation [12] where constraint checking facilities are available. Considering session states as graphs and state changes as graph transformations leads to a formal session model. Consistency constraints can be

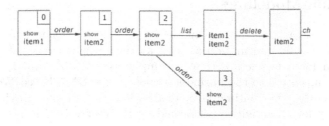

Fig. 1. On-line shopping scenario

formulated in OCL [15]. In this paper, we concentrate on the formulation of (a restricted form of) invariants. They can be used to express that certain safety conditions hold during a session. For constraint checking of the session model, the OCL constraints are translated into a set of graph formulae. Then, the formal model has to be validated by doing consistency checking on the initial state graph and showing that all rules preserve the consistency.

1.1 Running Example

Our running example is taken out of the area of e-commerce. An on-line customer can browse through the products presented in a catalog. When the customer finds an item to buy, it can be added to a shopping cart. Before purchasing the items they are listed. In that situation the customer still has the chance to change the list which means deleting items or adding new ones. No matter how the item list changes, the shopping cart has always to show the actual number of items as well as the right overall sum.

Let's consider the concrete scenario in Figure 1: The customer selects one item and then another one. The shopping cart shows that the number of items is equal to 2. Listing the items the customer deletes the first one. Choosing the second one for changing it, the cart is shown with a number of items equal to 1. Jumping back three pages, item 2 is shown together with the cart containing two items. Buying another copy of item 2 it is not clear how many items are altogether in the cart. Since the customer jumped back to a past state and continued that, the number of items is now 3. But there is also e-commerce software which would take into account that the customer deleted one item in between and would set the number of items to 2. In this case, buying an item would not increase the number of items. A behaviour which cannot be followed easily. But even when a number of items equal to 3 is depicted, it happens that the corresponding item list contains only two items. Thus we are running into a consistency problem: The number of items in the shopping cart does not always correspond with number of items selected.

In the following, we develop a session model for this example where we have the possibility to jump back to complete session states such that inconsistencies are avoided.

2   Session Modelling

In a general setting, the session model is part of the business model. It keeps the session specific information as long as the session persists. Each time the information for a business action is completely available and in a consistent form it is transferred to the business model. In the UML-based Web Engineering approach [9], the runtime layer deals with sessions and provides a history functionality for all activities performed by the user within a session. The user can browse through instances, modifiy them or is just inactive. Components such as pages are stored and have instantiations. Mapping instantiations to session objects and components to session classes, our session meta model could extend the UWE approach by versioning facilities. Completing the description by collaborations concerning typical session operations such as opening a session, creating a session object, editing or deleting it, etc. leads then to an executable model.

2.1   Session Modeling with Versioning Facilities

The session model keeps all user input data and user requested information which is essential for the running session. We use UML class diagrams to describe the structural aspects of the session model. session management, we consider a slight UML extension for the revision management in session models Figure 2 in the following: We consider a slight UML extension for the revision management in session models Figure 2 in the following:

- We consider a stereotype of objects, called session objects, which have a new tagged value for objects: isCurrent. It indicates the current state of a session object.
- Session objects of one type are collected in a session class, a stereotype of classes.
- There is a new dependency between session objects: <<revision>> which is a stereotype of dependency relations. This dependency is ordered to keep track of the order of revisions on one and the same session object.

In [9], a number of consistency constraints are given for sessions formulated in OCL. Here, we add two further ones focussing on the revision structure of session objects:

1. Each session object is the revision of at most one other session object.
   (1) `context SessionObject inv: self.history.origin.size() <= 1`
2. Exactly one session object in one revision tree is current.
   To formulate this constraint by OCL, we need the following additional operations:
   `allRevisions: Set(SessionObject);`
   `allRevisions = self.revision.derived→`
   `    union(self.revision.derived.allRevisions);`

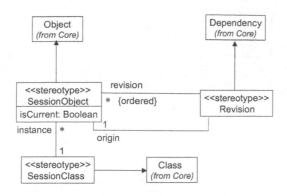

Fig. 2. Metamodel extension for session management

```
allOrigins: Set(SessionObject);
allOrigins = self.history.origin→
    union(self.history.origin.allOrigins);
```

(2) context SessionObject inv: ((self.isCurrent = true) xor
    self.allRevisions->select(isCurrent = true)→size = 1) xor
    self.allOrigins->select(isCurrent = true)→size = 1

The behavioural aspects of sessions are described by special collaborations which show the typical flow of activities. The activities are method calls. Here, we do not concentrate on the contents of the methods is simple, but only state important constraints which have to be valid after a method call. Actions which are important for the session such as user input actions or input dependent business computations (intermediate input validation or configuration settings) would cause revisions of corresponding session objects. Session actions such as backward and forward jumping do not cause new revisions but reset the current session state.

Example The session model has to keep track of all items in the shopping cart. Each time the user puts a new item into the cart, changes it or deletes it from the item list, this action has to be reflected in the session model. Thus, we have to keep track of the item list, all the items in the list and the cart. This results in three session classes in the class diagram of the session model in Figure 3. The session model is connected with the Web page model by server page `EditList` which delegates user actions changing an item or the item list, to the item list.

One specific session state is depicted in the object diagram in Figure 4. It shows the state of a session where the user has put one item into the cart and jumped back to the initial page showing e.g. a special offer. Performing then

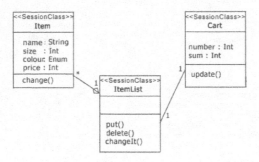

Fig. 3. Session model Class diagram

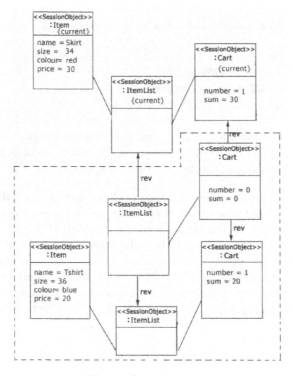

Fig. 4. A session state

another put action leads to a new revision branch. The framed part shows the session state before the last put action took place.

Putting an item into the cart, changing it or deleting it from the item list are user actions which cause new revisions in the session model. We model two of them by the collaborations in Figures 5 and 6. These are collaborations on

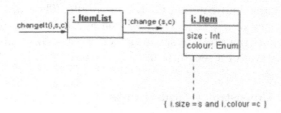

Fig. 5. Collaboration for changing an item

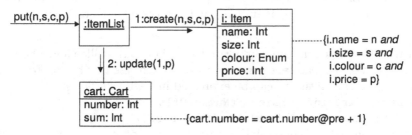

Fig. 6. Collaboration for putting an item into the list

the instance level and show the method flow within the session model. Main constraints are described using OCL.

3   Session Validation

After building up a session model we present how to use it to validate consistency properties. The validation of semantical consistency conditions is possible if we translate the session model which is given as UML model, into a semantic domain. We choose graph transformation as semantic domain, since there is the possibility to not only test consistency constraints but also validate the complete behaviour of the session model. Considering session states as graphs and state changes as graph transformations leads to a formal session model. State changes are caused by actions which are formalized to graph rules.

Besides syntactic consistency constraints given in the previous section, also semantic constraints can be formulated by OCL. Syntactic constraints are formulated for abstract syntax graphs being instances of the UML meta-model, while semantic constraints are formulated on one semantic model which is a graph transformation system in our case. Here, a state is modelled by a graph and all graphs have to satisfy the semantic constraints.

In general, checking of OCL constraints can mean twice: (1) to test if concrete instances (abstract syntax graphs or states) are consistent or (2) to validate if the set of all instances (abstract syntax graphs or states) is consistent.

The USE tool [11] and the Dresden OCL Toolkit [6] provide concepts and tool support to parse and compile OCL constraints. They are useful to check syntactic consistency of a UML model as well as to test semantic consistency. They do not support the proof of semantic consistency constraints, i.e. validate that all possible states of a semantic model are consistent.

3.1 Semantic Consistency Constraints

Similar to syntactic constraints also semantic consistency constraints can be formulated by OCL. In this section, we concentrate on the formulation of invariants within the session model and show sample invariants important for our running example. They describe the main two safety constraints in sessions.

Example One if not the most important invariant is the following: The number of items in the shopping cart has to correspond with the number of items in the selection list. This invariant can be expressed in OCL as follows:

```
context Cart inv consistentNumberOfItems:
  self.number = self.itemList.item→size()
```

The second important constraint states: The sum of items' costs in the shopping cart is the sum of items' prices in the list. The OCL formulation of this constraints is:

```
context Cart inv consistentSum:
  self.sum = self.itemList.item.price→sum()
```

3.2 A Semantic Model

Graphs Graphs are often used as abstract representation of diagrams, e.g. of UML diagrams. In the following, we consider typed attributed graphs. The manipulation of graphs is performed by the so-called double-pushout approach to graph transformation [4]. It was extended to the attributed case in [13].

In object-oriented modelling, the structural aspects can be described on two levels: the type level (modelled by class diagrams) and the instance level (modelled by object diagrams). Semantically, this coherence is mapped to typed graphs where a fixed type graph T serves as abstract representation of the class diagram. Its instances are mapped to graphs equipped with a structure-preserving mapping to the type graph, formally expressed by a graph homomorphism.

Example Translating the UML model of our running example to a graph transformation model, we first have to construct the type graph which is depicted in Figure 7. It contains a graph representation of the class diagram in the session model extended by revision edges and attributes isCurrent for all vertices which originate from the session model. These edges and attributes are the semantical representation of session classes as containers for session objects. Session states

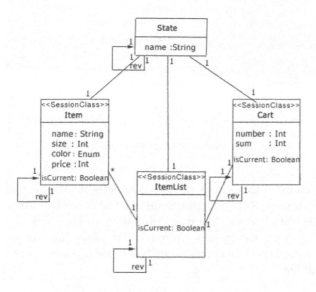

Fig. 7. The type graph

as the ones depicted in Figure 4 by object diagrams can be mapped straight-forward to instance graphs. If tagged value isCurrent is shown in a session object, the corresponding attribute has value true, otherwise it has value false. It is obvious that such instance graphs are typed over the type graph in Figure 7.

Graph Rules After having defined session states as typed attributed graphs, actions are formalized by graph rules describing how session states may change. Identities of vertices are represented by dashed arrows. Identities of edges are deduced from the vertices they are connecting and their types. We use the double-pushout approach [4] which is type graph compatible. Furthermore, multi-objects are needed. Rules with multi-objects represent rule schemes which expand to a countably infinite set of graph transformation rules, one for each legal multiplicity of the multi-object. When applying such a rule to a given graph, always the maximal rule is chosen among all applicable ones.

Example Actions changeIt and put described in collaborations in Figures 5 and 6 are translated to rules in Figures 8 and 9. The rule name and parameter list correspond to those of the corresponding method. Rule put produces new revisions of the item list and the cart. The resulting item list contains all previous items with the new one in addition. The previous items are depicted by a multi-

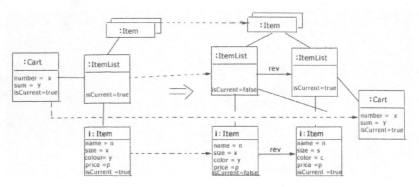

Fig. 8. Rule *changeIt(in i:Item, s:Int, c:Enum)*

object which is mapped to all item vertices connected to the current list vertex when the rule is applied. Rule changeIt changes attributes of an item which leads to a new revision of that item. The rule also has to create a new revision of the item list, although only the item is changed. But assuming that several items are changed one after the other, this order has to be recorded which is done by revising the item list.

Graph Transformation A graph transformation from a pre-state graph G to a post-state graph H describes changes on a concrete session state. The structural and behavioural aspects of a session model can be formally represented by a typed graph transformation system $GTS = (T, I, R)$ consisting of a type graph T, an initial graph I which is an instance graph typed over T, and a possibly infinite set R of graph rules with all left and right-hand sides typed over T. Infinite sets of rules are necessary, because multi-objects can occur and rules with multi-objects represent rule schemes.

Example Considering the session state in Figure 4 represented as graph, an application of rule put to the framed part would lead to the whole graph representating the follow-up session state.

The whole formal session model for our example consists of the type graph in Figure 7, the initial graph in Figure 10 showing the initial state of a session and a set of rules. This set comprises rule changeIt and put as depicted in Figures 8 and 9 as well as a rule delete not explicitly shown. These are the action rules in our formal session model. Moreover, it contains rules for forward and backward jumping which do not insert new revisions, but just change the isCurrent attribute along the revision structure. Backward jumping is specified by rules jumpBack1 and jumpback2 in Figures 11 and 12. Rule jumpBack1 is used when the previous action was a put or delete. In this case, the cart as well as the item list have been revised, but not the listed items. Rule jumpBack2 is used if the previous action was a change of an item. In this case, this item and the list, but not the cart have been revised. Rules for forward jumping look similar.

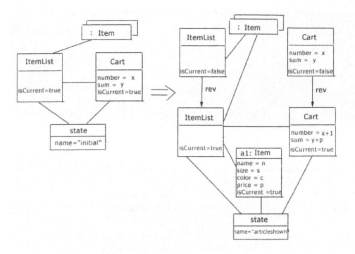

Fig. 9. Rule *put(in n:String, s:Int, c:Enum, p:Int)*

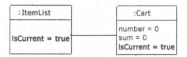

Fig. 10. The initial state

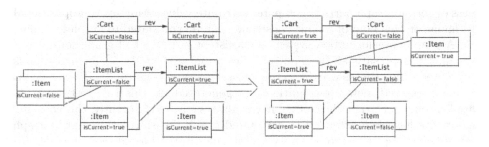

Fig. 11. Rule *jumpBack1*

3.3 Consistency Checking

In section 3.1, we showed how semantic session consistency can be stated by invariants in OCL. To validate a given session model, we have to translate these OCL invariant to a set of graph constraints.

In a first approach, this is possible, if the set of OCL is restricted in a way that the existence of certain object structures and attribute values is required or prohibited. This kind of atomic constraints can be directly translated to graph constraints which describes the existence or non-existence as well as the unique-

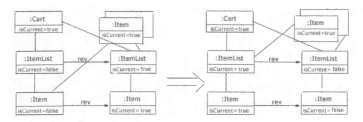

Fig. 12. Rule *jumpBack2*

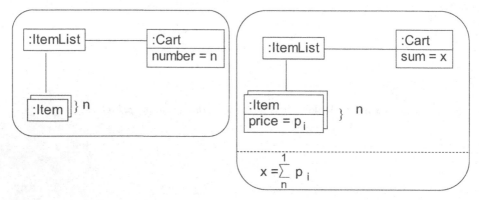

Fig. 13. Constraint *consistentNumberOfItems* and *consistentSum*

ness of certain graph parts or compares attribute values. Moreover, propositional formulae on this kind of atomic constraints can be translated to graph formulae.

A graph transformation system is consistent wrt. a set of graph formulae if the initial graph satisfies them and all rules preserve them. In [7], an algorithm is presented which checks whether a rule preserves graph constraints. If a constraint is not preserved, new pre-conditions are generated for this rule restricting its applicability such that the consistency is always ensured. Recently, this checking algorithm has been extended to attributed graph transformation and to graph formulae in [10]. It has to be applied to all rules in the formal session model. To validate a session model, the initial session state graph has to satisfy all graph constraints. Thereafter, we validate the action rules.

Often multi-objects are also useful when formulating graph constraints. If multi-objects occur in graph constraints, we do not just look for existence or non-existence of the corresponding graph structures, but look for its maximal occurrence similarly to the matching of a rule with multi-objects. Thus, having one multi-object in the constraint, we end up with set $\{c_i | i \in I\}$ of constraints where c_n contains n copies of the multi-object.

Example OCL constraints consistentNumberOfItems and consistentSum are translated to graph constraints. In Figure 13 these graph constraints are de-

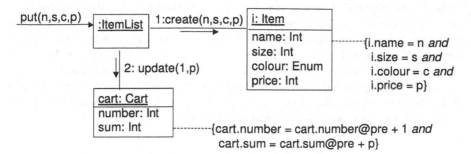

Fig. 14. Completed collaboration *put*

picted using both multi-objects. It is obvious that the initial state in Figure 10 fulfills these constraints. Checking the rules, we find out that rule changeIt fulfills both constraints, while rule put satisfies constraint consistentNumberOfItems but not constraint consistentSum. The algorithm would equip rule put with a precondition stating that $y = y + p$ which means that the rule can only be applied if $p = 0$. The problem lies in not updating attribute sum of the Cart vertex. Comparing rule put with its corresponding collaboration in Figure 6, we find out that constraint c.sum = c.sum@pre + p is missing as post-condition of applying method update. Compare the completed version of this collaboration in Figure 14.

Tool Support for Validation The validation of graph constraints in a formal graph transformation model is supported by the graph transformation engine AGG (see [5]). The tool provides several visual editors to view and edit graph transformation systems, an interpreter to simulate concrete scenarios, and a debugger. Recently, an initiative has been started to implement static analysis techniques for graph transformation such as consistency checking and critical pair analysis to determine conflicts and dependencies between different actions. Graph formulae as described above are supported, except of the usage of multi-objects within graph constraints. It is up to future work, to extend constraint checking for this case. To use AGG for session validation, a UML model which has been produced by some CASE tool has to be translated to the AGG input format such that the corresponding sentic model is constructed. Assuming that the UML model is given in XMI (XML Meta data Interchange) format [2], an XSL (Extensible Stylesheet Language) transformation has to be provided to produce a GGX (Graph Grammar Exchange) document as input for AGG. If the validation results in transformed rules, they have to be retranslated to XMI.

4   Conclusion

Session management is a major issue in the design of Web applications. In this paper, we presented a UML-based approach to session modelling which supports the description of session-specific data together with version management aspects. Future work on this approach has to further investigate the integration of this work into the comprehensive UML-based Web Engineering approach by Nora Koch [9]. Translating the UML model of a session to a formal model based on graph transformation, the semantic consistency of a session model can be validated. Having formulated consistency conditions within a restricted form of OCL, they can be translated to graph formulae which are then used to validate the formal session model. As pointed out this validation process leads to a formulation of important session-specific conditions in collaborations. Although there are a number of approaches to session management at the design level, our approach adds explicit version management issues and offers the possibility to validate semantic consistency constraints for session management in Web applications (based on UML).

Consistent session design is not only a key issue for the development of Web applications, but plays an important role when designing custom wizards in any application. Although presented within the setting of Web applications, our approach seems to be general enough to design consistent session management in any application. It is up to future work to show the general usability of the session modelling and validation approach we have presented.

References

[1] *The Apache Jakarta Projekt - Struts* http://jakarta.apache.org/struts. 16
[2] *XML Metadata Interchange* http://www.omg.org/technology/documents/formal/xmi.htm. 27
[3] J. Conallen. *Building Web Applications with UML*. Addison-Wesley, 2000. 16
[4] A. Corradini, U. Montanari, F. Rossi, H. Ehrig, R. Heckel, and M. Löwe. Algebraic approaches to graph transformation part I: Basic concepts and double pushout approach. In G. Rozenberg, editor, *Handbook of Graph Grammars and Computing by Graph transformation, Volume 1: Foundations*, pages 163–246. World Scientific, 1997. 22, 23
[5] C. Ermel, M. Rudolf, and G. Taentzer. The AGG-Approach: Language and Tool Environment. In H. Ehrig, G. Engels, H.-J. Kreowski, and G. Rozenberg, editors, *Handbook of Graph Grammars and Computing by Graph Transformation, volume 2: Applications, Languages and Tools*, pages 551–603. World Scientific, 1999. available at: http://tfs.cs.tu-berlin.de/agg. 27
[6] Frank Finger. Design and implementation of a modular ocl compiler. Master's thesis, Dresden University, Germany, 2000. 22
[7] R. Heckel and A. Wagner. Ensuring Consistency of Conditional Graph Grammars – A constructive Approach. *Proc. of SEGRAGRA'95 "Graph Rewriting and Computation"*, *Electronic Notes of TCS*, 2, 1995. http://www.elsevier.nl/locate/entcs/volume2.html. 26
[8] N. Koch and A. Kraus. The expressive power of uml-based web engineering. In O. Pastor, G. Rossi, and L. Olsina, editors, *Second International Workshop on Web-oriented Software Technology (IWWOST02)*, 2002. 16
[9] Nora Koch. *Software Engineering for Hypermedia Systems*. PhD thesis, LMU München, Germany, 2001. 16, 18, 28

[10] M. Matz. Konzeption und Implementierung eines Konsistenznachweisverfahrens für attributierte Graphtransformation. Master's thesis, TU Berlin, Fak. IV, 2002. 26

[11] M. Richters. *A precise Approach to Validating UML Models and OCL Constraints.* Monographs of the Bremen Institute of Safe Systems. University of Bremen, 2002. thesis at the university of Bremen. 22

[12] G. Rozenberg, editor. *Handbook of Graph Grammars and Computing by Graph Transformations, Volume 1: Foundations.* World Scientific, 1997. 16

[13] G. Taentzer, I. Fischer, M Koch, and V. Volle. Visual Design of Distributed Systems by Graph Transformation. In H. Ehrig, H.-J. Kreowski, U. Montanari, and G. Rozenberg, editors, *Handbook of Graph Grammars and Computing by Graph Transformation, Volume 3: Concurrency, Parallelism, and Distribution,* pages 269–340. World Scientific, 1999. 22

[14] *Unified Modeling Language – version 1.5.* Available at http://www.omg.org/uml. 16

[15] J. Warmer and A. Kleppe. *The Object Constraint Language: Precise Modeling with UML.* Addison-Wesley, 1998. 17

Specifying Pointer Structures by Graph Reduction*

Adam Bakewell, Detlef Plump, and Colin Runciman

Department of Computer Science, University of York, York YO10 5DD, UK
{ajb,det,colin}@cs.york.ac.uk

Abstract. Graph reduction specifications (GRSs) are a powerful new method for specifying classes of pointer data structures (shapes). They cover important shapes, like various forms of balanced trees, that cannot be handled by existing methods.

This paper formally defines GRSs as graph reduction systems with a signature restriction and an accepting graph. We are mainly interested in PGRSs — polynomially-terminating GRSs whose graph languages are closed under reduction and have a polynomial membership test.

We investigate the power of the PGRS framework by presenting example specifications and by considering its language closure properties: PGRS languages are closed under intersection; not closed under union (unless we drop the closedness restriction and exclude languages with the empty graph); and not closed under complement.

Our practical investigation presents example PGRSs including cyclic lists, trees, balanced trees and red-black trees. In each case we try to make the PGRS as simple as possible where simpler means fewer rules, simpler termination and closure proofs and fewer non-terminals. We show how to prove the correctness of a PGRS and give methods for demonstrating that a given shape cannot be specified by a PGRS with certain simplicity properties.

1 Introduction

Pointer manipulation is notoriously dangerous in languages like C where there is nothing to prevent: the creation and dereferencing of dangling pointers; the dereferencing of nil pointers or structural changes that break the assumptions of a program, such as turning a list into a cycle.

Our goal is to improve the safety of pointer programs by providing (1) means for programmers to specify pointer data structure shapes, and (2) algorithms to check statically whether programs preserve the specified shapes. We approach these aims as follows.

1. Develop a formal notation for specifying shapes (languages of pointer data structures); that is the main concern of this paper. We show how shapes can be defined by graph reduction specifications (GRSs), which are the dual of graph grammars in that graphs in a language are reduced to an accepting graph rather

---

* Work partly funded by EPSRC project *Safe Pointers by Graph Transformation*[1].

J.L. Pfaltz, M. Nagl, and B. Böhlen (Eds.): AGTIVE 2003, LNCS 3062, pp. 30–44, 2004.

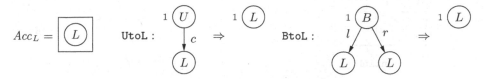

Fig. 1. A graph reduction specification of binary trees

than generated from a start graph. Polynomially terminating GRSs whose languages are closed under reduction (PGRSs) allow a simple and efficient membership test for individual structures, yet seem powerful enough to specify all common data structures.

2. The effect of a pointer algorithm on the shape of a data structure is captured by abstracting the algorithm to a graph rewrite system annotated with the intended structure shape at the start, end and intermediate points if needed. A static verifier then checks the shape annotations (see [3]).

Example 1 (Specifications of binary trees and full binary trees)
Fig. 1 gives a graph reduction specification of binary trees. The smallest binary tree is a leaf. We can draw it as Acc_L, the accepting graph, a single node labelled L. Trees may contain unary or binary branches. Therefore any other binary tree can be reduced to Acc_L by repeatedly applying the reduction rules UtoL and BtoL. These replace bottom-most branches, whose arcs point to leaves, by a leaf. The "1" indicates that any arcs pointing to the branch are left in place by the reduction rule. Full binary trees are specified by omitting the rule UtoL so that each node is either a leaf or a binary branch.

This reduction system only recognises trees because applying the inverse of its rules to any tree always produces a tree. Intuitively, forests cannot reduce to a single leaf as the rules do not break up graphs or connect broken graphs; no rule reduces a cycle; rules are matched injectively so BtoL cannot reduce a DAG with shared sub-trees; our signatures, introduced later, limit node outdegree so branches must be unary or binary.   □

Graph reduction is a very powerful specification mechanism, we show how it can be used to define various kinds of balanced binary trees. Some shapes are harder to specify than others; we categorise shapes according to whether their PGRS needs non-terminal node labels; the difficulty of proving termination and closedness under reduction are also indicative of shape complexity. Some difficult languages can be specified as the union or intersection of simpler languages; we consider how the power of single PGRSs compares with such combinations.

Although many of our examples are trees, a graph-based specification framework is essential because we need precise control over the degree of sharing. Term rewriting ignores this issue and algebraic type specifications are unable to guarantee that members of tree data types are trees. Previous work on shape specifications uses variants of context-free graph grammars, or certain logics,

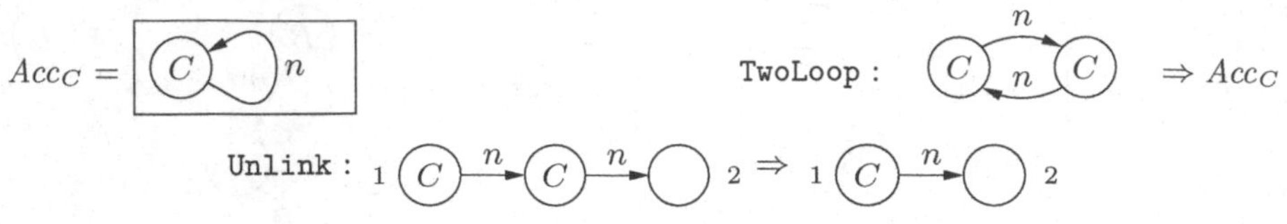

$$Acc_C = \boxed{\;C\;\circlearrowleft\; n\;} \qquad\qquad \mathtt{TwoLoop}: \;\; \overset{n}{C \underset{n}{\rightleftarrows} C} \;\Rightarrow Acc_C$$

$$\mathtt{Unlink}: \;\; {}_1\!\bigcirc\!{C}\xrightarrow{n}\bigcirc\!{C}\xrightarrow{n}\bigcirc\,{}_2 \;\Rightarrow\; {}_1\!\bigcirc\!{C}\xrightarrow{n}\bigcirc\,{}_2$$

Fig. 2. A graph reduction specification of cyclic lists

which are unable to express properties like balance [12, 15, 5, 14, 8]. GRSs are far more powerful than the syntactic type restrictions expressible in languages like AGG, PROGRES and Fujaba. The suitability of general-purpose specification languages like OCL for specifying and checking shapes is unclear. PGRSs can define shapes with sharing and cycles. Our second example presents cyclic lists.

Example 2 (Specification of cyclic lists)
Fig. 2 gives rules defining cyclic lists. A single loop, Acc_C, is a cyclic list and all other cyclic lists reduce to Acc_C. Two-link cycles are reduced by `TwoLoop`. Longer cycles are reduced a link at a time by `Unlink`.

Clearly a graph of several disjoint cycles will not reduce to a single loop; no rules reduce branching or merging structures, and acyclic chains cannot become loops.   □

The rest of this paper is organised as follows. Section 2 defines GRSs. Section 3 discusses polynomial GRSs (PGRSs) and their complexity for shape checking. Section 4 discusses power, showing when shapes are undefinable without non-terminals and demonstrating the closure properties of PGRS languages. Section 5 applies our theory to specify red-black trees. Section 6 discusses related work. Section 7 concludes. Proofs are omitted from this paper, they are given in the full technical report [2].

2   Graph Reduction Specifications

This section describes our framework for specifying graph languages by reduction systems. We define graphs, rules and derivations as in the double-pushout approach [10], and add a signature restriction to ensure that graphs are models of data structures and that rules preserve the restriction. The running example builds a specification of balanced binary trees (BBTs) — binary trees in which all paths from the root to a leaf have the same length.

Definition 1 (Signature)
A signature $\Sigma = \langle \mathscr{C}_V, \mathscr{C}_N, \mathscr{C}_E, \mathtt{type} : \mathscr{C}_V \to \wp(\mathscr{C}_E) \rangle$ consists of a finite set of vertex labels $\mathscr{C}_V$, a set of non-terminal vertex labels $\mathscr{C}_N$ such that $\mathscr{C}_N \subseteq \mathscr{C}_V$, a finite set of edge labels $\mathscr{C}_E$ and a total function $\mathtt{type}$ assigning a set of edge labels to each vertex label.   □

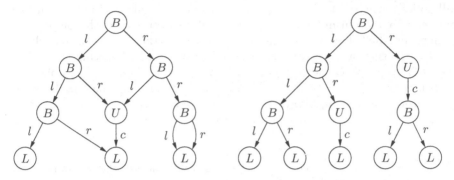

Fig. 3. Two Σ_{BT}-total graphs. The right one is a BBT, the left one is not

Intuitively, graph vertices represent tagged records. Their labels are the tags. Outgoing edges represent the record pointer fields of which each tag has a fixed selection defined by type. Edge labels in $\mathscr{C}_E$ correspond to the names of pointer fields. Non-terminal labels may occur in intermediate graphs during reduction but not in any graph representing a pointer structure. There is no need to restrict the permissible target node labels for edges in the signature because the reduction rules introduced below can encode any such restrictions. In the following, Σ always denotes an arbitrary but fixed signature $\langle \mathscr{C}_V, \mathscr{C}_N, \mathscr{C}_E, \text{type}\rangle$.

Example 3 (Binary tree signature)
Let $\Sigma_{BT} = \langle\{B, U, L\}, \{\}, \{l, r, c\}, \{B \mapsto \{l, r\}, U \mapsto \{c\}, L \mapsto \{\}\}\rangle$. Tree nodes are labelled B(inary branch), U(nary branch) or L(eaf). There are no non-terminals. Arcs are labelled l(eft), r(ight) or c(hild). Binary branches have left and right outgoing arcs, unary branches have a child and leaves have no arcs. □

Definitions 2, 3 and 4 below are consistent with the double-pushout approach to defining labelled graphs, morphisms, rules and derivations (see [10]; [11] considers graph relabelling). Fig. 3 shows two example graphs over Σ_{BT}.

Definition 2 (Graph)
A graph over Σ, $G = \langle V_G, E_G, s_G, t_G, l_G, m_G\rangle$ consists of: a finite set of vertices V_G; a finite set of edges E_G; total functions $s_G, t_G : E_G \to V_G$ assigning a source and target vertex to each edge; a partial node labelling function $l_G : V_G \to \mathscr{C}_V$ (a partial function $f : A \to B$ maps $\text{dom} f$, a subset of A, to B. We write $f(x) = \bot$ when $x \notin \text{dom} f$); and a total edge labelling function $m_G : E_G \to \mathscr{C}_E$. □

Definition 3 (Morphism, inclusion and rule)
A graph morphism $g : G \to H$ consists of a node mapping $g_V : V_G \to V_H$ and an edge mapping $g_E : E_G \to E_H$ that preserve sources, targets and labels: $s_H \circ g_E = g_V \circ s_G$, $t_H \circ g_E = g_V \circ t_G$, $m_H \circ g_E = m_G$ and $l_H(g_V(x)) = l_G(x)$

for all nodes x where $l_G(x) \neq \perp$ ($f(x) = \perp$ means f is undefined for x). An iso-morphism is a morphism that is injective and surjective in both components and maps unlabelled nodes to unlabelled nodes. If there is an isomorphism from G to H they are isomorphic, denoted by $G \cong H$. Applying morphism $g : G \to H$ to graph G yields a graph gG where: $V_{gG} = g_V V_G$ (i.e. apply g_V to each node in V_G); $E_{gG} = g_E E_G$; $s_G(e) = n \Leftrightarrow s_{gG}(g_E(e))$ and similarly for targets; $m_G(e) = m \Leftrightarrow m_{gG}(g_E(e)) = m$; $l_G(n) = l \Leftrightarrow l_{gG}(g_V(n)) = l$. A graph inclusion $H \supseteq G$ is a graph morphism $g : G \to H$ such that $g(x) = x$ for all vertices and edges x in G. Note that inclusions may map unlabelled nodes to labelled nodes.

A rule $r = \langle L \supseteq K \subseteq R \rangle$ consists of three graphs: the interface graph K and the left and right graphs L and R which both include K.   □

Intuitively, a rule deletes nodes in $L - K$, preserves nodes in K and allocates nodes in $R - K$. In [10] rules may merge nodes but we have no need for this more general formulation here. Our pictures of rules show the left and right graphs; the interface graph is always just the set of numbered vertices common to left and right. For example, the interface of BtoL in Fig. 1 consists of the unlabelled node 1. So BtoL deletes two leaf nodes and two arcs, and preserves node 1 which is relabelled as a leaf.

Definition 4 (Direct derivation)
Graph G directly derives graph H through rule $r = \langle L \supseteq K \subseteq R \rangle$ and mor-phism g, written $G \Rightarrow H$, $G \Rightarrow_r H$ or $G \Rightarrow_{r,g} H$, if there is an injective graph morphism $g : L \to G$ such that: 1. no edge in $G - gL$ is incident to a node in $gL - gK$ (the dangling condition); 2. $H \cong H'$ where H' is constructed from G as follows: (i) remove all vertices and edges in $gL - gK$ (and restrict s_G, t_G, l_G and m_G accordingly) to obtain a subgraph D of G, (ii) add disjointly all vertices and edges (and their labels) in $R - K$ to D to form H': so there is another injective morphism $h : R \to H'$ with $h(R - K) \cap D = \emptyset$; if the source of an edge $e \in R - K$ is $x \in V_K$ then $s_{H'}(h(e))$ is $g(x)$ otherwise it is $h(x)$; similarly for targets; for every vertex $x \in V_K$ if $l_L(x) \neq l_R(x)$, the label of $g(x)$ in H' becomes $l_R(x)$.   □

Injectivity of the matching morphism g means that BtoL in Fig. 1 is only applicable to a graph in which some B-labelled node has left and right arcs to distinct L-labelled nodes; the dangling condition means the L-labelled nodes must have no other in-arcs and the B-labelled node may have in-arcs.

If $H \cong G$ or H is derived from G by a sequence of direct derivations through rules in set $\mathscr{R}$ we write $G \Rightarrow^*_{\mathscr{R}} H$ or $G \Rightarrow^* H$. If no graph can be directly derived from G through a rule in $\mathscr{R}$ we say G is $\mathscr{R}$-irreducible. Definitions 2 and 3 are too general for modelling data structures because the outdegree of nodes is unlimited, and graphs and rules can disrespect the intentions of our signatures.

Example 4 (Unrestricted graph reduction is too general)
Fig. 4 shows a simple rule Rel which relabels a node, and an example derivation in which the relabelling results in a graph containing a leaf with a child. Un-

Rel : $_1$ (B) ⇒ $_1$ (L)   (B) —$\xrightarrow{r}$— (L) ⇒$_{Rel}$ (L) —$\xrightarrow{r}$— (L)

Fig. 4. A rule `Rel`, which does not respect the BT signature, and the effect of applying it to a graph which does respect the BT signature

---

restricted rules could make trees cyclic or give branches multiple left-children. This motivates the following restrictions.   □

Definition 5 (Outlabels and Σ-graph)
The outlabels of node v in graph G are the set of labels of edges whose source is v: $\text{outlabels}_G(v) = \{m_G(e) \mid s_G(e) = v\}$.
A graph G respects Σ, or G is a Σ-graph for short, if: (1) $\forall e, e' \in E_G \cdot s_G(e) = s_G(e') \Rightarrow m_G(e) \neq m_G(e') \vee e = e'$ and (2) $\forall v \in V_G \cdot l_G(v) \neq \bot \Rightarrow \text{outlabels}_G(v) \subseteq \text{type}(l_G(v))$. Note the set of Σ-graphs is closed under subgraph selection.   □

Every node has at most one outgoing edge with any given label, and the outlabels of a node labelled l form a subset of the type of l.

Definition 6 (Σ-total graphs)
A Σ-graph G is Σ-total if l_G is total and for every node $v \in V_G$, $\text{outlabels}_G(v) = \text{type}(l_G(v))$.   □

A Σ-total graph models a data structure: all its nodes are labelled and each node has a full set of outlabels. Apart from these restrictions nodes may be connected to others in the same graph arbitrarily. In this paper we do not model nil pointers. Alternatives are considered in [2]. Non-total Σ-graphs are used in rules where it is essential, or convenient, to have unlabelled nodes and missing outlabels.

Example 5 (Σ_{BT} and Σ_{BT}-total graphs)
In the right half of Fig. 4, the left graph respects Σ_{BT} and the right graph does not. In Fig. 3 both graphs are Σ_{BT}-total.   □

To prevent reduction rules breaking either the signature or the totality of graphs we define a simple restricted rule form: Σ-total rules.

Definition 7 (Σ-total rule)
A rule $\langle L \supseteq K \subseteq R \rangle$ is a Σ-total rule if L, R are Σ-graphs and for every node x:
1. $l_L(x) = \bot \Rightarrow x \in V_K \wedge l_R(x) = \bot \wedge \text{outlabels}_L(x) = \text{outlabels}_R(x)$.
That is, unlabelled nodes in L are preserved and remain unlabelled with the same outlabels.
2. $x \in V_K \wedge l_L(x) \neq \bot \wedge l_L(x) = l_R(x) \Rightarrow \text{outlabels}_L(x) = \text{outlabels}_R(x)$.
That is, labelled nodes in L which are preserved with the same label have the same outlabels in L and R.

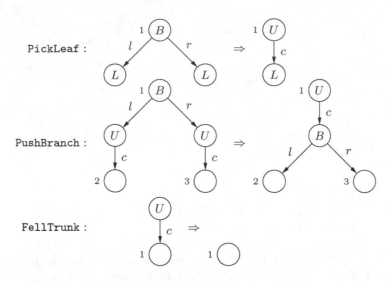

Fig. 5. BBT shape specification rules

---

3. $x \in V_K \wedge l_L(x) \neq \perp \wedge l_L(x) \neq l_R(x) \Rightarrow$
$l_R(x) \neq \perp \wedge$ outlabels$_L(x) = $ type $(l_L(x)) \wedge$ outlabels$_R(x) = $ type $(l_R(x))$.
That is, relabelled nodes have a complete set of outlabels in L and R. Nodes
may not be labelled in L and unlabelled in R, or vice versa.
4. $x \in V_L - V_K \Rightarrow$ outlabels$_L(x) = $ type $(l_L(x))$.
That is, deleted nodes have a complete set of outlabels.
5. $x \in V_R - V_K \Rightarrow l_R(x) \neq \perp \wedge$ outlabels$_R(x) = $ type $(l_R(x))$.
That is, allocated nodes are labelled and have a complete set of outlabels.   □

Example 6 (Rules specifying balanced binary trees)
Example 7 specifies BBTs with the Σ_{BT}-total rules $\mathcal{R}_{BBT} = \{$PickLeaf,
PushBranch, FellTrunk$\}$, given in Fig. 5. PickLeaf replaces a binary branch of
leaves by a unary branch of a leaf; PushBranch forces a binary branch of unary
branches one level down, it applies anywhere in a tree. Note that both rules
preserve height and balance. FellTrunk removes unary branches which are not
the target of any arcs, it preserves balance but decreases height.   □

Theorem 1 (Σ-total rules preserve Σ and Σ-totality)
Let r be a Σ-total rule and $G \Rightarrow_r H$ a direct derivation on graphs over Σ.
Then G is a Σ-graph iff H is a Σ-graph. Moreover, G is Σ-total iff H is Σ-total.
     □

Definition 8 (GRS, NT-free GRS)
A graph reduction specification (GRS) $S = \langle \Sigma, \mathcal{R}, \text{Acc} \rangle$ consists of a signature
Σ, a finite set of Σ-total rules $\mathcal{R}$ and an $\mathcal{R}$-irreducible Σ-total graph Acc, the

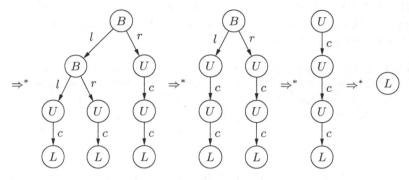

Fig. 6. A reduction of the right graph in Fig. 3. The steps in the four reduction sequences are: `PickLeaf`, `PickLeaf`; `PushBranch`, `PickLeaf`; `PushBranch`, `PushBranch`, `PickLeaf`; `FellTrunk`, `FellTrunk`, `FellTrunk`

---

accepting graph. The graph language of S is $\mathcal{L}(S) = \{G \mid G \Rightarrow^*_{\mathcal{R}} \text{Acc} \wedge l_G(V_G) \cap \mathcal{C}_N = \emptyset\}$. If $\mathcal{C}_N = \emptyset$ we say that S is NT-free.   □

Termination and closedness are discussed in Section 3. Note that Acc is Σ-total, so every graph in $\mathcal{L}(S)$ is Σ-total by Theorem 1.

Example 7 (Specification of balanced binary trees)
We define BBTs by the NT-free GRS BBT $= \langle \Sigma_{BT}, \mathcal{R}_{BBT}, \text{Acc}_L \rangle$, where $\mathcal{R}_{BBT}$ is defined in Example 6. That is, $\mathcal{R}_{BBT}$ reduces BBTs, and nothing else, to Acc_L. Fig. 6 shows an example reduction. The left graph in Fig. 3 is irreducible under $\mathcal{R}_{BBT}$, owing to the various forms of sharing it contains, and therefore is not a BBT (it is a balanced binary DAG); the right graph is a BBT.   □

Theorem 2 (BBT specifies balanced binary trees)
For every Σ_{BT}-graph G, $G \in \mathcal{L}(\text{BBT})$ iff G is a balanced binary tree.   □

3   Membership Checking

Graph reduction rules are just reversed graph-grammar production rules so reduction specifications can define every recursively enumerable set of Σ-total graphs (that exclude the empty graph, see [2]). This follows from Uesu's result that double-pushout graph grammars can generate every recursively enumerable set of graphs [18]. Consequentially, arbitrary reduction rules can specify languages with an undecidable membership problem.

For testing example structures we need specifications for which language membership can be checked — preferably in polynomial time. Therefore we will require that GRSs are polynomially terminating and their languages closed under reduction. Testing membership of such languages is simple: given a graph G, check that G only has terminal labels and apply the rules in $\mathscr{R}$ (nondeterministically) as long as possible; G belongs to $\mathscr{L}(S)$ iff the resulting graph is isomorphic to Acc. First we consider termination.

Definition 9 (Graph size, polynomially terminating, size-reducing)
Graph size is defined by $size(G) = \#V_G + \#E_G$ where $\#$ denotes set cardinality. A GRS $S = \langle \Sigma, \mathscr{R}, Acc \rangle$ is terminating if there is no infinite derivation $G_0 \Rightarrow_{\mathscr{R}} G_1 \Rightarrow_{\mathscr{R}} \cdots$. It is polynomially terminating if there is a polynomial p such that for every derivation $G \Rightarrow_{\mathscr{R}} G_1 \Rightarrow_{\mathscr{R}} \cdots \Rightarrow_{\mathscr{R}} G_n$, $n \leq p(size(G))$. It is size-reducing if $size(L) > size(R)$ for every rule $\langle L \supseteq K \subseteq R \rangle$ in $\mathscr{R}$. $\square$

The example specifications in this paper have linear reduction lengths; this is usually easily shown, but there is no general decision method, so new GRSs may require individual termination analysis. For example, BBT is size-reducing, while RBT (Section 5) reduces the natural number $size(G) + \#\{v \mid l_G(v) = B\}$ at each step. Now we consider closedness and complexity.

Definition 10 (Closedness, Confluence, PGRS)
A GRS $S = \langle \Sigma, \mathscr{R}, Acc \rangle$ is closed if for every direct derivation $G \Rightarrow_{\mathscr{R}} H$, $G \Rightarrow_{\mathscr{R}}^* Acc$ implies $H \Rightarrow_{\mathscr{R}}^* Acc$. S is confluent if for every pair of derivations $H_1 {}_{\mathscr{R}}^* \!\! \Leftarrow G \Rightarrow_{\mathscr{R}}^* H_2$ over Σ, there is a graph H such that $H_1 \Rightarrow_{\mathscr{R}}^* H {}_{\mathscr{R}}^* \!\! \Leftarrow H_2$. A polynomially terminating and closed GRS is a polynomial GRS (PGRS). $\square$

Confluence implies closedness (the converse does not hold). Confluence of a terminating specification can be shown by adapting the critical pair method of [17] to GRSs (see [2]). All examples in this paper are confluent by this method as all their critical pairs are strongly joinable. Two reduction rules form a critical pair if they can be applied to the same graph such that one rule removes part of the graph required to apply the other rule. Closedness can be tested by disregarding any critical pair which only occurs as part of non-language member graphs.

Theorem 3 (Complexity of testing membership)
If S is a PGRS then membership of $\mathscr{L}(S)$ is decidable in polynomial time. $\square$

We assume S is fixed, so the number of rules is fixed and the size of the largest left graph in $\mathscr{R}$ is a constant c. Checking whether any rule in $\mathscr{R}$ matches a graph G requires $O(size(G)^c)$ time. This is because there are at most $size(G)^c$ injective mappings $V_L \to V_G$ for any left graph L, and checking whether a mapping induces a graph morphism $L \to G$ and the dangling condition can be done in constant time if graphs are suitably represented. Given a match, rule application is constant time. Hence the procedure sketched in the introduction to this section runs in polynomial time. The procedure is correct as the closedness of S makes backtracking unnecessary.

4   Extensions and Closure Properties

NT-free PGRSs are powerful but there are still lots of shapes they cannot describe; PGRSs are more powerful and GRSs have the universal specification power of graph grammars. This section develops the idea of classifying the simplicity of shapes by showing whether they have an NT-free specification or not. We show that: intersection extends the range of shapes definable by NT-free (P)GRSs to all the (P)GRS-definable shapes, and that (P)GRSs are closed under intersection; union extends the range of shapes definable by NT-free PGRSs and PGRSs, but terminating and possibly non-confluent GRSs are closed under union (provided $A\propto \neq \emptyset$); complement extends the range of shapes definable by NT-free (P)GRSs and (P)GRSs.

Complete binary trees (CBTs) are BBTs where every branch is binary. Theorem 4 says they cannot be defined by an NT-free GRS. Lemma 1 presents a general method for showing that an NT-free GRS cannot define a given shape.

Lemma 1 (Proving graph languages are undefinable)
Graph language $\mathscr{L}$ cannot be defined by an NT-free GRS if:
$\forall k \in \mathbb{N}, \mathscr{R} \subseteq \mathscr{L} \times \mathscr{L} \cdot \max\{\delta(G, H) \mid (G, H) \in \mathscr{R}\} \geq k \vee \mathscr{R}^* \neq \mathscr{L} \times \mathscr{L}$
where $\delta(G, H) = \min\{\max\{\mathrm{size}(L), \mathrm{size}(R)\} \mid r = \langle L \supseteq K \subseteq R \rangle \wedge G \Rightarrow_r H\}$. $\square$

To use Lemma 1 we show that for every k there is a graph $G \in \mathscr{L}$ which cannot be rewritten to some other graph $H \in \mathscr{L}$ without a rule of size at least k.

Theorem 4 (CBTs cannot be defined by an NT-free GRS)
No NT-free GRS can specify complete binary trees. $\square$

We can often make a language specifiable by using non-terminals. Alternatively, we can take the intersection or union of two NT-free GRS languages. We show that using non-terminals is equivalent to using intersection and hence GRSs are closed under intersection. The following examples give non-terminal and intersection specifications of CBTs.

Example 8 (Specification of complete binary trees)
Let $CBT = \langle \Sigma_{BT} + \langle\{\}, \{U\}, \{\}, \{\}\rangle, \mathscr{R}_{BBT}, A\propto_B\rangle$. Hence CBTs are BBTs which do not contain any unary branches. $\square$

Example 9 (CBTs by intersection)
Let $\mathscr{L}(CBT) = \mathscr{L}(FBT) \cap \mathscr{L}(BBT)$. CBTs are full binary trees (left conjunct, Example 1). CBTs are balanced (right conjunct). Both GRSs are NT-free. $\square$

By Theorem 4 and Example 9, the languages of NT-free (P)GRSs are not closed under intersection. Theorem 5 shows that (P)GRSs and intersections of NT-free (P)GRSs have equivalent power. Theorem 6 shows that (P)GRSs are closed under intersection.

Theorem 5 (GRSs equivalent to intersections of NT-free GRSs)
1. If N is a GRS there are NT-free GRSs S and T s.t. $\mathscr{L}(N) = \mathscr{L}(S) \cap \mathscr{L}(T)$. Further, if N is a PGRS then so are S and T.
2. If S and T are NT-free GRSs there is a GRS N s.t. $\mathscr{L}(N) = \mathscr{L}(S) \cap \mathscr{L}(T)$. Further, if S and T are PGRSs then so is N.   □

Theorem 6 (Graph reduction languages closed under intersection)
If S and T are (P)GRSs, then $\mathscr{L}(S) \cap \mathscr{L}(T)$ can be defined by a (P)GRS N. □

Language union offers another way to compose specifications. It is easy to see that union extends the range of languages specifiable by PGRSs and NT-free PGRSs. For example, a GRS cannot define a finite language that includes the empty graph and some other graph, but such a language is easily specified as a union of PGRSs with no reduction rules whose accepting graphs are the language elements. Similarly, PGRSs are not closed under union for infinite languages with or without the empty graph. If we allow terminating but possibly non-confluent GRSs, we can show that they are closed under union, provided their languages exclude the empty graph (see [2]).

GRS languages are not closed under complement. This follows from the ability of reduction specifications to simulate Chomsky grammars.

5   Red-Black Trees

This section applies the theory to specify red-black trees (RBTs). Our specification in Definition 12 is interesting because it is an NT-free PGRS but is not size-reducing. Theorem 7 says that a size-reducing RBT specification needs non-terminals (using a simplification of Lemma 1; see [2] for the proof and a size-reducing specification with non-terminals).

Definition 11 (Textbook red-black tree definition [6])
Red-black trees are trees of binary-branches and leaves where branches are labelled red or black, children of red branches are black or leaves and all paths from root to leaf have the same number of black nodes.   □

Theorem 7 (A size-reducing GRS of RBTs needs non-terminals)
Red-black trees cannot be specified by a size-reducing NT-free GRS.   □

Definition 12 (Specification of red-black trees)
Let $\Sigma_{RBT} = \langle \{R, B, L\}, \{\}, \{l, r\}, \{R \mapsto \{l, r\}, B \mapsto \{l, r\}, L \mapsto \{\}\} \rangle$ and $\mathrm{RBT} = \langle \Sigma_{RBT}, \mathscr{R}_{RBT}, \mathrm{Acc}_L \rangle$, where Fig. 7 shows the reduction rules in $\mathscr{R}_{RBT}$ and Fig. 1 shows Acc_L.   □

Note that RBT is not size reducing but it linearly terminates as every reduction step reduces $\mathrm{size}(G) + \#\{v \in V_G \mid l_G(v) = B\}$.

Theorem 8 (Correctness of RBT)
For every Σ_{RBT}-graph G, $G \in \mathscr{L}(\mathrm{RBT})$ if and only if G is a red-black tree. □

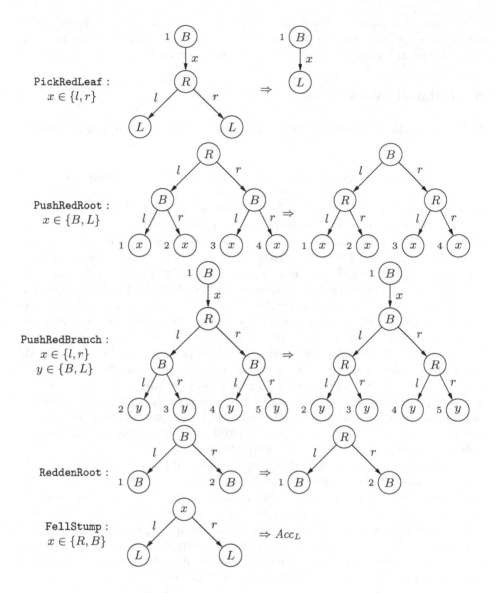

Fig. 7. Red-black tree reduction rules

Each rule preserves the red-black properties and produces either a smaller or a redder tree (therefore $\mathscr{R}_{RBT}$ terminates). The smallest RBT is a leaf. We can think of the tree reduction process as follows. `PickRedLeaf` can remove any red leaf-parent with a black parent. Any red node higher up the tree can be pushed by the tree by recolouring it and its children as in `PushRedRoot` or `PushRedBranch`, provided that its grandchildren are black or leaves. These rules alone produce a complete black tree. The root can be coloured by `ReddenRoot`, safely reducing the black height, and then pushed down and picked by the other rules. Eventually we reach a singleton which is rewritten to A cc_L by `FellStump`.

6   Related Work

In functional programming, Nested types can be used to specify perfect binary trees [13]; however, these are only complete balanced binary DAGs as they do not preclude sub-tree sharing.

The following papers specify shapes using variants of context-free graph grammars, or certain logics. They can all specify trees, but none tackle the problem of specifying non-context-free properties like balance. ADDS [12] specifies structures by a number of dimensions where arcs are restricted to point away from, or towards, the root in a specified dimension. It can also limit node indegree. The logic of reachability expressions [5] allows the reachability, cyclicality and sharing properties of pointer variables to be specified as logical formulae. It is decidable whether a structure satisfies such a specification (but the complexity is unclear) and the logic is closed under intersection, union and complement. In role analysis [15] the shapes of pointer data structures are restricted by specifying whether pointers are on cyclic paths and by stating which pointer sequences form identities. The number and kind of incoming pointers are also specified. An algorithm verifies programs annotated with role specifications. Graph types [14] are recursive data types extended with routing expressions which allow the target of a pointer to be specified relative to its source. In [16], graph types are defined by monadic 2nd-order logic formulae and a pointer assertion logic is used to annotate C-like programs with partial correctness specifications; a tool checks that programs preserve their graph type invariants.

Shape types [8, 9] are specified by context-free graph grammars. They can always be converted to equivalent GRSs [2], but the classes of context-free graph languages and PGRS languages are incomparable. However, PGRSs can specify context-sensitive shapes and we are not aware of any common data structure with a context-free specification and no PGRS. Shape types have a method for checking the shape-invariance of atomic transformations (individual pointer manipulations). GRSs have a similar method which can verify the shape safety of an algorithm; more detailed explanations are available in [3, 1]. Context-exploiting shapes [7] are generated by hyperedge-replacement rules extended with context; the precise relation to PGRSs is unclear. Membership checking is exponential but there is a restricted, decidable class of shaped transformation rules that preserve context-exploiting shapes.

For related work on graph parsing see [4] which discusses context-free graph grammars and layered graph grammars which are powerful but have an exponential membership algorithm

7 Conclusion and Future Work

Graph reduction specifications are a powerful formal framework, capable of defining data structures with non-context-free properties. The examples presented here show how they can specify complete binary trees (Section 4) and red-black trees (Section 5). Many other examples, including AVL trees and grids, are given in [2]. The GRS tool available from [1] implements GRS checking including confluence, membership and operation checks.

We intend to develop programming languages which offer safe pointer manipulation based on GRSs. We are investigating two approaches.

1. A new pointer programming paradigm. Algorithms will be described as operations on graphs with data fields; the shapes of intermediate structures will be specified or inferred and checked. Checking is undecidable in general; we plan to investigate its feasibility on practical examples, the method described in [3] is a starting point. For operations like insertion into red-black trees [6] a better checker will be required, and possibly more informative specifications, because the current checker is often non-terminating on non-context-free shapes.

2. An imperative programming language. Combining conventional pointer manipulation with types specified by GRSs: pointer algorithms will be abstracted and then checked as in the first approach. Here the main challenge is to fit the operational semantics of a garbage-collected imperative language to the semantics of double-pushout graph rewriting.

References

[1] Safe Pointers by Graph Transformation, project webpage. http://www-users.cs.york.ac.uk/~ajb/spgt/. 30, 42, 43

[2] A Bakewell, D Plump, and C Runciman. Specifying pointer structures by graph reduction. Technical Report YCS-2003-367, Department of Computer Science, University of York, 2003. Available from [1]. 32, 35, 37, 38, 40, 42, 43

[3] A Bakewell, D Plump, and C Runciman. Checking the shape safety of pointer manipulations. In *Proc. 7th International Seminar on Relational Methods in Computer Science (RelMiCS 7)*, LNCS. Springer-Verlag, 2004. 31, 42, 43

[4] R Bardohl, M Minas, A Schürr, and G Taentzer. *Handbook of Graph Grammars and Computing by Graph Transformation*, volume 2, chapter Application of Graph Transformation to Visual Languages, pages 105–180. World Scientific, 1999. 42

[5] M Benedikt, T Reps, and M Sagiv. A decidable logic for describing linked data structures. In *Proc. European Symposium on Programming Languages and Systems (ESOP '99)*, volume 1576 of *LNCS*, pages 2–19. Springer-Verlag, 1999. 32, 42

[6] T H Cormen, C E Leiserson, and R L Rivest. *Introduction to algorithms*. MIT Press, 1990. 40, 43

[7] F Drewes, B Hoffmann, and M Minas. Context-exploiting shapes for diagram transformation. *Machine Graphics and Vision*, 12(1):117–132, 2003. 42

[8] P Fradet and D Le Métayer. Shape types. In *Proc. Principles of Programming Languages (POPL '97)*, pages 27–39. ACM Press, 1997. 32, 42

[9] P Fradet and D Le Métayer. Structured Gamma. *Science of Computer Programming*, 31(2–3):263–289, 1998. 42

[10] A Habel, J Müller, and D Plump. Double-pushout graph transformation revisited. *Math. Struct. in Comp. Science*, 11:637–688, 2001. 32, 33, 34

[11] A Habel and D Plump. Relabelling in graph transformation. In *Proc. International Conference on Graph Transformation (ICGT 2002)*, volume 2505 of *LNCS*, pages 135–147. Springer-Verlag, 2002. 33

[12] L J Hendren, J Hummel, and A Nicolau. Abstractions for recursive pointer data structures: Improving the analysis and transformation of imperative programs. In *Proc. ACM SIGPLAN '92 Conference on Programming Langauge Design and Implementation (PLDI '92)*, pages 249–260. ACM Press, 1992. 32, 42

[13] R Hinze. Functional Pearl: Perfect trees and bit-reversal permutations. *Journal of Functional Programming*, 10(3):305–317, May 2000. 42

[14] N Klarlund and M I Schwartzbach. Graph types. In *Proc. Principles of Programming Languages (POPL '93)*, pages 196–205. ACM Press, 1993. 32, 42

[15] V Kuncak, P Lam, and M Rinard. Role analysis. In *Proc. Principles of Programming Languages (POPL '02)*, pages 17–32. ACM Press, 2002. 32, 42

[16] A Møller and M I Schwartzbach. The pointer assertion logic engine. In *Proc. ACM SIGPLAN '01 Conference on Programming Langauge Design and Implementation (PLDI '01)*. ACM Press, 2001. 42

[17] D Plump. Hypergraph rewriting: Critical pairs and undecidability of confluence. In M R Sleep, M J Plasmeijer, and M C van Eekelen, editors, *Term Graph Rewriting: Theory and Practice*, chapter 15, pages 201–213. John Wiley & Sons Ltd, 1993. 38

[18] T Uesu. A system of graph grammars which generates all recursively enumerable sets of labelled graphs. *Tsukuba J. Math.*, 2:11–26, 1978. 37

Specific Graph Models and Their Mappings to a Common Model

Boris Böhlen

RWTH Aachen University
Department of Computer Science III
Ahornstraße 55, 52074 Aachen, GERMANY
boehlen@cs.rwth-aachen.de

Abstract. Software engineering applications, like integrated development environments or CASE tools, often work on complex documents with graph-like structures. Modifications of these documents can be realized by graph transformations. Many graph transformation systems operate only in volatile memory and thus suffer from a couple of drawbacks.

In this paper, we present the graph model of the Gras/GXL database management system. Gras/GXL enables graph based applications to store their graphs persistently in commercial databases. Because these applications usually have their own graph model, a mapping from this graph model to the Gras/GXL graph model has to be realized. We will present mappings for the PROGRES, DiaGen, and DiaPlan graph models in this paper. For the PROGRES graph model we will also show its realization.

1 Introduction

Integrated development environments, visual language editors, and reengineering applications often use graphs or graph-like structures, as data structures for their documents. These graphs contain entities of different types on different levels of abstraction. Besides accessing these entities efficiently, applications perform complex queries and operations on the graphs — e.g. for ensuring consistency constraints, enumerating affected and related entities, or removing a net of related entities.

Nowadays, most of the applications mentioned before are still implemented manually. Specifying the application logic with graph transformations is a more elaborate approach. The logic is no longer implemented by hand but by using a textual or graphical specification. Now, the developer can concentrate on the solution of the problem and ignore implementation specific details. Moreover, specifying complex operations and queries visually is more convenient than implementing the operations manually, modifying them later, and keeping track of all dependencies.

Some graph transformation systems can even generate code for languages like C++ or Java from the specification. Applications are then built on top

J.L. Pfaltz, M. Nagl, and B. Böhlen (Eds.): AGTIVE 2003, LNCS 3062, pp. 45–60, 2004.

of the generated code. Many examples show that this approach performs well: Based on the graph transformation system PROGRES [21] and the prototyping framework UPGRADE [3], different groups at our department developed the AHEAD [11] and E-CARES [15, 16] prototypes, among others. Similar results were achieved by other groups: DiaGen [17] is used for realizing editors for visual languages like state chart diagrams [18]. Fujaba [12] is used for simulating production control systems.

Many graph transformation systems store their graphs in volatile memory. At each system start the graph is loaded and must be stored before the application exits. The drawbacks of this approach are obvious: Data are lost when the application crashes, the graph size is limited, database functionality must be reimplemented, etc. Let us discuss these problems in detail using two examples from our department.

The E-CARES reengineering environment analyses the source code, runtime traces, and other sources of information of a telecommunication system to recover its architecture. The initial steps of the analysis create huge graphs of about 300,000 nodes per subsystem[1]. For inter-subsystem communication, multiple subsystems have to be analysed. Storing these graphs in main memory has a couple of disadvantages: they are too huge, recovering lost information is very expensive[2], and system startup and shutdown times slow down significantly.

Another example is AHEAD, an environment for the administration of development processes. AHEAD features an agenda containing the tasks assigned to each developer. The developers use a distributed application to perform the tasks to which they have been assigned. Ensuring the data consistency of the distributed application is a very complicated task and many problems solved in database management systems before must be solved again. Both applications demand a graph database, such as GRAS.

When the development of the GRAS database management system started in 1984, common database management systems lacked many features required for the implementation of software development environments — for example, storing graphs efficiently, offering undo / redo of graph modifications, graph change events, etc. To obtain these features, problems solved in every database management system had to be solved again — transaction management, efficient data organization, concurrent access, data consistency, etc. PROGRES and all prototypes built with it, as well as the IPSEN environment [20], utilize the GRAS database.

However, databases have changed since the early days of GRAS, and the costs for maintaining and extending GRAS have increased. Moreover, our experience

---

[1] Because of GRAS' current 65,000 node limit only the most important informations are stored in the graph at the moment. A recent modification of GRAS — which supports roughly 1,000,000 nodes — reduces the problem only slightly because the overall storage space could not be increased.

[2] A recovery may require the re-analysis of a significant amount of documents which may have to be created again. Moreover, architecture modifications reflecting reengineering decisions may be lost.

in building prototypes with PROGRES and UPGRADE showed that the graph model of PROGRES is not sufficient for specifying some problems appropriately. As the PROGRES graph model is tightly coupled to the GRAS graph model, enhancements should be made to the GRAS graph model, which are not feasible. Thus, the implementation of the new GRAS database management system, Gras/GXL, has been decided. Whereas the proven useful functionality of GRAS must remain, more flexibility has to be added to the system.

Instead of a fixed graph model the new Gras/GXL database provides a graph model on top of which concrete graph models are realized. The advances in database management systems and standardization of service interfaces make it possible to reduce the development effort dramatically by building on third-party components, such as commercial database management systems or transaction managers. Besides a reduction in development efforts we expect increased reliability and stability, due to enhancements in the components we use. Because of standardized service interfaces, one component can be replaced by another more easily. The ability to replace components contributes to the scalability of Gras/GXL: For small prototypes a small and main memory database can be used for storing graphs. For applications which have to handle a huge amount of data — like E-CARES — commercial database management systems can be used.

The Gras/GXL database management system allows graph transformation systems to store their graphs persistently in different commercial database systems. Because virtually every graph transformation system has its own graph model, the predecessors of Gras/GXL could not be utilized by them. However, the Gras/GXL graph model is adopted to match exactly the graph model of the graph transformation system. The graph transformation system can now use a commercial database without undergoing major modifications.

In this paper, we present the graph model used by the successor of GRAS, the Gras/GXL database management system. Section 2 presents the specific graph models of DIAGEN, DIAPLAN, and PROGRES for which we will define a mapping in Section 4. Our discussion of the PROGRES graph model will show which situations cannot be specified appropriately and outline their influence on the specification and the prototype. In Section 3, we present the graph model, its origin, and relate it to the GXL graph model [10]. A mapping of the graph models discussed in Section 2 to the graph model of Gras/GXL is presented in Section 4. For the PROGRES graph model we will present the realization of the mapping in Section 5. Section 6 summarizes the presented results and sketches some open problems.

2 Graph Models of Different Graph Transformation Approaches

Before introducing the graph model of Gras/GXL we give a short overview on the graph models of DIAGEN, DIAPLAN, and PROGRES. Based on this overview a mapping for these graph models to the graph model of Gras/GXL will be presented in Section 4.

2.1   Graph Models of DiaGen and DiaPlan

DIAGEN generates diagram editors for a specific visual language from a textual specification. The resulting editors support free hand editing of diagrams in conformance with the corresponding visual language. If the resulting visual language is executable — like state charts — the diagrams can be animated. The formal background of DIAGEN are hypergraphs and hypergraph transformations.

The hypergraphs used by DIAGEN are a generalization of directed graphs as explained in [17]. A hypergraph consists of a set of labeled nodes, a set of directed, labeled edges, and a set of labeled hyperedges. The labels in hypergraphs correspond to types of ordinary programming languages. An edge connects, or visits, two nodes which may be identical. Hyperedges have a fixed number of labeled tentacles, which are determined by the label of the hyperedge. The tentacles connect nodes to the hyperedge. The order of the tentacles, and thus, the order in which the nodes are visited, is determined by the hyperedge's label. Obviously, edges can be represented by binary hyperedges. Thus, it is a matter of style whether edges are used or not.

DIAPLAN [7] is a visual rule-based diagram programming language that allows more complex animations than DIAGEN. The computational model of DIAPLAN are shapely nested hierarchical graph transformations. For DIAPLAN the graph model of DIAGEN has been extended by hierarchies. Edges as well as hyperedges may contain at most one hypergraph and are then called a frame. Of course, a hypergraph contained in a frame can contain other frames. Thus, the resulting hierarchy has a tree-like structure and there is always exactly one top graph. Hoffmann extended the role of nodes in [8] and [9]: nodes may contain hypergraphs as well, and special external nodes, called points, were introduced. Points are nodes at which a graph may be connected to other graphs and which can not contain a hypergraph. Note that edges crossing frame boundaries are forbidden. The integration of DIAPLAN into DIAGEN for generating diagram editors is planed.

2.2   Graph Model of PROGRES

PROGRES is used at our department to build the application logic of prototypical applications. The application domains of the prototypes range from conceptual design in civil engineering [13] to project management (AHEAD). The PROGRES graph model is based on directed attributed graphs. A graph consists of a set of labeled nodes and a set of directed labeled edges. Only the nodes are first class objects and may carry attributes.

Edges do not have an identity and serve as binary relationships only. Attributed edges, edges on edges, n-ary relations, or hierarchical graphs are not supported. But, our experience in building prototypes with PROGRES showed that these constructs are needed in virtually every specification. Fortunately, the PROGRES graph model is capable to emulate all constructs mentioned before. For example, an attributed edge can be emulated by an edge–node–edge construct. One edge points to the source of the attributed edge, the other edge points to the target. The node carries the attributes. Hierarchical graphs can be emulated by specially typed edges.

The influence of emulating attributed edges and other constructs on the specification style and prototype creation cannot be neglected. Because edge–node–edge constructs are harder to use in visual graph tests and graph transformations — even if path expressions are used — textual transformations dominate the specification. When a prototype is built based on the code generated from the specification, the edge–node–edge constructs have to be transformed at the UI level to an attributed edge. Otherwise, a user has difficulties understanding the diagrams presented by the prototype[3]. Mapping every single instance of an edge–node–edge construct to an attributed edge is very expensive and results in slower updates of the user interface. The Gras/GXL data model solves these problems when supported by PROGRES and the UPGRADE framework as we will see in Section 4.

3   Graph Model of Gras/GXL

The previous discussion on PROGRES, GRAS, and the prototypes built at our department outlined the disadvantages of the present GRAS version — an inflexible graph model, limited graph size, and maintenance problems.

In this section, we present the graph model of Gras/GXL. One major difference between GRAS and Gras/GXL is that GRAS defines a single graph model for all applications. In contrast, the Gras/GXL graph model is never used directly by an application. On top of the Gras/GXL graph model, a specific graph model is defined — for example a DIAGEN graph model — which is then used by an application.

Before introducing the Gras/GXL graph model, we give a short overview on the GXL graph model, as it served as the starting point of our graph model development.

3.1   GXL Graph Model

GXL has been widely accepted as a standard format for exchanging graphs. It is used by many reengineering applications, like Rigi [19] and Gupro [14], and graph visualization tools, like JGraph [1] and UPGRADE [3].

The graph model of GXL is shown in Figure 1. A GXL document stores several graphs. Graphs may be typed and can have an arbitrary number of attributes. Attribute types are defined by a sublanguage which is not covered in this article.

---

[3] Imagine a UML class diagram where the roles and cardinalities are nodes connected to the association by edges instead of labels.

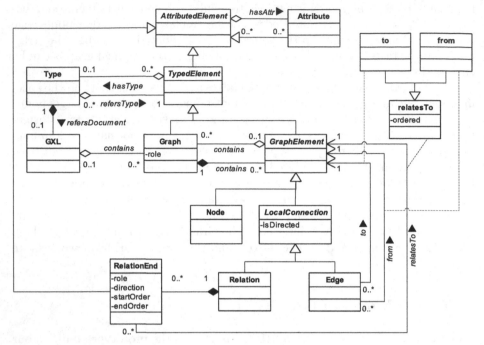

Fig. 1. GXL graph model (from [6])

Every graph belongs to exactly one GXL document and can be identified by a user defined identifier, called a role. A graph contains an arbitrary number of graph elements (nodes, edges, and relations), where each graph element belongs to exactly one graph. Relations represent n-ary relationships between graph elements. They can also be used to express nodes (unary relations) and edges (binary relations). But, this representation of nodes and edges is not convenient for most users. Therefore, the creators of GXL decided to provide nodes and edges as primitive — or atomic — constructs. Edges as well as relations can be ordered. Relation ends cannot be typed but they may carry attributes. Just like graphs, relation ends can be distinguished from each other by user defined roles. Hierarchical graphs are created by placing an arbitrary number of graphs inside a graph element. Like graphs, graph elements can have a type and attributes.

Graph elements (i.e., nodes, edges, and relations), but not graphs, can be connected by edges and — through relation ends — relations. Applications that require edges between graphs have to emulate these inter-graph edges — for example, by using nodes which contain graphs. Another limitation, which is not expressed in the graph model, is that only elements within the same GXL document can be inter-connected[4].

---

[4] The common super class of edges and relations, *LocalConnection*, can be regarded as a realization of this constraint.

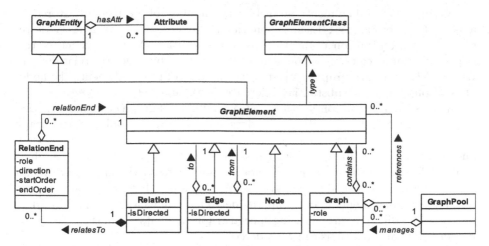

Fig. 2. Graph model of Gras/GXL

Referencing graph elements or graphs within a graph is not allowed, yet. A change request that addresses this open issue is still pending. At the moment, references to graph elements can only be realized by using edges that have special user-defined types. Also, references to graphs can not be realized directly because the GXL graph model does not permit edges to graphs.

Another option for realizing references to graphs or graph elements are attributes. But, as the attribute sublanguage of GXL does not support attributes with graphs or graph elements as values, these references are user-defined.

Both issues are addressed by the upcoming GXL release 1.1 for which a preliminary document type definition (DTD), but no formal graph model, is available. This release provides graph references as well as graph-valued attributes. References to graph elements will not be supported.

Another limitation that has not been addressed so far is direct nesting of graphs. The GXL graph model demands that graphs can only be contained in a graph element or the surrounding GXL document.

3.2 Gras/GXL Graph Model

The discussion of the GXL graph model made clear that this model provides a good starting point for the graph model of the Gras/GXL database management system. However, the discussion also showed that some features, which may be used by certain graph models, are not supported, yet. In the following, we will discuss the Gras/GXL graph model (see Figure 2) in detail and present its differences compared to the GXL graph model. Despite the extensions and modifications the Gras/GXL graph model is able to express the same graph classes as the GXL graph model.

As in a GXL document, an arbitrary number of graphs are stored in a Gras/GXL graph pool Each graph can be identified by its role. GXL regards graphs and graph elements as different entities. As a drawback of this approach, graphs can not be connected to each other by edges (or relations). In Gras/GXL we regard graphs as just a special kind of graph element, like nodes or relations. This is a substantial difference as we shall see.

A Gras/GXL graph contains an arbitrary number of graph elements — nodes, edges, relations, and graphs. In GXL, edges and relations can only connect graph elements within the same document. The Gras/GXL graph model permits edges and relations connecting graph elements stored in different graphs, even in physically different databases[5], which is not supported by GXL but important for connecting elements in different graphs. As explained before, graphs are just ordinary graph elements in our graph model. Thus, they can be visited by edges and relations directly without using special graph elements. As in the GXL graph model, edges and relations can be ordered.

Neither the current nor the upcoming GXL version support references to graph elements. References to graphs are only supported by the upcoming GXL 1.1 version. Our graph model allows a graph to reference other graph elements. We decided not to restrict them to graphs — as GXL does — for two reasons: (1) references to graph elements allow for a much greater flexibility and (2) as graphs are a special kind of graph element this comes naturally.

Hierarchical graphs are created either by a containment relationship or by graph-valued attributes. The containment relationship is used if a graph should be contained in another graph. Graph-valued attributes are used in all other situations, for example if a relation should contain a graph. The use of graph-valued attributes together with the containment relationship allows us to create arbitrary hierarchical graphs. Although this approach seems to be complicated it allows us to handle even complex situations uniformly — like hierarchies of graphs stored in different databases — which is more important than convenient usage of the graph model. The result is a clean and efficient realization of graph hierarchies.

Graph elements and relation ends may have an arbitrary number of attributes, just as in GXL. We do not offer a separate language for attribute definitions. Instead, the graph schema of the application will be defined using Java classes. Hence, attributes can be of any serializable Java type.

A graph element (including graphs) must have a type, whereas in GXL graphs and graph elements may be typed. Although untyped graphs are common in some application domains, we do not support them in the graph model for the following reasons: First, complex schema-based operations can be realized with the help of the underlying database more efficiently. Second, a concrete graph model can create an implicit schema for schema-less graph models.

Summarizing, the graph model of the Gras/GXL database system solves the problems identified for GRAS before. As our graph model is not targeted towards a particular graph model, only some consistency checks can be performed.

---

[5] Not yet supported by the existing implementations of the graph model.

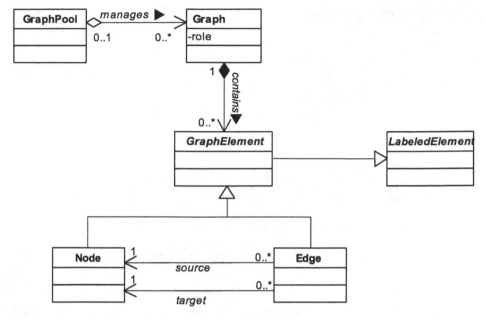

Fig. 3. Class diagram for the PROGRES graph model

Most consistency checks — for example, the number of subgraphs allowed within a graph element — must be performed by the implementation of a concrete graph model. The next section will explain how concrete graph models are realized based on Gras/GXL's graph model.

4   Mapping Different Graph Models to Gras/GXL

One aim of the Gras/GXL database management system is the support of different graph models. The implementation of these graph models and the applications that use them can rely on the infrastructure provided by Gras/GXL — for example, incremental attribute evaluation or graph queries. As mentioned before, a drawback of this approach is that the graph model can only perform limited consistency checks — existence checks for types or graph elements, etc. More complex checks (like cardinality checks) or prohibiting the use of certain graph elements — for example, references to graph elements — have to be shifted to the realization of a concrete graph model.

4.1   Mapping for the **PROGRES** Graph Model

The PROGRES graph model is simple enough to present the steps which are necessary to map a certain graph model to the Gras/GXL graph model in a paper. In some cases, only a mathematical or textual description of the graph

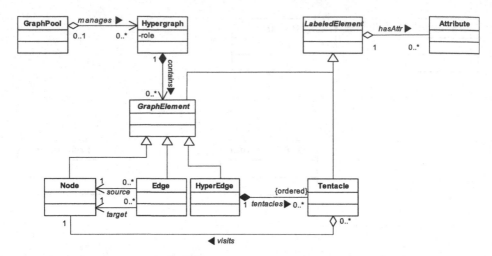

Fig. 4. Class diagram for DiaGen's graph model

model is available. Then, the first step is the creation of a class diagram based on this description. Figure 3 presents the PROGRES graph model we deduced from the description in Section 2.2.

The next step should be to add methods to the class diagram which are used to create and modify graphs. But, we will skip this step for the sake of brevity.

Now, we define the mapping of the PROGRES graph model to the Gras/GXL graph model. In this case a simple one–to–one mapping is sufficient: Graphs are mapped to graphs, nodes to nodes, and edges to edges. The implementation of all classes — for example, `Graph` — is backed by the corresponding classes of the graph model.

Of course, creating the mapping is not enough. The constraints defined for the PROGRES graph model must be ensured by the realization of the mapping. For example, before an edge connects two nodes the realization must check if the types of the nodes are compatible with the types allowed for the source and target node of the edge. In Section 5 we will discuss the realization of the PROGRES mapping in more detail.

4.2 Mapping for the DiaGen Graph Model

Next we define a mapping for DiaGen's graph model to Gras/GXL's graph model. We use the textual description from Section 2.1 to deduce the class diagram shown in Figure 4. A `Hypergraph` contains an arbitrary number of labeled `GraphElement`s — Nodes, Edges, and `Hyperedges`. We do not consider tentacles as graph elements that should be contained in the hypergraph, because they are part of a hyperedge. Thus, the class `Tentacle` is not a sub-class of `GraphElement`, but of `LabeledElement`, because tentacles have a label. The order of tentacles is of importance, which is expressed by the constraint {`ordered`}

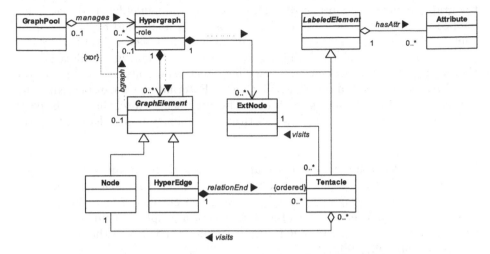

Fig. 5. Class diagram for DIAPLAN's graph model

for the association **tentacles** between **Hyperedge** and **Tentacle**. The association **visits** expresses that a tentacle visits exactly one node and that a node
can be visited by several tentacles. The same applies to the associations **source**
and **target** between the classes **Node** and **Edge**. In addition, we introduce the
class **GraphPool** which contains an arbitrary number of hypergraphs. Thus, our
database can store more than just one hypergraph.

After deducing a class diagram which corresponds to the description of
DIAGEN's graph model, we can define the methods necessary for modifying
hypergraphs. We consider only one operation as an example in this paper.
The class **Hypergraph** provides the method **createEdge(source node, target
node, edge label)** to creates an edge with the specified label from the source
node to the target node.

After all methods have been defined we can proceed to the next step, the
definition of a mapping of DIAGEN's graph model to our graph model. We begin
with the easy steps: Hypergraphs are mapped to graphs, nodes to nodes, and
edges to edges. Hyperedges are mapped to relations and tentacles to relation
ends. The order of the tentacles is ensured by the order attribute of the relation
ends. The label of a tentacle is mapped to the role of the corresponding relation
end. The method **createEdge** should connect edges with nodes, but not with
other edges. This can be ensured by using appropriate parameters for the source
node and target node parameters.

4.3   Mapping for the DiaPlan Graph Model

In Section 2.1 we introduced the graph model of DIAPLAN as an extension of
the DIAGEN graph model. Based on the discussion of this extension we deduced

the graph model shown in Figure 5. DIAPLAN's graph model has two major differences compared to the DIAGEN graph model: (1) nodes and hyperedges may contain a hypergraph, and (2) a new kind of node has been introduced, called point or external node. These differences lead to the following modifications of Figure 4: We introduce a new class ExtNode which represents external nodes (or points) which are contained in a hypergraph. External nodes can be visited by hyperedges — just like nodes — through their tentacles. Note, that we dropped the Edge class, because it can be substituted by binary hyperedges. The association subgraph between the classes *GraphElement* and Hypergraph denotes that any graph element, i.e. nodes and hyperedges, may contain at most one hypergraph. Because external nodes are not allowed to contain a hypergraph, they are not a subclass of *GraphElement*. Remember, that a hypergraph can only be a top-level graph or contained in a graph element. But, a hypergraph can not be a top-level graph and a subgraph of a graph element at the same time, which is denoted by the OCL [23] constraint xor between the associations contains and subgraph.

With the class diagram shown in Figure 5, a mapping to the Gras/GXL graph model can be defined by extending the mapping defined for DIAGEN's graph model before: external nodes are mapped to nodes. The hierarchies are realized by using graph-valued attributes. Of course, the implementations of all methods must ensure the consistency constraints of the graph model — for example, no boundary crossing edges, existence of a top-level hypergraph, etc.

The mappings for the PROGRES, DIAGEN, and DIAPLAN graph model show that it is possible to map these graph models onto the graph model of Gras/GXL. Our mapping of the DIAPLAN graph model showed that graph hierarchies can be realized with the help of graph-valued attributes.

5   Realizing the PROGRES Mapping

The previous section showed how mappings of specific graph models to the graph model of Gras/GXL are realized on the conceptual level. In this section we will focus on the concrete realization of the mapping for the PROGRES graph model based on a few examples.

We have seen that the Gras/GXL graph model is richer than the graph models which are mapped on it, i.e. features not required by the PROGRES graph model — like n–ary relations — are provided by the Gras/GXL graph model. From a conceptual point of view a much simpler graph model would be sufficient, for example one that just provides attributed nodes and binary edges. Any other graph model could be mapped on such a simple graph model. However, such a mapping could be quite complicated and its realization would not be very efficient.

Instead, the Gras/GXL graph model offers a superset of the features found in existing graph models. Thus, the mapping of a specific graph model to the Gras/GXL graph model is quite simple, as we have seen before, and the imple-

```
public ProgresEdgeID createEdge(ProgresNodeID source,
                                ProgresNodeID target,
                                ProgresEdgeTypeID type)
  throws EntityNotFoundException, GrasGXLException \{
  if (!contains(source) || !contains(target)) \{
    throw new EntityNotFoundException("...");
  \}
  ProgresEdgeType edgeType = schema.getEdgeTypeByID(type);
  if (!source.instanceOf(edgeType.getSourceClassID()) ||
      !target.instanceOf(edgeType.getTargetClassID())) \{
    throw new GrasGXLException("...");
  \}
  EdgeID edgeID = graph.createEdge(source.getBackingNodeID(),
                   target.getBackingNodeID(),
                   type.getBackinEdgeTypeID()).getEdgeID();
  return new ProgresEdgeID(edgeID, source, target, type);
\}
```

Fig. 6. Source code for creating an edge in the PROGRES graph model

mentation of the mapping can be realized efficiently, as well see in the following examples.

Our first example deals with the creation of an edge to connect two nodes. The PROGRES graph model requires that the types of the source and target node matches the types of the source and target node in the definition of the edge type. Of course, the source and target node must exist in the graph and the edge type must have been declared in advance. Our implementation of the mapping must check the following things before the edge can be created: existence of source and target node, existence of the edge type, and the type conformance of the source and target node. After ensuring that these conditions hold, the edge can be created between the two nodes. Since the application should not be aware of the existence of the Gras/GXL graph model we create an identifier which is specific for the PROGRES graph model and return it to the application. The source code in Figure 6 shows exactly these steps.

The realization of the method is straightforward and not very surprising. Besides its simplicity, the benefit of having a rich graph model is not obvious.

The following example illustrates the benefits of having a rich graph model like the Gras/GXL graph model. In this example we implement a method which returns all nodes in the graph which are an instance of a node class, including all sub-classes. Because the Gras/GXL graph model is aware of graph schemas with node classes and inheritance, we can realize this method easily and efficiently. Our realization can directly use the method provided by the Gras/GXL graph model for this task, which is able to delegate most of parts of the query to the underlying database — e.g., a couple of joins in the case of a relational database. The only task which is left to our realization, shown in Figure 7,

```
public Collection getAllNodesOfClass(ProgresNodeClassID nc)
  throws GrasGXLException \{
  Collection retSet = new HashSet();
  Collection set=graph.getAllNodesOfClass(nc.getNodeClassID());
  for (Iterator i = set.iterator(); i.hasNext(); ) \{
    NodeID nid = (NodeID) iter.next();
    String typeName = nid.getNodeClassID().getName();
    ProgresNodeTypeID ntid=schema.getNodeTypeIDByName(typeName);
    retSet.add(new ProgresNodeID(nid, ntid));
  \}
  return retSet;
\}
```

Fig. 7. Source code for retrieving all nodes matching a specific node class

is the conversion of Gras/GXL node identifiers to PROGRES node identifiers. Without Gras/GXL's rich graph model the realization would be much more complicated and less efficient, because we would have to implement our own schema management.

The experience gained during the realization of the mapping for PROGRES graph model led us to the conclusion that it should be able to generate the mapping based on an UML class diagram, constraints, and possibly other informations. At the moment we are investigating this issue. General criteria for a good mapping have not been defined up to now and are also part of our future work.

6   Conclusion and Future Work

In this paper we presented the Gras/GXL database management system and its graph model. We explained how different graph models can be implemented on top of the graph model — namely the graph models of PROGRES, DiaGen, and DiaPlan. The realization of these graph models can be used to add persistency capabilities to these graph transformation systems.

The Gras/GXL database management system is still in its early stages. Currently, the implementation of the graph model for the PostgreSQL DBMS and the in-memory implementation have been finished. An implementation for the FastObjects OODBMS is under construction. A transaction management service based on the OpenORB transaction manager as well as the rule engine for triggering and handling events have been implemented. The two finished implementations of the graph model utilize these services and are fully functional, tested with the help of unit test cases [2]. To compare the different realizations we just begun the implementation of a benchmarking application.

The implementation of a PROGRES graph model has been half done, the implementation of the DiaGen and DiaPlan graph model has just started. After finishing the implementation of the PROGRES graph model, we will extend the PROGRES environment to generate code from PROGRES specifications for our database management system instead of GRAS. In combination with the implementation of a new base layer for the UPGRADE framework, it will be possible to use Gras/GXL for PROGRES prototypes. Hereafter, we will focus on the development of a graph query language and how to utilize it for graph matching and defining views on graphs. In addition, we will examine how Gras/GXL can be used as the foundation of an engine for executing graph transformations exchanged as GTXL [22] documents. Another open issue is the generation of mappings based on UML class diagramms and OCL constraints.

References

[1] Gaudenz Alder. *Design and Implementation of the JGraph Swing Component*, 2002. 49

[2] Robert V. Binder. *Testing Object-Oriented Systems: Models, Patterns, and Tools*. Object Technology Series. Addison Wesley, Reading, 2000. 58

[3] Boris Böhlen, Dirk Jäger, Ansgar Schleicher, and Bernhard Westfechtel. UP-GRADE: A framework for building graph-based interactive tools. In Tom Mens, Andy Schürr, and Gabriele Taentzer, editors, *Electronic Notes in Theoretical Computer Science*, volume 72 of *ENTCS*. Elsevier Science Publishers, October 2002. 46, 49

[4] Andrea Corradini, Hartmut Ehrig, Hans-Jörg Kreowski, and Grzegorz Rozenberg, editors. *Proc. 1st Intern. Conf. on Graph Transformation (ICGT '02)*, volume 2505 of *LNCS*. Springer, October 2002. 59, 60

[5] H. Ehrig, C. Ermel, and J. Padberg, editors. *UNIGRA 2001: Uniform Approaches to Graphical Process Specification Techniques*, volume 44 of *ENTCS*. Elsevier Science Publishers, April 2001. 60

[6] GXL Web Site. http://www.gupro.de/GXL, 2003. 50

[7] B. Hoffmann. Shapely hierarchical graph transformation. In *Proc. Symp. on Visual Languages and Formal Methods (VLFM '01)*, pages 30–37. IEEE Comp. Soc. Press, September 2001. 48

[8] B. Hoffmann. Abstraction and control for shapely nested graph transformation. In Corradini et al. [4], pages 177–191. 48

[9] Berthold Hoffmann and Mark Minas. Towards rule-based visual programming of generic visual systems. In *Proc. 1st Intern. Workshop on Rule-Based Programming (RULE '2000)*, pages 111–125, September 2000. 48

[10] Richard Holt, Andreas Winter, and Andy Schürr. GXL: Towards a standard exchange format. In *Proc. 7th Working Conf. on Reverse Engineering (WCRE '00)*, pages 162–171. IEEE Comp. Soc. Press, November 2000. 47

[11] Dirk Jäger, Ansgar Schleicher, and Bernhard Westfechtel. AHEAD: A graph-based system for modeling and managing development processes. In Mandred Nagl, Andy Schürr, and Manfred Münch, editors, *Proc. AGTIVE'99*, volume 1779 of *LNCS*, pages 325–339. Springer, September 2000. 46

[12] H. J. Köhler, U. Nickel, J. Niere, and A. Zündorf. Integrating UML diagrams for production control systems. In Carlo Ghezzi, Mehdi Jazayeri, and Alexander L. Wolf, editors, *Proc. 22nd Intern. Conf. on Software Engineering (ICSE '00)*, pages 241–251. ACM Press, June 2000. 46

[13] Bodo Kraft, Oliver Meyer, and Manfred Nagl. Graph technology support for conceptual design in civil engineering. In M. Schnellenbach-Held and Heiko Denk, editors, *Advances in Intelligent Computing in Engineering, Proc. 9th Intern. EG-ICE Workshop*, pages 1–35. VDI Düsseldorf, 2002. 48

[14] Bernt Kullbach and Andreas Winter. Querying as an enabling technology in software reengineering. In Paolo Nesi and Chris Verhoef, editors, *Proc. 3rd Europ. Conf. on Software Maintenance and Reeng.*, pages 42–50. IEEE Comp. Soc. Press, March 1999. 49

[15] André Marburger and Dominikus Herzberg. E-CARES research project: Understanding complex legacy telecommunication systems. In *Proc. 5th Europ. Conf. on Software Maintenance and Reengineering*, pages 139–147. IEEE Comp. Soc. Press, 2001. 46

[16] André Marburger and Bernhard Westfechtel. Graph-based reengineering of telecommunication systems. In Corradini et al. [4], pages 270–285. 46

[17] Mark Minas. Concepts and realization of a diagram editor generator based on hypergraph transformation. *Science of Computer Programming*, 44(2):157–180, 2002. 46, 48

[18] Mark Minas and Berthold Hoffmann. Specifying and implementing visual process modeling languages with DIAGEN. In Ehrig et al. [5]. 46

[19] Hausi A. Müller, Mehmet A. Orgun, Scott R. Tilley, and James S. Uhl. A reverse engineering approach to subsystem structure identification. *Journal of Software Maintenance*, 5(4):181–204, December 1993. 49

[20] Manfred Nagl, editor. *Building Tightly Integrated Software Development Environments: The IPSEN Approach*, volume 1170 of *LNCS*. Springer, 1996. 46

[21] Andy Schürr, Andreas J. Winter, and Albert Zündorf. PROGRES: Language and environment. In Hartmut Ehrig, Gregor Engels, Hans-Jörg Kreowski, and Grzegorz Rozenberg, editors, *Handbook on Graph Grammars and Computing by Graph Transformation*, volume 2, pages 487–550. World Scientific, Singapore, 1999. 46

[22] Gabriele Taentzer. Towards common exchange formats for graphs and graph-transformation systems. In Ehrig et al. [5]. 59

[23] Jos Warmer and Anneke Kleppe. *The Object Constraint Language — Precise Modeling with UML*. Addison Wesley, Reading, 1999. 56

Transforming Graph Based Scenarios into Graph Transformation Based JUnit Tests

Leif Geiger and Albert Zündorf

University of Kassel, Software Engineering Research Group
Department of Computer Science and Electrical Engineering
Wilhelmshöher Allee 73, 34121 Kassel, Germany
{leif.geiger,albert.zuendorf}@uni-kassel.de
http://pm-pc1.pm.e-technik.uni-kassel.de/se/

Abstract. This paper describes how the Fujaba CASE tool supports
a semi-automatic transformation of usecase scenarios specified by so
called story boards into automatic test specifications and test implementations. A story board is a sequence of graph snapshots showing the
evolution of a graph based object structure during a typical example execution of an usecase. From such an example execution we automatically
derive a test specification that executes the following three basic steps:
First, a graph transformation is generated that creates an object structure serving as the test bed for the following steps. Second, we generate
an operation that invokes the core method realizing the corresponding
usecase. Third, we generate a graph test with a left-hand side corresponding to the graph structure described as result in the story board.
On test execution, this graph test validates whether the object structure resulting from the usecase execution matches the results modeled in
the usecase scenario. Support for this approach has been implemented
within the Fujaba case tool. The approach has been validated in a major
research project and in several student projects.

1 Introduction

Many modern software development approaches propose a so-called usecase
driven process, e.g. the Rational Unified Process RUP, [4]. In these approaches,
requirements are analyzed using usecase diagrams and textual scenario descriptions. During the analysis phase these textual scenario descriptions are refined
using UML behavior diagrams like sequence diagrams or collaboration diagrams.
In the design phase, the program structure is defined using e.g. class diagrams
and the program behavior may be modelled using e.g. statecharts or in our case
using graph grammar specifications. During these steps and during ongoing system maintenance, it is a major problem to ensure that the program behavior
matches the behavior outlined in the usecase scenarios.

There are two solutions to this behavioral consistency problem. First, there
are a number of approaches that generate program behavior from a number of example scenarios. This works especially for sequence diagrams and statecharts, cf.

J.L. Pfaltz, M. Nagl, and B. Böhlen (Eds.): AGTIVE 2003, LNCS 3062, pp. 61–74, 2004.

[10, 11, 8, 5, 7]. Following such an approach guarantees consistency between scenarios and program behavior. However, these approaches are not yet very mature and they require a large number of very elaborated scenarios in order to work well. In addition, there are still unsolved problems if the generated statecharts are further modified during design and maintenance.

In this work we propose an alternative idea that we have implemented as part of the Fujaba case tool project, cf. [3, 6]. We turn example scenarios into test cases. From each usecase scenario we semi-automatically derive a JUnit test operation. This test operation checks whether given the example situations modelled by the corresponding usecase description, the program behave as described with respect to some observable results. Thereby, we have a simple means to do some plausibility testing for the consistency of usecase scenarios and usecase implementation. To be able to achieve this, our approach relies on scenario specifications using sequences of UML collaboration diagrams, i.e. graph snapshots, for the analysis of usecases. Then we turn certain graph snapshots into graph transformations used by our test operations. Since graph transformations are conceptually just pairs of graphs, this derivation of graph transformations merely requires the copying of certain parts of the scenario descriptions.

The following chapter introduces our running example. This is followed by a description of our software development process. Section 4 outlines our approach for test case generation. We close with some experiences and some future work.

2   Running Example

As running example for this paper, we use the ISILEIT project funded by the German Research Society (DFG Schwerpunktprogramm) at University of Paderborn, cf. [9]. Figure 1 shows a simplified schematic view of such a transportation system. This example employs a number of shuttles that transport goods and material between different robots and assembly lines. Traditionally, such systems are controlled by a single programmable logic device (PLD). However, to become more flexible and to be able to scale to an arbitrary number of shuttles, we proposed to model the different shuttles as autonomous agents. Each such agent is in charge of a specific transportation task. Different agents may execute different tasks at the same time. We modelled the behavior of each agent using our Fujaba environment and we were able to provide a simulation environment allowing to test the interaction between the different agents.

3   The Fujaba Development Process

The FUjaba development Process FUP extends the ideas of [6] by a more elaborated process and by the explicit derivation of tests from scenarios. The FUP is an iterative process starting with requirements elicitation based on usecases. For each usecase FUP requires at least one textual scenario description with

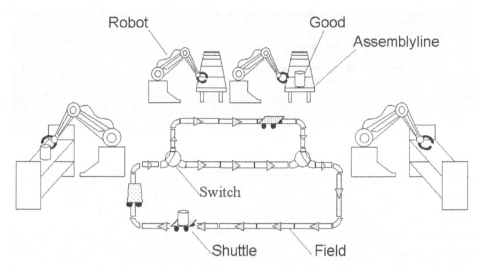

Fig. 1. The ISILLEIT automatic transportation system

a predefined structure. To allow this, we extended the Fujaba case tool with an HTML based text editor with embedded editing of UML diagrams, see Figure 2.

Figure 2 shows a usecase diagram for our transportation system and the textual description of a standard scenario for usecase `load shuttle`. Note, in FUP each textual usecase description has a description of the start situation, a description of the invocation of that usecase, a number of steps outlining the execution of the usecase and a description of the result situation.

In the next step, a special command derives a so-called story board from the textual scenario description. Initially, this story board is just an activity diagram with one activity for each element of the textual scenario. These activities contain the original textual descriptions as a comment. Now the developer models each step by a collaboration diagram that is embedded in the corresponding story board activity, cf. Figure 3. We call this phase story boarding.

The first activity of Figure 3 models the start situation of the `load shuttle` scenario with a shuttle in front of an assembly line that owns a good of type key. The developer modelled this situation as an object diagram/as a graph consisting of a shuttle object `s` and an assembly robot object `ar` that are located at the same field `f`. In addition, there is a good object `g` attached to `ar`.

The second activity shows a collaboration diagram modelling the invocation of the corresponding usecase. In FUP, each usecase is finally realized by a method of some object. Thus, the second activity of the story board just shows one collaboration message naming the operation that corresponds to the usecase. In our case this is the operation `loadShuttle`.

The following activities correspond to the scenario execution steps. They are modelled by collaboration diagrams/graph transformations that typically show a method invocation and the object structure modifications caused by this

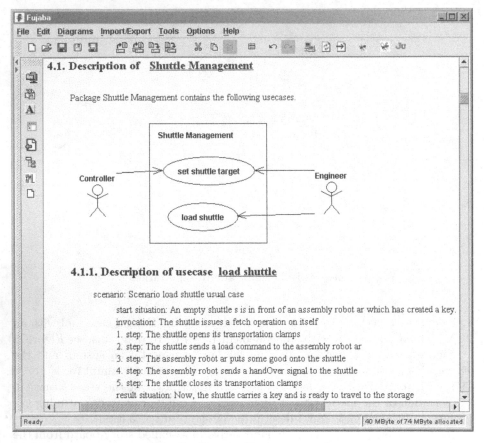

Fig. 2. Usecase diagram example

operation. For example, in our simulation, the `openClamps` operation in the third activity causes an assignment to the `clamps` attribute of the shuttle object `s`.

Note, in Fujaba the left-hand and right-hand side of a graph transformation are shown as a single graph with «destroy» and «create» markers identifying elements that are only contained in the left-hand or only in the right-hand side, respectively. See for example how the `holds` link is replaced by a `carries` link in activity 5 in order to model that the good is loaded onto the shuttle.

In our approach, the last activity of a story board always models an object diagram/a graph representing the result of the usecase execution with respect to the corresponding start situation. Thus, the object diagram of the start situation may also be interpreted as one possible pre-condition for the execution of the usecase and the object diagram of the corresponding result situation may be interpreted as the post-condition for this scenario that have to be ensured by the implementation of the operation that realizes the usecase functionality. This is the operation that is called within the second story board activity.

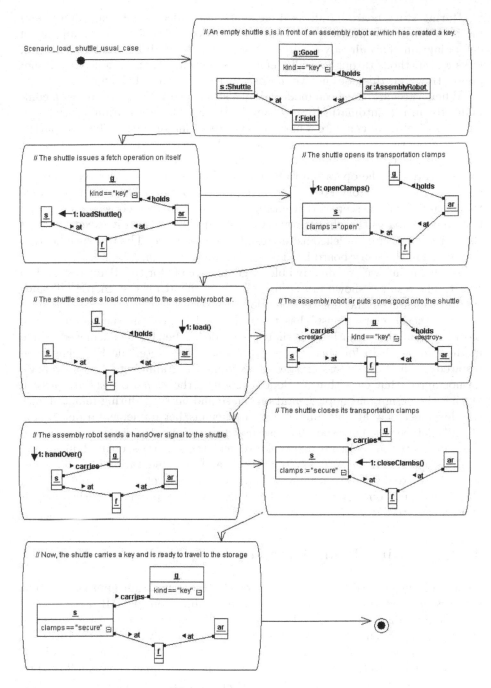

Fig. 3. Story board example

During story boarding all used elements like objects, links, attributes and methods have to be provided with appropriate declarations in an accompanying class diagram. This already ensures a consistent use of object kinds, attributes, links and methods throughout all scenarios and even within the following design phase. However, this does not yet cover consistency at the behavioral level.

When a scenario has been modelled by a story board, Fujaba provides a command to turn it automatically in a simple JUnit test specification for this scenario. Basically, we generate a method with three major parts. The first part is derived from the modelled start situation of the story board. The generated test operation just creates a similar object structure at runtime. The second part is the invocation of the operation to be tested. The third part is derived from the last activity of the story board that is supposed to model the object structure that results from the scenario execution. We turn this into an operation/a graph test that compares the object structure resulting from the test execution with the object structure modelled as the result of the scenario. The test specification derived from the story board in Figure 3 is shown in Figure 4.

At this point, we are already able to generate code for the JUnit test and to run it. The result is shown in Figure 5. For now the JUnit test should fail since the implementation has not yet been done.

Now the developer "just" has to design and implement the methods employed within the scenario such that the modelled behavior is achieved. Some methodological help for this step is proposed in [2]. Note, the FUP has been inspired by the Test-First-Principle of eXtreme Programming [1]. However, we do not agree, that everything is done as soon as the scenario tests are passed. During the design and implementation phase and later on during maintenance, the developer may at any time check whether his/her implementation already or still matches the corresponding usecase descriptions. Due to our experiences this simple approach already helps a lot to keep the scenarios, created during the analysis phase, and the actual system design and implementation in a consistent state. Note, we do not claim to provide a thorough approach for system testing. Our focus is just on keeping the system documentation provided by the analysis scenarios up-to-date.

4   Generating JUnit Testcases

As already mentioned, the basis for the our test generation approach is that the usecase scenarios are modelled using a combination of UML activity and collaboration diagrams, so called story boards. To facilitate the test generation, these story boards must have a certain structure. They always have to start with a model of the Start situation, i.e. with a graph representing the initial object structure. This describes the precondition for the story board and is followed by the Invocation step which is a collaboration diagram containing an invocation of the method that implements the functionality of the corresponding usecase. The last step of the story board is the Result Situation showing the final object structure. The Result Situation is the postcondition of the story board. This

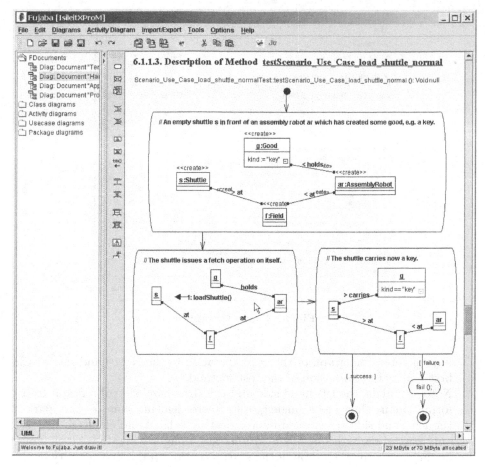

Fig. 4. The generated test

specific story board structure is supported by our tool as every story board is derived from a textual description which already has this structure.

Creating tests is now simply copying graphs into graph transformations and generating code from these graph transformations. To create a test case specification, the following steps are automatically executed by the Fujaba tool:

1. Fujaba creates a JUnit test class for each usecase and adds a test method for each scenario of this usecase to this class. In addition, this new JUnit test class is added to a JUnit test suite for the whole usecase diagram. In the example, a class Scenario_Use_Case_load_shuttle_normalTest is created containing a method testScenario_Use_Case_load_shuttle_normal
   Note, we have chosen the JUnit test framework just because it is simple and popular. Our approach does not depend on this framework.

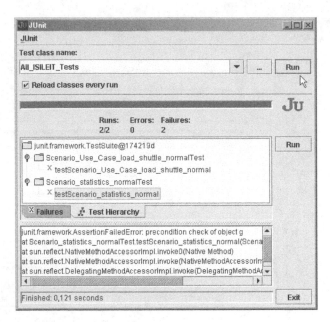

Fig. 5. Example test run

2. Fujaba copies the graph of the start situation to the right-hand side of the first graph transformation of the test method.

   Note, in Fujaba the left-hand side and the right-hand side of a graph transformation are shown as a single graph, where elements that are only part of the left-hand side are marked by a «destroy» label and elements that are only part of the right-hand side are marked using a «create» label. Thus, actually, we copy the graph of the start situation to the first activity of the test method and then we add «create» labels to all elements of the graph, cf. the first activity of Figure 3 and Figure 4.

   As a result, on test execution, the first graph transformation of the test method creates exactly the object structure, that has been modelled as the starting point of the corresponding usecase scenario.

3. Fujaba copies the graph of the Invocation to the left-hand and to the right-hand side of the second graph transformation of the test method.

   Actually, in the Fujaba notation, we need only one copy of the graph without «destroy» and «create» labels, cf. the second activity of Figure 4. At execution time, the resulting graph transformation will not modify the given graph, but it just re-computes the match between graph transformation and host graph. In Fujaba, this matching can be facilitated by using so-called bound objects. Bound objects are rendered as boxes containing just the object name but not the object type. For the Java code generator of the Fujaba environment, a bound object indicates that for this object the match of a previous graph transformation step shall be reused. We do not need to compute the match again. Thus, we copy the graph of the invocation activity to the test method and we mark all objects as bound in order to facilitate the test execution.

Note, as already discussed, we assume that the start situation of a story board models the whole object structure that is a pre-condition for the considered scenario. In addition, the invocation step shall just show the method-calls that trigger the execution of the scenario but it should not introduce new graph elements that are not already mentioned in the start situation. Thus, the object structure employed in the invocation step should be a subgraph of the object structure modelled in the start situation. During test generation this restriction allows us to mark all objects of the invocation activity as bound which facilitates the test execution. If the invocation step would introduce new graph elements, the invocation activity of the test method would have to create these elements. This means, we would have to identify such newly introduced graph elements and to mark them with ≪create≫ labels. This is not yet implemented.

As already mentioned, the invocation step of the story board should contain a method invocation triggering the execution of the corresponding usecase. This method invocation is copied from the story board to the test method, too. In Fujaba, such a method invocation is shown as a collaboration message. Fujaba graph transformations execute such a collaboration message after the execution of the rewrite step. Thus, at test execution time, the second activity of the test method just invokes the method that should implement the corresponding usecase. In our example, method loadShuttle is issued on the shuttle object.

4. Fujaba copies the graph of the Result Situation to the left- and right-hand side of the third graph rewrite rule. As already discussed, in Fujaba this requires just a plain copy. In addition, we mark all objects of the result situation that are already part of the start situation as bound in order to facilitate the test execution. In our example, all objects of the result situation are already known from the start situation, thus all objects of the third acitivty are marked as bound.

   At test execution time, the resulting graph transformation compares the object structure resulting from the method invocations of the previous step with the object structure modelled as result situation in the original story board.

   Note, although our example employs only bound objects, i.e. for all objects we reuse the match stemming from the start situation, our result situation differs from our start situation with respect to the holds and carries link. Thus, in our example the final test activity checks, if the invocation of method loadShuttle in the previous step actually achieved that object s has a carries link to the object g and if its kind attribute has the value "key", cf. Figure 4).

5. Fujaba creates two outgoing transitions from the result activity of the test method that signal the success or failure of the test.

If the graph test of the result activity is successfully applied we provide a [success] transition leading to a stop activity. At test execution time, this will just terminate the test and JUnit will flag the test as passed, cf. Figure 4. If the graph test of the result activity does not match, we provide a [failure] transition leading to a so-called statement activity that executes the JUnit command fail() and then we terminate the method. At test execution time, this causes JUnit to flag that the test has not been passed.

6. Fujaba generates code for the test class.
   The Fujaba code generator is able to translate such (test-) method specifications into usual Java code. This usual Java code may be employed together with manually coded program parts, it may utilize existing libraries and it may be embedded into existing frameworks like i.e. the JUnit framework.

7. Fujaba compiles the code with a standard Java compiler and the JUnit library.

8. Fujaba runs the test within the JUnit Framework. Figure 5 shows that the test execution has failed as in our example the loadShuttle() method has not yet been implemented.

9. If the test fails, Fujaba debugs the test execution and uses our Dynamic Object Browsing System DOBS in order to visualize the result graph stemming from the test execution and to compare it with the result graph modelled in the story board/test specification, cf. Figure 6.

To summarize, we derive a JUnit test method from a story board modelling some example execution of some usecase, that performs three major steps. First, the object structure is created that models the start situation of the considered scenario. Second, we invoke the method that implements the usecase. Third, we compare the object structure resulting from the method invocation with the object structure modelled as result situation within the story board.

Note again, such simple tests may only be used to check whether the implementation of a usecase matches the corresponding usecase descriptions. They do not provide thorough system tests. We have no experience whether this approach can be extended for the purpose of system testing or not.

5   Conclusions and Future Work

Our approach of generating test methods out of the scenarios has been used with good success in several student projects with altogether about 60 students. Within these projects we have made the following observations.

Since we use scenarios, i.e. story boards, as input for test and code generation, our approach requires relatively elaborated scenario diagrams. However, we don't consider this as a flaw but as a feature. Without the subsequent test generation, in earlier student projects we observed that frequently the analysis scenarios missed important details. In these earlier projects, the scenarios frequently covered only a small fraction of the important design aspects while many important questions were post-poned to the design and implementation phase. Using the test generation mechanisms, in the new projects the students were forced to come up with an object structure that covered almost all important design aspects of our example applications. We saw much better scenarios and analysis documents.

In addition, we observed, that the students were much more ready to invest work into better scenarios since they knew that this work pays back through the generated test specifications.

Note, although we require a level of detail that allows code generation, it is still possible to model the scenarios at a very high level of abstraction. The scenarios only have to cover the relevant parts of the application domain. They still do not need to deal with implementation details. In the student projects, such details were typically added during the design and implementation of the usecase realization methods. In that phase, the students introduced supporting object structure details and they dealt with algorithmic and GUI problems and so on.

All of our students followed the test driven software development approach proposed by the Fujaba Process. Thus, they fist generated the JUnit tests and they then began to realize the scenarios step-by-step while using the tests as executable requirements checks. Due to our impressions, this approach worked very well. The student were very focused in their design and implementation work and trying to fulfill the test conditions was very motivating for them. We had some concerns that the students would come up with very specific implementations covering only the cases that were tested. However, this happened only a few times when time pressure was very high at the end of the project. Most of the time, the student came up with very general methods that covered general cases, too, and that would work in all plausible usage scenarios.

Overall, the desired behavioral consistency between analysis scenarios and design and implementation was very well achieved. Due to our experiences, the resulting scenarios provided a much better documentation of the implemented systems than the documentations that have been produced in earlier projects without our test generation approach.

Besides this encouraging results, we observed a number of improvement possibilities. First of all, there is some learning curve. Our students needed some time to understand how the start situation, the invocation step and the result situation have to be modelled in order to fit for our test generation. Frequently, the start situation was not showing all objects and graph elements that were required in order to allow the usecase method to work without problems. The students tended to introduce some of these objects during subsequent scenario steps. In the future we will address this problem more explicitly in our lectures. In addition, the Fujaba tool might check subsequent scenario steps for not yet introduced graph elements and either mark them with errors or it may even try to extend the start situation with such elements automatically during test generation.

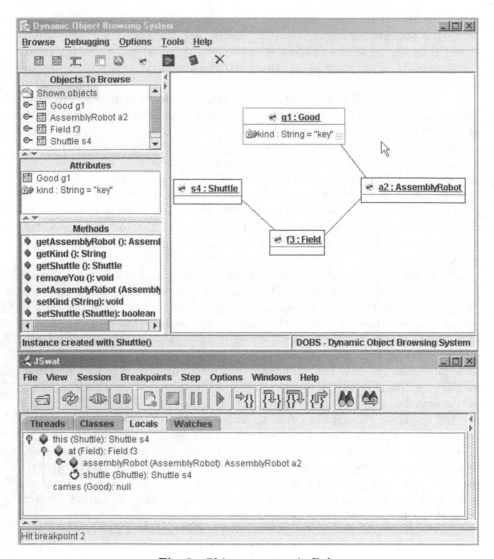

Fig. 6. Object structure in Dobs

Similarly, we frequently had the problem that the students misused destroy and create labels and they made mistakes in using bound objects in the scenario steps. Unfortunately, these faults often remained undetected until test generation or even until test execution time. Thus, the students were always uncertain whether the test failed due to an erroneous implementation or due to such a mistake within the scenario. This problem may be addressed by extending Fujaba with a compile time checker for scenarios.

If the test fails due to an incorrect implementation, most of the time it was very difficult for the students to find the cause of the problem. JUnit just reported an `AssertionFailedError` without further hints, cf. Figure 4. In this situation, the students had to use a debugger and our Dynamic Object Browsing System DOBS to execute the test stepwise and to inspect the runtime object structure and to compare it with the object structure expected by the result situation graph test, cf. Figure 6. This debugging was very tedious. To improve this, as a first step it should be reported which parts of the result situation graph test were successful and which parts did not match and Dobs should be started automatically allowing the comparison between expected and actual result graph. This might give a hint, which object or link is missing or which attribute has a wrong value. So, the problem would be much easier to understand than with just the cryptic JUnit exception outputs and debugging would be facilitated.

To complete the implementation it could also be useful to know up to which point the test execution fits the modelled scenario and where it starts to differ such that the postcondition check finally fails. In our example it could happen, that, e.g., in the fifth graph of the story board (Figure 3) the `holds` link is destroyed but the `carries` link is not created. To help identifying such problems, the Fujaba tool should report that the test execution meets the modelled scenario until scenario step four and that scenario step five is never reached. This may be achieved by two different techniques. First, the developer could add some "control points" into the implementation, where at test runtime it is checked if the object structure matches a certain scenario step. It should be possible add such a mechanisms to Fujaba. Another idea is to derive detailed traces from the test runs and to compare these execution traces to the story board. However, this approach faces the problem that the test trace probably shows all kinds of implementation details while the story board scenario is modelled at the domain level of abstraction. We need to find out, how to filter the scenario relevant steps from a test trace. Generally, we would like to make more use of the intermediate steps of a story board.

References

[1] K. Beck: Extreme Programming Explained: Embrace Change; Addison-Wesley, ISBN 0-201-61641-6, 1999. 66
[2] I. Diethelm, L. Geiger, T. Maier, A. Zündorf: Turning Collaboration Diagram Strips into Storycharts; Workshop on Scenarios and state machines: models, algorithms, and tools; ICSE 2002, Orlando, Florida, USA, 2002. 66
[3] Fujaba Homepage, Universität Paderborn, http://www.fujaba.de/. 62
[4] I. Jacobson, G. Booch, J. Rumbaugh: The Unified Software Development Process; Addison-Wesley, ISBN 0-201-57169-2, 1999. 61
[5] K. Koskimies, T. Männistö, T. Systä, J. Tuomi: SCED - An environment for dynamic modeling in object-oriented software construction; Proc. Nordic Workshop on Programming Environment Research '94 (NWPER '94), Lund, Department of Computer Science, Lund Institute of Technology, Lund University, pp. 217-230, 1994. 62

[6] H. Köhler, U. Nickel, J. Niere, A. Zündorf: Integrating UML Diagrams for Production Control Systems; Proc. of ICSE 2000 - The 22nd International Conference on Software Engineering, June 4-11th, Limerick, Ireland, acm press, pp. 241-251, 2000. 62

[7] E. Mäkinen, T. Systä: MAS: An Interactive Synthesizer to Support Behavioral Modeling in UML; Proc. 23rd International Conference on Software Engineering (ICSE 2001), Toronto, Canada, acm press, 2001. 62

[8] T. Maier, A. Zündorf: The Fujaba Statechart Synthesis Approach; 2nd International Workshop on Scenarios and State Machines: Models, Algorithms, and Tools (ICSE 2003), Portland, Oregon, USA, 2003. 62

[9] U. Nickel, W. Schäfer, A. Zündorf: Integrative Specification of Distributed Production Control Systems for Flexible Automated Manufacturing; M. Nagl, B. Westfechtel (eds.): Modelle Werkzeuge und Infrastrukturen zur Unterstützung von Entwicklungsprozessen, Wiley-VCH, 2003. 62

[10] Somé S.: Beyond Scenarios: Generating State Models from Use Cases; Scenarios and state machines: models, algorithms, and tools (ICSE 2002 Workshop), Orlando, Florida, USA, 2002. 62

[11] Whittle J. and Schumann J.: Statechart Synthesis From Scenarios: an Air Traffic Control Case Study; Scenarios and state machines: models, algorithms, and tools (ICSE 2002 Workshop), Orlando, Florida, USA, 2002. 62

[12] A. Zündorf: Rigorous Object Oriented Software Development, Habilitation Thesis, University of Paderborn, 2001.

On Graphs in Conceptual Engineering Design

Janusz Szuba[1,3], Agnieszka Ozimek[2], and Andy Schürr[3]

[1] Institute of Fundamental Technological Research
Polish Academy of Sciences
ul. Swietokrzyska 21, 00-049 Warsaw, Poland
jszuba@ippt.gov.pl
[2] Institute of Computer Modelling
Cracow Technical University
ul. Warszawska 24, 31-155 Cracow, Poland
aozimek@pk.edu.pl
[3] Real-Time Systems Lab
Darmstadt University of Technology
Merckstr. 25, D-64283 Darmstadt, Germany
{janusz.szuba,andy.schuerr}@es.tu-darmstadt.de

Abstract. This paper deals with the subject of knowledge-based computer aided design. A novel method, giving additional support for conceptual design, is presented. Using this method, a designer first specifies the functional requirements and the structure of the object to be designed, based on use cases and function graphs. A prototype design is then derived from these requirements. Subsequently, the designer checks the fulfilment of certain consistency rules and engineering norms by the application of a constraint checker. This checker uses background knowledge stored in graph structures and the reasoning mechanism provided by the graph rewriting system PROGRES. An example of designing a swimming pool illustrates the proposed methodology.

1 Introduction

Designers, especially architects, very frequently use graph structures to represent functional and spatial relations of the object to be designed. Based on this observation, a new conceptual design method has been created, in which use cases and scenarios, as well as functional requirements and constraints, are specified by using graph structures. At present, we restrict our considerations to architectural design, but it seems to be the case that the developed method could also be applied to other engineering domains such as Machine Building or Electrical Engineering. The method is presented here for designing a swimming pool, but it has already been applied to and is suitable for other types of buildings. The design tool GraCAD [25] currently being elaborated, which utilises this method, can be seen as a conceptual pre-processor for a new generation of CAD-tools. For prototyping the graph part of GraCAD we use the graph-rewriting system PROGRES developed at the RWTH Aachen [21], i.e. we use a kind of graph transformation-based approach for knowledge representation purposes.

J.L. Pfaltz, M. Nagl, and B. Böhlen (Eds.): AGTIVE 2003, LNCS 3062, pp. 75–89, 2004.

Pioneered by N. Chomsky [8], the linguistic (grammar-based) approach to world modelling has been applied in many areas. The core idea in this methodology is to treat certain primitives as letters of an alphabet and to interpret more complex objects and assemblies as words or sentences of a language based upon the alphabet. Rules governing the generation of words and sentences define the grammar of the considered language. In terms of words, modelling such a grammar generates a class of objects that are considered to be plausible. Thus, grammars provide a very natural knowledge representation formalism for computer-based tools that should aid design.

Since G. Stiny [22] developed shape grammars, many researchers have shown how such grammars allow the architect to capture essential features of a certain style of buildings. However, the primitives of shape grammars are purely geometrical, which restrict their descriptive power. Substantial progress was achieved after graph grammars were introduced and developed (cf. e.g. [20]). Graphs are capable to encode much more information than linear strings or shapes. Hence, their applicability for CAD-systems was immediately appreciated [12].

A special form of graph-based representation used for design purposes has already been developed by E. Grabska [13] in 1994. Later on, Grabska's model served as the basic knowledge representation scheme in research reported at conferences in Stanford [14], Ascona [4], and Wierzba [5]. It turned out that by introducing an additional kind of functionality graphs into Grabska's model, conceptual solutions for the designed object can be conveniently reasoned about. The additional functionality analysis of houses, as the starting point for the conceptual design, has been proposed by several researchers (compare, e.g. [6], [9]). Such a methodology allows the designer to detach himself from details and to consider more clearly the functionality of the designed object incorporating the constraints and requirements to be met, and the possible ways of selecting optimum alternatives.

The results described in sections 2 and 3 can be viewed as the further development of the research reported previously in [23], [24] and [25], i.e. of combining, for the first time, graph transformation techniques with a conceptual design approach for buildings, based on functionality analysis. As a result of co-operation with architects, we were able to apply our method to a more complex, realistic example than in previous publications and verify the usefulness of this method. According to one architect's suggestions (2nd author of this paper), our method has been supplemented with a new requirements elicitation phase (cf. Section 2). UML activity diagrams [3] are now used to capture the behaviour of prototypical users of the designed object, in the form of use cases. Apart from this extension of our conceptual design method, the paper studies different ways of replacing the previously used 'procedural' way of implementing constraint checks, by using the new built-in graph constraint checking mechanisms of the language PROGRES (cf. Section 3). We will see that these new constructs are still difficult to use and should be improved in various ways.

2   Graph Technology Supporting Conceptual Design

This section concerns a very early phase of engineering design called conceptual design. The main aim of conceptual design is to specify the functional requirements and constraints resulting from the conversation with a customer, and then to create a prototype design that fulfils the specified requirements. In this phase the designer operates on high-level concepts. Details that would distract him from conceptual thinking are omitted. Rapid specifying of consistent functional requirements and the creation of a prototype design meeting these requirements are very important in the process of designing. Such a prototype facilitates communication with the customer and is the basis for further discussion about the designed object. Such a discussion facilitates the understanding of the customer's intentions, preferences and may result in modifications to the requirements or the prototype. After reaching an agreement with the customer, the designer begins working out the detailed design. Below, a novel method supporting conceptual design activities of architects is presented.

Architects very frequently use graphs to depict the functional and spatial relations of designed objects. Furthermore, they use control flow graphs, similar to UML activity diagrams, to show the order of activities performed in the considered design object; i.e. — similar to software engineers — architects follow a use case driven approach for requirements elicitation purposes (cf. [16], [19]). Based on these observations, we have created a method that addresses the conceptual phase of architectural design, in which the functional requirements and constraints for designed buildings are specified in the form of graph structures. In this method, UML use case and activity diagrams are integrated with so-called area and room graphs, which are then translated into a prototype design (cf. Fig. 1). One of the main advantages of the graph-based design approach introduced thereby is the possibility to specify domain-specific design rules and norms on area and room graphs, in the form of constraints on a very high level, and to derive the corresponding consistency checking code automatically by using the graph transformation system PROGRES.

In the following, the new design method is explained in more detail using the running example of a swimming pool. The architect starts with a functional analysis of the building, based mainly on conversations with an investor (cf. Fig. 1):

- First, the architect specifies some basic parameters. For the swimming pool example these parameters include the type of the swimming pool (recreational, sports, learner) and an approximate number of users.
- Then the architect identifies different sorts of users (stakeholders) of the constructed building (swimming pool client, lifeguard, technical staff, cleaning service, administration staff etc.).
- The following step is to identify use cases for these users: 'swimming' for swimming pool clients, 'observing' and 'rescuing swimmers' for lifeguards (further considerations are restricted only to the client and the lifeguard).
- The next step is to define for each use case a scenario in the form of an activity graph (UML activity diagram), which explains in more detail the

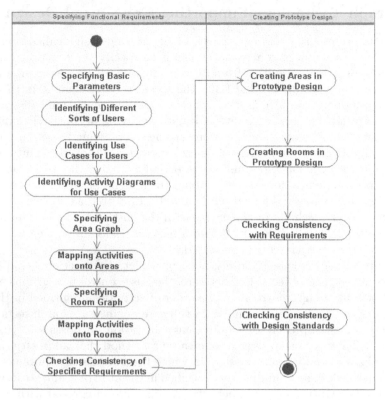

Fig. 1. Aiding conceptual phase of architectural design

planned interaction between the user and the designed object. These activity graphs model the most frequent and important behaviours of the users. By creating scenarios in the form of activity graphs for various types of users, the functionality of the object is considered from various points of view (from the perspective of the client of the swimming pool, the lifeguard — cf. Fig. 2, etc.).

– Then the architect has to decompose the designed object into areas. For this purpose, the architect creates an area graph in which area nodes and relations between them like accessibility, visibility, adjacency are specified. Area nodes correspond to functional areas of the designed building. Fig. 3 presents an area graph; area nodes are surrounded by dashed lines; edges between area nodes are not displayed, to preserve the clarity of our example.[1]

– Afterwards, activity nodes from activity graphs have to be mapped onto area nodes. Mapping activities onto a given area node means that these activities are performed in the assigned area. This mapping could be skipped

---

[1] For some types of buildings, especially small or medium ones, like single storey family houses, the specification of an area graph could be skipped, for big buildings, like an airport the decomposition into areas is recommended.

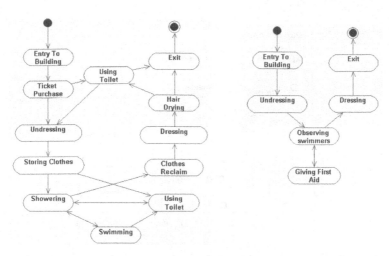

Fig. 2. Activity graphs/diagrams a) for client of swimming pool b) for lifeguard

for small and medium size buildings (mapping onto areas is not displayed in our example).

- Next, a room graph has to be specified, i.e. the decomposition of the designed object at the level of rooms. Room nodes, relations between rooms, and assignment rooms to areas are specified. The relations most frequently used by architects at the level of rooms are accessibility, adjacency and visibility, but other relations could be introduced as well.
- Finally, activities from activity graphs are transferred from their assigned areas to the corresponding rooms. Mapping an activity onto a given room node means that the activity is performed in the room.

By the way of this method, a functional requirements' graph of the building, consisting of areas, rooms, activities, and edges between them, is created (cf. Fig. 3). Lower level decomposition elements like changing cabins, cabinets, urinals, showers etc. could be introduced into the graph model, but they have been excluded from our example.

Functional requirements defined in this way may be checked with respect to quite a number of general or domain specific consistency rules. For instance, the following consistency rules have been identified for activities:

- Check if the order of activities in the activity graph is appropriate: for many types of architectural objects it is possible to specify the permitted order of activities for the considered subdomain in advance and independent from the design process of a specific object of this type. For instance, in a swimming pool after showering the next activity might be swimming.
- Check if for every activity an appropriate room and area is assigned: Start and Stop activities are exceptions, which are used to mark beginning and ending of a scenario. All other activities are first assigned to specific areas and then to the rooms related to these areas.

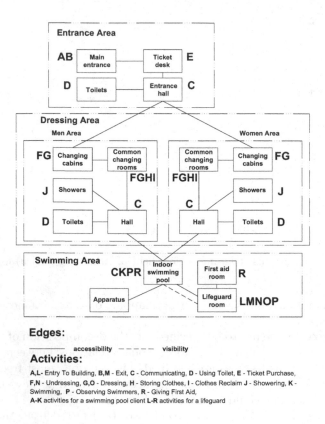

Fig. 3. Swimming pool graph

– Check whether users are able to perform the activities comfortably in the order imposed by the activity graph if the building (at the level of rooms) has a structure matching the defined room graph.
– ...

In Section 3, the PROGRES implementation of the graph checker for activity-related consistency rules explained above is presented.

After specifying consistent functional requirements, the architect creates a prototype design/floor layout of the building. He/she operates on high level geometrical concepts like zones, rooms and maps them to area nodes and room nodes from area/room graphs (cf. Fig. 4 and [25]). Due to this mapping it is possible to check whether the design meets functional requirements specified in the graph. Based on geometrical elements of the prototype — like walls, doors, and windows — the relations between conceptual elements are computed by the checker and compared with the structure of the graph of functional requirements. An example of such a check is verifying whether room nodes connected by the

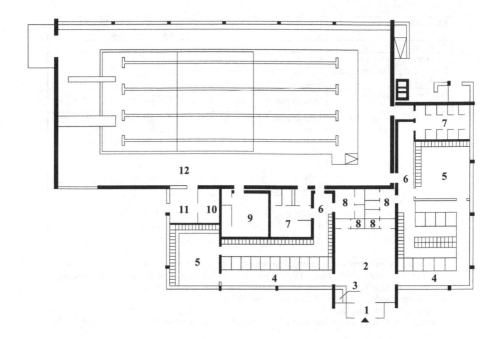

1) main entrance 2) entrance hall 3) ticket desk 4) changing cabins 5) common changing rooms 6) hall 7) showers 8) toilets 9) apparatus 10) first aid room 11) lifeguard room 12) indoor swimming pool

Fig. 4. Prototype design of swimming pool — Hallenbad Espelkamp/D, architect: A.Wagner(1963), published in [10], p. 132–133

accessibility edge in the room graph correspond in the prototype design/floor layout to rooms that are adjacent and accessible via appropriately placed doors.

The next issue worth checking is the agreement between the layout proposed and the architectural norms. In particular we have to check whether:

– each area, room, and other design element has adequate spacing
– the location of each room with regard to geographical considerations is correct (for example whether the swimming pool is placed with its longer wall towards the South)
– windows provide rooms with enough light (the minimal window area for rooms intended for prolonged occupancy is equal to 1/8 of the floor area)
– the layout is optimal for the supply of essential services (water, gas, electricity) and ventilation, etc.

Fig. 5 shows standards for the swimming pool design cf. ([10], [19], [26]) that could be used in architectural norm checkers.

After creating the prototype design and checking its consistency with functional requirements and architectural norms, the prototype is transformed to

Elements	Indicators	Quantities for 25 x 12,5 m. swimming pool
Pool water area	-	312,5 m^2
Surface of pool surrounds	60 – 70 % of pool water area	190 – 230 m^2
Number of places for storing clothes	1 place per 1,5 m^2 of pool water area	210 places
Number of places for changing clothes	60% of places for storing clothes	120 places
Max. number of users at the same time including:	1 person per 1,5 m^2 pool water area	210 people (70 people in the water, 140 – in pool surrounds, changing room, cafe, etc.)
men	55%	115 men
women	45%	95 women
Number of showers	1 shower per 5-7 users	24-17
Number of toilets for: men	1 toilet per 50 men 1 urinal per 30 men	2 toilets 4 urinals
women	1 toilet per 25 women	4 toilets

Min. swimming hall length - length of the swimming pool + 2 x 3m (3m. on both sides of the pool)
Min. swimming hall width - length of the swimming pool + 2 x 2m. (2m. on both sides of the pool)
Most common lengths for sports swimming pools: **51m**, **50m**, 25m, 33 $^1/_3$ m.
Most common widths for sports swimming pools: 12,5m, 15m:
Min. width of swimming lane: 2m.
Most common dimensions for learner pools: 10,0 x 4,5 m, 12,5 x 6,0 m, 12,5 x 8,0 m, 16 $^2/_3$ x 8 $^1/_3$

Fig. 5. Standards for swimming pools

the standard CAD environment (ArchiCAD, Architectural Desktop), which is more suitable for the detailed, technical design. In this environment the architect continues designing but at a much more detailed level than in the conceptual phase.

3   Swimming Pool Checkers — PROGRES Specification

For prototyping (constraint) checkers for the above mentioned consistency rules we now use (in contrast to previously published papers about the related activities) the recently introduced mechanism of local constraints and repair actions available in the graph rewrite system PROGRES. A local constraint declaration introduces an integrity constraint, which is an arbitrary first-order predicate logic formula and which is (incrementally) checked at run-time. The constraint-defining Boolean expression uses path expressions to navigate from the considered node self to related nodes in its neighbourhood. The constraints are, therefore, a kind of derived Boolean attributes, which have to be true at the end of so-called safe transactions or productions. (A transaction is a parameterised subprogram, which calls an arbitrary number of tests and productions to implement the desired graph transformation.) The keyword safe marks those productions and transactions which take a consistent graph as input and produce a consistent graph as output with respect to all relevant integrity constraints. All productions without this prefix may produce intermediate graph states which violate some integrity constraints. The violation of a constraint at the end of safe graph transformations causes immediate termination of a running execution process or the activation of a so-called repair action. The repair action is a graph transformation which should be successful and eliminate the detected inconsistency; otherwise, the PROGRES execution machinery stops with throwing a final error message.

The following part of this section shows an example of a consistency checker specified with local constraints and repair actions of PROGRES. This checker verifies whether the order of activities in an activity graph is appropriate and whether every activity has an adequate room assigned. The most important class in our example is an abstract class Activity that represents the activity performed in the designed object.

```
node class Activity is a Object
  meta
    followingActivities : type in Activity [0:n] := Activity;
    activityRooms : type in Room [0:n] := Room;
  constraint
    checkFollowingActivities
      = for all activity : Activity := self.-next-> ::
          activity.type in self.followingActivities
        end
      else
        for all act : Activity := self.-next->
        do
          choose
            when (not (act.type in self.followingActivities))
            then
              removeNextEdge ( self, act )
            else
              skip
          end
        end
      end                                          ;
    checkActivityRooms
      = for all room : Room := self.-activityRoom-> ::
          room.type in self.activityRooms
        end                                        ;
end;
```

Activity is the base class for Stop, Start classes

```
node type Start : Activity
  constraint
    checkIncomingEdges = empty ( self.<-next- );
  redef meta
    activityRooms := nil ;
end;

node type Stop : Activity
  redef meta
    followingActivities := nil ;
    activityRooms := nil ;
end;
```

that exist in the specifications for all kinds of buildings, and the base class for the building (swimming pool) specific activities like Dressing, Undressing (for the sake of clarity we show only these two building specific classes).

```
node type Dressing : Activity
   redef meta
     followingActivities := HairDrying ;
     activityRooms :=
        LifeGuardRoom or ChangingCabins or CommonChangingRooms ;
end;

node type Undressing : Activity
   redef meta
     followingActivities := StoringClothes ;
     activityRooms :=
        LifeGuardRoom or ChangingCabins or CommonChangingRooms ;
end;
```

Activity contains two meta attributes: followingActivities and activityRooms. (A meta attribute is not a node attribute, but a node type attribute. Its value may not be changed at run-time.) The first meta attribute is a set of node types derived from the type Activity, indicating types of activities that could be linked by next edges with the activity of the considered type. The next relation is defined between two Activity classes and indicates the order of activities performed in the designed building. The followingActivities attribute is set in the Activity class to the Activity type and could be redefined in Activity subclasses (cf. Dressing, Undressing, Stop classes). The second meta attribute is a set of node types derived from the type Room, indicating types of rooms that could be linked by activityRoom edges with the activity of the considered type. The activityRoom relation is defined between Activity and Room classes. If an activity a is linked by an activityRoom edge with a room r, it means that the activity a is performed in the room r. The activityRoom attribute is set in the Activity class to the Room type and could be redefined in Activity subclasses (cf. Dressing, Undressing, Start, Stop classes). These meta attributes are used in checkFollowingActivities and checkActivityRooms constraints. The constraint checkFollowingActivities verifies whether every activity linked by the next edge with the considered activity is of a type included in the followingActivities set. The constraint checkActivityRooms verifies whether every room linked by activityRoom to the considered activity is of a type included in its own activityRooms set. In addition to these two constraints, checkIncomingEdges is defined for the Stop class. It checks whether the set of incoming next edges is empty for a given Stop activity. The constraint checkFollowingActivities contains a user defined repair action. This action finds all inconsistent next edges and removes them with the removeNextEdge transaction.

The mechanism of constraints and repair actions seems to be very useful for the specification of checkers. The main disadvantage of the current PROGRES implementation is that it either deactivates constraint checking completely or enforces all constraints, but gives you no option:

1. to activate only selected constraints
2. to display error messages about constraint violation
3. to activate only selected repair actions

The following part of this section shows ad-hoc solutions to these problems. For simplicity we restrict our considerations only to the checkFollowingActivities constraint. The Activity class has been modified as follows:

```
node class Activity is a Object
  intrinsic
    followingActivitiesCheckActivated : boolean := true;
    followingActivitiesRepairActionActivated : boolean := true;
    errorMsg : string := "";
  meta
    followingActivities : type in Activity [0:n] := Activity;
    activityRooms : type in Room [0:n] := Room;
  constraint
    newCheckFollowingActivities
      = (not self.followingActivitiesCheckActivated)
        or ((oldCheckFollowingActivities ( self )
              and (self.errorMsg = "")            )
            or (not oldCheckFollowingActivities ( self )
                  and (self.errorMsg # "")              ))
      else
        choose
          when (self.followingActivitiesRepairActionActivated
                  and not oldCheckFollowingActivities ( self ) )
          then
              oldRepairAction ( self )
              & setErrorMsg ( self, "" )
          else
            setErrorMsg
            ( self,
              "Activity contains not recomanded outgoing next edge." )
          end
        end                                              ;
end;
```

The constraint checkFollowingActivities was replaced in the modified Activity class by newCheckFollowingActivities. In the newCheckFollowingActivities constraint the function oldCheckFollowingActivities is invoked.

```
function oldCheckFollowingActivities :
  ( activity : Activity) -> boolean =
  for all nextActivity : Activity := activity.-next-> ::
    nextActivity.type in activity.followingActivities
  end
end;
```

This function returns true if every activity linked by the nextedge with the activity passed as a parameter has a type included in the activity.followingActivities set. In other words, oldCheckFollowingActivities does the same as checkFollowingActivities, but is defined as a function. In the repair action of the newCheckFollowingActivities constraint, oldRepairAction is invoked.

```
transaction oldRepairAction( activity : Activity) [1:1] =
  for all nextActivity : Activity := activity.-next->
  do
    choose
      when
        (not (nextActivity.type in activity.followingActivities))
      then
        removeNextEdge ( activity, nextActivity )
      else
        skip
      end
  end
end;
```

The oldRepairAction PROGRES transaction has the same functionality as the repair action for the checkFollowingActivities constraint. In the new Activity class, three intrinsic attributes have been added:

- the followingActivitiesCheckActivated Boolean attribute is responsible for activation/deactivation of following activities checking for single node instances. If the value of this attribute is true, checking is activated.

- the followingActivitiesRepairActionActivated Boolean attribute is responsible for activation/deactivation of a repair action for a considered node instance. If the value of this attribute is true, inconsistencies are eliminated by the execution of oldRepairAction, otherwise inconsistencies are marked only by assigning error messages to the errorMsg attribute.
- the errorMsg string attribute is used for marking inconsistencies with error message texts.

In the newCheckFollowingActivities constraint, first the Boolean condition (not self.followingActivitiesCheckActivated) is evaluated. If this condition is true, then the evaluation of the considered constraint is finished, otherwise the remaining part ((oldCheckFollowingActivities (self) and (self.errorMsg = "")) or (not oldCheckFollowingActivities (self) and (self.errorMsg # ""))) is executed. This remaining part means that one of the conditions below has to be fulfilled:

- the function oldCheckFollowingActivities returns a true value for the considered self activity and errorMsg for the self activity equals the empty string.
- the function oldCheckFollowingActivities returns a false value for the considered self activity and the attribute errorMsg for the self activity is a non-empty string.

A violation of both conditions invokes the repair action. In the repair action a guarded choose statement checks the condition (self.followingActivitiesRepairActionActivated and not oldCheckFollowingActivities (self)) i.e. it is checked if the repair action is activated and oldCheckFollowingActivities returns a false value for the considered activity. If the condition is fulfilled the inconsistencies are eliminated with the oldRepairAction transaction and errorMsg is set to the empty string, otherwise the errorMsg attribute is set to the message that describes the error.

The specification above solves problems 1–3. However, a very low level, difficult to understand PROGRES code has to be written. A general PROGRES mechanism:

- allowing activation/deactivation of selected constraints
- marking inconsistencies by means of appropriate error messages
- activation/deactivation of a repair actions for selected constraints

for all nodes or specific nodes of a given node class/type would simplify writing PROGRES specifications for checkers considerably. Future versions of PROGRES or other graph transformation environments with similar features should be extended appropriately or refrain totally from giving any language support for constraint checking purposes.

4   Summary

The research concerning conceptual design, in particular the requirements elicitation phase and the phase of creating the prototype design for the specified requirements, seems to be worth to continue. To the best of our knowledge, the CAD tools for architects developed until now ([1], [2]) do not support the elicitation phase. We are also not aware of the usage of UML activity diagrams for this purpose. Based on this observation, the method supporting conceptual design, previously presented in [23], [24] and [25], has been supplemented with a new requirements elicitation phase. UML activity diagrams are now used to capture the behaviour of prototypical users of the designed object in the form of use cases. In the method, use cases and scenarios as well as functional requirements and constraints for the considered object are specified using graph structures. The consistency of the specified requirements is verified by graph checkers. In other works concerning the usage of graph rewriting system PROGRES in the area of conceptual design of building ([17], [18]), the elicitation phase is skipped and the consistency of a designed building is verified based on the parametrizable graph knowledge specified by a knowledge engineer. In our case it is checked if the object to be designed fulfils the graph requirements specified by designer in the elicitation phase. The combination of those two complementary approaches will be considered in the future.

Graph transformation systems like PROGRES appear to be appropriate tools for prototyping the software implementing such a method. In our graph specification we used the new built-in graph constraint checking mechanisms of the PROGRES language for that purpose in order to considerably simplify the construction of these specifications. But unfortunately, experience so far shows that these new constructs are still difficult to use and should be improved in various ways. In spite of this fact, PROGRES seems to be useful in communication between an architect (a domain expert) and a knowledge engineer, while defining and implementing architectural domain knowledge. The specification presented in section 3 was created in co-operation with architects and the most of the PROGRES statements used in this specification were clear for them. Moreover, the architects made an interesting remark that the formalism of hierarchical graphs and transformations on them would be useful, because the structure of all, or almost all, buildings is hierarchical.

The PROGRES specification for a swimming pool is still in the process of construction. Based on this specification we are going to create a prototype application integrated with an existing CAD tool for architects and verify whether the presented method is useful in practice. In our work on this prototype we are going to use the commercial tool for architects ArchiCAD 8.0 [1]. An add-on extending the functionality of ArchiCAD will be implemented with the use of ArchiCAD API Development Kit 4.3 and Microsoft Visual C++ 6.0. It is planned to integrate this add-on with a graph editor for specifying the functional requirement of the building to be designed. The graph editor will be built on the basis of PROGRES UPGRADE Framework [7]. The other interesting task to consider would be a transformation from an ArchiCAD representation into a graph representation; however, such a reverse engineering step seems to be quite difficult (or even impossible) because of the lack of semantic information about the rooms' purpose in the standard ArchiCAD model.

In some of the design tasks, where engineering rules are strict, it is possible to find a solution to the design problem by the means of graph generation strategies. An example of such a strategy would be the generation of a furniture layout in an office, based on the function of a considered room. The creation of such a strategy in co-operation with architects is planned in the future (cf. [14]).

Finally, we would like to emphasize that the presented method seems to be worth considering in other engineering domains like Machine Building or Electrical Engineering as well.

Acknowledgement

Many thanks to Adam Borkowski and Pawel Ozimek for fruitful discussions and for help in the paper preparation.

This research was partly supported by the Federal Ministry of Education and Research (BMBF) under the joint Polish-German research project 'Graph-based tools for conceptual design in Civil Engineering' coordinated by M. Nagl (RWTH, Aachen), European Research Training Network 'SegraVis' and Computational Engineering Center of Darmstadt University of Technology.

References

[1] ArchiCAD 8.0 Reference guide, Graphisoft, Budapest, 2002
[2] Autodesk Architectural Desktop 3.3 User's Guide, Autodesk, 2002
[3] Booch, G., Rumbaugh, J., Jacobson, I.: The Unified Modeling Language User Guide. Addison Wesley Longman, Reading(1999)
[4] Borkowski, A., Grabska, E.: Converting function into object. In: I. Smith, ed., Proc. 5th EG-SEA-AI Workshop on Structural Engineering Applications of Artificial Intelligence, LNCS 1454, Springer-Verlag, Berlin (1998), 434–439
[5] Borkowski, A. (ed.): Artificial Intelligence in Structural Engineering, WNT, Warszawa (1999)
[6] Borkowski, A., Grabska, E., Hliniak, G.: Function-structure computer-aided design model, Machine GRAPHICS & VISION, 9, Warszawa (1999), 367–383
[7] Böhlen, B., Jäger, D., Schleicher, A., Westfechtel B.: UPGRADE: A Framework for Building Graph-Based Interactive Tools, Proceedings International Workshop on Graph-Based Tools (GraBaTs 2002), Barcelona, Spain, Electronic Notes in Theoretical Computer Science, vol. 72, no. 2 (2002)
[8] Chomsky, N.: Aspects of Theory of Syntax, MIT Press, Cambridge (1965)
[9] Cole Jr., E. L.: Functional analysis: a system conceptual design tool, IEEE Trans. on Aerospace & Electronic Systems, 34 (2), 1998, 354–365
[10] Fabian, D., Bäderbauten: Handbuch für Bäderbau und Badewesen: Anlage, Ausstattung, Betrieb, Wirtschaftlichkeit (Aquatic buildings), Verl. Georg D. W. Callwey, München (1970)
[11] Flemming, U., Coyone, R., Gavin, T., Rychter, M.: A generative expert system for the design of building layouts - version 2, In: B. Topping, ed., Artificial Intelligence in Engineering Design, Computational Mechanics Publications, Southampton (1999), 445–464

[12] Göttler, H., Günther, J., Nieskens, G.: Use graph grammars to design CAD-systems! 4th International Workshop on Graph Grammars and Their Applications to Computer Science, LNCS 532, Springer-Verlag, Berlin (1991), 396–410

[13] Grabska E.: Graphs and designing. In: H. J. Schneider and H. Ehrig, eds., Graph Transformations in Computer Science, LNCS 776, Springer-Verlag, Berlin (1994), 188–203

[14] Grabska, E., Borkowski, A.: Assisting creativity by composite representation, In: J. S. Gero and F. Sudweeks eds., Artificial Intelligence in Design'96, Kluwer Academic Publishers, Dordrecht (1996), 743–760

[15] Grabska, E., Palacz, W.: Floor layout design with the use of graph rewriting system Progres, In: M. Schnellenbach-Held, H. Denk (Eds.), Proc. 9th Int. Workshop on Intelligent Computing in Engineering, 180, VDI Verlag, Düsseldorf (2002), 149–157

[16] Korzeniewski, W.: Apartment Housing — Designers Guide (in Polish), Arkady, Warszawa (1989)

[17] Kraft B., Meyer O., Nagl M.: Graph technology support for conceptual design in Civil Engineering, In: M. Schnellenbach-Held, H. Denk (Eds.), Proc. 9th Int. Workshop on Intelligent Computing in Engineering, 180, VDI Verlag, Düsseldorf (2002), 1–35

[18] Kraft B., Nagl M.: Parameterizable Specification of Conceptual Design Tools in Civil Engineering, to appear in this proceedings

[19] Neufert, E.: Bauentwurfslehre, Vieweg & Sohn, Braunschweig-Wiesbaden (1992)

[20] Rozenberg, G. (ed.): Handbook of Graph Grammars and Computing by Graph Transformation, World Science, Singapore (1997)

[21] Schürr, A., Winter, A., Zündorf, A.: Graph grammar engineering with PROGRES. Proc. 5th European Software Engineering Conference (ESEC'95), W. Schäfer, P. Botella (Eds.), LNCS 989, Springer-Verlag, Berlin (1995), 219–234

[22] Stiny, G.: Introduction to shape and shape grammars, Environment and Planning B: Planning and Design, 7, 1980, 343–351

[23] Szuba, J., Grabska, E., Borkowski, A.: Graph visualisation in ArchiCAD. In: M. Nagl, A. Schürr, M. Münch, eds., Application of Graph Transformations with Industrial Relevance, LNCS 1779, Springer-Verlag, Berlin (2000), 241–246

[24] Szuba, J., Borkowski, A.: Graph transformation in architectural design, Computer Assisted Mechanics and Engineering Science, 10, Warszawa (2003), 93–109

[25] Szuba, J., Schürr, A., Borkowski, A.: GraCAD — Graph-Based Tool for Conceptual Design, In: A. Corradini, H. Ehrig, H.-J. Kreowski, G. Rozenberg eds., First International Conference on Graph Transformation (ICGT 2002), LNCS 2505, Springer-Verlag, Berlin (2002), 363–377

[26] Wirszyllo R. (ed.): Sport equipment. Designing and building. (in Polish), Arkady, Warszawa (1966)

Parameterized Specification of Conceptual Design Tools in Civil Engineering*

Bodo Kraft and Manfred Nagl

RWTH Aachen University, Department of Computer Science III
Ahornstrasse 55, 52074 Aachen, Germany
{kraft,nagl}@i3.informatik.rwth-aachen.de

Abstract. In this paper we discuss how tools for conceptual design in civil engineering can be developed using graph transformation specifications. These tools consist of three parts: (a) for elaborating specific conceptual knowledge (knowledge engineer), (b) for working out conceptual design results (architect), and (c) automatic consistency analyses which guarantee that design results are consistent with the underlying specific conceptual knowledge. For the realization of such tools we use a machinery based on graph transformations.

In a traditional PROGRES tool specification the conceptual knowledge for a class of buildings is hard-wired within the specification. This is not appropriate for the experimentation platform approach we present in this paper, as objects and relations for conceptual knowledge are due to many changes, implied by evaluation of their use and corresponding improvements.

Therefore, we introduce a parametric specification method with the following characteristics: (1) The underlying specific knowledge for a class of buildings is not fixed. Instead, it is built up as a data base by using the knowledge tools. (2) The specification for the architect tools also does not incorporate specific conceptual knowledge. (3) An incremental checker guarantees whether a design result is consistent with the current state of the underlying conceptual knowledge (data base).

1 Introduction

In our group, various tools for supporting development processes have been built in the past, for software engineering [14], mechanical engineering, chemical engineering, process control [10], telecommunication systems [13], and authoring support [7], some of them are presented at this workshop. This paper reports about a rather new application domain, namely civil engineering.

For all tools mentioned above, we use a graph-based tool construction procedure: internal data structures of tools are modeled as graphs, changes due to command invocations are specified by graph rewriting systems. Then, there are two different branches for constructing tools, a research-oriented and an industry-oriented one. In this paper we restrict ourself to the research-oriented branch.

* Work supported by Deutsche Forschungsgemeinschaft (NA 134/9-1)

J.L. Pfaltz, M. Nagl, and B. Böhlen (Eds.): AGTIVE 2003, LNCS 3062, pp. 90–105, 2004.
© Springer-Verlag Berlin Heidelberg 2004

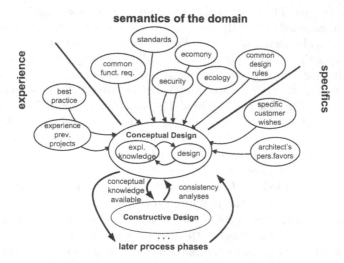

Fig. 1. Different areas of conceptual design

There, we derive tools automatically from specifications, using the PROGRES system [16] for specification development, a code generator for producing code out of the specification, and the UPGRADE visual framework environment [2] into which the code is embedded. The resulting tools are efficient demonstrators for proof of concept purposes. Theses tools are based on our academic development infrastructure, having been developed in the last 15 years.

Conceptual Design in civil engineering means that design results are elaborated on a coarse and abstract level without regarding details which are later included in constructive design (in other disciplines called detail engineering) [12]. The main goal of conceptual design is to take the various levels of semantics for a design problem into consideration(cf Fig. 1): (a) domain specific knowledge, as standards, economy rules, security constrains, or common and accepted design rules, (b) experience knowledge in form of best practice or of using previous design results and, finally, (c) specific user behavior knowledge or wishes, where users are customers or architects, respectively.

The essentials of our conceptual design approach are that (i) explicit knowledge can be formulated, enhanced, or used, (ii) change support is specifically supported, where changes can happen on the level of knowledge as well as for design results, (iii) a lot of consistency checks are included in order to report errors as soon as possible, and (iv) a smooth connection to constructive design is aimed at. The approach specifically pays off, if (v) specific classes of buildings are regarded and, within a class, different designs for buildings and different variants thereof.

We realize a graph-based demonstrator by which a senior architect (knowledge engineer) can specify knowledge by tools. The knowledge is specific for a class of buildings. For the usual architect, there are further tools for developing conceptual designs. These designs are immediately checked against the underlying specific knowledge. For the realization of these tools, we use the enhanced machinery already sketched above. We call this demonstrator the conceptual knowledge experimentation platform as it allows to experiment with concepts without being forced to change the realization of tools.

In this paper we take a certain class of buildings as an example namely one-floor medium-size once buildings. The example is simplified with respect to breadth and depth. The paper also gives no details of the implementation of tools, only the graph transformation specifications are presented here. Tool functionalities and user interface style of the experimentation platform are given in a separate demo description [11].

The paper goes as follows: In section 2 we give a specification of architect tools in a traditional form, were the building type specific knowledge is fixed within the specification. This motivates the different specification method presented in this paper. In section 3 we discuss the specification for the knowledge engineer tools. Furthermore, we give an example of a host graph which can be produced by interactively using these tools. This graph, called domain model graph describes the characteristics of a class of buildings (here office-buildings). Section 4 gives a specification of the architect's tools by which conceptual designs for buildings can be elaborated. In section 5 we discuss the specification for analyses. Section 6 emphasizes the difference of the two specification methods, the traditional and the parameterized one, summarizes the main ideas of this paper, and discusses related literature.

2   A Traditional Tool Specification

There are many projects in the group using graph technology. The specific knowledge of the appropriate domain usually is hard-wired in the schema and the transaction part of a PROGRES tool specification. In this paper, we apply a different specification method.

The reason is, that the knowledge engineer will not be able to learn the PROGRES language and to use the realization machinery, adequate for a tool builder. Furthermore, the knowledge should be easily modifiable, as we are experimenting to find suitable object and relation types, restrictions, and rules for conceptual design in civil engineering.

To illustrate the difference between a traditional specification and the parameterized one described here, we briefly introduce an example specification which shows how tools for architectural design of an office building would be described in the traditional way.

The schema part of our example is shown in Fig. 2. It shows the abstract node class ROOM with a comment attribute. Nodes of that class can be related to each other by Access and Contains edges. The node class is specialized into five different node types representing different room types we want to model. Therefore, the relations can connect rooms of all specific types. The node class ROOM evidently expresses the similarities of different room node types.

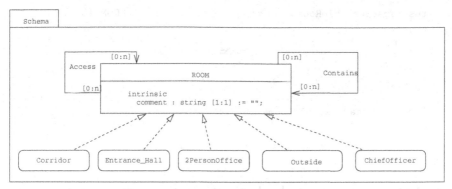

Fig. 2. Schema of an office building

production Create_EntranceHall(**out** newHall : Entrance_Hall [1:1])
=

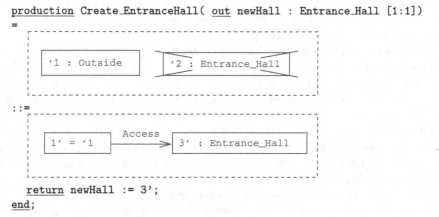

return newHall := 3';
end;

Fig. 3. Example of a graph production, inserting a node and an edge

The transaction part determines how different graphs of that graph class are built. Fig. 3 shows a sample production for our example. The graph pattern to be searched requires an **Outside** node to exist and no **Entrance_Hall** node to be already present (negative application condition). If this pattern is found when applying the production, a new **Entrance_Hall** node is created and connected with the outside node by an **Access** edge. So, the application of the production guarantees that the **Entrance_Hall** is always directly accessible from outside.

Thus, each graph of our example specification models the structure of an office building floor plan. A **ROOM** node without an **Access** relation stands for an inaccessible room. The test shown in Fig. 4 finds such rooms. It searches the graph for **ROOM** nodes which are not connected with the outside node by a path containing **Access** or **Contains** relations. The result is a possibly empty node set. In this way we formally define the meaning of inaccessibility.

We see that the knowledge about the building type "office building" is fixed within the specification. There are room types **Entrance_Hall** or **2PersonOffice**

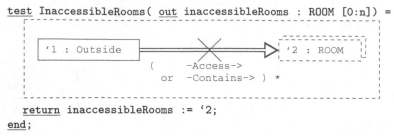

```
test InaccessibleRooms( out inaccessibleRooms : ROOM [0:n]) =
```

```
return inaccessibleRooms := '2;
end;
```

Fig. 4. Example of a graph test, finding inaccessible rooms

room defined as node types. Evidently, it is not possible to create new room types, like coffee kitchen with certain accessibility constrains, without changing the PROGRES specification (schema, transactions, tests). Using PROGRES this way means that the knowledge engineer and the specificator (and later on the visual tool builder) are the same person.

Our request is to keep these jobs separate. The specificator develops a general specification, which does not contain specific application knowledge. This knowledge is put in and modified by the knowledge engineer, as a database, here called domain model graph where he is using tools derived from the general specification. In the same way, there is an unspecific specification for the architect tools, from which general tools are derived. The architect tools now use the knowledge domain model graph interactively elaborated by the knowledge engineer. Thereby, the design results are incrementally checked against the underlying specific knowledge.

It is obvious, that this approach has severe implications on how a specification is written, where the domain knowledge is to be found, and where it is used. The different approaches to fix domain knowledge in the specification or to elaborate it in a host graph are not PROGRES specific, they are different ways to specify.

3 Specification of the Knowledge Engineer Tools

In this section we describe the specification for the knowledge engineer tools. Using the corresponding tools, specific domain knowledge for a class of buildings is explicitly worked out (domain model graph). This knowledge is used to restrict the architecture tools to be explained in the next section.

The upper box of Fig. 5 depicts the PROGRES schema part of the knowledge engineer specification. This schema is still hard-wired. It, however, contains only general determinations about conceptual design in civil engineering. Therefore, it is not specific for a certain type of building.

The node class m_Element (m_ stands for model) serves as root of the class hierarchy, three node classes inherit from it. The class m_AreaType describes "areas" in conceptual design. Usually, an area is a room. It may, however, be a part of a room (a big office may be composed of personal office areas) or

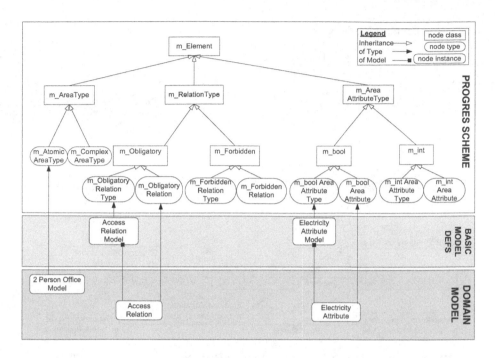

Fig. 5. Schema of knowledge engineering tool and specific model

a grouping of rooms (a chief officer area may contain a secretariat, a personal office for the chief officer, and a meeting room).

From the class **m_AreaType** two node types are defined; **m_AtomicAreaType** represents an area not further decomposed, **m_ComplexAreaType** a complex area composed of several areas. In the same way, classes **m_Obligatory** and **m_Forbidden** describe obligatory and forbidden relations. As we model knowledge about a building type, optional relations are not modeled explicitly. Everything what is not forbidden or obligatory is implicitly optional. The reader may note that attributed relations are represented in PROGRES as nodes with adjacent edges. Finally, attributes may appear as constituents of areas and relations. So, we have again node classes and types to represent the attributes, here only for integer and Boolean values. Note again that attributes have to be defined as nodes, as they are defined by the knowledge engineer.

Fig. 5 in the lower box shows some nodes which stand for kinds of concepts to be used for our office building example. We call these kinds m odels. These models appear in the host graph, interactively produced by the knowledge engineer by using the tools the specification of which we regard. The 2PersonOfficeModel node represents a corresponding room kind, the **AccessRelation** node an accessibility relation between rooms, the **ElectricityAttribute** an attribute node needed to describe a property of an office room. These nodes are schematic information (information on type level) for the class of buildings to be described.

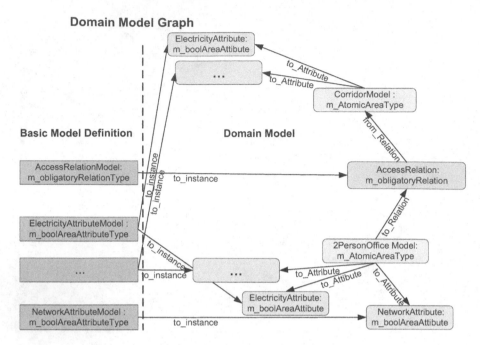

Fig. 6. Cutout of a Domain Model Graph (specific for a type of buildings)

As, however, this type info is not static but interactively elaborated, we call it model.

An attribute, such as for electric access, may appear in several room models. So, it has to be defined before being used several times. In the same way, accessibility occurs between different room models. Furthermore, it is up to the knowledge engineer, which attribute and relation concepts he is going to introduce. By the way, these definitions may be useful for several types of buildings. Therefore, there is a basic model definition layer in the middle of Fig. 5.

Summing up, Fig. 5 introduces a 3 level approach for introducing knowledge. The PROGRES schema types are statically defined. They represent hard-wired foundational concepts of civil engineering. The other levels depend on the knowledge tool user. Thereby, the middle layer defines basics to be used in the specific knowledge which is dependent on the type of building. So, the static layer on top defines invariant or multiply usable knowledge, whereas the middle layer and the bottom layer are specific for a type of building. The host graph built up by knowledge engineer tools contains information belonging to the middle and the bottom layer.

Fig. 6 shows a cutout of this graph structure the knowledge engineer develops, which we call domain model graph. On the left side, basic attribute and relation models are depicted. They belong to the level 2 of Fig. 5. On the right side their use in a specific domain model is shown. This right side shows the area models

```
production m_CreateBoolAreaAttribute( attributeModelDescr : string;
            attributeValueDescr : boolean;
            areaModel : m_AtomicAreaType)
[0:1] =
```

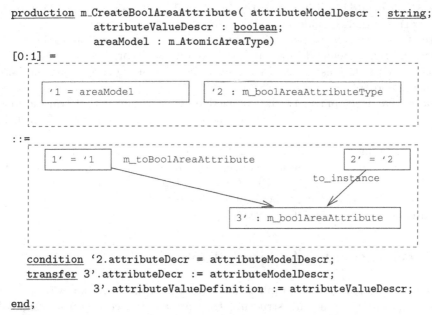

```
condition '2.attributeDecr = attributeModelDescr;
transfer 3'.attributeDecr := attributeModelDescr;
         3'.attributeValueDefinition := attributeValueDescr;
end;
```

Fig. 7. Creating an instance of an attribute model

2PersonOfficeModel and CorridorModel. The 2PersonOfficeModel has two attributes to demand network sockets and electricity to be available. Between the two area models, an access relation is established, to demand an access from all 2 person offices to the corridor, in the graph realized through an edge-node-edge construct.

In Fig. 5 we have introduced a three level "type" system. Any lower level is an instance of an upper level. A node, however, can be an instance of the static type and a dynamically introduced basic type as well. We can see that the electricity attribute is an instance of the static type m_boolAreaAttribute and of the dynamic basic type ElectricityAttributeModel. This is realized by giving the electricity attribute node a string attribute denoting the dynamic basic type and an edge to_instance from the basic type to the attribute node. Tests and transactions guarantee the consistency between these static or dynamic types of instances.

Fig. 7 shows a production to create an attribute assigned e.g. to a 2Person-OfficeModel. Please note that the model is represented by a node with the denotation areaModel of the static type m_AtomicAreaType which has a PRO-GRES node attribute storing the dynamic type 2PersonOfficeModel. Input parameters are the attribute model description as a string, an attribute value, and the model node representing the 2 person room concept. Node '2 on the left side represents an attribute model node. By the condition clause we ensure that it corresponds to the input parameter attributeModelDescr. Only if an attribute model (node '2) with this description exists, a new attribute

(node 3') is created and linked to the 2PersonOfficeModel (node 1') and to
the attribute model (node 2'). The model description is stored in a string attribute of node 3', just as the attribute value. The inverse operation, to delete an
attribute is trivial. Before deleting an attribute model all corresponding instances
have to be deleted. This is done by a transaction executing several productions
in a specific order.

Interactive development by the knowledge engineer means that transactions
modifying the domain model graph are now invoked from outside. Then, this
domain model graph is built up containing two levels as shown in Fig. 6. Thereby,
the corresponding to_instances, to_attribute, to_Relation, and from_Rela-
tion edges are inserted. Any new concept is represented by a node of a static
type (to be handled within the PROGRES system), of a dynamic type, with
bordering nodes for the corresponding attributes which belong to predefined
attributes of the basic model definition layer.

4   Specification for Architect Tools

Whereas the domain model graph is used to store conceptual knowledge, the
design graph provides a data structure to represent the conceptual design of
a building. The specification of the designer tools directly uses the runtime-
dependent basic domain knowledge (layer 2 of Fig. 5). So, the consistency of
a design graph with this basic knowledge can be obeyed. The consistency of the
design graph with the building type specific knowledge of layer 3 is guaranteed
by other analyses. Both analyses are described in the next section.

The design graph allows to specify the structure and the requirements of
a building in an early design phase, above called conceptual design. To design
a building without any layout and material aspects allows the architect to concentrate on the usage of this building on a high abstraction level. During the
constructive design, this design can be matched with an actual floor plan to
discover design errors. This is not further addressed in this paper.

The design graph again is the result of the execution of a PROGRES specification, where transactions are interactively chosen. The 3 level "type" system,
which is similar to that of Fig. 5, is shown in Fig. 8. The essential difference
is that we now model concrete objects, relations, both with corresponding attributes and not knowledge describing how such a design situation has to look
like. This is denoted by the prefix d_, which stands for classes and types for
design.

Another difference is the d_Notification node class with three corresponding node types. The nodes of these types are used to represent warnings, errors,
and tips to be shown to the architect. Furthermore, there are now concrete relation nodes between design objects and not rules that certain relations have
to exist or may not exist. Finally, the design graph nodes now are instances,
and not nodes describing types for instances as it was the case on layer 3 of the
knowledge engineer tools.

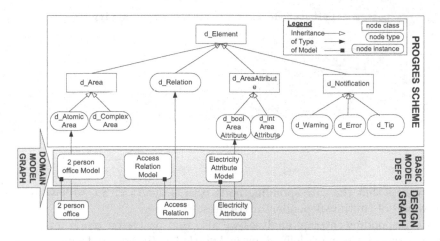

Fig. 8. Scheme of the design graph

The instantiation of attributes, areas, and relations works in the same way as described in Fig. 7 for models. In the design graph we find instances of concepts with a static and dynamic type with bordering instances of attributes and relations both being applied occourences of the corresponding basic models introduced on layer 2 of Fig. 5. As this basic model layer is again needed on the design graph level we just import it from the domain model graph.

5   Consistency Analyses

In this section we present two different forms of consistency analyses. The first form is part of the domain model graph specification. So, these analyses are executed when the knowledge engineer tool is running, to keep the dynamic type system consistent. Corresponding internal analyses can be found for the design graph, respectively. The second form of analyses shows how the consistency between the domain model graph and the design graph is checked.

Let us start with the first form of analyses built in the domain model graph specification. Fig. 9 shows a test being part of the analyses to guarantee the consistency of the dynamic type system. Each basic model has to be unique. So, if the knowledge engineer tries to create a model that already exists, the enclosing transaction should fail. The test m_AttributeModelExists gets as input parameter the model description, e.g. ElectricityAttributeModel. If the model already exists, then a node of type m_boolAreaAttributeType exists, whose attribute attributeDescr has the value of the input parameter.

These analysis transactions work as usual in PROGRES specifications. They guarantee that certain structural properties of a graph class (here domain model graph) are fulfilled. In the above example this means that a basic model definition occurs only once. The difference to traditional PROGRES specifications,

```
test m_AttributeModelExists( modelDescr : string)
[0:1] =
```

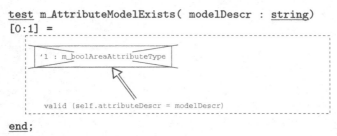

```
end;
```

Fig. 9. Test if a model already exists

however, is that the corresponding node type is dynamic. So we have to check the values of runtime-dependent attributes.

Corresponding internal analyses we also find on design graph level, for the consistency between predefined basic knowledge (imported from the domain knowledge graph) and the current form of the design graph. As they work in the same way as the internal analyses of the domain model graph, we skip them.

The second form of analyses check whether there are violations of the predefined specific knowledge within the design graph. For this, we have to find out inconsistencies between the design graph and the domain model part of domain model graph (cf. Fig. 6). The attributes of an **area model** prescribe the usage of an **area** in the design graph. In an office block, there should be network sockets in all offices, but not in the corridor. This rule is defined in the domain model graph by the Boolean attribute `NetworkAttribute` whose value can be true or false. If the architect constructs a network socket in the corridor, by connecting the area `Corridor` with the attribute `NetworkAttribute`, the design graph is in an inconsistent state.

Tools immediately report such inconsistencies. However, we allow the architect to violate rules and do not stop the design process, because we do not want to hinder his creativity.

Fig. 10 shows an example production, which checks whether the value of an attribute, defined in the model graph, corresponds to the attribute value in the design graph. Whereas the nodes '1 and '2 describe an **area model** and an **attribute** defined in the domain model graph, the nodes '3 and '4 describe an **area** and an **attribute** defined in the design graph. The first two lines of the condition clause ensure that only these nodes of the design graph (node '3 and '4) are found, which correspond to the **area model** (node '1) and its **attribute** (node '2). The next two lines of the condition clauses demand the attributes to be `false` in the domain model graph (node '2) and to be `true` in the design graph (node '4). So, an inconsistency between the domain model graph and the design graph is found. In this case, on the right side of the production, the new node 5' is inserted to mark this inconsistency and to store a specific error message.

<u>production</u> d_CheckAreaAttribute(AreaModel : m_AtomicAreaType ;
            Attribute : m_boolAreaAttribute)

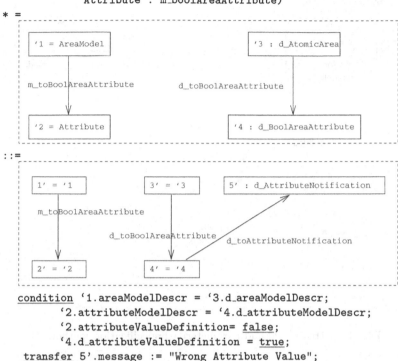

<u>condition</u> '1.areaModelDescr = '3.d_areaModelDescr;
        '2.attributeModelDescr = '4.d_attributeModelDescr;
        '2.attributeValueDefinition= <u>false</u>;
        '4.d_attributeValueDefinition = <u>true</u>;
    <u>transfer</u> 5'.message := "Wrong Attribute Value";
<u>end</u>;

Fig. 10. Analysis to check the consistency of bool attributes

6 Conclusion and Discussion

6.1 Summary and Discussion

In this paper we introduced a specification method for tools in the domain of civil engineering. Different tools provide support for knowledge engineering and conceptual design, respectively. Analyses within either the knowledge engineer or the architecture tool guarantee internal consistency with the basic knowledge interactively introduced. Furthermore, analyses guarantee the consistency of a design result with the building type specific knowledge. Correspondingly, the specifications are split into three parts. The interactively elaborated domain knowledge consists on the one side of a basic part which is useful for several classes of buildings. The specific part on the other side represents the knowledge about one class of buildings.

The specification of the knowledge engineering tools allows to introduce basic model nodes for attributes and relations. Furthermore, the specific knowledge is elaborated by model instance nodes for areas, relations and attributes. The complete information, dependent on the input of the knowledge engineer, is kept

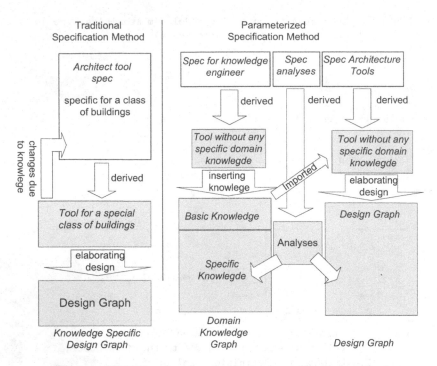

Fig. 11. Traditional and parameterized specification method

in the domain model graph. This information is used by the specification for the designer tools, namely by invoking the analyses between designer results and specific domain knowledge.

So, resulting tools are parameterized. In the same way, the (architecture) tool specification is parameterized in the sense that it depends on specific knowledge to be put in, altered, or exchanged. More specifically, it uses a host graph produced by the knowledge engineer specification. The interactively determined knowledge information can be regarded as dynamic type information.

Fig. 11 shows both approaches, namely the traditional and parameterized one, to specify tool behavior. In the traditional way (left side) the specific knowledge is contained in the specification of the architecture tool. Whenever the knowledge changes, the specification has to be changed and an automatic tool construction process has to be started. On the right side there is a sketch of the parameterized approach presented in this paper. The knowledge engineer tool has, in its initial form, no specific knowledge. This is interactively elaborated. The resulting host graph (domain model graph) acts as typing information for the architecture tool. The basic knowledge information is imported by the design tool. The specific knowledge information is used for building type-dependent analyses of a concrete design result.

6.2 Related Work in Civil Engineering

Both specification methods have pros and cons. If the knowledge is fixed, then the traditional way is advantageous. More checks can be carried out at specification elaboration time, which is more efficient. If the underlying knowledge changes, as it is the case with our experimentation platform, the parameterized method is not only better but necessary. Here, changes of the underlying knowledge need no modification of tools. The price is to have more and more complicated checks at tool runtime due to levels of indirectness which are more costly. Furthermore, the specifications do contains less structural graph information and, therefore, are more difficult to read and write.

Let us now compare the results of this paper with other papers in the area of conceptual design in civil engineering on one side, and with other graph specification approaches on the other. Let us start with the design literature and concentrate on those which also use graphs. There are several approaches to support architects in design. Christopher Alexander describes a way to define architectural design pattern [1]. Although design pattern are extensively used in computer sciences, in architectural design this approach has never been formalized, implemented and used. In [8] Shape Grammars are introduced to support architectural design, e.g. the design of Queen Ann Houses [6]. The concept of shape grammars is related to graph grammars. However this approach rather supports a generation of building designs than an interactive support while designing, what we propose.

Graph technology has been used by [9], to build a CAD system that supports the design process of a kitchen. In contrast to our approach, the knowledge is hard-wired in the specification. In [4] [3] graph grammars are used to find optimal positions of rooms and to generate an initial floor plan as a suggestion for the architect. Formal concept analysis [18] and conceptual graphs [17] describe a way to store knowledge in a formally defined but human readable form. The TOSCANA systems [5] describes a systems to store building rules.

6.3 Comparison to Other GraTra Specification Approaches

Finally, we are going to relate our graph specification method to others in the area of graph technology. We concentrate on those papers where typical and different tool specification methods are applied. In the AHEAD project [10], a management system for development processes is developed. AHEAD distinguishes between a process meta model, to define the general knowledge hardwired in the specification, and the process model definition to represent domain specific knowledge, which can be elaborated or changed at runtime. Nevertheless, the tool construction process has to be run again to propagate changes to the AHEAD prototype.

In the ECARES project [13] graph-based tools are developed to support the understanding and restructuring of complex legacy telecommunication systems. The specific domain knowledge consists in this case e.g. of the formal definition of the underlying programming language to be found in a specific specification.

As result of a scanning and parsing process a host graph is automatically created representing a system's structure. Changing the specific knowledge, the parser and the specific part of the PROGRES specification have to be adapted and the tool construction process has to restart.

In the CHASID project [7] tools are introduced to support authors writing well-structured texts. Its specification method resembles to the one presented in this paper. The specific domain knowledge is here stored in so called schemata, they are again elaborated at runtime. In contrast to our approach, however, the defined schemata are directly used to write texts and not to be checked against a text to uncover structural errors. So, the main advantage of the new specification method of this paper is a gain in flexibility!

References

[1] Christopher Alexander, Sara Ishikawa, Murray Silverstein, Max Jacobson, Ingrid Fiksdahl-King, and Shlomo Angel. *A Pattern Language*. Oxford University Press, New York, NY, USA, 1977. 103

[2] Boris Böhlen, Dirk Jäger, Ansgar Schleicher, and Bernhard Westfechtel. UP-GRADE: A framework for building graph-based interactive tools. In Tom Mens, Andy Schürr, and Gabriele Taentzer, editors, *Electronic Notes in Theoretical Computer Science*, volume 72 of *Electronical Notes in Theoretical Computer Science*, Barcelona, Spain, October 2002. Elsevier Science Publishers. 91

[3] A. Borkowski, E. Grabska, and E. Nikodem. Floor layout design with the use of graph rewriting system progres. In Schnellenbach-Held and Denk [15], pages 149–157. 103

[4] A. Borkowski, E. Grabska, and J. Szuba. On graph based knowledge representation in design. In A.D. Songer and C.M. John, editors, *Proceedings of the International Workshop on Information Technology in Civil Engineering*, Washington, 2002. 103

[5] D. Eschenfelder and R. Stumme. Ein Erkundungssystem zum Baurecht: Methoden und Entwicklung eines TOSCANA Systems. In Stumme and G. Wille [18], pages 254–272. 103

[6] U. Flemming. More than the Sum of Parts: the Grammar of Queen Anne Houses, Environment and Planning B . *Planning and Design*, (14), 1987. 103

[7] Felix Gatzemeier. Patterns, Schemata, and Types — Author support through formalized experience. In Bernhard Ganter and Guy W. Mineau, editors, *Proc. International Conference on Conceptual Structures 2000*, volume 1867 of *Lecture Notes in Artificial Intelligence*, pages 27–40. Springer, Berlin, 2000. 90, 104

[8] J. Gips and G. Stiny. Shape grammars and the generative specification of painting and sculpture. In *Proceeding of the IFIP Congressn 71*, pages 1460–1465, 1972. 103

[9] H. Göttler, J. Günther, and G. Nieskens. Use of graph grammars to design cad-systems. In *Graph Grammars and their application to Computer Science*, pages 396–409, LNCS 532, 1990. Springer, Berlin. 103

[10] Dirk Jäger, Ansgar Schleicher, and Bernhard Westfechtel. AHEAD: A graph-based system for modeling and managing development processes. In Mandred Nagl, Andy Schürr, and Manfred Münch, editors, *AGTIVE'99*, volume 1779 of *Lecture Notes in Computer Science*, pages 325–339, Kerkrade, The Netherlands, September 2000. Springer, Berlin. 90, 103

[11] B. Kraft. Conceptual design tools for civil engineering, demo description, this workshop, 2003. 92

[12] Bodo Kraft, Oliver Meyer, and Manfred Nagl. Graph technology support for conceptual design in civil engineering. In Schnellenbach-Held and Denk [15], pages 1–35. 91

[13] André Marburger and Dominikus Herzberg. E-CARES research project: Understanding complex legacy telecommunication systems. In *Proceedings of the 5th European Conference on Software Maintenance and Reengineering*, pages 139–147, Lisbon, Portugal, 2001. IEEE Computer Society Press, Los Alamitos, CA, USA. 90, 103

[14] Manfred Nagl, editor. *Building Tightly Integrated Software Development Environments: The IPSEN Approach*, volume 1170 of *Lecture Notes in Computer Science*. Springer, Berlin, 1996. 90

[15] M. Schnellenbach-Held and Heiko Denk, editors. *Advances in Intelligent Computing in Engineering, Proceedings of the 9th International EG-ICE Workshop*, Darmstadt, Germany, 2002. 104, 105

[16] Andy Schürr. *Operationales Spezifizieren mit programmierten Graphersetzungssystemen*. PhD thesis, RWTH Aachen, DUV, 1991. 91

[17] F. J. Sowa, editor. *Conceptual Structures*. Addison Wesley, Reading, MA, USA, 1984. 103

[18] R. Stumme and E. (ed.) G. Wille, editors. *Begriffliche Wissensverarbeitung*. Springer, Berlin, 2000. 103, 104

Design of an Agent-Oriented Modeling Language Based on Graph Transformation*

Ralph Depke, Jan Hendrik Hausmann, and Reiko Heckel

Faculty of Computer Science, Electrical Engineering and Mathematics
University of Paderborn, D-33095 Paderborn
{depke,hausmann,reiko}@upb.de

Abstract. The use of UML extension mechanisms for the definition of an Agent-Oriented Modeling Language only fixes its syntax. But agent concepts demand an appropriate semantics for a visual modeling language. Graphs have been shown to constitute a precise and general semantic domain for visual modeling languages. The question is how agent concepts can be systematically represented in the semantic domain and further on be expressed by appropriate UML diagrams. We propose a language architecture based on the semantic domain of graphs and elements of the concrete syntax of UML. We use the proposed language architecture to define parts of an agent-oriented modeling language.

1 Introduction

Agents and related concepts have been shown to be useful abstractions that extend standard Object-Orientation. To employ these concepts in practical software development they need to be represented in the models preceeding the system implementation. Thus, there is a need for a modeling language that incorporates agent-based features. Building such a language only upon the standard OO modeling language for the software industry, the UML, proves to be difficult. UML and the underlying language architecture (MOF) focus on the definition of syntax only and neglect semantic issues (see also Sect. 2.4 for details).

Yet, a precise denotation of the semantics of added features is important for several reasons: The definition of a new language is motivated by new semantic concepts. For example, an agent-oriented modeling language is based on the concept of an agent which is different from other concepts like that of an object (cf. [7]). A precise semantics of the language is necessary for consistently reflecting agent concepts like roles, protocols, and goals. Moreover, unique requirements for the implementation of an agent based system rely on the precise description of the structure and the behavior of a system.

The question arises whether a new modeling language is necessary at all. In section 3 we will show that the available features of UML, without extension and semantics, are not sufficient to support the agent concepts of role and protocol

---

* Work supported in part by the EC's Human Potential Programme under contract HPRN-CT-2002-00275, [Research Training Network SegraVis].

J.L. Pfaltz, M. Nagl, and B. Böhlen (Eds.): AGTIVE 2003, LNCS 3062, pp. 106–119, 2004.

appropriately. Thus, UML is tailored to a specific agent-oriented modeling profile with an appropriate semantics. In this way, the syntactically correct usage of the language ensures the correct application of agent concepts. Because the syntax of the language is defined using the extension mechanisms of UML we do not introduce a completely new syntax and the acceptance of the language is enhanced. In a nutshell, we aim at defining a UML-based agent-oriented modeling language in a better way.

In order to introduce a precise semantics of the agent-oriented modeling language (AML) we rely on the semantic domain of graphs. We propose a language design with three different levels of description, see Sec. 2. On the instance level we have instance graphs which represent states of the modeled system. On the model level we use typed attributed graph transformation systems (TAGTS) which consist of type graphs and graph transformation rules. They are used for expressing structural and dynamical models of systems. On the meta level we use a meta type graph for restricting the type graphs of TAGTS to desired elements of the language.

Orthogonal to the structure of the semantic domain is its relationship to the syntax of a modeling language. A solution to the problem is to relate the concrete syntax to the abstract syntax of the language and to transform the abstract syntax to the semantic domain. This solution has the advantage that the abstract syntax usually offers a homogeneous representation of the language which abstracts from aspects like the visual presentation of language elements. The next step is to relate the abstract syntax to the semantic domain.

UML provides such a homogeneous abstract syntax. The meta model is given by class diagrams and user models are just instances of the classes. The concrete syntax is specified rather informally by example diagrams and descriptions of the mapping to the meta model. UML can be tailored for specific purposes of use by the profile mechanism. Thus, there are means to define the syntax of specific modeling languages. In the next section we present a language design which provides a mapping of the UML based syntax of a modeling language to the semantic domain of graphs being stratified in three levels of description.

In section 3 we show how the proposed language architecture can be used to express the agent concepts of role and protocol in an agent-oriented modeling language.

2   A Graph Based Language Architecture of AML

The structure of the agent-oriented modeling language (AML) is determined by its syntax and semantics. The concepts of a language essentially reflect in the semantic domain of the language which we deal with in this section. We choose graph transformation as semantic domain because it has been shown to be appropriate for visual languages (cf. [8]). The semantic domain of graph transformation is structured in three different abstraction levels (see Fig. 1). We proceed to explain the different levels in more detail.

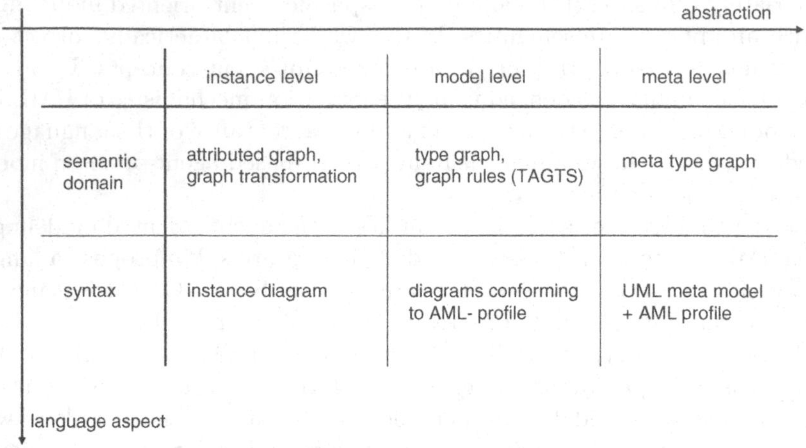

	instance level	model level	meta level
semantic domain	attributed graph, graph transformation	type graph, graph rules (TAGTS)	meta type graph
syntax	instance diagram	diagrams conforming to AML- profile	UML meta model + AML profile

abstraction

language aspect

Fig. 1. Language architecture

2.1   The Semantic Domain of Graph Transformation

The semantic domain of graphs represents instances, types, and meta types. The relationships among different levels of abstraction are formalized by notions from graph theory. The operational semantics of models is given in terms of rule-based graph transformation.

More precisely, we use typed attributed graph transformation systems as a semantic domain. Graphs are attributed if their vertices or edges are coloured with elements of an abstract data type (like Strings or Integers) [12, 4]. In our context, entities which carry attributes will only be represented as vertices. Thus, graphs with coloured vertices are sufficient for us.

Mathematically, abstract data types are represented as algebras over appropriate signatures. A many sorted signature consists of type symbols (sorts) and operation symbols. For each operation its arguments' types and its result type are defined. An algebra for a signature consists of domain sets for the type symbols of the signature and it defines for each operation symbol of the signature an operation with respect to the domain sets. The integration of graphs and algebras results in attributed graphs [12]. Attribute values are elements from a domain of the algebra. In an attributed graph they are contained as data vertices in the set of vertices. Attributes are considered as connections among non data entity vertices and data vertices, i.e., attribute values. The introduction of attribute values from an algebra ensures that graph transformations change values only according to the laws of the algebra.

On the instance level system states are given by attributed graphs. Entity vertices are used to represent agents, roles, messages, etc. as members of a system state defined by an attributed graph. Data vertices are attribute values for agents, roles, etc. Each of these data vertices is connected by an edge to the entity vertex which possesses the belonging attribute.

An attributed graph morphism is a mapping between two graphs which respects the connections between vertices and edges [12]. Additionally, data vertices are only mapped to data vertices. An attributed type graph is an attributed graph in which the elements of the algebra represent the type symbols (sorts) of the data vertices. Now, an attributed graph is typed by an attributed type graph if there exists an attributed graph morphism from the former to the latter graph.

As usually, a graph transformation rule consists of a left hand side and a right hand side. The rule is typed if both sides are typed by the same type graph. An attributed type graph together with a set of typed graph transformation rules form a Typed Attributed Graph Transformation System (TAGTS) [12]. The operational semantics of a TAGTS depends on the chosen graph rewrite mechanism. We choose the DPO approach in which every change of the state must be explicitly specified in the rules (cf. [7]).

We use three different levels of description: on the model level structural and dynamic properties of a system are described. Such models must conform to the language description on the meta level. On the instance level system states evolve according to the system description on the model level.

On the model level typed attributed graph transformation systems (TAGTS) provide the semantic domain of agent-oriented models, see Fig. 1. The attributed type graph of a TAGTS is used for typing attributed graphs on the instance level, i.e., system states.

The attributed type graph of a TAGTS is itself typed by the attributed meta type graph on the meta level, see Fig. 2. This graph determines the kinds of graph elements and admissible graphs on the model level.

The set of graph transformation rules of a TAGTS determines the dynamics of the system. The graph rules of the TAGTS must conform to the attributed type graph of the TAGTS. On the instance level subsequent system states are generated by graph transformations that result from the application of graph transformation rules to an admissible system state.

We will discuss the syntactical aspects of the language in section 2.2. Next, we use the presented language architecture to define the structure of an agent-oriented modeling language.

We regard essential elements of the agent-oriented modeling language AML and relate them in a meta type graph, see Fig. 2. The structure of this graph is motivated by the agent concepts and their dependencies. In section 3, the concepts of role and protocol are discussed in more detail. Next, we will motivate parts of the meta type graph roughly. A detailed discussion of the requirements for an agent-oriented modeling language can be found in [7].

The initial element is the **Agent** which contains attributes and operations. A **Data** element is considered as data container which contains attributes and links to other **Data** elements. **Roles** are attached to agents in order to make agents interact. **Roles** contain **attributes**, **operations** and **messages**. They are able to send or receive messages through the execution of one of their operations. A message is named and it has a list of parameter data entities. A **protocol** comprises roles

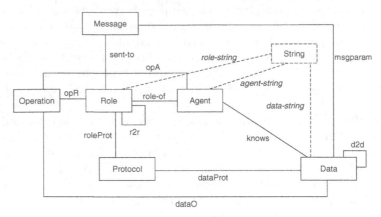

Fig. 2. The meta type graph of structural elements of AML

and data. A subset of the set of data entities may be marked as parameters of the protocol. These parameters are substituted by data entities of the context in which the protocol is used. Attributes are given by nodes denominated with their sort. In Fig. 2 the sort String is introduced. It is connected to all entities which posses attributes of type String.

The structure of the semantic domain influences the syntactic representation of models which we discuss now.

2.2   The Syntax of Models

We deal with the syntax of models by introducing a profile for the agent-oriented modeling language AML, see Fig. 3. For every vertex in the meta type graph there is introduced a stereotype in the profile. For each stereotype its base meta class is designated. A description explains the stereotype and a constraint restricts its application. The constraints of the stereotypes transfer semantic information to the syntax of the language. In Fig. 3, the constraints establish an (incomplete) restriction of UML language elements in order to represent agent concepts appropriately. A more formal description of constraints results from using a constraint language like OCL. The formalization is omitted here.

On the model level the semantic domain consists of a TAGTS which comprises a typed attributed graph and a set of graph transformation rules. The typed attributed graph is represented by an agent class diagram, see Fig. 5. The typing of the model elements is expressed by using stereotypes from the profile of the language AML to the graph elements' identifiers. In this way the attributed type graph morphism from the TAGTS to the attributed meta type graph is represented syntactically.

A graph transformation rule is represented by a stereotyped package diagram in which the left and right hand sides are distinguished. Two packages named Left and Right are inserted in the rule package and contain diagrams for the left

Stereotype	Base Class	Parent	Description	Constraint
<<Agent>>	Class	N/A	An <<Agent>> owns attributes and operations.	self.isActive = true
<<Role>>	Class	N/A	A <<Role>> has attributes, operations, and messages. It performs the interaction to other roles for an agent.	Must have at least one <<Message>> and one <<OpRule>>
<<Data>>	Class	N/A	<<Data>> is a container for attributes.	Contains only attributes.
<<role-of>>	Association	N/A	By the <<role-of>> association a <<Role>> is bound to an <<Agent>>	Same constraints as composition
<<OpRule>>	Operation	N/A	An <<OpRule>> names a local graph transformation rule of an <<Agent>> or a <<Role>>	The visibility is private
<<Message>>	Signal	N/A	A <<Message>> has a name and parameters.	parameters are of class <<Agent>>, <<Role>>, or <<Data>>
<<Protocol>>	Package	N/A	A <<Protocol>> is a template package which may contain <<Agent>>s, <<Role>>s, and <<Message>>s	A protocol contains at least two <<Role>>s
<<Rule>>	Package	N/A	A <<Rule>> contains two packages: one for the left and one for the right hand side	Identical elements in the left and right hand side are defined once and then referred on the other side
<<Protocol Rules>>	Package	N/A	This package contains all <<Rule>>s of a protocol	

Fig. 3.   The UML profile of AML

and right hand sides of the rule which conform to the AML profile. An example is given in Fig. 4. The diagrams represent the left and right hand sides of a graph transformation rule which are both typed by the attributed type graph of the TAGTS. The correct typing is ensured by the use of the AML profile. We prefer this new notation because the left hand side and the right hand side of a rule are clearly separated. This is different to notations relying on UML collaboration diagrams. Also, the labels of the packages accurately indicate the context of the rule.

Before we deal with the semantics of agent interaction protocols we compare our approach to some related work.

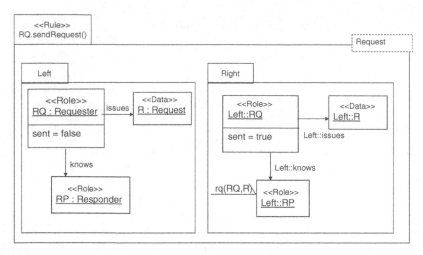

Fig. 4. A <<Rule>> package diagram

2.3 Related Approaches to Agent-Oriented Modeling

Yet, there exist some approaches to agent-based modeling. We regard AgentUML, Tropos and MaSE which introduce agent concepts or use UML.

AgentUML is a prominent proposal for the extension of UML with language elements for agent concepts [1]. But there are two problems with AgentUML. First, the extension of UML neither does use the extension mechanisms of UML explicitly nor does it describe the extension at least on the level of abstract syntax, i.e., the meta model. Second, AgentUML lacks a formal semantics which reflects agent concepts precisely.

Tropos is another approach to agent-oriented modeling which is intended to support different activities of software development [10]. Different diagrams are introduced for capturing actors and their respective dependencies in an organizational model. Goals and their dependencies are depicted in diagrams of a specific non-UML type. In the design activity diagrams use elements of AgentUML [1]. Altogether, Tropos supports visual modeling of agent-based systems to some degree and it barely relies on UML.

Deloach et al. [5] introduce the Multi-agent Systems Engineering (MaSE) methodology for developing heterogeneous multiagent systems. MaSE uses a number of graphically based models to describe system goals, behaviors, agent types, and agent communication interfaces. The visual models are only roughly based on UML. Thus, the approach lacks a precise syntax, e.g., given by a UML profile. The semantics of the models is not defined.

Next, we discuss related work concerning the definition of modeling languages.

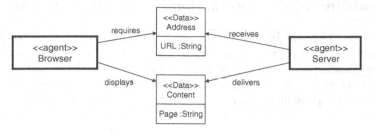

Fig. 5. An agent class diagram

2.4   Related Meta Modeling Approaches

The most common way of defining modeling languages (at least the one standardized by the OMG) is the Meta Object Facility (MOF) [13]. The MOF introduces 4 layers of models. The abstract syntax of each layer is defined by providing a class diagram on the level above it. As models are thus used to define modeling languages, the term meta-modeling has been coined for this approach. Each layer of the MOF may define concrete syntax representations for the elements it introduces. For instance, the UML, the best known MOF-based luguage, introduces lots of diagrams which all provide their own special notation but are defined in terms of the MOF meta-model. The widely known shortcomings of the MOF approach are that it does not supply facilities to define the semantics domain and that the instance relation between the different levels is not properly defined (see e.g. the various discussions in the precise UML group). Our approach addresses these problems by providing an explicit semantics domain (graphs) and by using well-known structures in this domain (typegraph-relations) to precisely define the meaning of level-spanning instantiation.

Other approaches have also set out to extend the MOF to improve its preciseness. A feasibility study for IBM [3] suggests a recursive structure of definition levels in which each level is used to define the structure of the level beneath it as well as the instantiation concept that this level should provide. This instantiation concept is also the (denotational) semantics since the purpose of each level is to define the constructs on the lower levels. The semantics of behavioral constructs is not addressed in this approach.

Another approach [2] proposed a mathematical notation based on sets for the precise definition of UML's semantics. Sets are also underlying the approach in [14] which suggests M odel T ransform ation System s, a graph-transformation-based approach extended by high-level control constructs to provide a description of the dynamic semantics. Our approach avoids the use of such additional constructs and uses standard graph transformations only.

Graph Transformations have furthermore been used to provide operational semantics for the UML in an interpreter-like style [8] and in a compiler-like translation approach [11]. Our approach uses the same basic ideas but takes a more fundamental view and investigates the impact that graph transformations as a semantic domain have for the whole language architecture.

3   Modeling Agent Interaction Protocols with AML

The modeling of agent based systems must take properties and specific behavior of agents into account. Typical properties of agents are their autonomous behavior, structured interaction and proactivity (cf. [7]). We concentrate on autonomy and interaction. Autonomy emphasizes the fact that agents have control over their own operations: They are not called from outside like methods of an object but are only invoked by the agent itself. Thus, agents own separate threads of control, they have only little (and explicitly stated) context dependencies, and they interact in an asynchronous way.

We aim at developing a modeling language which captures the interaction behavior of agents in the semantic domain. Therefore, the afore proposed language architecture will be applied.

In order to support the general reuse of often occurring interaction patterns the interaction between agents is often described in terms of protocols, which represent templates of coordinated behavior (see, e.g., [9]). Instead of making reference to the agents involved in an interaction, reusability requires that we reduce the description of the participating agents to such features and properties that are relevant to the protocol. For this purpose, the concept of role will be used. In the protocols roles interact on behalf of the agents. From this idea there result specific requirements to roles (cf. [7]). We regard only some of the requirements.

Unlike interfaces in UML roles are instantiable because state information is necessary if roles are protocol participants that have to store the state of an interaction. A role instance is bound to an agent in order to make him interact. Roles can be dynamically attached to and retracted from an agent. If a role is attached to an agent then the agent's state must change during the interaction. Otherwise, the interaction would not have any effect on the agent.

Roles are specified in terms of attributes, operations and messages. Attributes store the state of an interaction and operations of roles are able to send and receive messages. A role becomes used if it is bound by the role-of relationship between the role and a base agent. In order to establish an interaction between some agents the role-of relationship has to carry a suitable semantics which fulfills the above mentioned requirements. Considering the behavior of a role bound to a base agent, the operations of a role must also affect the state of its base agent. Next, we will discuss how this specific semantics of roles can be modeled precisely.

Now, we discuss how the requirements can be fulfilled in UML. UML offers sequence diagrams which can be used to describe the message exchange between instances. But sequence diagrams do not allow the description of internal state changes of the concerned instances when a message is sent or received. The same problem exists with collaboration diagrams which offer very similar features. Besides the collaboration diagrams on the instance level there are also collaboration diagrams on the specification level. They allow the definition of behavioral patterns for some context of collaboration roles and the interaction within this context. The use of a pattern demands substitution instances for the

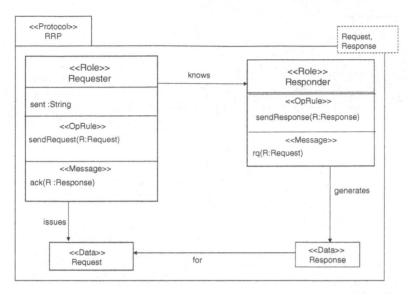

Fig. 6. The template package for the structure of the request-response protocol

collaboration roles. In our case an agent had to be substituted for a collaboration role. This kind of "binding" is not useful in our case because an agent is able to play different roles in different interactions. Instead, we use the template mechanism of UML. By template packages we will define both the structure of a protocol and its behavior given by graph transformation rules.

An example shows how protocols are given in a notation for agent-specific model elements. We use a trivial request-response protocol which is a typical example of a connector between agents enabling interaction (see [9]). Note, that the protocol is rather incomplete. Especially, the behavior of the responder must be refined in order to be useful. Anyhow, we are able to demonstrate how protocols are used in our approach.

The structure of the request-response protocol is given in Figure 6. The protocol is given by an agent class diagram which is contained in a template package. The parameters Request and Response of the template refer to two different data classes within the package.

The operations sendRequest and sendResponse are specified by use of graph transformation rules, see Figure 7 and Figure 4. Rules are given by a pair of instance diagrams which represent the left and right hand sides of the rule. In the rules, the delivery of a message is represented by an arrow whose head is attached to the receiver of the message. All rules of the protocol are integrated in one package for the protocol rules, see Figure 7.

The data entity Request is used as a parameter of the operation sendRequest which is used by the role Requester in order to send a request message rq to a receiver. The role Responder uses its operation sendResponse to answer the

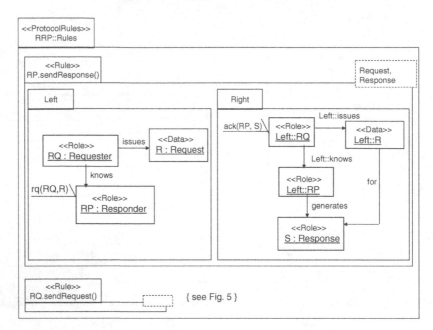

Fig. 7. The rules of the request-response protocol

request. It attaches a message ack with parameter data Response to the role Requester.

The binding of a protocol to a given model is a pure syntactic mechanism of macro expansion: The parameters of the protocol template are filled in with model elements from a given agent class diagram (shown in Fig. 5). Then, the model elements of the protocol are inserted in the agent class diagram and the role classes are attached to agent classes by the role-of relationship.

The graph transformation rules of the protocol are expanded in the following way: For each role in a rule an agent is inserted in the rule. The role and the agent are connected by the role-of relationship. The correlation of roles and agents must conform to the expanded agent class diagram. Then, model elements which have been assigned to protocol parameters are inserted in the rule. Now, these expanded graph transformation rules reflect the change of the system behavior through the expanded protocol.

We show the expansion of the request-response protocol. Syntactically, a role-of relationship is depicted as an arrow with a filled in triangle head. In Figure 8 the binding of a protocol is shown with respect to the two agents browser and server which interact by playing the roles Requester and Responder. The data entities of type Address and Content act as Request and Response.

In Figure 9 the binding of the rule sendRequest is depicted in a rule diagram. The rule is extended by instances of the base agents to which the roles of the protocol are bound according to the agent class diagram in Figure 8. The rule is

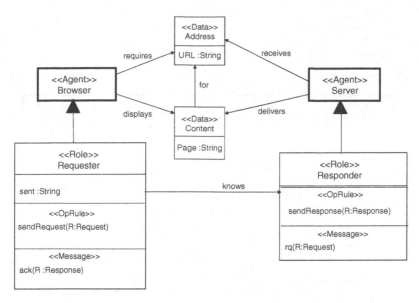

Fig. 8. Two agents use the request-response protocol (agent class diagram)

also extended by an instance of a protocol parameter (**Address**) in order to affect a part of the (system) state which is related to the base agent.

In the context of our proposed language architecture the resulting agent class diagram and the rule diagrams map to a corresponding type graph and to graph transformation rules of a TAGTS. Thus, the modification of the agent class diagram and rule diagrams becomes semantically effective.

The intended semantics of agent interaction by protocols is precisely enabled by a combination of a pure syntactical expansion mechanism and a mapping to the semantic domain of graph transformation. In the context of protocol expansion graph transformation is an appropriate semantic domain because the necessary change of system behavior is achievable by pure structural modification of graph transformation rules. It is not necessary to change the operational semantics of the TAGTS in order to enable protocol-based interaction.

Thus, the proposed language architecture fits well for the syntax and semantics of elements of an agent-oriented modeling language.

4   Conclusion

In this paper we introduced a language architecure that relies on visual elements from UML and on the semantics of graph transformation. The abstract syntax of diagrams of the new language is tailored to the needs of the semantic domain of graphs. We have shown how concepts for elements of an agent-oriented modeling language can be described precisely within the architecture. Based on the

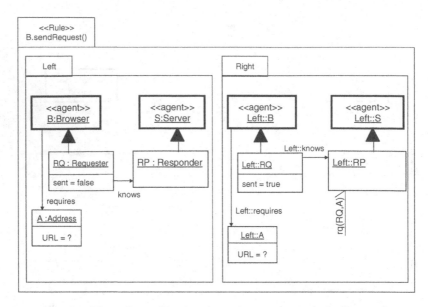

Fig. 9. The rule sendRequest bound to an agent (rule diagram)

operational semantics of the DPO approach we demonstrated the semantics of protocol based interaction of agents.

In the next step more elements of an agent-oriented modeling languages are to be integrated. The semantic relation of the different elements is to be described in more detail. Further on, the use of the modeling elements in a software development process must be clarified.

The precisely defined agent-oriented models can be used to reason about properties of agent based systems. For example, by applying techniques like model checking it becomes possible to decide whether an agent is able to reach a certain state. With respect to agents this is an important question because particular states of an agent often characterise goals which an agent tries to reach. First results regarding these aspects are contained in [6].

Acknowledgement

We thank the referees for their valuable comments.

References

[1] B. Bauer, J.P. Müller, and J. Odell. Agent UML: A Formalism for Specifying Multiagent Software Systems. In P. Ciancarini and M.J. Wooldridge, editors, *Proc. 1st Int. Workshop on Agent-Oriented Software Engineering (AOSE 2000), Limerick, Ireland, June 2000*, volume 1957 of *LNCS*, pages 91–104. Springer-Verlag, Berlin, 2001. 112

[2] R. Breu, U. Hinkel, C. Hofmann, C. Klein, B. Paech, B. Rumpe, and V. Thurner. Towards a formalization of the Unified Modeling Language. In M. Aksit and S. Matsuoka, editors, *ECOOP'97 – Object-Oriented Programming, 11th European Conference*, volume 1241 of *LNCS*, pages 344–366. Springer, 1997. 113

[3] S. Brodsky, T. Clark, S. Cook, A. Evans, and S. Kent. A feasibility study in rearchitecting UML as a family of languages using a precise OO meta-modeling approach, 2000. 113

[4] I. Claßen and M. Löwe. Scheme evolution in object oriented models: A graph transformation approach. In *Proc. Workshop on Formal Methods at the ISCE'95, Seattle (U.S.A.)*, 1995. 108

[5] S. A. DeLoach, M. F. Wood, and C. H. Sparkman. Multiagent systems engineering. *Int. Journal of Software Engineering and Knowledge Engineering*, 11(3), 2001. 112

[6] R. Depke and R. Heckel. Modeling and analysis of agents' goal-driven behavior using graph transformation. In H.D.Ehrich, J.J.Meyer, and M.D.Ryan, editors, *Objects, Agents and Features - Structuring Mechanisms for Contemporary Software*, Revised Papers, LNCS, Int. Seminar, Dagstuhl Castle, Germany, Feb. 16 – 21, 2003. Springer-Verlag. To appear. 118

[7] R. Depke, R. Heckel, and J. M. Küster. Roles in agent-oriented modeling. *Int. Journal of Software Engineering and Knowledge Engineering*, 11(3):281–302, 2001. 106, 109, 114

[8] G. Engels, J.H. Hausmann, R. Heckel, and St. Sauer. Dynamic meta modeling: A graphical approach to the operational semantics of behavioral diagrams in UML. In A. Evans, S. Kent, and B. Selic, editors, *Proc. UML 2000, York, UK*, volume 1939 of http://www.springer.de/comp/lncs, pages 323–337. Springer-Verlag, 2000. 107, 113

[9] Foundation for Intelligent Physical Agents (FIPA). Agent communication language. In *FIPA 97 Specification, Version 2.0*, http://www.fipa.org. 1997. 114, 115

[10] F. Giunchiglia, J. Mylopoulos, and A. Perini. The tropos software development methodology: processes, models and diagrams. In *Proc. of the 1st Int. Conference on Autonomous Agents and Multiagent Systems*, pages 35–36. ACM Press, 2002. 112

[11] M. Gogolla, P. Ziemann, and S. Kuske. Towards an Integrated Graph Based Semantics for UML. In Paolo Bottoni and Mark Minas, editors, *Proc. ICGT Workshop Graph Transformation and Visual Modeling Techniques (GT-VMT'2002)*. Electronic Notes in Theoretical Computer Science (ENTCS), Elsevier (October 2002), 2002. 113

[12] M. Löwe, M. Korff, and A. Wagner. An algebraic framework for the transformation of attributed graphs. In M.R. Sleep, M.J. Plasmeijer, and M.C. van Eekelen, editors, *Term Graph Rewriting: Theory and Practice*, chapter 14, pages 185–199. John Wiley & Sons Ltd, 1993. 108, 109

[13] Object Management Group. Meta object facility (MOF) specification, September 1999. http://www.omg.org. 113

[14] D. Varró and A. Pataricza. Metamodeling mathematics: A precise and visual framework for describing semantics domains of UML models. In J.-M. Jézéquel, H. Hussmann, and S. Cook, editors, *Proc. Fifth Int. Conference on the Unified Modeling Language – The Language and its Applications*, volume 2460 of *LNCS*, pages 18–33, Dresden, Germany, September 30 – October 4 2002. Springer-Verlag. 113

Specification and Analysis of Fault Behaviours Using Graph Grammars*

Fernando Luis Dotti[1], Leila Ribeiro[2], and Osmar Marchi dos Santos[1]

[1] Faculdade de Informática, Pontifícia Universidade Católica do Rio Grande do Sul
Porto Alegre, Brazil
{fldotti,osantos}@inf.pucrs.br
[2] Instituto de Informática, Universidade Federal do Rio Grande do Sul
Porto Alegre, Brazil
leila@inf.ufrgs.br

Abstract. In this paper we make use of formal methods and tools as means to specify and reason about the behavior of distributed systems in the presence of faults. The approach used is based on the observation that a fault behavior can be modeled as an unwanted but possible transition of a system. It is then possible to define a transformation of a model M_1 of a distributed system into a model M_2 representing the behavior of the original system in the presence of a selected fault. We use a formal specification language called Object Based Graph Grammars to describe models of asynchronous distributed systems and present, for models written in terms of this language, the transformation steps for introducing a set of classical fault models found in the literature. As a result of this process, over the transformed model(s) it is possible for the developer to reason about the behavior of the original model(s) in the presence of a selected fault behavior. As a case study, we present the specification of a pull-based failure detector, then we transform this model to include the behavior of the crash fault model and analyze, through simulation, the behavior of the pull-based failure detector in the presence of a crash.

1 Introduction

The development of distributed systems is considered a complex task. In particular, guaranteeing the correctness of distributed systems is far from trivial if we consider the characteristics open systems, like: massive geographical distribution; high dynamics (appearance of new nodes and services); no global control; faults; lack of security; and high heterogeneity [21]. Among other barriers, in open environments (e.g. Internet) it is hard to assure the correctness of applications because we cannot be sure whether a failure is caused by a fault in the system under construction itself or by the environment in which it runs. It is therefore necessary to provide methods and tools for the development of distributed

---

* This work is partially supported by HP Brasil - PUCRS agreement CASCO, ForMOS (FAPERGS/CNPq), PLATUS (CNPq), IQ-Mobile II (CNPq/CNR) and DACHIA (FAPERGS/IB-BMBF) Research Projects.

J.L. Pfaltz, M. Nagl, and B. Böhlen (Eds.): AGTIVE 2003, LNCS 3062, pp. 120–133, 2004.
© Springer-Verlag Berlin Heidelberg 2004

systems such that developers can have a higher degree of confidence in their solutions. We have developed a formal specification language [4], called Object Based Graph Grammars (OBGG), suitable for the specification of asynchronous distributed systems. Currently, models defined in this formal specification language can be analyzed through simulation [1] [5]. Moreover, in our activities we are also working on an approach to formally verify, using model checking tools, models defined in OBGG. Besides the analysis of distributed systems specified in OBGG, we can also generate code for execution in a real environment, following a straightforward mapping from an OBGG specification to the Java programming language. By using the methods and tools described above we have defined a framework to assist the development of distributed systems. The innovative aspect of this framework is the use of the same formal specification language (OBGG) as the underlying unifying formalism [6].

The results achieved so far, briefly described above, have addressed the development of distributed systems without considering faults. In order to properly address the development of distributed systems for open environments, in this paper we present a way to reason about distributed systems in the presence of selected fault behaviors. Thus, we revise our framework in order to consider some fault models found in the literature [9]. The rationale is to bring the fault behavior to the system model and reason about the desired system in the presence of faults. To achieve that, we show how to introduce selected fault behaviors in a model written in terms of the formal specification language OBGG.

As stated in [10], a fault can be modeled as an unwanted but possible state transition of a system. Moreover, it states that the transformation of an original model M_1 into a model M_2 considering a kind of fault consists on the insertion of virtual variables and statements that define the fault behavior. We show how to perform this transformation process for models written in OBGG, such that the transformation from M_1 into M_2 preserves that all possible computations of M_1 and adds the desired fault behavior. Having the transformed model M_2, the developer can reason about its behavior (currently through simulation tools, but formal verification using model checking tools is being integrated to the framework). Code for execution in a real environment can be generated using the model M_1. While running in a real environment, if the environment exhibits the fault behavior corresponding to the one introduced in M_2, then the system should behave as expected during the analysis phase.

This paper is organized as follows: Section 2 discusses related work; Section 3 presents the formal specification language OBGG and introduces a case study used throughout the paper; Section 4 shows how to transform a specification to incorporate selected fault behaviors; Section 5 briefly analyzes the case study; finally, Section 6 brings us to the conclusions.

2   Related Work

Concerning related works, we have surveyed the literature trying to identify approaches that allow developers to reason about distributed systems in the presence of faults. As far as we could survey, we could not identify other approaches based on specification transformations, providing various fault models and supporting analysis based on tools.

The SPIN (Simple Promela INterpreter) model checker [12] enables a developer to specify a system (using the formal specification language PROMELA (PROcess/PROtocol MEta LAnguage)) and formally verify, using model checking. The similarity is that SPIN allows one to check models using channel abstractions that may loose messages. The fault behavior is provided by the supporting tool. Other fault behaviors are not provided.

Another work, but related to the development of mobile systems, is presented in [8]. The Distributed Joint Calculus is a process calculus that introduces the notions of locations and agents, where locations may be nested. The calculus provides the notion of crash of a location, whereby all locations and agents in the crashed location are halted. A crash of a location changes the behavior of basic characteristics hindering any kind of reaction, like for communication or process migration. Systems that are represented with this calculus can be formally verified, but using theorem proving, requiring better skilled developers.

In [17] I/O-Automata are proposed as means for specifying and reasoning about distributed algorithms for asynchronous environments. I/O-Automata were designed to have nice features, such as the composition of I/O-Automata resulting in another I/O-Automata. Also, it is possible to derive properties of the composed automata from the analysis of its component automata. In [16] a rich set of distributed algorithms was modeled with I/O-Automata. In many of these, faults are taken into account and new versions of the algorithms are proposed. In this paper we have focused on fault representation, and not on how to handle a specific kind of fault in a given system or distributed algorithm. Considering the techniques around I/O-Automata, however, we could not identify an approach such as the proposed in this paper, whereby a fixed transformation step to embed a selected fault behavior can be carried out in the same way for different models. Representation of fault behavior with I/O-Automata is carried out by manually extending state and transitions of the model, for each case.

3   Object Based Graph Grammars

Graphs are a very natural means to explain complex situations on an intuitive level. Graph rules may complementary be used to capture the dynamical aspects of systems. The resulting notion of graph grammars generalizes Chomsky grammar from strings to graphs [20, 7]. The basic concepts behind the graph grammars specification formalism are:

- states are represented by graphs;
- possible state changes are modeled by rules, where the left- and right-hand sides are graphs; each rule may delete, preserve and create vertices and edges;
- a rule have read access to items that are preserved by this rule, and write access to items that are deleted/changed by this rule;

- for a rule to be enabled, a match must be found, that is, an image of the left-hand side of a rule must be found in the current state;
- an enabled rule may be applied, and this is done by removing from the current graph the elements that are deleted by the rule and inserting the ones created by this rule;
- two (or more) enabled rules are in conflict if their matches need write access to common items;
- many rules may be applied in parallel, as long as they do not have write access to the same items of the state (even the same rule may be applied in parallel with itself, using different matches).

Here we will use graph grammars as a specification formalism for concurrent systems. The construction of such systems will be done componentwise: each component (called entity) is specified as a graph grammar; then, a model of the whole system is constructed by composing instances of the specified components (this model is itself a graph grammar). Instead of using general graph grammars for the specification of the components, we will use object-based graph grammars (OBGG) [4]. This choice has two advantages: on the practical side, the specifications are done in an object-based style that is quite familiar to most of the users, and therefore are easy to construct, understand and consequently use as a basis for implementation; on the theoretical side, the restrictions guarantee that the semantics is compositional, reduce the complexity of matching (allowing an efficient implementation of the simulation tool), as well as eases the analysis of the grammar. Basically, we impose restriction on the kinds of graphs that are used and on the kind of behaviors that rules may specify.

Each graph in an object-based graph grammar may be composed by instances of the vertices and edges shown in Figure 1. The vertices represent entities and elements of abstract data types (the time stamps tm in and tm ax are actually special attributes of a message, we defined a distinguished graphical representation for these attributes to increase the readability of the specifications). Elements of abstract data types are allowed as attributes of entities and/or parameters of messages. Messages are modeled as (hyper)arcs that have one entity as target and as sources the message parameters (that may be references to other entities, values or time stamps). Time stamps describe the interval of time in which the message must occur in terms of the minimum and maximum time units relative to the current time. These time stamps (tm in and tm ax) are associated to each message (if they are omitted, default values are assumed).

For each entity, a graph containing information about all attributes of this entity, relationships to other entities, and messages sent/received by this entity is built. This graph is an instantiation of the object-based type graph described above. All rules that describe the behavior of this entity may only refer to items defined in this type graph.

A rule must express the reaction of an entity to the receipt of a message. A rule of an object-based graph grammar must delete exactly one message (trigger of the rule), may create new messages to all entities involved in the rule, as well as change the values of attributes of the entity to which the rule belongs.

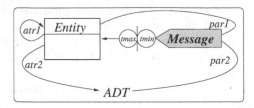

Fig. 1. Object-Based Type Graph

A rule shall not delete or create attributes, only change their values. At the right-side of a rule, new entities may appear (entities can be dynamically created). Besides, a rule may have a condition, that is an equation over the attributes of its left- and right-hand sides. A rule can only be applied if this condition is true.

An Object-Based Graph Grammar consists of a type graph, an initial graph and a set of rules. The type graph is actually the description of the (graphical) types that will be used in this grammar (it specifies the kinds of entities, messages, attributes and parameters that are possible – like the structural part of a class description). The initial graph specifies the start state of the system. Within the specification of an entity, this state may be specified abstractly (for example, using variables instead of values, when desired), and will only become concrete when we build a model containing instances of all entities involved in this system. As described above, the rules specify how the instances of an entity will react to the messages they receive.

According to the graph grammars formalism, the computations of a graph grammar are based on applications of rules to graphs. Rules may be applied sequentially or in parallel. Before going into more details on the way computations are built, we will briefly discuss how time is dealt with in our model.

Time stamps of messages describe when they are to be delivered/treated. In this way, we can program certain events to happen at some specific time in the future. As rules have no time stamps, we assume that the application of a rule is instantaneous. The time unit must be set in the specification. This will be used as a default increment for messages without a time stamp as well as to postpone messages. Time stamps of messages are of the form: $\langle tmin, tmax \rangle$, with $tmin \leq tmax$, where $tmin$ and $tmax$ are to be understood as the minimum/maximum number of time units, starting from the current time, within which the message shall be treated. As the time stamps are always relative to the current time, it is not possible to send messages that must be applied in the current time. The semantical model used by now is a hard-time semantics: if a message is sent, it must be treated within its specified time interval; if this is not possible, the computation fails. In this paper, we will not use the maximum time limit for messages, that is, in our case, there is no failure due to time constraints.

Each state of a computation of an OBGG is a graph that contains instances of entities (with concrete values for their attributes), messages to be treated, and the current time. In each execution state, several rules (of the same or different entities) may be enabled, and therefore are candidates for execution at that instant in time. Rule applications only have local effects on the state. However, there may be several rules competing to update the same portion of the state. To determine which set of rules will be applied in each time, we need to choose a set of rules that is consistent, that is, in which no two or more rules have write access to (delete) the same resources. Due to the restrictions imposed in object-based graph grammars, write-access conflicts can only occur among rules of the same entity. When such a conflict occurs, one of the rules is (non-deterministically) chosen to be applied. This semantics is implemented in the simulation tool PLATUS [2, 1, 5].

3.1 Pull-Based Failure Detector

Now we introduce a pull-based failure detector modeled using OBGG. In this example, every pull-based failure detector of the model is modeled as an entity, called PullD etector. The type graph and rules of this entity are shown in Figure 2. Some internal attributes of the PullD etector entity are abstract data types (M ap and List). The M ap abstract data type is used to handle a collection of tuples. The List abstract data type is the implementation of a chained list. The GS entity is the group server for a particular group being monitored. Through the GS the pull-based failure detectors can obtain references to communicate with other participants of the group being monitored. The detailed definition of the abstract data types M ap and List, and the GS entity can be found in [19]. The initial graph of this entity was omitted due to space limitations (in Section 5 we show a specification that has some instances of this initial graph).

In OBGG there is no embedded notion of local or remote communications. However, a developer may specify minimum delays for messages. This notion allows us to express some differentiation between remote and local communication by assigning different minimum delay times. By default, all messages that do not have minimum delays specified are treated as having the lowest minimum delay possible. In the case study presented in this section we have explicitly specified only the minimum remote communication times (m rct).

Figure 2 shows the rules that define the behavior of the PullD etector entity. A PullD etector entity monitors the activity of a known group of PullD etector entities (attribute group_server) using liveness requisitions. The monitoring behavior is started at the beginning of the system, where a Q liveness message is sent to each PullD etector entity of the system. The Q liveness message indicates the beginning of an interval of d1 units of time (rule ReqLiveness1). During this interval other entities (processes) will be asked if they are alive. To do that, a PullD etector has an associated GS (group server) which records all the other PullD etectors of the group. A PullD etector then asks GS these PullD etectors (rule ReqLiveness2), receives References to them, and sends messages AYA live for each one (rule AreYouA live). Responses to AYA live messages, via ImA live messages (rule IamA live), are accepted until the interval d1 has not expired. Upon reception of an ImA live message during the valid interval time, a PullD etector entity records every entity that confirms its situation (rule A liveReturn1). If the period for the response of monitored entities has finished, for

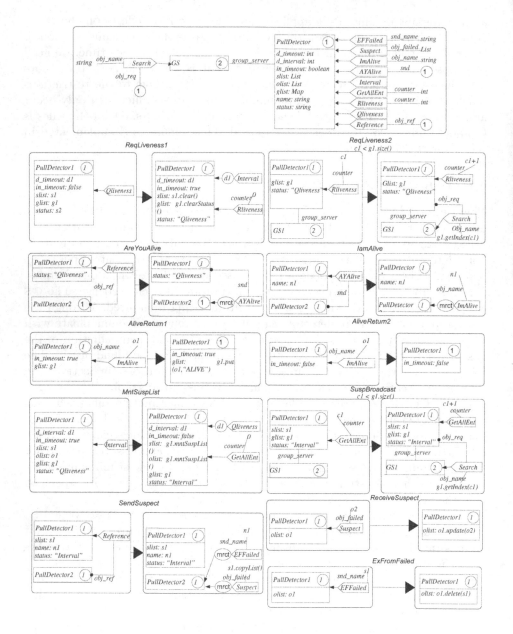

Fig. 2. Type Graph and Rules of the *PullDetector* Entity

every response message received after the interval time, the PullDetector entity does nothing (rule AliveReturn2). After the time interval for receiving ImAlive responses finishes, a PullDetector starts a new monitoring cycle, in an interval of d1 units of time, with a Qliveness message (rule MntSuspList). At the beginning of this pause the PullDetector entity builds a local list of suspected entities, located in the slist attribute (rule MntSuspList) and sends this list to every entity on the monitored group, using a Suspect message (rules SuspBroad-cast and SendSuspect). Besides, it sends a message EFFailed to every member of the monitored group. Thus, every entity in the monitored group receives (rule ReceiveSuspect) the local suspect list of the PullDetector that originated the Suspect message, and raises the union of this list with its own local suspect list, originating a unified list (without repeated entries). The reception of the message EFFailed causes the originator to be excluded from the suspect list.

4   Transforming Specifications to Represent Fault Models

In this section we explain the methodology we have used to represent a subset of the classical fault models found in the literature using the OBGG formalism. After that, we show how to use (insert) these selected models in an already defined model of a desired system.

Traditionally, terms like fault, error and failure are used as in [15]. Here we try to avoid the terms error and failure, and adopt a formal definition, given in [10], for fault. As stated in [3] a system may change its state based on two types of events: events of normal system operation and fault occurrences. Based on this observation, a fault can be modeled as an unwanted (but possible) state transition of a system [10]. Thus a fault behavior of a system is just another kind of a (programmable) behavior. These unwanted state transitions can be modeled through the use of additional virtual[1] variables, acting like guards to activate specific commands (guarded commands). In this case, a group of guarded commands represents a specific fault, i.e. the manner in which the system will exhibit the fault behavior, being activated whenever its associated guard is satisfied, by the assignment of a true value to it [9].

The addition of virtual variables and guarded commands can be viewed as a transformation of a model M_1 into a model M_2 that contains in its state space the behavior of a selected fault model [10]. Here, we adopted these concepts. Since our specification formalism supports implicit parallelism and is declarative, it is very suitable to represent guarded commands: the left-side of a rule corresponds to the guard of the command; applying the rule transformation (according to the right side) corresponds to executing the guarded command. Thus we can model fault behaviors for an OBGG specification inserting virtual variables (used for guards) and messages (used to activate a fault behavior) in every entity of the model. Besides, we need both to create rules representing the fault behavior introduced and, depending on the fault model, to change original rules defined for the entities that appear in the model. Depending on the fault model, different rule transformations occur. These transformations are explained in more detail in the next sections.

---

[1] The term virtual is used to qualify variables that are not part of the desired system itself, but part of the introduced fault behavior.

For the selection of fault models to be described using our formalism we have adopted the fault classification found in [13]. There, fault models are classified in: crash, omission, timing, and Byzantine. Specifically, with respect to the omission model, we have used the classification of [9], where the omission model is splitted up in: send omission, receive omission, and general omission. In the next sections we present the modeling of: crash, send omission, receive omission, and general omission fault models. Also, we show how to introduce a selected fault behavior in an existing system model. We have not yet modeled the timing and Byzantine fault models. In the timing fault model a process might respond too early or too late [13]. In the Byzantine fault model a process may assume an arbitrary behavior, even malicious [14]. In the future we intend to model these fault models as well. In this paper we concentrated our efforts in the modeling of the previous cited fault models, which are very often used in the fault tolerance literature (e.g. the crash fault model is commonly considered for distributed systems).

4.1   Crash Fault Model

In the crash fault model a process fails by halting. The processes that communicate with the halted process are not warned about the fault. Below it is shown how to transform a model M_1 (without fault behavior) into a model M_2 that incorporates the behavior of a crash fault model.

To add the behavior of a crash fault model we perform the following transformation on each entity $GG = (T, I, Rules)$ (with type graph T, initial graph I and set of rules $Rules$): insert a message Crash and an attribute in the type graph of the entity (depending on the value of this variable, the entity may exhibit the fault behavior or not); insert the same message in the initial graph of the entity, in order to activate the fault behavior; create a new rule that activates the fault behavior of the entity; create new rules whose left-sides are replicas of the original rules left-sides and right-sides specify no activity (let the state unchanged), representing the fault behavior once the guard is true (down is true); modify all the rules of an entity with the insertion of a guard (down:false), meaning that the entity will exhibit the original behavior only if it is not crashed. These modifications generate a new entity $GG' = (T', I', Rules')$ in which the components are illustrated by schemes in Figure 3: it shows how the type graph and initial graph are transformed, how each rule r generates two rules r' and r'' in $Rules'$ and shows the rule to be added $EntCrash$.

An example of these rule transformations is presented in Figure 4, showing the transformation of the rule IamAlive previously defined for the PullDetector entity. Figure 4 (a) shows the normal behavior of the entity when the variable down is set to false (the fault behavior is not active). Once the fault behavior is activated, the rule defined in Figure 4 (b) will occur. This rule simply consumes the message used to trigger the rule but does nothing (neither change the state of the entity nor create new messages), in this way we incorporate the behavior of a crash fault model.

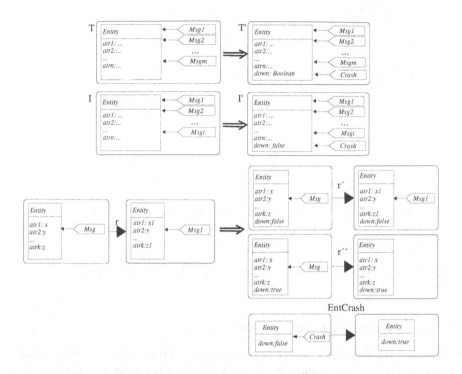

Fig. 3. Transformation over a model to represent a crash fault model

4.2   Receive Omission Fault Model

In the receive omission fault model a faulty process receives only a subset of the messages sent to it [11]. The receive omission fault model is analogous to the send omission fault model. The main difference is that in this fault model only a subset of the total messages sent to the fault process are received, differently from the send omission where only a subset of the total messages sent by the process are actually sent.

To add the behavior of a receive omission fault model we perform the following transformation on each entity $GG = (T, I, Rules)$: insert an attribute in the entity's type graph (depending on the value of the variable, the entity may exhibit the fault behavior); insert a R cv0 m it message in the entity's type graph; create a new rule that activates the fault behavior of the entity; create new rules whose left-sides are replicas of the original rules left-sides and right-sides specify no activity (let the state unchanged), representing the fault behavior once the guard is true (rcv_om itted is true); let unchanged the rules of the entity (guards are not inserted), since in the receive omission fault model a process may fail to receive only a subset of the total set of messages sent to it. That is why we do not insert a guard on the original rules, leaving the choice (once the guard is true)of

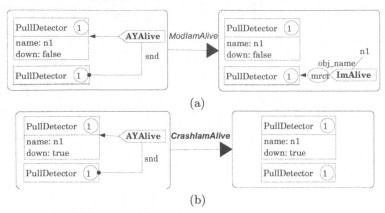

Fig. 4. (a) Rule *IamAlive* without fault behavior (b) Rule *IamAlive* with fault behavior

which rule to apply in a non-deterministic way. These changes are illustrated in Figure 5.

4.3   Other Omission Fault Models

Due to space restrictions we do not discuss the send omission and general omission fault models at the same level of detail, but rather discuss them in terms of the ideas already presented.

A process in a send omission fault fails to transmit a subset of the total messages that it was supposed to send [11]. The transformation of a model to incorporate the send omission behavior is analogous to the previous transformations. However, in this case, some messages that should be sent will simply be ignored. Again, in this model we use the non-deterministic choice of rules offered by OBGG to model the fact that a message may be sent or not.

The general omission model states that a process may experience both send and receive omissions [18]. Using these concepts we model the general omission model as a merge of the previously discussed send and receive omission fault behaviors.

5   Analysis of the Pull-Based Failure Detector

In this section we exemplify the introduction of fault models described in the previous sections using the example of a pull-based failure detector defined in Section 3.1. We have applied the transformation for the crash fault model defined in Section 4.1, in order to reason about the system in presence of crash faults.

First, we generated the entity PullDetectorCrash following the construction given in Section 4.1. This transformation of the PullDetector model presented in Section 3.1 considering the crash fault model generated 23 rules out of the

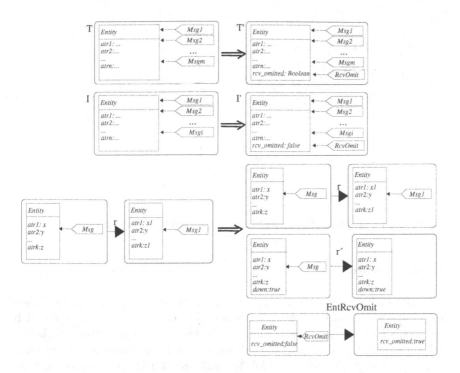

Fig. 5. Transformation over a model to represent a receive omission fault model

11 original ones. From these 23 rules, 11 describes the behavior of the model without crash, other 11 rules describe the behavior of a crashed pull detector and 1 rule serves for the activation of the crash model.

Second, we build the specification of a system containing three pull detectors (three instances of the entity PullDetectorCrash), instantiating in two of them the tmin of message Crash as MaxInt, and as 110 in the other (see Figure 6). This models the fact that pull detectors 1 and 3 will not crash, and pull detector 2 will have the possibility to crash after 110 time units. When this Crash message is received, Pdetector2 entity will halt, assuming a behavior where neither messages will be sent nor received, and its state will not change. Two of the three Qliveness messages, used for the activation of the entities, in the system have also been delayed, such that the entities do not start the detection cycle at the same time. Note that the initial graph shown in this figure is just a part of the initial graph of the system being modeled, the instantiations of the GS entities are not shown. In the execution of this scenario both entities Pdetector1 and Pdetector3 will detect that Pdetector2 has failed, and will put the entity in their unified suspect list (olist variable). As presented in Section 1, currently we have, as means to analyze the behavior of OBGG formal specifications, a simulation tool. We use this tool to generate a log of an execution of the system.

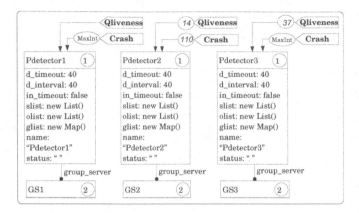

Fig. 6. Initial graph for PullDetector entities considering a crash fault model

6 Final Remarks

In this paper we have presented an approach for the specification and analysis of distributed systems considering the presence of faults. We have shown the definitions of crash and omission fault models, the last one being further specialized in send, receive and general omission. Moreover, we have shown how to combine the behavior of a given system model M_1 and one of the fault models defined, achieving a model M_2. Model M_2 can be used to reason about the behavior of the original model M_1 in the presence of the considered fault.

The formal specification language we have used to represent models of systems as well as to describe fault models is a restricted form of graph grammars, called Object Based Graph Grammars. The fact that this formalism is declarative and supports the notions of implicit parallelism and non-determinism resulted that transformations from M_1 to M_2 were rather simple to define (see Section 4). Since models described according to the used formalism can be simulated, we have then a method and a supporting tool to help reasoning about distributed systems in the presence of faults. Simulation can be used as means for testing the behavior of models in the presence of faults, as well as for performance analysis. Since we aim also to prove properties over the models we define, we are analyzing an approach for model checking models written in OBGG. This approach is based on the mapping of OBGG specifications to a description language that could serve as input to a model checker.

References

[1] B. Copstein, M. Mora, and L. Ribeiro. An environment for formal modeling and simulation of control system. In *Proc. of the 33rd Annual Simulation Symposium*, pages 74–82. SCS, 2000. 121, 125

[2] B. Copstein and L. Ribeiro. Compositional Construction of Simulation Models using Graph Grammars. In *Application of Graph Transformations with Industrial Relevance (AGTIVE'99)*, LNCS 1779, pages 87–94. Springer, 2000. 125

[3] F. Cristian. A rigorous approach to Fault-Tolerant programming. *IEEE Transactions on Software Engineering*, 11(1):23–31, 1985. 127

[4] F. Dotti and L. Ribeiro. Specification of mobile code systems using graph grammars. In S. Smith and C. Talcott, editors, *Formal Methods for Open Object-based Systems IV*, pages 45–64. Kluwer Academic Publishers, 2000. 121, 123

[5] F.L. Dotti, L.M. Duarte, B. Copstein, and L. Ribeiro. Simulation of Mobile Applications. In *Communication Networks and Distributed Systems Modeling and Simulation Conference*, pages 261–267. SCS International, 2002. 121, 125

[6] F.L. Dotti, L.M. Duarte, F.A. Silva, and A.S. Andrade. A Framework for Supporting the Development of Correct Mobile Applications based on Graph Grammars. In *6th World Conference on Integrated Design & Process Technology*, pages 1–9. SDPS, 2002. 121

[7] H. Ehrig, G. Engels, H.-J. Kreowski, and G. Rozenberg, editors. *Handbook of Graph Grammars and Computing by Graph Transformations, Volume 2: Applications, Languages and Tools*. World Scientific, 1999. 122

[8] C. Fournet, G. Gonthier, J.-J. Lévy, L. Maranget, and D. Rémy. A Calculus of Mobile Agents. In *7th International Conference on Concurrency Theory*, pages 406–421. Springer, 1996. 122

[9] F.C. Gärtner. Specifications for Fault Tolerance: A Comedy of Failures. Technical Report TUD-BS-1998-03, Darmstadt University of Technology, Department of Computer Science, Germany, 1998. 121, 127, 128

[10] F.C. Gärtner. Fundamentals of Fault-Tolerant Distributed Computing in Asynchronous Environments. *ACM Computing Surveys*, 31(1):1–26, 1999. 121, 127

[11] V. Hadzilacos and S. Toueg. A Modular Approach to Fault-Tolerant Broadcasts and Related Problems. Technical Report TR94-1425, Cornell University, Department of Computer Science, Ithaca, NY, USA, 1994. 129, 130

[12] G.J. Holzmann. The Model Checker SPIN. *IEEE Transactions on Software Engineering*, 23(5):279–295, 1997. 122

[13] P. Jalote. *Fault Tolerance in Distributed Systems*, pages 51–53. Prentice Hall, Englewood Cliffs, NJ, USA, 1994. 128

[14] L. Lamport, R. Shostak, and M. Pease. The Byzantine Generals Problem. *ACM Transactions on Programming Languages and Systems*, 4(3):382–401, 1982. 128

[15] J.-C. Laprie. Dependable Computing and Fault Tolerance: Concepts and Terminology. In *15th International Symposium on Fault-Tolerant Computing*, pages 2–11. IEEE Computer Society Press, 1985. 127

[16] N. Lynch. *Distributed Algorithms*. Morgan Kauffman, 1996. 122

[17] N. Lynch and M. Tuttle. An Introduction to Input/Output Automata. *CWI-Quarterly*, 3(2):219–246, 1989. 122

[18] K.J. Perry and S. Toueg. Distributed Agreement in the Presence of Processor and Communication Faults. *IEEE Transactions on Software Engineering*, 12(3):477–482, 1986. 130

[19] E.T. Rödel. Modelagem Formal de Falhas em Sistemas Distribuídos envolvendo Mobilidade (Formal Modeling of Faults in Distributed Systems with Code Mobility). Master's thesis, Pontifícia Universidade Católica do Rio Grande do Sul, Porto Alegre, RS, Brasil, 2003. 125

[20] G. Rozenberg, editor. *Handbook of Graph Grammars and Computing by Graph Transformations, Volume 1: Foundations*. World Scientific, 1997. 122

[21] F.A. Silva. *A Transaction Model based on Mobile Agents*. PhD thesis, Technical University Berlin - FB Informatik, Berlin, Germany, 1999. 120

Integrating Graph Rewriting
and Standard Software Tools[*]

Uwe Aßmann and Johan Lövdahl

Research Center for Integrational Software Engineering (RISE)
Programming Environments Lab (PELAB)
Linköpings Universitet, Sweden
{uweas,jolov}@ida.liu.se

Abstract. OptimixJ is a graph rewrite tool that can be embedded easily
into the standard software process. Applications and models can be developed in Java or UML and extended by graph rewrite systems. We
discuss how OptimixJ solves several problems that arise: the *model-
ownership problem*, the *embedded graphs problem*, the *library adaptation
problem*, and the *target code encapsulation problem*. We also show how
the tool can be adapted to host language extensions or to new host
languages in a very simple way, relying on the criterion of *sublanguage
projection*. This reduces the effort for adapting OptimixJ to other host
languages considerably.

1 Introduction

How can we get more people to employ graph rewriting in their daily modelling
and programming tasks? One of the standard answers is: "by better tooling".
However, there are two major problems with this answer. Firstly, most graph
rewrite tools have employed their proprietary DDL[1], basing the graph rewrite
specifications and the generated code on this DDL. However, this principle forces
applications to use the DDL of the tool and, hence, to base their code on the
tool: the tool owns the model However, in today's model-based development
processes, models are maintained in the language of the application or in UML.
And UML tools do not know nor care about other generators (model-ownership
problem, Figure 1).

The second problem is a person-power problem. Graph rewrite tooling is lagging behind the development in standard software engineering, mainly because

---

[*] Work partially supported by European Community under the IST programme -
Future and Emerging Technologies, contract IST-1999-14191-EASYCOMP. The authors are solely responsible for the content of this paper. It does not represent the
opinion of the European Community, and the European Community is not responsible for any use that might be made of data appearing herein.
[1] In the following, we distinguish data definition languages (DDL), which are used
for type declarations, from data manipulation languages (DML), which are used for
procedural and rule-based behavior description.

J.L. Pfaltz, M. Nagl, and B. Böhlen (Eds.): AGTIVE 2003, LNCS 3062, pp. 134–148, 2004.
© Springer-Verlag Berlin Heidelberg 2004

the graph rewrite tool developer community is rather small and has not so much person-power available as e.g., the community that develops UML tools. So, the tool gap keeps on increasing. It almost looks like an inverted Zenon's paradox: every time the graph rewrite tool developer has realized a step towards a tool, the rest of the world has done already two steps into other directions.

This paper presents solutions for both problems. It presents a tool, OptimixJ, which works seamlessly together with Java and jUML[2]. The tool does not superimpose its own graph model to the standard languages, but refers to the graph models in the users' specification, enabling reuse of standard models. Hence, the tool does not force users to change their software process when working with graph rewriting. Changing to OptimixJ can be done step-by-step, in a piece-meal growth; users can start with small additional graph rewrite specifications that work on standard models, and, being more experienced, go on with larger specifications. We believe that this philosophy is somewhat more realistic than the philosophy of current tools: graph rewriting is a powerful technique for several application fields, but not for all of them. Users need flexibility and do not want to be constrained by the full-metal jacket of a monolithic graph rewrite tool. Nevertheless, they want to be able to decide when they use graph rewriting.

However, even this integration feature will not be sufficient, because new modelling and programming languages are evolving all the time. For instance, Java 1.5 will integrate generic data structures. A graph rewrite tool that is integrated with Java has to be extended to generic data types, which can be a very laborious process. Hence, we have to require for a successful tool that it must be extensible to new versions of the host language[3] very quickly and easily. This is the case for OptimixJ. We show this by investigating its architecture in more detail. In fact, OptimixJ exploits a novel sublanguage projection concept to adapt its rewrite specification to a host language. Using the adapter, graph rewrite systems can be type-checked against the host data model, enabling reuse of standard models. Additionally, when such an adapter is exchanged, a new host language can be connected to the tool. Instead of writing a full new compiler for host language plus graph rewriting extension, only specific parts of the tool's architecture need to be reimplemented. This simplifies the construction of new tools enormously.

Apart from these main problems, several other minor problems hobble the use of graph rewrite systems in the modern development process. Firstly, many UML tools implement associations with neighbor sets, i.e., implement graphs not in a closed form, but embed them into the nodes (embedded graph problem). Usually, the graph rewrite tools cannot handle this. Secondly, if the model is owned by the UML tool, the GRS generator should be able to know, which node attributes belong to graphs and which ones don't. In particular, the UML tools

---

[2] *jUML* is assumed to be the variant of UML that maps directly to Java. For instance, class diagrams should not employ multiple inheritance.

[3] The *host language* is the language to which all specifications and programs should be translated.

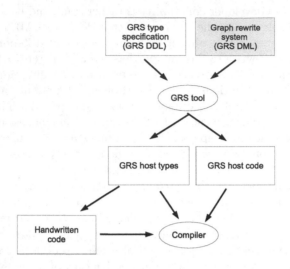

Fig. 1. The standard scenario of graph rewrite system (GRS) tools in the software process. The tool maintains its own DDL, forcing the application to use generated type definitions, or proprietary graph classes

use data types of the host language library to type the associations. Usually, the GRS tool cannot adapt to arbitrary host language libraries, because it does not know whether a class has a graph-like semantics or not (host library adaptation problem).

Next, modern object-oriented languages enforce encapsulation of code and data in classes. While, in general, this is a well-accepted structuring principle for systems, the encapsulation prevents that classes are extended later on (target code extension problem). Hence, a generator has either full control over a class, or none at all. It is no longer possible, as in the target language C, to augment a module stemming from the application model with additional procedures generated by the GRS tool.[4] Often, this problem is solved by delegation: graph nodes and objects from the application are realised in different physical objects, delegating work to each other. However, delegation is subject to object schizophrenia [9] and slows down navigations in large graphs ([3], Chapter 9).

OptimixJ can be fully integrated with standard software engineering languages and tools, such as Java and jUML. When using OptimixJ, the model is owned by the major development tool and the graph rewrite system only enriches the generated code (Section 2). Models from jUML structure diagrams can be used, legacy Java implementations can be augmented with GRS code. Hence,

---

[4] This problem does not occur in all object-oriented languages. For instance, C++ has an extension operator :: that allows for extending a class remotely, and this operator can be used to augment the classes generated from the model.

the graph rewrite systems adapt to the data model of the application, which is a major advantage. OptimixJ allows for embedded graph representations that are used by UML tools. OptimixJ is adapted to the JDK, all graph types of the JDK can be employed. New Java container libraries can be integrated easily. Since OptimixJ provides executable GRS, it can be used as an executable form of OCL and we investigate the commonalities and differences (Section 4). Due to sublanguage projection, OptimixJ is easy to port to other DDL (Section 5). Finally, we compare to related work (Section 6). We believe that the tool and its underlying concepts can provide a valuable contribution to an intensified use of graph rewrite systems in the industrial software development.

2 OptimixJ in Action

OptimixJ is a graph rewrite tool that can generate implementations of algo-rithmic attributed graph rewrite systems with 1-context [2]. It is an extension of Optimix [1], but seamlessly integrated with the host language Java. Hence, OptimixJ can be employed very easily in a standard object-oriented software process.

We start with a simple example. Consider the following Java class `Person`, which models relationships of persons in families (`father`, `mother`, `children`, a.s.o.). The main program builds up a little database of persons, and starts some processing.

```
public class Person {
 public Person father;        public Mother mother;
 public Person[] parents;     public Person[] children;
 public Vector ancestors;     public Vector childrenAbove18;
 public Vector oldChildren; public String id;
 public int age;
 public Person(String id) {
  father = new Person() ;      mother = new Vector();
  parents = new Person[10] ;   children = new Person[10];
  ancestors = new Vector();    this.id = id;
  childrenAbove18 = new Vector();
 }
 public static void main(String[] s) {
  Person gDad = new Person("grand_dad");
  Person gMom = new  Person("grand_mom");
  Person child1 = new Person("child1");
  Person child2 = new Person("child2");
  Person father = new Person("father");
  Person mother = new Person("mother");
  child1.age = 23;             child2.age = 13;
  child2.setFather(father);    child2.setMother(mother);
  child1.setFather(father);    child1.setMother(mother);
  father.setFather(gDad);      father.setMother(gMom);
  Vector persons = new Vector();  persons.add(father);
```

```
    persons.add(child2);                persons.add(child1);
    Person[] dummy = new Person[1];
    makeChildrenConsistent((Person[])persons.toArray(dummy));
    calculateOldChildren((Person[])persons.toArray(dummy));
  }
}
```

This class is incomplete because the methods `makeChildrenConsistent` and `calculateOldChildren` are realized by graph rewrite systems. We store it in a file `Person.jox`, indicating its incompleteness. The rewrite systems can be defined in a separate specification `Person.ox`. OptimixJ generates their implementation and inserts them into `Person.jox`, producing the final `Person.java`.

```
module Person;
grs MakeChildrenConsistent(persons:Person[]) {
  range C <= persons;
  rules
  // Datalog syntax: <conclusion> :- <premise>.
  // (1) consistency of relations
  parent(C,M) :- mother(C,M);
  parent(C,F) :- father(C,F);
  children(P, C) :- parent(C, P);
}
grs calculateOldChildren(persons:Person[]) {
  range parent <= persons;
  rules
  // Rule syntax: if <premise> then <conclusion>;
  if parent in child.parents then parent in child.ancestors;
  if parent in child.parents and ancestor in parent.ancestors
    then ancestor in child.ancestors;
  if child in parent.children,
    child matches Person{age=>Val}, Val >= 18
    then child in parent.childrenAbove18;
  if child in parent.childrenAbove18
    // (2) Building up a new node domain
    then add g:GrownUpChild,
    // (3) Put into new relationship
      g in parent.oldChildren;
}
end Person
```

While `makeChildrenConsistent` inverts several relations in the model, `calculateOldChildren` computes a transitive closure and allocates new nodes of a new type `GrownUpChild`. ¿From this GRS definition, OptimixJ generates two methods fragments, which are embedded into class `Person`, stored in file `Person.java`:

```
public class Person {
  public static void makeChildrenConsistent(
    persons:Person[]) {...}
```

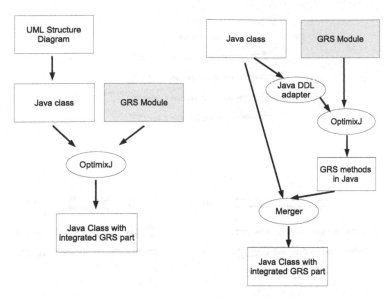

Fig. 2. With OptimixJ, graph rewrite specifications enrich Java classes

```
public static void calculateOldChildren(
    Person[] persons) {...}
    ... other members as above ...
}
```

This class is a standard Java class, declarationally complete, and can be compiled to byte code. Figure 2 summarizes how OptimixJ is integrated with Java: graph rewrite specifications enrich standard Java models. And of course, `Person.jox` could also be generated from a jUML model.

3   What OptimixJ uses From Java

The JavaDDL Adapter. To find out about the graphs in the model, OptimixJ contains an adapter to Java type definitions (DDL adapter). This adapter parses the type declarations in the Java classes. Parsing all of Java is not necessary. OptimixJ restricts itself to JavaDDL, the data definition sublanguage of Java (classes, fields, method signatures). The adapter translates JavaDDL to the internal format of OptimixDDL, the graph type language of Optimix itself. In this way, OptimixJ finds out about graph node types, inheritance relations between them (extends and implements), and association relations. During compilation, it type-checks the graph rewrite systems against this model.

Embedded Graphs. Graphs are assumed to be implemented with Java container types. In the most simple case, set container types are employed for one direction

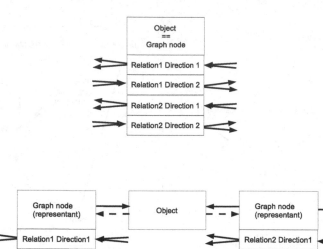

Fig. 3. Embedded and separate graphs

of a relationship type. This is similar to the way UML tools realize relations (embedded graphs, implemented by neighbor sets, Figure 3). In the example, the parent-child relation is represented by fields

```
public Person[] parents;
public Person[] children;
```

Alternatively, a graph can be realized by a closed data structure in which every object is represented by a graph node, and referenced by the representant (Figure 3). If the host language tool and the graph rewrite tool are different in this respect, they are hard to integrate.

Many GRS tools do not employ embedded graphs. For instance, the AGG tool ships a Java library for graphs, nodes, and edges [10]. To these nodes, arbitrary Java objects can be attached as attributes. When an object should be integrated into several graphs, objects either are referenced only by the graph representants, or have to be extended with a link to the graph representant. This has at least two disadvantages. Firstly, without back-references, graph node representants have to be looked up dynamically in graphs. This costs additional overhead during graph navigations and can slow down applications considerably [3]. Secondly, the back-references have to be typed with the graph node type of the library, but cannot be typed with a type from the application model. Embedded graphs are faster for navigations and can be statically typed.

It should be noted that also other tools use DDL adapters. The UML validation wizard of AGG transforms XMI, which was exported from an XML tool, to the AGG external representation GGX [10]. Then, AGG will read the UML dia-

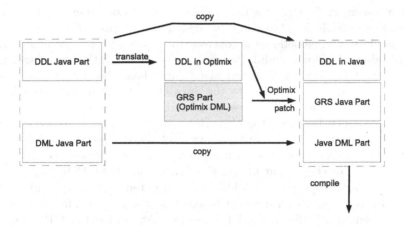

Fig. 4. The internal architecture of OptimixJ. Several parts of the Java models are copied to the output file. Additionally, the output file is extended with the code generated from the graph rewrite systems

grams, and can work on them. In the validation wizard, the adapter is employed to check inconsistencies of UML specificiations.

The JDK Library Adapter. Java and its libraries develop. To be able to incorporate new Java container types for neighbor sets, OptimixJ offers to create an adapter by a functor declaration. In general, functors are functions over types; in our case they denote container data type specifications for OptimixJ. Appendix A contains a functor declaration for a container type of a Java library. A functor declaration consists of a feature structure (associative array) with the following parts:

1. Name of the container class
2. Source type and target type of the functor (`Object` is also possible)
3. Specification of graph features: bidirectional, homogeneous, etc.
4. Specification of optimization features: OptimixJ can optimize certain algorithms over certain graphs, based on features of the functors
5. Names of functions in the container class for adding and deleting nodes and edges

It is important to see that functor declarations make OptimixJ independent of Java libraries. If a container class has a list- or set-like structure, a functor can be declared such that OptimixJ can generate code for it. If a new library appears, new functor declarations have to be written, and the generator will work with it. This makes OptimixJ a very flexible tool useful in many different application scenarios.

The Code Merger. To treat the target code extension problem, i.e., to overcome Java class encapsulation, the generated code must be inserted into an appropriate Java class. The embedding is controlled by class and module names. In our example above, Optimix module **Person** is embedded into the Java class **Person**. Whenever a GRS module is named like a Java class, OptimixJ will embed the generated graph rewrite code into the Java class. Conceptually speaking, the tool uses module and class names as joinpoints [6] and weaves graph rewrite code and ordinary Java code together. Hence, with OptimixJ, graph rewrite systems can be seen as class extensions. As a whole, the process is illustrated in Figure 4. OptimixJ copies the Java classes, type declarations and method implementations, to the output file (in the figure marked by DDL Java Part and DML Java Part, respectively). Additionally, it extends the output file with the code generated from the graph rewrite systems (the grey box in the figure).

After generating the method fragments that realize the GRS, OptimixJ merges them into the Java class of the application.

4   Comparing the Roles of OptimixJ and OCL

Since embedded graph representations of host languages can be handled by OptimixJ, since the Java DDL is understood by the language adapter, since library container data types are handled by the library adapter, and since the encapsulation problem is solved by the merger, OptimixJ can play the role of an executable constraint language for UML. Graph rewrite systems in OptimixJ can play a similar role as constraint specifications in OCL (object constraint language), both conceptually and pragmatically. However, OptimixJ specifications are executable, OCL specifications need not be.

Conceptually, OCL specifications are relative to a class diagram, i.e., they form an aspect of the specification. Constraints without classes do not make sense. Similarly, OptimixJ graph rewrite rules without a Java data model cannot be translated to executables.

Pragmatically, OptimixJ and OCL are used for similar purposes, however, OptimixJ is constructive while OCL is checking. Our example contains several consistency rules, for instance, rule (1) in the above example describes that two relations are inverse to each other. In OCL, such a rule is interpreted declaratively, i.e., the rule checks whether the model is consistent. In OptimixJ, the rule is interpreted constructively, i.e., when the rule system is executed, the model is consistent by construction. Hence, OptimixJ can be used to build up consistent new information, starting from arbitrary Java models. For more details about OptimixJ's code generation, please consult [2].

Secondly, graph rewrite systems can play the role of filters that can be applied to object domains. In the above example, relation **childWithAgeOver18** is filtered from **children** with a boolean condition. Again, OptimixJ interprets the rule constructively.

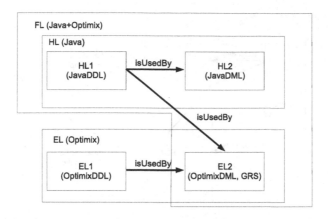

Fig. 5. Languages that should be integrated. FL is HL extended by EL2

As a logic, OptimixJ only supports binary typed Datalog, in its constructive form of edge-addition rewrite systems (EARS) [2]. OCL provides full first-order logic. On the other hand, OptimixJ allows for transformations on the model, i.e., it can be specified how certain parts of the model are built from other parts. For instance, GRS can build up new object domains (see rule (2) in the above example) and span up new relations between the new node domains (rule (3)).

As can be seen in system `calculateOldChildren`, OptimixJ allows for a simple business rule syntax. Logic-like specifications in the form of binary Datalog are possible, but we presume that business rule like syntax will be accepted by more users. Already today, such rules are extensively used in software processes for business systems [12], and this community is likely to accept a notation looking less formal.

There remains one final problem for the tool to be useful on the long run. Because OptimixJ lets the standard development tool own the model of the application, it must be adapted to extensions of the host language or to other ones rather quickly. How can that be achieved with minimal costs?

5   Sublanguage Projection - A New Architecture for Compilers of Extended Languages

OptimixJ is not only easy to adapt to new Java libraries, its internal architecture is also easy to extend to new modelling languages or new extended versions. We estimate that an adaptation can be done in about 1-3 personmonths. Actually, OptimixJ itself demonstrates this, since it is an extension of an earlier version of

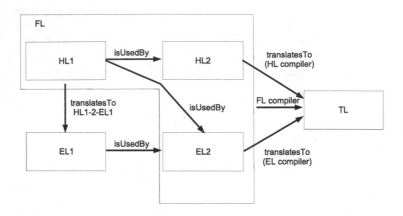

Fig. 6. The separation of the FL compiler into DDL mapper, EL compiler, and HL compiler

Optimix and was constructed in about 8 person weeks. This simple extensibility relies on a specific tool architecture that exploits a systematic feature of Java and the Optimix language, sublanguage projection. If this criteria is fulfilled for another host language, it can be easily extended with graph rewrite features.

The criterion we develop assumes a scenario in which a graph rewriting language (extending language, EL, e.g., Optimix) should be integrated with a host language (HL), e.g., Java, to a full language FL (Java+Optimix, Figure 5). To this end, a compiler for FL has to be created that translates FL programs to a target language (TL). Building such a compiler usually costs several person years, but under certain conditions, the full language can be split into several sublanguages, which can be treated in separation. In particular, the compiler for the graph rewriting language EL can be reused, although the host language changes. And this saves many person months of development time.

The central idea of a system architecture relying on sublanguage projection is that the compiler for the host language (HL compiler) and the compiler of the extending language (EL compiler) should be reused as components in the compiler for the full language (Figure 6). To this end, we assume that both HL and EL can be compartmented into at least two sublanguages each, HL1, HL2, EL1 and EL2, where the translation of HL2 only depends on HL1 but not vice versa. EL2 must be independent of HL2, i.e., both sublanguages can be translated or interpreted independently of each other. Then, to reuse the EL compiler for FL in the context of HL, it need to understand HL1, but not HL2. And instead of constructing an FL compiler directly, we develop a compiler from

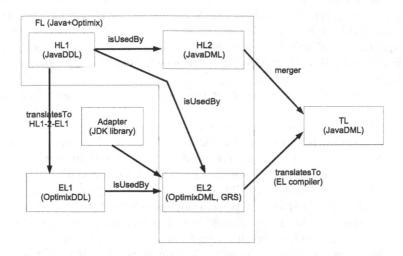

Fig. 7. The separation of the OptimixJ compiler into JavaDDL mapper, Optimix compiler, and JDK adapter. Since HL = TL, the HL compiler is the identity

HL1 to EL1. This compiler should be easier to construct than a compiler for FL, because HL1 is a subset of HL and independent of HL2.

Hence, the central idea of OptimixJ's system architecture is that the compiler for a Java+Optimix language can be split into a Java compiler, an Optimix compiler, and some additional tools (Figure 7, cf. Figure 4). To do the translation of the OptimixDML, the Optimix compiler must understand the DDL sublanguage of Java (JavaDDL adapter) and its ways how to represent graphs (Java Library adapter), but it need not recognize full Java. The Java adapter is a JavaDDL-2-OptimixDDL compiler (HL1-2-EL1), while the JavaDDL adapter consists of a specification that assert graph features for JDK container classes. With these adapters, type definitions in JavaDDL (HL1) can be evaluated from the Optimix compiler (EL), when translating OptimixDML (EL2) to Java (TL). This takes advantage of the fact that JavaDDL is much simpler than full Java; Optimix need not know about JavaDML (HL2), since OptimixDML and JavaDML are independent. Hence, the language adapter enables us to reuse the EL compiler (Optimix), at least to translate EL2 to TL. In the general case, we can employ the standard HL compiler to translate all HL parts of an FL program to TL. In our case, HL and TL are both Java, so we only need to integrate the result of both compilers into one consistent TL specification (m erging). The result is a full compiler for the extended language FL (Java+Optimix), constructed with much less effort since we reused both compilers for HL and EL.

It could be said that the independence of the language components HL2 and EL2 enable us to separate out the concerns of the host language and the graph rewrite language. This leads to the reuse of the compilers. In other words, the independence allows us for projecting FL into two languages, HL and EL (sublanguage projection). The projections are performed by HL1-2-EL1 and by FL-2-HL. Projection is a form of decomposition; merging TL components is a form of composition, and hence, we have employed a form of divide-and-conquer for compiler construction, in particular for the extension of GRS languages to host languages.

It could also be said that, in order to compile FL, we change domains. The algorithm class domain transformation is well known [4]. It solves problems by transforming them into domains in which simpler solution algorithms exist. Then, the solution is transformed back into the original domain. In our case, we transform Java+Optimix into Java and Optimix, for both of which compilers exist. The transformations are the JavaDDL-2-OptimixDDL compiler and the identity function (since HL=TL). The reverse transformation consists the merging of the generated Java code components. In case the transformations are simple to construct, the domain transformation will solve the original problem (a compiler for Java+Optimix) in a simple way.

We expect that the sublanguage projection principle will permit us to port OptimixJ to future versions of Java, or languages such as C#. Since every typed language has a DDL and DML part, the principle can easily be used for other combinations of graph rewriting languages and host languages.

6   Related Work

In agile processes, such as XP, schema evolution in the host language plays a major role. Often, semantics-preserving restructurings are required to bring the software into a better shape, before a new functionality is introduced (refactoring). While the first semi-automatic refactoring tools appear, they cannot be employed if the model is owned by the GRS tool (refactoring problem). Building a refactoring tool for a full-blown GRS language costs several person years [7], and the GRS community doesn't have the ressources to do that. Refactoring of OptimixJ applications, however, is simpler: the model is done in Java or jUML and can be refactored with standard tooling. Graph rewrite specifications have to be refactored by hand, but usually, they are smaller than the model so that this requires less effort. It is an interesting question for future research, whether a refactoring tool for the GRS language can be built in a similar way as sublanguage projection suggests.

To our knowledge, Fujaba is the only tool that integrates UML with graph rewrite systems [11]. This integration seems to work well, however, many of the person-power problems apply also to Fujaba.

XML namespaces are a language component model that does not allow for dependencies between the components: all components are independent in their interpretation, i.e., for interpretation, the language constructs of different language modules do not relate to each other. Immediately, XML languages are candidates for sublanguage and compiler partitioning.

Aspect languages are candidates for language layering [6]. By definition, an aspect is relative to a core [3], which implies a layered language architecture. Given an aspect compiler (a weaver) for aspects (EL) and core language HL, it can be reused with another core language HL' if a mapper HL'-2-EL is written. Hence, the sublanguage projection principle should also simplify the construction of weaver tools.

7 Conclusion

OptimixJ solves several problems of current graph rewrite tools. In development, the application model need not be done in OptimixJ, but can be done in a standard language. OptimixJ understands models in JavaDDL, also in jUML, allows for embedded graph representations, and can be adapted to new library container classes. Hence, it solves the model-ownership problem, the embedded graphs problem, and the library adaptation problem. OptimixJ invasively embeds the generated code into application classes, breaking Java encapsulation in a controlled way (target code encapsulation problem). We also have discussed how such a tool can be adapted to host language extensions or to new host languages in a very simple way, by sublanguage projection. This could simplify the architecture of future GRS tools considerably.

At the moment, OptimixJ is used in two interesting software engineering applications, and their authors have also been involved in the OptimixJ implementation. The SWEDE Semantic Web Ontology Development Environment [8] is a toolset that compiles OWL, the ontology description language of the W3C [13] to Java class hierarchies and OptimixJ. OptimixJ evaluates the constraints of the ontology. The GREAT aspect-oriented design framework [5] is a transformation framework that transforms abstract and platform-independent UML design models into more concrete implementation models. Both applications show that OptimixJ and its underlying concepts should provide a valuable contribution to the use of graph rewriting with modern software tools.

References

[1] Uwe Aßmann. OPTIMIX, A Tool for Rewriting and Optimizing Programs. In *Graph Grammar Handbook, Vol. II*. Chapman-Hall, 1999. 137

[2] Uwe Aßmann. Graph rewrite systems for program optimization. *ACM Transactions on Programming Languages and Systems*, 22(4):583–637, 2000. 137, 142, 143

[3] Uwe Aßmann. *Invasive Software Composition*. Springer-Verlag, February 2003. 136, 140, 147

[4] G. Brassard and P. Bratley. *Algorithmics: Theory and Practice*. Prentice-Hall, 1988. This book contains a very nice chapter on probabilistic algorithms for a variety of problems such as numerical integration, sorting, and set equality. 146

[5] Alexander Christoph. GREAT - a graph rewriting transformation framework for designs. *Electronic Notes in Theoretical Computer Science (ENTCS)*, 82(4), April 2003. 147

[6] Gregor Kiczales, John Lamping, Anurag Mendhekar, Chris Maeda, Cristina Lopes, Jean-Marc Loingtier, and John Irwin. Aspect-oriented programming. In *Proceedings of the European Conference on Object-Oriented Programming (ECOOP 97)*, volume 1241 of *Lecture Notes in Computer Science*, pages 220–242. Springer, Heidelberg, 1997. 142, 147

[7] Andreas   Ludwig.   The   RECODER   refactoring   engine. http://recoder.sourceforge.net, September 2001. 146, 148

[8] Johan Lövdahl. An Editing Environment for the Semantic Web. Master's thesis, Linköpings Universitet, January 2002. 147

[9] Clemens Szyperski. *Component Software: Beyond Object-Oriented Programming*. Addison-Wesley, New York, 1998. 136

[10] AGG Team. Agg online documentation. http://tfs.cs.tu-berlin.de/agg, 2003. 140

[11] Fujaba Team. Fujaba UML tool home page. http://www.uni-paderborn.de/ fujaba, May 2003. 146

[12] Barbara von Halle. *Business Rules Applied*. Wiley, 2001. 143

[13] W3C. OWL ontology web language. http://www.w3c.org/sw/2001, July 2003. 147

A   Functor Declarations

The following is an example for the definition of a list functor `ProgramElement-List` over a type `ProgramElement`, for the RECODER refactoring tool [7].

```
ProgramElementList: FUNCTOR OxSetFunctor(
    name              => ProgramElementList,
    TargetType        => ProgramElement,
    /* Graph features */
    IsHomogeneous     => true,
    IsBidirectional   => false,
    HasExplicitEdges  => false,
    IsOrdered         => true,
    /* Optimization features */
    IsUnionOptimizable          => false,
    CheapCopyingPossible        => false,
    /* Names of graph methods */
    AllocFunc                => ProgramElementList,
    AddEdgeFunc              => addElement,
    DelEdgeFunc              => removeElementAt,
    AddNodeFunc              => addElement,
    AddNode2Func             => addElement,
    DelNodeFunc              => removeElementAt,
    //... and some more
);
```

Expressing Component-Relating Aspects with Graph Transformations

Alon Amsel and Dirk Janssens

University of Antwerp, Middelheimlaan 1, B-2020 Antwerp
{alon.amsel,dirk.janssens}@ua.ac.be

Abstract. Aspect Oriented Programming (see [3]) is an attempt to deal with so-called cross-cutting concerns and the tangled code that often results from them. The aim of this paper is to explore the possibility of developing a lightweight rule-based representation of aspects, enabling one to discuss and reason about aspects at a high level of abstraction. It is shown, for a concrete example, that aspects can be represented by graph transformation systems, and that the extension of a base program by an aspect can be viewed as a composition of graph transformation systems. We focus on aspects that do not modify the base program's behavior, but that can only allow or disallow its execution. We elaborate on an aspect concerning synchronization, after which we discuss ways to generalize our approach to other aspects, and to situations where several aspects are combined.

1 Introduction

1.1 Aspect-Oriented Programming

The aim of this paper is to explore the use of graph rewriting for the definition and visualization of aspect-oriented software. In many software systems, readability and understandability become poor as some modules are full of tangled code and requirements are scattered all throughout the program. Aspect-oriented programming (AOP) [3] provides a solution for this kind of problems. Aspects basically are concerns that cross-cut any possible object-oriented modularity, and therefore appear on several different places in the code. AOP languages such as AspectJ [4], provide the possibility to implement those concerns separately from the base program in modular non-hierarchical entities. However, aspects need to be incorporated in an object-oriented design without support from a higher-level design.

Different aspect-oriented programming languages use distinct mechanisms to define and to weave aspects. AspectJ uses a join point mechanism. Join points are program points such as method calls, method returns, constructor calls, where aspect code can be inserted. AspectJ provides the possibility to define the desired join points in pointcuts. The code that an aspect adds to a join point is called advice. This code can be inserted before, after or around (i.e. instead of) a join point. Since there are practically no limitations to the use of around

J.L. Pfaltz, M. Nagl, and B. Böhlen (Eds.): AGTIVE 2003, LNCS 3062, pp. 149–162, 2004.

advice, safety problems can arise. In this paper, we abstract from possibly unsafe aspects.

Inspired by the terminology of Aßmann and Ludwig [9] who use similar terms to categorize graph rewriting rules for aspect weaving but not for the aspects themselves, we distinguish between three kinds of aspects, according to their relationship to the base program: component-relating, component-additive and component-modifying aspects. Component-relating aspects essentially are conditions that should be checked at some stage of the program execution, and that allow the base program to run if the conditions are met or prevent it from running if they are not. They only tolerate unilateral parameter passing from objects to an aspect, but not in the opposite direction. Component-additive aspects add extra functionality to the base program or provide some extra information about the progam. Logging [12] and tracing [13],[14] are the most common examples of this kind. Both these categories retain the base program's entire functionality. Finally, component-modifying aspects, that include the largest variety of cross-cutting concerns, may bilaterally interact with objects, by accessing and modifying variables or method returns and they may even take over entire control over the base program. They can also modify the class or type hierarchy of the base program, for instance using AspectJ's introduction [16], [18].

1.2 Graph Rewriting

Graph rewriting has been proposed as a conceptual tool for the description and formalization of a variety of programming concepts, such as modularity [7], distribution [8], the semantics of various sublanguages of the UML [11], security policies [15], and many others. The basic idea is that system configurations are represented by graphs, a program is viewed as a set of graph rewriting rules, and sequences of rule applications model the program execution.

In this paper the use of graph transformations for expressing aspects is explored. Only component-relating aspects are considered, and it is shown for a concrete case that such aspects can be represented by sets of graph rewriting rules. Thus, since the base program is also represented by a set of rewriting rules, the complete program (the base program equipped with aspects) can be viewed as a composition of three sets of rules. The required composition operations are on the one hand the union of sets of graph rewriting rules, and on the other hand the synchronous composition of these rules: in the latter operation several rules are glued together to form a more complex rule — thus this operation is a variant of the amalgamation operation already studied in the context of algebraic graph rewriting [2]. A few other techniques, such as process algebras [1], have previously been proposed to serve as a formal basis for aspect-oriented programming languages.

We focus on synchronization, a component-relating aspect. Section 2 introduces our running example, a police case management program, and explains the synchronization problem that arises. Section 3 shows how we model the aspect separately from the base program, using graph transformations. Section 4 explains how different rules are glued together. In Section 5 we study the effect of adding other component-relating aspects to a program and consider how several aspects are combined with each other and with a base program. Finally, we briefly discuss related and future work.

2   Running Example: Police Management Software

2.1   Attributes and Operations

Consider a network of police officers, including managers (chiefs of police) and agents (police officers). Their case management is implemented in a Java program, where managers are responsible for assigning agents to cases and for hiring agents. One agent can be assigned to several cases and several agents can be assigned to one case. Each agent has his unique Agent Identification Number (AId). Each case has a unique Case Identification Number (CId). Agents and cases are objects. They contain their identification number and possible references to other objects. The manager class contains nine methods that are relevant to this paper, which we call operations.

Table 1 shows an overview of the operations that managers can perform on agents and cases and Table 2 shows the essential parts of the manager class implementation. The sets of agents and cases are implemented as arrays of booleans. The value of a boolean is true if an agent or case with that identification number exists. Edges are items of a two-dimensional array of booleans.

2.2   Base Program Graph Representation

A program configuration is represented by a labeled directed graph (V, E, lab_V, lab_E). The data concerning agents and cases are represented by a twofold graph: there are two kinds of nodes (distinguished by their labels) representing agents and cases, respectively. The fact that an agent a is assigned to a case c is represented by an edge from a to c. Moreover, there are nodes representing pending calls and results. More information about these, such as the name of the method called or the value of a result, is encoded into the node labels.

Table 1. Manager Operations

	Method Name	Comments
1	AssignAgent(AId,CId)	Assign an agent to a case
2	UnassignAgent(AId,CId)	Remove an agent from a case
3	NewAgent()	Appoint a new agent to the staff
4	NewCase()	Define a new case
5	SearchCasesOfAgent(AId)	Show overview of the cases AId works on
6	SearchAgentsForCase(CId)	Show overview of the agents working on CId
7	ViewAll()	Display all cases and agents
8	FireAgent(AId)	Permanently remove an agent
9	CloseCase(CId)	Permanently remove a case

Table 2. Manager Class Code

```
01 class Manager{
02  private boolean cases[];
03  private boolean agents[];
04  private boolean edges[][];
05  private char busy_write;
06  private int busy_read;
07
08  Manager(int max_agents, int max_cases){
09    boolean agents[]= new boolean[max_agents];
10    boolean cases[]=new boolean[max_cases];
11    boolean edges[][]=new boolean[max_agents][max_cases];
13  }
14
15  public void AssignAgent(int AId,int CId)
16  ...
17  public void CloseCase(int CId)
18 }
```

For instance, if SearchCasesOfAgent(x) or SearchAgentsForCase(x) has been executed, the label of the matching result node contains the names of the nodes adjacent to the one x refers to. We denote this information by $N(a)$ or $N(c)$ if x refers to an agent a or a case c.

In a first approach, the fact that a call cl contains the AId x of an agent a is represented by an x-labeled edge from the call node labeled cl to a. The presence of a CId is represented in a similar way. All throughout this paper, we depict calls, agent and case nodes with circles, and results with vertically oriented ellipses. Using these conventions, we show a possible configuration of the base program graph in Fig. 1.

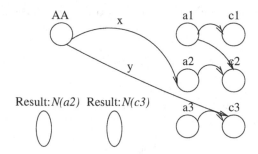

Fig. 1. Possible configuration of the base program graph. Two result nodes show that read information was requested twice. There are three agents, three cases and one pending call of AssignAgent

We consider program executions starting from a configuration (graph) that consists of the agent-case graph and a number of nodes representing pending method calls. The execution of such a call corresponds to the application of a graph rewriting rule; in this application the corresponding call node as well as part of the agent-case graph is rewritten. The graph rewriting rules used in this paper consist of a left-hand side (the subgraph to be replaced), a right-hand side (the subgraph that replaces the removed occurrence of the left-hand side) and an embedding mechanism that specifies the way the newly created occurrence of the right-hand side gets connected to the remaining part of the original graph. The embedding mechanism used in this paper is very simple: each rule is equipped with a set of pairs (u, v), where u is a node of the left-hand side and v is a node of the right-hand side. If (u, v) belongs to the embedding mechanism, this means that each edge connecting u to a node outside the rewritten occurrence of the left-hand side has to be redirected to v. The pairs of the embedding mechanism are represented as dashed lines in the figures. The rewriting rules for the Assign-Agent and SearchCasesOfAgent methods are depicted in Fig. 2.

2.3 Synchronization Aspect

The first aspect we want to add to the base program concerns synchronization: assume that one introduces concurrency into the system, and that one needs to impose constraints on the system behavior in order to avoid conflicts. In the example system, assume that several method calls can be executed in parallel. We distinguish two kinds of operations in Table 1: write and read operations. Write operations (1,2,3,4,8 and 9 in Table 1) are those that introduce nontrivial changes in the agent-case graph, while read operations (5,6 and 7) create a result node, but do not change anything about agent or case nodes.

In order to avoid any possible conflicts, we would like to prohibit two or more simultaneous write operations, or simultaneous write and read operations. On the other hand, several (i.e. an unlimited number of) concurrent read operations should be allowed since they cannot be in conflict. We need to explicitly enforce

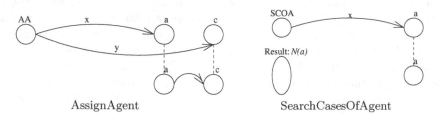

AssignAgent   SearchCasesOfAgent

Fig. 2. Base program rules. Operations are schematically denoted by their initials. The left figure shows how AssignAgent(x,y) adds an edge between two nodes and the right figure shows how SearchCasesOfAgent(x) produces a result node. In both figures, the left hand side is shown on top and the right hand side on the bottom. Dashed lines connect the pairs of the embedding mechanism

Table 3. Java-code for AssignAgent

```
01 public void AssignAgent(int AId,int CId){
02  if(busy_read==0 && busy_write=='0'){//checking synch
03  busy_write:='X';
04    edge[AId,CId]:=true;
05  busy_write:='0';
06 }
```

this synchronization property. Including it in the code of every manager operation will result in tangled code, so we define this property separately from that
code, in the graph context.

Table 3 shows a possible Java implementation of AssignAgent equipped with
the synchronization aspect. The integer busy_read indicates the number of reads
being processed. If busy_write is set to the symbol 'X', a write operation is being
processed. Every write operation contains the exact same lines 02, 03, 05 and
06. Only line 04 takes care of the assignment of an agent to a case. Lines 02, 03
and 05 handle the synchronization property [1].

When a manager wants to perform a write operation, the synchronization
aspect has to check whether no other read or write operation is going on. If
another operation is being executed, the requested operation will not be performed and the manager will be notified [2]. Similarly, when a manager wants to
perform a read operation, the synchronization aspect has to check whether no
write operation is currently being executed. If these conditions are not fulfilled,
the operation cannot be executed. Before the property is checked, we say that
the operation is a request.

3   Expressing Synchronization by Graph Rewriting Rules

The configurations that were described in the previous section, represent the
base program. From now on, we will refer to these configurations as the base
program view. In order to represent a component-relating aspect in our graph,
we introduce a few terms. A new node, called aspect node, is added to our graph.
Its label consists of the name of the aspect, a colon, and all the information that
is necessary to evaluate the aspect's condition. This information forms the aspect
state. Beside the base program view, an aspect view consists of the aspect node,
its adjacent nodes, and all the edges between them. Finally, the global view
consists of the entire configuration.

---

[1] The synchronization property is described as in [5].

[2] This notification can be considered another aspect or it can be viewed as part of the
synchronization aspect. In any case, it can be argued to be component-additive and
no longer component-relating.

In order to express constraints expressed by aspects, like synchronization, one has to modify the way the program is represented. The possibility that several method executions can happen concurrently cannot be expressed when method executions are described as atomic — one needs to express the fact that a method execution has started but not yet terminated, thus one rewriting rule for each operation doesn't suffice anymore. A method execution will now be described by a sequence of two rule applications, instead of a single one: the first one models the decision whether or not to start the execution, and if that decision is positive, the second one models the completion of the execution. Thus, let B PR be a rewriting rule of the base program. B PR is replaced by a pair (A, B) of sets of rules of the extended program: A contains rules that model the test, B contains rules that model the completion of the operation modeled by B PR. Each application of B PR determines a concrete pair (ACR, EBR) of rules, where ACR (for aspect-checking rule) belongs to A and EBR (for extended base rule) belongs to B. Clearly ACR and EBR are to be applied consecutively.

The first of these rule applications boils down to the evaluation of the synchronization constraint and thus requires information about the number of read and write operations that have started. The synchronization aspect state acts like a generalized counter: its value will either be any integer (the number of read operations being processed), or the special symbol X (indicating an ongoing write operation).

The second rule to be applied, EBR, which models method execution in its original form (i.e. as in the base program), is obtained by synchronizing the rewriting rule given by the base program with a graph rewriting rule AUR (for aspect-updating rule) that updates the aspect state. We describe this mechanism in detail in Section 4.

Since the rules of A model the checking of the synchronization constraint, there are two variants of them: one describing a positive result of the check, and one describing a negative result, in which case the requested operation is made unrealizable, and EBR becomes the identity rewriting rule. Both variants may only modify nodes and edges that occur in the aspect view. The aspect requires each manager request to have an outgoing edge to the aspect node. Our way to represent the effect of an aspect-checking rule is by reversing the edge between call node and aspect node, labeling it with 'OK', for a positive result, or 'NOK', for a negative result. Once a call node has an incoming edge labeled 'OK', we say that it is active.

Figures 3 and 4 depict these aspect-checking rules for write and read operations respectively, first the cases where the constraint is satisfied, then the two cases yielding a negative result. Aspect nodes are visualized by horizontally oriented ellipses. Write operations on the one hand and read operations on the other hand require different aspect-checking rules, depending on the tolerated aspect states.

After the application of an aspect-checking rule, the base program rewriting rule must be applied and the aspect state must be updated so as not to make subsequent calls fail unnecessarily. Finished read operations decrement the in-

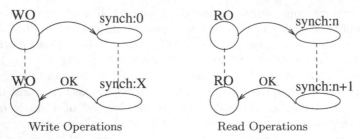

Fig. 3. Aspect-checking rules for the synchronization aspect. The left figure shows the aspect-checking rule for a write operation, changing the aspect state to X and the right figure shows the aspect-checking rule for a read operation, incrementing the aspect state by one. Synchronization is abbreviated to 'synch'. The labels WO and RO stand for any write operation and read operation in Table 1, respectively

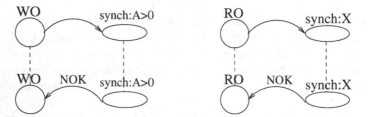

Fig. 4. Aspect-checking rules for write(left) and read(right) operations yielding an error

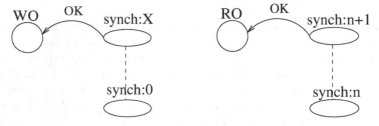

Fig. 5. Aspect-updating rules for write(left) and read(right) operations

teger value and finished write operations set the aspect state back to 0. Both remove the call node. Like the aspect-checking rule, this rule may only modify nodes and edges that occur in the aspect view. Fig. 5 depicts the aspect-updating rule for the example.

4   Gluing Graph Rewriting Rules

The entire program, now composed of the base program and an aspect, manipulates the data of the base program as well as the aspect node. Changes caused by aspect-updating rules on the one hand and by base program rules on the other hand must be such that the consistency of the data in ensured. Thus a mechanism is required to synchronize them. We do this by gluing the aspect-updating rule to the base program rewriting rule over common nodes and edges. Technically, the way these rules are glued together is a variant of the amalgamation construction introduced by Corradini et al. (see e.g. [2]): the two rewriting rules are glued over an interface, which is a rule manipulating the overlaps. We refer to the resulting combined rules as extended base rules An example of this gluing is given in Fig. 6, for the AssignAgent operation.

In general, each gluing is uniquely determined by a span of graph rewriting rules $AUR \leftarrow interface \rightarrow BPR$. Since the embedding of the extended base rule can completely be specified in the base program rewriting rule and the aspect-updating rule, there is no need to specify it again in the interface rule. Thus the interface rule can be viewed as just a pair of graphs, and the arrows in

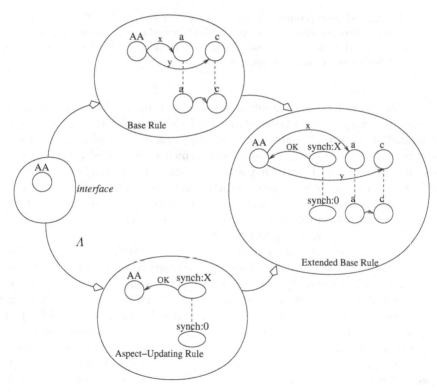

Fig. 6. Gluing of an aspect-updating rule to a base program rewriting rule (AssignAgent). The top spline contains the base program rewriting rule and the bottom spline contains the aspect-updating rule. The leftmost spline represents the gluing interface, in which Λ denotes the empty graph. Finally, the rightmost spline depicts the extended base rule

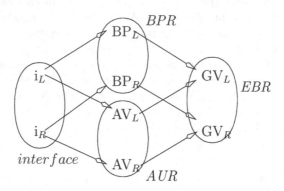

Fig. 7. Generalized gluing contsruction. AV stands for aspect view, GV for global view and BP for base program view. Subscripts indicate whether a graph belongs to the left hand side (L) or to the right hand side (R) of a rewriting rule. Arrows represent graph morphisms. Both top and bottom squares form pushouts

the span are pairs (one for the left hand side and one for the right hand side) of graph morphisms. The extended base rule is then obtained by completing the pushout for both left hand sides and right hand sides of a given span, as schematically shown in Fig. 7. To specify which aspects and aspect-updating rules correspond to which base program rewriting rules, one may use a set of such spans. This set then serves as our pointcut language. For our running example, this set contains a span for each base program rewriting rule, describing the way they must be synchronized with the first (for write operations) or the second (for read operations) aspect-updating rule of Fig. 5.

We are now ready to take a closer look at a sequence of rule applications. Fig. 8 shows the evolution of the global view for the processing of three requests. First an AssignAgent node becomes active (aspect-checking rule). Next, in turn, the UnassignAgent node is stopped (ACR with negative result), AssignAgent is completed (EBR), a SearchCasesOfAgent node becomes active (ACR) and, finally, SearchCasesOfAgent is completed (EBR), creating a result node. Note that our synchronization aspect prevents an operation to be executed forever. An alternative approach would be to make the program wait until the aspect state allows that operation to run. Only minor modifications to our model are required to adapt it to this approach: it is sufficient to change the embedding mechanism in such a way that outgoing edges labeled 'NOK', are changed into unlabeled edges pointing in the opposite direction, so the condition can be checked again.

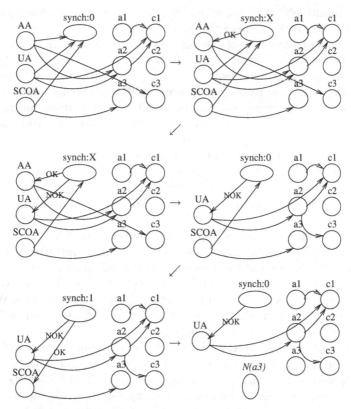

Fig. 8. Global view of possible sequence of applications. The arrows between the graphs indicate the order of the transformations. The labels of edges between call nodes and agents or cases are omitted for better readability

5   Adding More Aspects

Usually, base programs will be composed with more than just one component-relating aspect. For instance, in our running example, one could add an aspect checking whether an identification number refers to an existing agent or case. Every additional aspect has its own aspect-checking rules and aspect-updating rules. The global view will then consist of the base program view combined with several aspect views. Gluing several aspects to a base program simultaneously can prove to be quite tricky, since nothing guarantees that the interfaces between the base program and different aspects consist of the same rewriting rule. In general, this is not the case.

Thanks to our description of aspect views in Section 3 we can now define independent aspects. We say that two component-relating aspects A_1 and A_2 are independent iff the aspect node of A_1 does not occur in A_2's aspect view and vice versa. Intuitively, this means that the condition expressed by either aspect does not depend on the state of the other aspect.

For two independent aspects A_1 and A_2, one may glue the aspect-updating rule of A_1 to the base program rewriting rule and subsequently glue the aspect-updating rule of A_2 to the result. The choice of which aspect is added first does not affect the final result (up to isomorphism). This follows from the fact that component-relating aspects cannot modify the base program, so any node occurring in an aspect view as well as in the base program view is manipulated in the same way by the base program and the aspect-updating rule. Any edge that occurs in A_1's aspect view and in the base program cannot occur in A_2's aspect view, because otherwise both aspect nodes would be adjacent and hence the aspects would not be independent. Therefore, an aspect-updating rule cannot affect any node or edge that appears in the aspect view of another independent aspect.

If aspect A_2 does depend on aspect A_1, then the order in which the aspect-updating rules must be glued should be obvious from an intuitive point of view, because in that case one of the aspects, say A_2, refers to the data of A_1, and hence A_2 has no meaning in the absence of A_1. Thus we will glue A_1's aspect-updating rule to the base program first, and subsequently glue A_2's. We leave it up to future work to formalize this and to study the effect of adding larger sets of dependent aspects to a base program.

If a base program is composed with more than one component-relating aspect, call nodes connect to every aspect node. Obviously, several aspects also induce distinct aspect-checking rules, which must all be applied before the execution of the base program. Since the conditions expressed by all the aspects must be satisfied, call nodes can only become active when they only have incoming edges, all labeled 'OK'. For independent aspects, it doesn't matter which aspect-checking rule is applied first or whether they are applied simultaneously.

6   Conclusion and Future Work

We have shown that component-relating aspects can be expressed and combined in a simple way in the framework of graph rewriting: the base program is represented by a set of graph rewriting rules transforming graphs that represent system configurations. When aspects are added, the configuration graphs are extended by adding aspect nodes. The aspects can then be represented by sets of rules: on one hand, aspect-checking rules that represent the decision whether the constraint associated to the aspect is satisfied or not, and on the other hand aspect-updating rules that manage the aspect state. The program containing the aspects is a set of rewriting rules on the extended graphs that results from introducing the aspect-checking rules and from gluing the base program graph rewriting rules to aspect-updating rules. Thus each base program rewriting rule is mapped to a preliminary aspect-checking and a concluding extended base rule. Each additional aspect is represented by a supplementary aspect node. The base program's behavior is preserved and can be extracted from our model through the base program view. In this way one obtains a high-level rule-based representation of aspects, as well as of the composition of aspects with a base program.

Our approach is different from Aßmann's [9] in that we do not model the weaving process, but the effect of aspects on a base program. However, we adopt the idea of splitting up the large class of aspects into three categories. In that paper, the three categories apply to graph rewriting rules and describe three different ways to weave aspects in the base program.

There are also similarities to the work of Koch and Parisi-Presicce [15], where policies are described separately from a base program within a framework of graph transformations, while Colcombet and Fradet [10] introduce instrumented graphs with automaton structure to enforce trace properties on a program. Technically, the way aspect-updating rules are composed with rewriting rules of the base program is a variant of the amalgamation construction introduced by Corradini et al. (see e.g. [2]).

The work presented in this paper is obviously only a first step towards the development of a framework for the high-level description of component-relating aspects. Such a framework should be applied to, e.g., visualization of aspect-oriented programs and aspect-oriented code generation. Especially the ways programs or aspects, represented as sets of graph rewriting rules, can be composed, need further consideration. Furthermore, we need to improve our pointcut language, allowing parametrizations or use of wildcards, which can be very useful, especially if aspect-updating rules differ from one base program rewriting rule to another only by the label of one node. Graphs are known to be an effective and flexible modeling technique, and moreover a significant amount of results [6] concerning their use in software engineering are available. A hierarchical graph representation has been developed for the modeling of object-oriented refactorings [17]. This model may be extensible to include aspects and serve as a relatively broad software engineering model. In the future, we would also like to include component-additive and component-modifying aspects in our model, using similar techniques. We intend to study the interaction between aspects of different categories, as well as weaving techniques for our model — in terms of graph rewriting, meaning that techniques are required to compress or simplify the sets of graph rewriting rules that result from a straightforward modeling of the separate aspects.

References

[1] J. Andrews. Process-algebraic foundations of aspectoriented programming, 2001. 150

[2] A. Corradini, U. Montanari, F. Rossi, H. Ehrig, R. Heckel and M. Löwe, Algebraic approaches to graph transformation, Part 1, in *Handbook of Graph Grammars and Computing by Graph Transformation*, Vol.1, pp 163 - 245, World Scientific, 1997 150, 157, 161

[3] Gregor Kiczales et al: Aspect Oriented Programming, *European Conference on Object-Oriented Programming* (ECOOP 1997), Springer Verlag LNCS1241 149

[4] AspectJ website: http://www.eclipse.org/aspectj 149

[5] Ada 95 Quality and Style Guide, Chapter 6: Concurrency, 6.1.1: Protected Objects 154

[6] *Handbook of Graph Grammars and Computing by Graph Transformation*, Volumes 1-3, World Scientific 161

[7] Hans-Jörg Kreowski and Sabine Kuske, Graph transformation units with interleaving semantics, *Formal Aspects of Computing*, 11(6), pp 690-723, 1999 150

[8] G. Taentzer, M. Koch, I. Fischer and V. Volle, Distributed graph transformation with application to visual design of distributed systems, in *Handbook of Graph Grammars and Computing by Graph Transformation*, Vol.3, pp 269 - 340, World Scientific, 1999 150

[9] Uwe Aßmann, Andreas Ludwig: Aspect Weaving by Graph Rewriting, *Generative and Component-Based Software Engineering* (GCSE 2000) pp 24-36 150, 161

[10] Thomas Colcombet, Pascal Fradet: Enforcing Trace Properties by Program Transformations, *27th Symposium on Principles of Programming Languages* (POPL 2000) pp 54-66 161

[11] G. Engels, R. Heckel, and J. M. Küster: Rule-Based Specification of Behavioral Consistency based on the UML Meta-Model, In M. Gogolla and C. Kobryn, editors, *Proceedings of the 4th International Conference on the Unified Modeling Language* (UML'2001), pp 272 -287, LNCS 2185, Toronto, Canada, October 2001 150

[12] Mohamed M. Kandé, Jörg Kienzle, Alfred Strohmeier: From AOP to UML: Towards an Aspect-Oriented Architectural Modeling Approach, *Technical Report IC 200258* 150

[13] Tina Low: Designing, Modelling and Implementing a Toolkit for Aspect-oriented Tracing (TAST), *Workshop on Aspect-Oriented Modeling with UML at the First International Conference on Aspect-Oriented Software Development* (AOSD 2002) 150

[14] Eric Wohlstadter, Aaron Keen, Stoney Jackson and Premkumar Devanbu: Accomodating Evolution in AspectJ, *Workshop on Advanced Separation of Concerns, at the ACM SIGPLAN Conference on Object-Oriented Programming, Systems, Languages and Applications* (OOPSLA 2001) 150

[15] Manuel Koch and Francesco Parisi-Presicce: Describing Policies with Graph Constraints and Rules, *International Conference on Graph Transformations* (ICGT 2002) pp 223-238 150, 161

[16] Ouafa Hachani, Daniel Bardou: Using Aspect-Oriented Programming for Design Patterns Implementation, *Workshop on Reuse in Object-Oriented Information Systems Design* (OOIS 2002) 150

[17] Niels Van Eetvelde, Dirk Janssens: A Hierachical Program Representation for Refactoring, *Uniform Approaches to Graphical Process Specification Techniques (UniGra) workshop at the European Joint Conferences on Theory and Practice of Software (ETAPS 2003)* 161

[18] Maximilian Störzer, Jens Krinke: Interference Analysis for AspectJ, *Foundation of Aspect-Oriented Languages (FOAL) Workshop at the Second International Conference on Aspect-Oriented Software Development* (AOSD 2003) pp 35-44 150

Modeling Discontinuous Constituents with Hypergraph Grammars

Ingrid Fischer

Lehrstuhl für Informatik 2
Universität Erlangen–Nürnberg
Martensstr. 3, 91058 Erlangen, Germany
Ingrid.Fischer@informatik.uni-erlangen.de

Abstract. Discontinuous constituents are a frequent problem in natural language analyses. A constituent is called discontinuous if it is interrupted by other constituents. In German they can appear with separable verb prefixes or relative clauses in the Nachfeld. They can not be captured by a context–free Chomsky grammar. A subset of hypergraph grammars are string-graph grammars where the result of a derivation must be formed like a string i.e. terminal edges are connected to two nodes and are lined up in a row. Nonterminal edges do not have to fulfill this property. In this paper it is shown that a context–free string-graph grammar (one hyperedge is replaced at a time) can be used to model discontinuous constituents in natural languages.

1 Discontinuous Constituents in Phrase Structure Grammars

Discontinuity is a phenomenon that can be found in various natural languages. It means that two parts of a constituent of a sentences are not found next to each other in this sentence. Take for example the following sentence:[1]

(1) Der Mann hat der Frau   das Buch gegeben, das   fleckig ist.
    The man   has the woman the book given   that   dirty is.
    The man has given the woman the book, that is dirty.

The relative clause das ◼eckig ist (that is dirty) belongs to the noun phrase das Buch (the book) though these phrases are not placed next to each other. The past participle gegeben (given) is in the last but one position, the relative clause is moved to the end in the so called Nachfeld [4]. A phrase structure tree modeling the syntactic structure of Example (1) is shown in Figure 1. Crossing links occur in this tree showing that it can not be derived with the help of a context–free Chomsky grammar [9].

---

[1] The German examples are first translated word by word into English to explain the German sentence structure and then reordered into a correct English sentences.

J.L. Pfaltz, M. Nagl, and B. Böhlen (Eds.): AGTIVE 2003, LNCS 3062, pp. 163–169, 2004.
© Springer-Verlag Berlin Heidelberg 2004

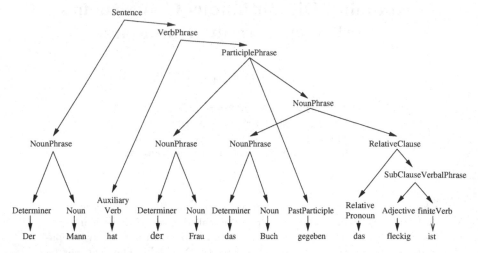

Fig. 1. The phrase structure tree for *Der Mann hat der Frau das Buch gegeben, das fleckig ist. (The man has given the woman the book, that is dirty.)*

The situation even gets worse for sentences like:

(2) Er kauft den Hund ein,   der beißt.
    He buys the dog   (verbal prefix) that bites.
    He buys the dog that bites.

The German verb einkaufen (buy) is split in two parts in present tense kauft ....
ein. The verbal prefix ein is separated and moved to the last but one position,
the finite verb part is in the second position. In this example the noun phrase
with its relative clause and the verb with verbal prefix form two discontinuous
constituents.

Discontinuous examples can be also found in English, but not as often:

(3) Ann, John told me, he had seen.   ,
(4) Peter's brewing of the last season I am in hopes will prove excellent.[2]

In (3) Ann is the direct object of the verb seen and should be part of the
corresponding verb phrase. To stress, that it was Ann that was seen, the proper
name was moved to first position leading to crossing edges in the phrase structure
tree. In example (4) one clause is moved into the second. This second clause
Peter's brewing of the last season will prove excellent is part of the verbal phrase
am in hopes.

Of course it is possible to write context–free Chomsky grammars for these
examples. But they do not reflect which discontinuous parts belong together,
leaving out the linguistic evidence. Several proposals have been made on how

---

[2] Thomas Jefferson to Joseph Miller, March 11, 1817.

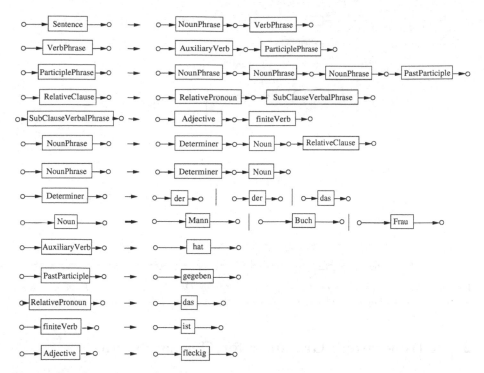

Fig. 2. Simple string-graph grammar generating the sentence *Der Mann hat der Frau das Buch, das fleckig ist, gegeben. (The man gave the woman the book, that is dirty.)* containing no discontinuous constituent

to extent context–free grammars. Chomsky himself suggested additionally to grammars transformations, that move around constituents [3]. Other approaches based on phrase structure grammars separate word order from the necessary constituents [2]. An overview for HPSG grammars also having a backbone of context–free Chomsky grammars can be found in [12]. When using functional uncertainty all information found is collected and after the parsing process it is checked if everything fits together [10].

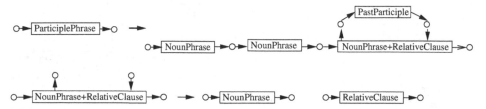

Fig. 3. Hyperedge replacement rules to connect a relative clause in the Nachfeld to a noun phrase

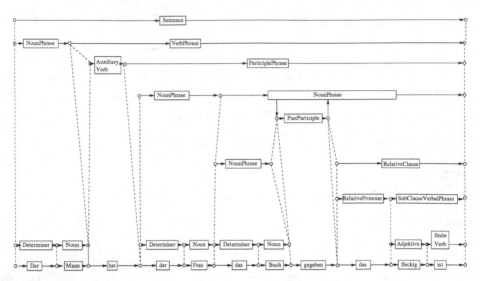

Fig. 4. Rules applied when parsing the sentence *Der Mann hat der Frau das Buch gegeben, das fleckig ist. (The man has given the woman the book, that is dirty.)*

2   A Hypergraph Grammar for German Syntax

In context–free hypergraph grammars one hyperedge is replaced in one derivation step by a hypergraph. Hypergraph grammars can also represent strings. A letter (word) of a string (sentence) can be seen as terminal hyperedge with two nodes that connect a letter to the letters resp. hyperedges on the left and right side. In a string generating context–free hypergraph grammar a nonterminal hyperedge

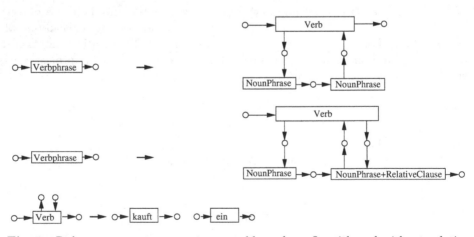

Fig. 5. Rules necessary to generate separable verb prefix with and without relative clause in the Nachfeld

may be connected to several points in the growing string, not just two as in
the case of ordinary context–free grammars. It is only the result of a derivation
that models a string. This subset of hypergraph grammars is called string-graph
grammars. In [8] it is shown that the context–sensitive language $a^n b^n c^n$ (in
Chomsky's sense) can be modeled with a context–free string-grammar. In [8], [5]
context-free hypergraph grammars are described in great detail. More information on string-graph grammars can be found in [7], [6]. These ideas can now be
used to describe discontinuous constituents in natural languages.

In Figure 2 the rules of a simple context–free Chomsky grammar are shown
as string-graph grammar.[3] Every terminal and nonterminal symbol of the Chomsky grammar is translated into a hyperedge with two tentacles. This grammar
is divided into two parts: first parts-of-speech such as noun, determiner, etc.
are mapped to words. Then constituents are constructed. With the help of this
grammar the following sentence can be parsed containing no discontinuous elements.

(5) Der Mann hat der Frau   das Buch, das fleckig   ist, gegeben.
    The man   has the woman the book that dirty   is   given.
    The man has given the woman the book, that is dirty.

When moving the relative clause to the Nachfeld of the sentence the noun phrase
rules generating das Buch, das ◼eckig ist (the book that is dirty) must allow for
a gap in between. This can be accomplished when taking hyperedges with four
nodes instead of two nodes for the noun phrase. The new rules are shown in
Figure 3. This can not be done with ordinary Chomsky grammars. The most important feature is the new nonterminal hyperedge NounPhrase+RelativeClause
connected to four nodes. The arrows between the hyperedge and these nodes
show the way the final sentence is taking "through" this nonterminal hyperedge. When the NounPhrase+RelativeClause is on the left hand side of a rule,
these rules breaks the nonterminal hyperedge apart into hyperedges that fit into
a hypergraph for strings.

Applying this rules in Figures 2 and 3 to generate sentence 1 can be done as
shown in Figure 4. In this Figure the dotted lines mark which nodes are mapped
onto each other indicating the rules that have been applied. It is easy to see that
now the discontinuous sentence was generated with the help of a context–free
grammar.

Extending the grammar with separable verb prefixes can be done by adding
the rules shown in Figure 5. Here a verb can consist of two parts. Two rules for
this new verb-hyperedge are given depending on whether a relative clause can
be found in the Nachfeld or not.

What is missing up to now is the free word order in German in the Mittelfeld
between the finite and infinite part of the verb. Several others authors in the
area of phrase structure grammars also refer to this phenomenon as discontinuous constituents. This is often handled with the help of sets [12]. In other syn-

---

[3] The two *der (the)* are different, one is supposed to be feminine accusative, the other
   masculine nominative.

tactic environments discontinuous constituents and free word order are strictly separated [1].

3   Conclusion and Outlook

In this paper simple context–free string-graph grammar rules have been used to generate discontinuous constituents in German. Of course the rules as shown do not work on their own. A lot of information was left out on purpose to keep the rules simple. It was not tested whether nouns and verbs resp. determiners and nouns agreed in gender, person, number and case. It was also not tested if the correct cases were used with the verbs. This can be done as usual when hyper-edges are extended with feature structures that are combined with unification as done in several natural language processing systems, e.g. PATR2 [14].

First a parser must be developed, that is able to handle string-graph gram-mars. The parser for hypergraph grammars of Diagen [11] is able to parse string-graph grammars for natural languages. At the moment an Earley-based parser [13] for string-graph grammars is being developed and extended. This parser has feature structures, unification and rules with probabilities. A small semantic formalism is also attached.

References

[1] Norbert Bröker. *Eine Dependenzgrammatik zur Koppelung heterogener Wis-sensquellen*. Niemeyer, 1999. 168

[2] Harry Bunt. *Discontinuous Constituency*, chapter Formal Tools for the Descrip-tion and Processing of Discontinuous Constituents. Mouton De Gruyter, 1996. 165

[3] Noam Chomsky. *Syntactic Structures*. The Hague: Mouton, 1957. 165

[4] Erich Drach. *Grundgedanken der deutschen Satzlehre*. Diesterweg, Frankfurt, 1937. 163

[5] Frank Drewes, Annegret Habel, and Hans-Jörg Kreowski. Hyperedge replacement graph grammars. In G. Rozenberg, editor, *Handbook of Graph Transformations. Vol. I: Foundations*. World Scientific, 1999. 167

[6] J. Engelfriet and L. Heyker. Context–free hypergraph grammars have the same term–generating power as attribute gramamrs. *Acta Informatica*, 29:161–210, 1991. 167

[7] J. Engelfriet and L. Heyker. The string generating power of context-free hyper-graph grammars. *Journal of Computer and System Sciences*, 43:328–360, 1991. 167

[8] Annegret Habel. *Hyperedge Replacement: Grammars and Languages*, volume 643 of *Lecture Notes in Computer Science*. Springer-Verlag, Berlin, 1992. 167

[9] D. Jurafsky and J. Martin. *SPEECH and LANGUAGE PROCESSING: An Intro-duction to Natural Language Processing, Computational Linguistics, and Speech Recognition*. Prentice Hall, 2000. 163

[10] Ronald M. Kaplan and John T. Maxwell. An Algorithm for Functional Uncer-tainty. *COLING-88*, pages 297–302, 1988. 165

[11] Mark Minas. Diagram editing with hypergraph parser support. In *Visual Languages*, pages 230–237, 1997. 168

[12] Stefan Müller. Continuous or Discontinuous Constituents? A Comparison between Syntactic Analyses for Constituent Order and their Processing Systems. *Research on Language and Computation*, 1(4), 2002. 165, 167

[13] Ulrike Ranger. Ein PATR-II–basierter Chart Parser zur Analyse von Idiomen. Master's thesis, Universität Erlangen–Nürnberg, Lehrstuhl für Informatik 2, 2003. 168

[14] Stuart M. Shieber. *Constraint–Based Grammar Formalisms*. MIT Press, 1992. 168

Authoring Support Based on User-Serviceable Graph Transformation*

Felix H. Gatzemeier

RWTH Aachen University
Department of Computer Science III
Ahornstrae 55, 52074 Aachen, Germany
fxg@i3.informatik.rwth-aachen.de

Abstract. Conceptual authoring support enables authors to model the content of their documents, giving both constructive and analytical aid. One key aspect of constructive aid are schemata as defined in the context of conceptual graphs. These are proven structures the author may instantiate in the document. Schemata may also be defined by an author to capture local experience. To support reuse by an author, a schema also includes documentation of the whole and the parts.

We present *weighted schemata*, in which the elements are assigned weights (optional, important, and crucial) indicating their importance. Schemata may be instantiated in variants depending on the usage context. There is also functionality for exporting parts of the document graph to a schema description, whose format allows further editing.

1 Authoring Problems and Conceptual Authoring Support

Creating well-structured documents is a complex activity. The author has to solve the problem of conveying a topic, usually in a given amount of space, building on some presupposed knowledge, and serving some interest. Key tasks in achieving this are devising a coherent concept structure of the document and maintaining that structure during edits [4]. Conventional authoring environments, however, concentrate on capturing, manipulation, and formatting text as a string of characters.[1] Little or no help is available for building the structure.

To explore ways to alleviate this, we are working on a prototype providing conceptual authoring support named CHASID (Consistent High-level Authoring and Studying with Integrated Documents, [6]). It operates not only on the presentation (the formatted string of characters) and the document hierarchy (the formal hierarchy of sections and subsections), but also on a content model.

By explicitly storing and editing this model, flaws in it or in the structure of its presentation can be detected. CHASID can also help to build the document

---

* This work is funded by Deutsche Forschungsgemeinschaft grant NA 134/8-1

[1] For simplicity, we restrict ourselves here to textual documents. The ideas and implementation presented here do not rely on specific media types.

J.L. Pfaltz, M. Nagl, and B. Böhlen (Eds.): AGTIVE 2003, LNCS 3062, pp. 170–185, 2004.
© Springer-Verlag Berlin Heidelberg 2004

by offering ready-made substructures to be integrated, thus providing some formalised authoring experience.

Conceptual graphs (CGs, [12]) lend themselves well to express the content model used by CHASID. The graph model has the appropriate expressiveness and flexibility. CG schemata in particular correspond well with the proposing character of the substructures, as they remain editable. They thus establish a level of modelling between type systems and graph instances.

However, the classical CG schema definition aims at mathematical proofs, not interactive tools. So, this definition is extended to provide documentation for the author and information for consistency checks. Later analysis attaches warnings if crucial or important parts of instances are removed [7]. Some required conditions given in type definitions are enforced.

Schemata are thus named graphs with documentation explaining their purpose and usage whose nodes and relations have weights attached which control their further modifiability. This definition employs author-level concepts, so we expect authors to be able to (and want to) define schemata themselves. CHASID supports this through functionality that exports a graph selection to a schema. This improves on [7], as its schemata existed only as a collection of graph transformation rules and external HTML documentation pages.

This document starts out with a longer example on how schemata are used and defined in section 2. Background information is deferred to section 3.1, which presents schemata and their role in editing and diagnostics more systematically. Some notes on the implementation in PROGRES (PROgrammed Graph REwriting Systems, [10, 11][2]) are also given.

In the discussion in section 4, benefits and limitations of schema-based authoring are discussed and put into the context of related work. Section 5 wraps up the document and gives an outlook on medium-term plans.

2   A Usage Example

To illustrate the use of schemata, we present some stages in the development of an article about an investigation into determining actual working groups. These may well differ from the organizational working groups, but are harder to determine. Two approaches at this problem have been explored, namely conventional questionnaires and automated E-mail analysis. If both approaches yield similar results, they may be considered to be cross-validated, allowing the much cheaper E-mail analysis to be used instead of costly questionnaires.

Figure 1 shows an initial document graph and a schema browser of the CHASID prototype. The document hierarchy part of the graph is displayed as a tree on the left-hand side, while the content model is a boxes-and-lines graph on the right. That graph contains another rendition of the document hierarchy to visualise the import and export structure. This structure is the connection between divisions and content: A division may import a concept, in which case

---

[2] http://www-i3.informatik.rwth-aachen.de/research/progres

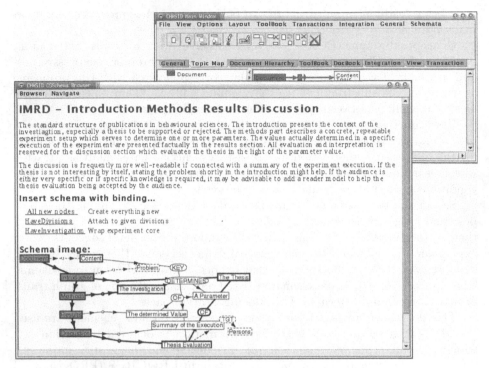

Fig. 1. Sample CHASID window with the description of the Introduction-Methods-Results-Discussion (IMRD) schema in the schema browser (front) and the initial empty document graph (back)

at least a working understanding of the concept is required for a reader to fully grasp the content of the division. In the other direction, a division may export a concept, which means that a reader is expected to have at least some idea of the concept after reading the division.

There is not much text in the document window because CHASID works as an extension to existing authoring applications. This allows authors to create the media content in an environment they are familiar with and relieves CHASID of major implementation burdens. In this discussion, we will not address media content creation or integration issues.

The starting point for the article is the Introduction - Methods - Results - Discussion (IMRD) schema, as shown in the schema browser. This has been chosen to achieve a clear distinction between descriptive and evaluative parts. This construction is very common for empirical reports, up to being required in behavioural science [3]. In its adaptation as a schema, different weights have been assigned to its elements, as visible in the preview in the schema browser: less important elements are dashed, more important ones bear a heavier outline.[3]

---

[3] Weights are discussed in detail in section 3.2.

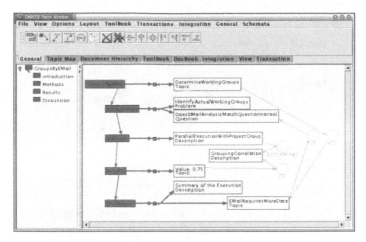

Fig. 2. First stage of article planning: the IMRD instance

In a first step, the author instantiates the schema and edits the generic labels for the specific document. The result, a framework for the article, is shown in figure 2. To fill this framework, an investigation is to be described that yields a grouping correlation factor between the methods. Only if that factor is high, the two investigations are cross-validated. As a part of that investigation, groups have to be determined in both questionnaires and E-mail. So, this is what the author does next: by generic editing actions extend the model to describe first one of the sub-investigations.

The result of these edits is shown in figure 3. There is another parameter to be determined (GroupStructure), whose values (QGroups) are the grounds (GRND) on which the overall parameter (GroupingCorrelation) is determined. Both in the methods and the results parts, the sub-investigations are described (Questionnaire). Our hypothetical author is quite content with this model and expects to reuse it, first for the other investigation to be discussed in the article, but maybe also in other IMRD articles that are to compare sub-studies. So, he goes about defining this as a schema.

The first step in defining a schema is selecting the parts of the graph that are to be the schema graph. The author has already done this in figure 3; the selection is highlighted by black corner marks. The selection does not just contain new nodes, but also nodes that had been added by instantiating the IMRD schema (and are thus already a part of an instance). In addition to nodes, also some edges are selected. Generally, the export selection is a selection like any other and defines what will be exported — with one exception: the first element selected must be a node (either a division or a topic). This will be the main reference point of the schema instance, serving for example as an anchor to attach warnings to.

The author now invokes the "Export as CG Schema..." command and supplies some documentation in the export dialog, as shown in figure 4. The dialog

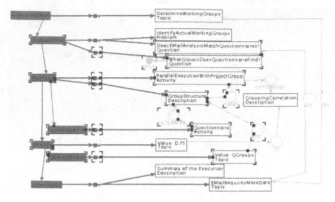

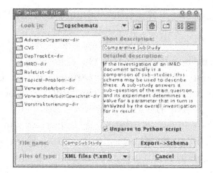

Fig. 3. Document graph with added sub-investigation

Fig. 4. Save Dialog with additional documentation input fields

offers a convenient way to attach indispensable information to the schema instance: Apart from the file name (which is also used as the schema name), there
is a short description and a longer description. The short one is used for the
schema index, while the long one is the initial documentation. By clicking on
'Export ↔ Schema', the author triggers the actual export.

The resulting user-defined schema is shown in figure 5 in two stages of refinement. First, there is the unedited result of exporting with no further user
interaction: the name and the short description make up the headline, followed
by the text of the long description. The first subsection consists of one insertion
binding. The second section displays a preview whose layout follows the location
of the schema elements in the document graph. At the bottom of the page, there
are some maintenance links and a collection of metadata. For macros that an
author has just exported as a permanent clipboard, this documentation may be
entirely sufficient. In our case, however, the questionnaire accent in the nodes
and the distribution of weights is unsatisfactory.

So, the author edits the description of the schema, which is stored as a single
XML file, from which commands and documentation are derived. He reduces

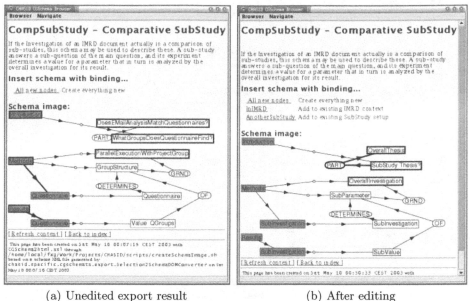

(a) Unedited export result   (b) After editing

Fig. 5. The author-defined schema

the weight of the division nodes to be just important so the schema may be applied in documents structured differently. He reformulates the names of divisions and topics to match the intended generality of the schema. As the schema is usually applied to extend an existing document, the default insertion binding that creates new instances for each schema element, is unsatisfactory. For different instantiation situations, the author also defines the insertion bindings 'InIMRD' (extending an IMRD document as it is done between figures 2 and 3) and 'AnotherSubStudy' (extending an IMRD document that already contains a CompSubStudy instance). An example difference is the latter offering the Group-Structure node as a parameter instead of creating it anew. To make these variants more useful, he defines labels for the nodes, which does not show in the browser. He then has all derived files (the preview, the documentation, and the instantiation scripts) rebuilt using the maintenance links at the bottom of the description page, resulting in the second window of figure 5.

After this excursion into defining a schema for reuse, the author returns to the article at hand. He instantiates the CompSubStudy schema again, this time for the EMail analysis study. For this, the 'AnotherSubStudy' is the instantiation variant of choice. He triggers that by activating the corresponding link in the browser window, then selects the context nodes as he would for any other graph transformation in CHASID, and has it finally executed. The result is shown in figure 6.

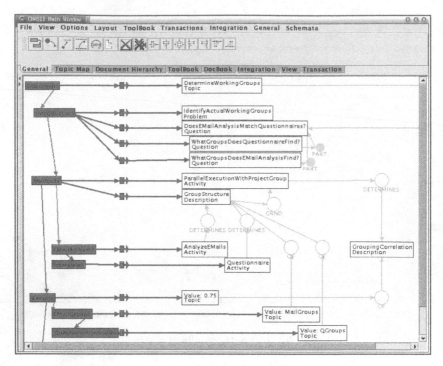

Fig. 6. Document graph with second investigation

At this point, we leave the author. If he were later to remove, for example, the topic 'ParallelExecutionWithProjectGroup', at least the three schemata instantiated here would trigger warnings about that being a central part.

3   Operations, Diagnostics, and Implementation

This section characterises kinds of editing operations available, ways to design warning functionality and some aspects of the implementation. It does not contain a singled-out definition of a weighted schema but discusses its aspects where appropriate. For a more formal definition of a weighted schema, see [7].

3.1   Editing Operations

The editing operations are classified here into three groups according to their origin: generic, schema instantiation, and hand-implemented.

Generic operations are the fundamental operations to create and remove concepts and relations, and to change their attributes. Concepts may be **merged**, which is a combination of the conceptual graphs restriction and join operations with relaxed conditions, where known types, type labels, and names do not have to fulfil any conditions. An example for a generic operation is 'Create Topic'.

Schema-related operations are concerned with instantiation, instance maintenance, and exporting. Instantiation operations are offered in the a schema browser under the header "Insert Schema with binding..." and vary in the nodes from the graph that are to be re-used in the new instance. There is always an "All new nodes" **binding**, while for other bindings, the author has to indicate the nodes to be used in the schema instance.

In exporting, a selection in the document, with additional documentation, is saved as a schema. The first selected element is the core node of the instance, that is, warnings about missing instances will be attached to it. The author usually then edits the schema description, mostly in the documentation, weighting, and bindings parts using an external editor (see section 3.3).

Finally, there is a group of specialised commands building on the known types. For example, the author may move the EMailGroups division behind the QuestionnaireGroups division instead of using generic removal and insertion. The command uses the known types of 'Division', 'First-descendant' and 'Next-sibling' in its implementation.

3.2   Diagnostics and Weighted Schemata

The content model and document hierarchy are evaluated according to information derived from schema instantiation and hand-implemented rules. The result of the evaluation are warning markers that are handled by generic operations.

Weighted Schemata Warnings.   Each element (concept or relation) of a schema has one of three **weights**.

OPTIONAL: The element may be removed without consequences.
IMPORTANT: The element may only be removed in special circumstances.
CRUIAL: If the element is removed, the schema does not apply any more.

The different weights are shown in the graphical notation in the schema browser by different border styles as in figure 1. CRUIAL elements are solid bold, IMPORTANT ones solid plain and OPTIONAL ones dashed plain.

"Removal" covers both deletion of concepts as well as disconnection of concepts, as that means removal of a relation. There is currently no provision for unwanted elements, whose addition (rather than removal) would trigger warnings. Elements may be modified (given new type labels or names) in any case without entailing warnings.

During schema instantiation, the weights of the schema elements are recorded for their counterpart in the host graph. Thus, graph elements now have weights specific to the schema contexts in which they are used. By storing the weights apart from the graph elements, each one may have several independent weights in several schema contexts. For example, the Introduction division is CRUIAL in the IMRD context and OPTIONALin two CompSubStudy contexts.

Hand-Implemented Warning Rules are patterns of unwanted graphs. To describe these, the programmer may use the full set of the underlying PRO-GRES facilities, most frequently nodes (single nodes or node sets) and edges that must or must not match, attribute restrictions, and edge path expressions. For example, there are warning rules to find document hierarchy sections that contain just one subsection, or a section that imports a concept that comes before a section which exports it.

Warnings are attached to the graph as additional nodes with edges to all concerned concepts and relations and a textual attribute containing the warning message. They show up as non-concept nodes in the graphical display and as annotations in the externally edited conventional document. If the condition causing the warning is fixed, the warning and all its representations are removed.

All warning messages can be manipulated with these generic operations:

- Suppress a specific warning, confirming the change that triggered this warning. The warning's representations disappear.
- Suppress all warnings of a type, indicating that this type does not apply to the current document. The representations of all warnings of this type disappear.
- Resurrect all suppressed warnings about one concept or some context,[4] to re-check whether these warnings were intended for suppression. The representations of the warnings re-appear as they were when suppressed.

Generic operations helping to cure warnings about crucial or important elements missing from schema instances are:

- Re-connect a missing important or crucial element and
- Dissolve the schema instance, leaving the concepts and relations as plain graph elements.

3.3   Implementation Issues

CHASID is implemented using the high-level graph transformation language PROGRES, whose graph model is the directed, node- and edge-typed graph with multiple inheritance and attributes defined on node types. Edges connect exactly two nodes; there are no hyper-edges. Graph manipulations are specified within PROGRES in graphical productions or textual transactions.[5] Concepts and relations are implemented as nodes, arcs as edges.

Generic and hand-implemented operations are arbitrary productions and transactions. The characteristic trait of generic operations is using only very abstract types and changing a very limited part of the document graph.

---

[4] The exact context is determined by the individual command, for example, a subtree in the document hierarchy.

[5] While the terms production and transaction are used consistently here, understanding the difference should not be required to understand the text.

From the PROGRES specification, C Code is generated, which is used in the Java-based UPGRADE framework in which the productions of the specification can be executed interactively [8]. UPGRADE is extended with custom Java code for specialised views and integration with the ToolBook multimedia authoring system and the psgml XML mode of the XEmacs editor. A feature of UPGRADE especially useful for schema instantiation is its PROGRES binding for Python scripts. This allows scripts to be used as a kind of run-time defined transaction.

Exporting a Schema. The schema export presents itself as a graph-saving operation to the user, but is internally more complicated. The export goes through the following stages:

1. The author selects the subgraph to use as a schema and activates the export operation;
2. The author names and describes the schema in an extended save dialog;
3. The export operation writes the schema description as an XML ole. This includes the already-mentioned parts name, short and longer descriptions, the edges and nodes, their types, and their attributes, and their current position in the visualisation. Additionally, metadata such as the creation time and user are recorded. Elements get the default weight IMPORTANT (unless a little heuristic determines otherwise);
4. The export operation uses various tools to generate the schema browser HTML page (XSL), an image of the schema for that page (XSL, PostScript, ImageMagick) and a default insertion script (Java, XSL).
5. The export operation updates the schema index (shell script, XSL).

Editing a Schema. There is currently no separate schema editor, so editing is restricted to changing the XML file and regenerating the derived files. Changes that may reasonably performed here are

- Changing the weight of an element (change the value of an enumerated type attribute)
- Adding or changing the tip label to be used in warnings (change the value of a string attribute)
- Changing the documentation (change content of an element in a very small HTML subset format)
- Adding or changing instantiation bindings (create an **binding** element with a unique ID and add that to IDREF attributes at the elements to be used for the binding)
- Change attribute values of schema elements (change string attributes)

A schema editor could offer assistance in adding or removing elements, repositioning them, and in defining bindings. It could check consistency constraints such as whether the graph is contiguous.

Instantiating a Schema. The implementation instantiates and joins a schema in several steps. These steps are illustrated for a trivial example schema (figure 7(a)).

1. The author is asked for joining concepts from the host graph;[6] in the example, the author selects the Middle node in the initial graph of 7(b).
2. New concepts and relations are created in the host graph for the schema elements not mentioned in the binding or not identified by the author;[7] in the example (7(c)), the new Target and a second Relation have been created.
3. For the entire instance, an instance marker node is created. For each element in the instance (just created or indicated by the author), a stub node is created that stores role it plays in the context of this schema. The stubs are connected to the instance marker, recording the weight in the edge type. The head node is connected with another edge.

   In the example, 7(d) shows these usually hidden nodes to the right of the document graph. A first column marks the nodes that are part of the instance, a second one the edges. The instance marker is the rightmost node.[8] The Middle node plays the role of the source node, which has weight CRUIAL. It is also the head.

Step 1 is performed through an user interface element generated from binding description. Steps 2 and 3 are the template-specific core. They are performed by the Python Script generated from the schema description.

With this procedure, a shadow of the expected instance is added to the document graph. On the one hand, the information required to check the instance cannot be attached directly to the nodes in the instance, as it is in danger of being removed with the node by the author. On the other hand, putting this information into the graph only once for all instances would introduce considerable administrative overhead.

Analysing Schema Instances. When schema instance elements are removed, the stub nodes remain. Warnings are then created by warning patterns that test for crucial or important stub nodes that have no corresponding node (and no such warning). The warning message states the role of the missing element and suggests operations based on the weight stored in the stub. If crucial nodes have been removed, it recommends removing the schema instance record, consisting of the instance node and the stubs, as the remaining nodes no longer represent

---

[6] These are the nodes from the host graph that are to be used in the new schema instance, thus providing a kind of embedding. It is only a *kind of* embedding, since that term usually denotes a context determined by an algorithm, not the user.

[7] This is a change from the process described in [7], where a complete instance was always created. This simplification is now possible because Python offers the matching control structures.

[8] In a previous more complex example, 'DoesEMAilAnalysisMatchQuestionnaires' plays the role of 'The Thesis' in the IMRD schema instance, and twice the role of 'OverallThesis' in the 'CompSubStudy' instances.

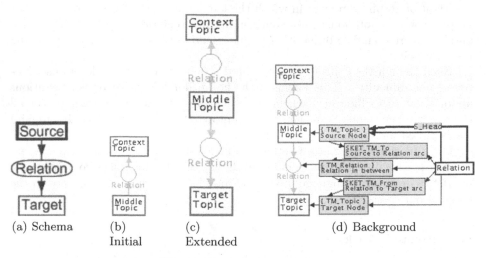

Fig. 7. Steps in instantiation

the original schema. This can be done with a generic production, since the instance/stub subgraphs are unambiguous. If merely important nodes have been removed, the stub may be removed to remove the warning. If optional nodes are removed, the stub is quietly removed. The warning also disappears if the author reconnects an appropriate element.

4 Discussion

The instantiation and weighting mechanisms presented here allow creation of graph models with parts of discriminated importance with some guidance. Giving reference points at instantiation time is an intuitive and efficient way to integrate schemata into graphs under construction. Weights are an useful addition to avoid less-than-helpful warnings about lost fringe elements, and to post warnings about severe mutilations where appropriate.

A shortcoming of the interactive join is that 'obvious' referents are not picked up automatically. For example, some parts of the context of CompSubStudy in the 'AnotherSubStudy' variant are redundant: there is the overall thesis and the section that exports it. That section should be the one that also exports the newly added sub-thesis. Grasping this obviousness formally could take the form of path expressions to be added to the schema definition. This, however, would make matching more complicated.

The current implementation detects only missing parts; a schema cannot inhibit relations being attached to it. This may be added by incorporating negated nodes and relations, for which additional warning patterns would have to be implemented that check negation instead of stubs.

These potential extensions lead to the point of graph grammar engineering concerns. Schemata, already in the form as presented here, may be regarded as a variant of graph grammars in which the left-hand side is not removed and the graph grows monotonously. So, some of the inherent problems of writing such grammars are bound to bite authors. This would only be worsened by extending the expressivity.

Even though the concept of a weighted schema is fairly simple, its use does pose some difficulties. This begins with the author having to be very cautious about what to include in a schema or a parameter binding in order to get all the expected elements. Some heuristics, a more robust user interface design, and a schema editor should be able to improve this.

On a more general note, CHASID is limited to what the author models explicitly. It does not attempt to infer anything from the document content. Some heuristics or natural language processing may be useful here, but are not in the focus of our research.

4.1   Related Work

CHASID being an application (a program aimed at supporting users in solving real-world problems) based on (Conceptual) Graphs, it may be interesting to classify it according to [2] into natural language processing, information retrieval, knowledge acquisition, requirements engineering, or miscellaneous. CHASID is, however, not really an artificial intelligence application, as it does not use the document's model (which would be the model of the natural language text or the basis of document retrieval, or the acquired knowledge) for deduction — unless evaluating the model is regarded as deducing warnings. It merely supports the author in building a good structure, avoiding structures that are known as problematic. So, it would be a dedicated system of the 'Miscellaneous' class, on the fringes of NLP.

The idea of using structures with attached weights is based on CIDRE (Cooperative & Interactive Document Reverse Engineering, [1].) To support a document hierarchy recognition process, certain hierarchical arrangements in SGML are given a probability rating. This uses a logical extension of the proven optical character and text recognition technique of statistical n-gram modelling. The weights are derived from a body of structured documents, which should allow this technology to scale well to large document management systems, as it does not require human intervention. This is, however, chiefly possible on the basis of the formal nature of the structure. The principle accounts well for the fact that formal structure definitions are bound to cater for exotic cases in order not to be too restrictive, so that the regular cases may get lost.

Well-documented document type descriptions, such as DocBook [14], provide a host of technical type and schema definitions. These remain on the technical level and do not address the problem of content structuring. If the element types of DocBook were to be integrated into CHASID, this would chiefly yield a detailed document structure and a number of domain topic types like 'Function' or 'Parameter'. The relationship of instances of these types already are not addressed by the standard. Special conventions nonwithstanding, it is not possible to determine wether a 'Function' is exported or imported by the division in which it occurs.

The SEPIA project [13] is concerned with cooperative, argumentation-based authoring. It builds an issue-based information schematic structures. It does not address the integration with existing authoring environments the way CHASID does and has later turned more into the field of cooperative work.

The term 'schema' is also used by other authoring applications. Schema-Text [9][9] uses it in the 'type system' sense. This schema may be edited in tandem with also graph-structured instance documents, providing support for schema evolution. The documents are then transformed using customisable patterns to yield audience- and medium-specific presentations. This generation step corresponds to CHASID's integration between document hierarchy and topic map. So, SchemaText is more suitable for managing large amounts of documents, while CHASID addresses more the single authoring process.

5   Conclusions

Conceptual authoring support systems have to cope with the markup dilemma. While an explicit structure pays off for the author in the long run, average authors do not look so far ahead. Patterns and schemata help in this respect by guiding the author through planning the document, which can speed up this process. Especially schemata help in providing proven modules to be instantiated in the document. By recording instantiation traces, CHASID is able to check schema instances for completeness. This is an improvement over the templates currently offered in authoring, which are overwritten during editing. Using weights in schemata offers a way to encode knowledge about problems (parts that are crucial for a text to work, but tend to be removed) in the schema instance scope easily. This allows non-programmers to extend the authoring environment significantly by defining own schemata.

This option of defining schemata requires them to be stored as data, which in turn leads to the question of the expressiveness of the data model. The static description of a schema as weighted elements with documentation trades user serviceability with advanced help such as determining 'obvious' contexts.

The usefulness of graphs in this context rests on their uniform model of complex structures. This allows the schema-related operations to be defined fairly generally, paying attention only to the restrictions defined in the graph model.

Given the current tools, a content model of a 15-page article may realistically be handled in terms of amount of data. The writing span of such a document is so short, however, that this will frequently not pay off. When considering a more coarse-grained model of a book, a benefit becomes more likely: the time spent writing increases, making it easier to get lost. There are, however, even less accounts of discernible reusable structures in books than there are in articles.

---

[9] www.schema.de

References

[1] Rolf Brugger, Frédéric Bapst, and Rolf Ingold. A DTD Extension for Document Structure Recognition. In Roger D. Hersch, J. André, and H. Brown, editors, *Electronic Publishing, Artistic Imaging, and Digital Typography*, volume 1375 of *Lecture Notes in Computer Science*, page 343ff. Springer, 1998.

[2] Michel Chein and David Genest. CGs Applications: Where Are We 7 Years after the First ICCS? In Ganter and Mineau [5], pages 127–139.

[3] Deutsche Gesellschaft für Psychologie (German association for psychology), Göttingen. *Richtlinien zur Manuskriptgestaltung (Guidelines for manuscript design)*, second edition, 1997.

[4] Linda S. Flower and John R. Hayes. The dynamics of composing: Making plans and juggling constraints. In Lee W. Gregg and Erwin R. Steinberg, editors, *Cognitive Processes in Writing*, pages 31–50. Lawrence Erlbaum, Hillsdale, NJ, 1980.

[5] Bernhard Ganter and Guy W. Mineau, editors. *Proc. International Conference on Conceptual Structures 2000*, volume 1867 of *Lecture Notes in Artificial Intelligence*. Springer, 2000.

[6] Felix H. Gatzemeier.   Patterns, Schemata, and Types — Author Support through Formalized Experience.   In Ganter and Mineau [5], pages 27–40. `http://www-i3.informatik.rwth-aachen.de/research/publications/by_year/2%000/PatScheTy-online.pdf`

[7] Felix H. Gatzemeier.   Authoring operations based on weighted schemata. In Guy W. Mineau, editor, *Conceptual Structures: Extracting and Representing Semantics — Contributions to ICCS 2001*, pages 61–74, August 2001.   `http://sunsite.informatik.rwth-aachen.de/Publications/CEUR-WS/Vol-41/Ga%tzemeier.pdf`

[8] Dirk Jäger. Generating tools from graph-based specifications. *Information and Software Technology*, 42:129–139, 2000.

[9] Christoph Kuhnke, Josef Schneeberger, and Andreas Turk.   A schema-based approach to web engineering.   In *Proceedings 33rd Hawaii International Conference on System Sciences*, volume 6, Maui, Hawaii, January 2000. `http://www.computer.org/proceedings/hicss/0493/04936/04936069.pdf`

[10] Manfred Nagl, editor. *Building Tightly Integrated Software Development Environments: The IPSEN Approach*, volume 1170 of *Lecture Notes in Computer Science*. Springer, Heidelberg, 1996.

[11] Andreas Schürr. *Operationelles Spezifizieren mit programmierten Graphersetzungssystemen (Operational specification with programmed graph rewriting systems)*. PhD thesis, RWTH Aachen, Wiesbaden, 1991.

[12] John F. Sowa. *Conceptual Structures: Information Processing in Mind and Machine*. The Systems Programming Series. Addison-Wesley, Reading, MA, USA, 1984.

[13] Norbert A. Streitz, Jörg M. Haake, Jörg Hannemann, Andreas Lemke, Wolfgang Schuler, Helge Schütt, and Manfred Thüring.   SEPIA – A Cooperative Hypermedia Authoring System.   In D. Lucarella, J. Nanard, M. Nanard, and P. Paolini, editors, *Proceedings of ECHT'92, the Fourth ACM Conference on Hypertext*, pages 11–22, Milano, Italy, November 1992. ACM. `ftp://ftp.darmstadt.gmd.de/pub/concert/publications/SEPIApaper.ps.Z`

[14] Norman Walsh and Leonard Muellner.   *DocBook: The Definitive Guide*. O'Reilly, October 1999. `http://www.oasis-open.org/docbook/documentation/reference/html/docbook.%html`

Re-engineering a Medical Imaging System Using Graph Transformations*

Tobias Rötschke

Technische Universität Darmstadt
Institut für Datentechnik, Fachgebiet Echtzeitsysteme
Merckstraße 25, 64283 Darmstadt, Germany
tobias.roetschke@es.tu-darmstadt.de
http://www.es.tu-darmstadt.de

Abstract. This paper describes an evolutionary approach to reengineer
a large medical imaging system using graph-transformations. The solution has been integrated into the ongoing development process within the
organizational and cultural constraints of a productive industrial setting.
We use graph transformations to model the original and the desired architecture, the mapping between architectural, implementation and design
concepts, as well as related consistency rules. Violations are calculated
every night and provided as continuous feedback to software architects
and developers, so they can modify the system manually according to
established procedures. With only a very limited global migration step
and moderate changes and extensions to the existing procedures it was
possible to improve the software architecture of the system, while new
features still could be implemented and released in due time. Although
this solution is dedicated to a concrete case, it is a starting point for
a more generic approach.

1   Introduction

In its early days, about 25 years ago, the medical imaging technique realized by
our system could only provide rather crude three-dimensional images of motion-
less human tissue. Being less invasive, but more expensive than other imaging
techniques, much effort has been spent improving image quality and performance
during the years, and nowadays the technique is on the verge of providing highly
detailed real-time images of a beating heart under surgery conditions.

Accordingly, the number of possible applications and hence the number of features has increased dramatically. As of today, the system has about 3.5 MLOC,
growing at a rate of several 100 KLOC each year. The trust in the stability of the
system depends mainly on the organizational and cultural framework established
around the development process. However, various technological and economical
reasons required a major reorganization of the software, while at the same time,

---

* This research was performed at Philips Medical Systems in Best, the Netherlands
  when the author was employee of Philips Research in Eindhoven, the Netherlands.

J.L. Pfaltz, M. Nagl, and B. Böhlen (Eds.): AGTIVE 2003, LNCS 3062, pp. 185–201, 2004.

more and more features had to be added to compete in the market. Among other things like porting the system to another hardware platform with different operating system, introducing a more object-oriented programming language, and integrating more third-party components, the reorganization included the introduction of different architectural concepts. To deal with the increasing complexity in an appropriate way, an overall refactoring of the system according to these concepts has been performed.

When the author became involved, the migration process had already begun. The new[1] architectural concepts called "Building Blocks" had been described in a document written in natural language. Only remotely related to the "Building Blocks" defined in [10], these concepts describe hierarchical units of responsibility. They correspond to the functional organization of the department, the scope of product related specifications, and an encapsulated piece of software. There were ideas about the visibility rules, but there was no formal specification, and the details were still subject to change.

The task of the author was to specify the new concepts using the site-specific terminology in a way that could be understood easily by software architects, designers and developers, and modified at least by software architects. Next to the specification, tool support was required to allow the software architects to monitor the progress of building UML models as part of new design specifications and of refactoring the source code according to the new concepts. Automated code transformations had to be avoided because they would affect virtually all source files at once, circumventing the established quality assurance mechanisms. This was considered a serious risk for the continuity of the ongoing development process.

In this paper, we focus on the evolutionary migration from the existing source code according to an implicit module concept to the desired situation where the implementation is consistent with newly written design models with explicit architectural concepts. In section 2, we provide an a-posteriori-specification of the original module concepts. Next, we define the desired concepts (section 3) and specify adequate consistency rules (section 4). The separate steps to achieve the desired situation are identified in section 5 and a migration plan is set up. Finally, section 6 describes how the migration is realized in a way that allows optimal integration into the ongoing development process.

2   Initial Concepts

The original architecture is a flat decomposition of the system into *modules*. Each module realizes an isolated piece of functionality. All modules are treated equally and may use functionality implemented by other modules as needed. In the source code, a module is represented by a directory. Modules and directories are mapped on each other using naming conventions according to a coding standard that has been defined as part of the development process.

---

[1] In this context we use the term "new" as opposed to the old concepts of the system, not at all "new to the world".

To organize intra- and inter-module connections, local and global header files are identified by naming conventions. Local header files are only visible in the module they are defined in. To allow the export of functions and data types to other modules, a global scope is defined. It consists of all global header files defined in the system. Every module can gain access to this global scope. By using different include paths for each module, this scope can be restricted to avoid functionally unrelated modules from using each other. All identifiers declared in global header files must be unique to avoid naming conflicts.

Over the years, developing the system using these original module concepts has become increasingly difficult as functionality has grown tremendously. This produced an increasing number of modules and interconnections. As a consequence, the system is harder to maintain, and individuals are no longer able to keep an overview over the system as a whole. So it takes more effort to add new features to the system without hampering the work of other developers.

Having more than two hundred software developers working together on a product is virtually impossible if there is no way to distribute them over smaller teams where they can operate (almost) independently from each other. For example, one team could be concerned with the adoption of a new version of the operating system. But this should not hamper the activities of other teams.

3   Desired Situation

The driving idea behind our new module concepts has been information hiding [14] and reducing dependencies. Every module should only see resources (data types and functions) that it actually needs. The import scope of every module should be defined explicitly. Available resources should be concentrated on few and unique locations to reduce ambiguity as well as dependencies between modules.

In the desired situation, the architecture consists of an aggregation hierarchy of so-called (building) blocks. Building blocks are architectural units like modules, but they also define an area of responsibility and ownership and a scope for appropriate documentation.

As opposed to the initial concepts, we define three meta models for the different kind of concepts we deal with:

- The architecture model describes the site-specific architecture concepts which, among other things, are used to define abstract visibility rules between different parts of the system.
- The design model is a subset of the UML meta model describing the UML concepts used to define concrete visibility rules between concrete architectural concepts in Rational Rose.
- The implementation model defines the relevant file system and language artifacts representing concrete architecture concepts.

Our long-term goal is to achieve mutual consistency between these models. This is a typical integration problem that can formally be described by triple graph

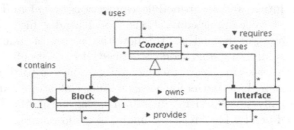

Fig. 1. Architecture Concepts

grammars [22]. As we decided to perform all modifications to the source code manually, if possible, we do not aim to achieve consistency by generating one model from one of the others. We rather perform a correspondence analysis of between each pair of models.

3.1   Architecture Concepts

The basic architecture concept is a Block as depicted in figure 1. A Block contains other blocks. All blocks are treated in a way that the same rules can be applied on different system levels and blocks can be moved easily. The system is represented by a root block, which transitively contains all other blocks.

A Block owns an arbitrary number of Interfaces. This relationship defines who is responsible for the Interface, i.e. the owner of a Block is also responsible for owned Interfaces. The provides dependency defines the export interface of a block. The block does not necessarily need to own this interface, but may forward an interface from another building block, typically a parent block. A block owning an interface automatically provides it.

Each concept has a sees dependency to all visible interfaces. This dependency is statically defined by the building block hierarchy. The requires dependency defines the import interface of a block or interface and is explicitly modeled by a software engineer in the UML model. A block providing an interface automatically requires it. A requires dependency is valid only if there is an adequate sees dependency.

Finally, the uses relationship represents an implemented dependency derived from the source code. A uses dependency is valid only if there is an adequate requires dependency in the UML model.

3.2   Design Concepts

The dependencies between different building blocks, i.e. the concrete visibility rules, are defined in so-called interface specifications as part of the UML model. So the relevant design concepts are defined by a partial UML meta model [21]. Figure 2 shows the ModelElements that are used by our approach.

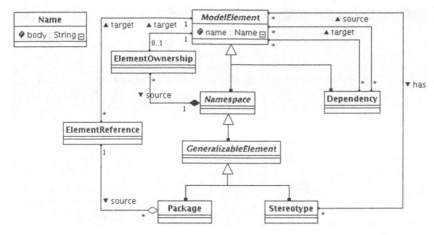

Fig. 2. Design concepts (Partial UML meta model)

Building blocks are represented as Packages. The hierarchy of blocks is mapped on a hierarchy of packages, which relates to the ElementOwnership relationship in the meta model. Interfaces are represented by Packages with a Stereotype "interface". The "owns"-relationship of our architecture model is represented by ElementOwnership relationship between adequate packages. The "provides"- and the "requires"-relationships are explicitly modeled by means of Dependencies. The Name of the dependency indicates the type of relationship.

Figure 3 shows an example of an interface specification of a building block called "acqcontrol". The export interface of the building block is defined by the "provides"-dependencies, the import interface by the "requires"-dependencies.

3.3   Implementation Concepts

The three major implementation concepts analyzed by our approach are directories, files, and include directives. Several subtypes are distinguished in the class diagram of the model as presented in figure 4.

3.4   The Correspondence Graph

As we want to achieve mutual consistency between architecture, design and implementation, we use the concept of triple graph grammars [22] to define consistency rules between two documents. These are referred to as source and target graph. Triple graph grammars define consistency by simultaneously creating an increment of the source graph, the appropriate increment of the target graph and an increment of a so called correspondence graph. The latter maps elements of the source graph to corresponding elements of the target graph.

So we need to define a fourth graph schema to define the node and edge types of the correspondence graph. In Figure 5, ADC denotes architecture-design

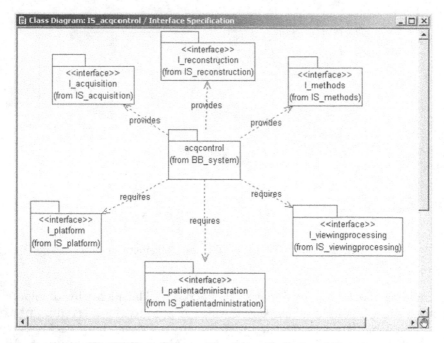

Fig. 3. Interface specification with Rational Rose

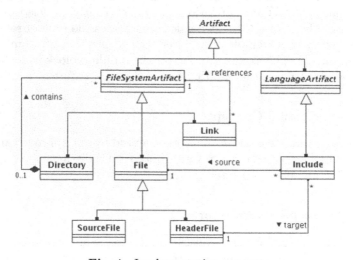

Fig. 4. Implementation concepts

consistency, A IC architecture-implementation consistency and D IC design-implementation consistency.

The correspondence nodes have specializations which are used to express the mapping between particular concepts. A B lockD irectory, for instance, is a sub-

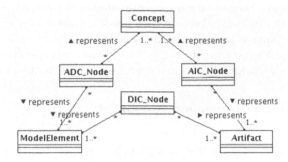

Fig. 5. Simplified graph schema for the correspondence graph

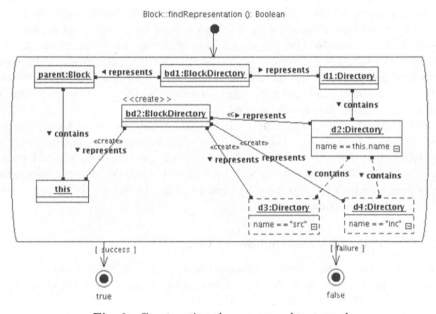

Fig. 6. Constructing the correspondence graph

type of an AIC-Node which relates the architecture concept Block to the corresponding implementation concept Directory. Figure 6 provides an example how this correspondence node is used. Note that we use n:m-correspondences, e.g. a Block corresponds to a set of three Directories.

4   Consistency Management

We use FUJABA to generate our consistency checking tool from a UML-like specification. The specification consists of a graph schema written as UML class diagram (see section 3.2) and graph transformations defined by story diagrams [5, 8]. Control flow is specified as UML activity diagram and productions are defined as UML collaboration diagrams inside the activities of the activity diagram. Left- and right-hand side of each production are integrated into a single diagram. Stereotypes attached to nodes and edge indicate whether an element occurs on both sides (no stereotype), left-hand side only (destroy) or right-hand side only (create). Optional nodes and edges are represented by dashed elements.

4.1   Creating the Correspondence Graph

As our version of Fujaba did not directly support triple graph grammars, we have to use standard graph transformations to simulate the triple rules. In figure 6, the nodes this and parents are architecture concepts (source graph), d1 to d4 implementation concepts (target graph) and bd1 and bd2 correspondence nodes.

To read this rule, start at the this node which is of the type Block. Follow the contains edge to the parent Block which has already been processes. So the should be a correspondence via the correspondence node bd1 to an appropriate directory d1. Otherwise the rule fails. This directory should contain another directory d2 which has the same name as the this Block. Optionally, this directory could contain subdirectories d3 and d4 named "src" and "inc". If this pattern is matched, the correspondence node bd2 is created and links to the related nodes this, d2, d3, and d4 are created as well.

If this rule is applied width-first, all BlockDirectory correspondences are created correctly. In a similar fashion, other correspondences are constructed. As the architecture document (a text file) and the existing source code are given, we can traverse the related graph representations and construct the correspondence nodes accordingly. The algorithm to visit all nodes in the correct order is part of our handwritten framework.

4.2   Checking the Correspondence Graph

Having constructed the correspondence graph based on the information from parsed documents, we check the consistency of these documents. Figure 7 shows a rule for checking whether an include directive found in a C-File is a valid RemoteInclude, i.e. an include from a file inside a block to an interface owned by (usually) another block.

Again, the rule is divided into three parts. The this node is an implementation concept. Via a correspondence node ri1 the corresponding block b1 and interface i1 are found. The rule succeeds if there is a requires link from b1 to i1 and fails otherwise. If the rule fails, there is one violation of the rule "Remote includes are valid", i.e. either the "requires" dependency in the UML model is still missing, or the source code contains an include directive that is not allowed.

Our handwritten tool framework contains an algorithm that checks the correctness of all elements in our model and counts the number of violations per architectural unit and kind of rule. The results are kept in a database for later retrieval. Our current tool implementation checks 18 different consistency rules.

The next section describes how these consistency checks fit in the overall re-engineering process.

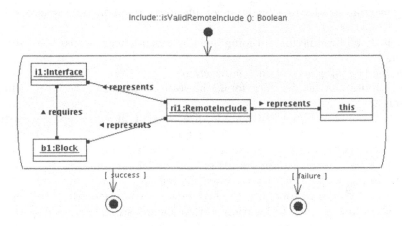

Fig. 7. Checking the consistency

5   Migration Plan for the New Software Architecture

When we started our re-engineering effort, the new building block hierarchy had already been set up. The original modules had been integrated as leaves of the building block hierarchy. By generating makefiles with suitable include paths, the original include directives could be preserved. Thus, no files had to be changed.

The major challenge was to introduce the new visibility concept based on explicitly defined interfaces, without disturbing the ongoing development process. Although the project members had agreed on the abstract visibility rules, the interfaces and concrete visibility relationships still had to be specified. To implement this specification, hundreds of files had to be moved, thousands of include dependencies had to be relocated, and tens of thousands of function calls had to be renamed to reflect the modification with respect to the naming conventions.

Using the normal development procedures including planning, documenting and inspecting each step would be far too complex and time consuming. A fully automated transformation of the complete code base, circumventing the trusted procedures was considered too risky. Besides, specifying all rules before actually implementing them would provide very late feedback on the effect of the new rules.

As a compromise, we decided to re-engineer the top-level building blocks (called subsystems) first and to perform all necessary code transformation. This should be done considering the actual include dependencies as given and improving them later. Once the subsystems work according to the new architectural concepts, we have enough feedback to re-engineer the rest of the system. Because the interfaces of each subsystem hide the internals from the rest of the system, further re-engineering can be done locally within single subsystem without effecting the work on other parts of the system.

The following steps have been identified for the first phase of the migration as described above:

1. Identify all header files hit by incoming include dependencies across subsystem borders.
2. Identify all files having outgoing include dependencies across subsystem borders.
3. Define the interfaces of each subsystem.
4. Distribute the header files found in step 1 over the interfaces defined in step 3.
5. Determine the changes to include directives of files in step 2 according to the new distribution of header files (step 4).
6. Define a scope directory (section 6.3 for each building block which serves as include path when building a file of that particular building block.
7. Create the scope directories defined in step 6.
8. Move all global header files to their new location according to step 4 and create a link at the old location so that includes inside the subsystems still work. This includes checking out and in affected directories in the ClearCase archive.
9. Change all files affected according to step 5. This includes checking out and in affected files in the ClearCase archive.

6 Migration Tools Support

While step 3 and step 4 must be performed manually, all other steps can be automated. Graph rewriting rules defined in FUJABA can help us to perform step 1, step 2, step 5, and step 6 and finally generated Perl scripts to perform the remaining steps.

6.1 Finding Includes Across Subsystem Borders

Using our graph model of the system, we could analyze the existing source code and calculate the impact on the system, assuming that we only redirect include dependencies across subsystem borders. We found out, that in total 14160 of the 44624 include dependencies would be modified. In 2386 of the 8186 files include directives needed to be changed. 521 header files were actually hit and thus had to be distributed manually over the interfaces defined in step 3.

6.2 Calculating Include Directives

We used the convention `#include "<MODULE>/inc/<HEADERFILE>"` to include header files from other modules in the original architecture. The new include directives look like `#include "<BLOCK>/<INTERFACE>/<HEADERFILE>"` to match the desired architecture. Using the graph model of the system, we can calculate the necessary changes and generate a script to checkout each affected file, modify the include directives, and check them in afterwards.

This step circumvents the regular development procedures, but it would be virtually impossible to perform these modifications by hand in a consistent way. However, this transformation is well understood and changes only a single aspect of the system many times. So the risk was considered minimal.

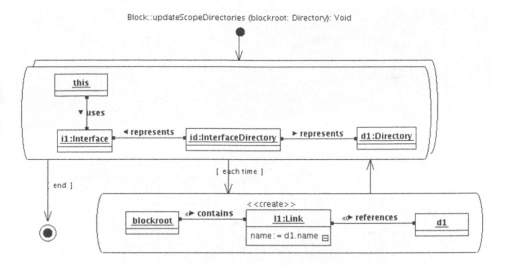

Fig. 8. Updating the scope directory

6.3 Calculating Scope Directories

To implement the visibility rules of the interfaces, we use different so called scope directories. For each building block the include path points to a dedicated directory. This directory contains a subdirectory for each building block that may be used. Such a subdirectory is called "scope directory" and contains links to directories representing available interfaces. This structure can modeled by the graph rewriting rule in figure 8. From this model we generate a script to perform the transformations in the file system.

We tested all the generated scripts on a clone of the repository until it worked flawlessly. Before applying the script to the productive repository, we asked all developers to check in their modifications before a weekend. During the weekend, we performed the transformation. When the next working week began, everybody started with a consistent version of the new architectural concepts.

6.4 Monitoring Re-engineering Progress

Figure 9 shows the tool chain of our approach. The source code is parsed by a reverse engineering tool called Code and Module Architecture Dashboard (CMAD) [20]. It allows to calculate simple code metrics per architectural unit each day and to keep the history of these metrics in a SQL repository. After selecting the dates of interest, the trends of these metrics can be observed with a web browser. Hyper-linked tables allow to navigate through the building block hierarchy.

The Interface Management Tool (IMAN) deals with the consistency of architecture, design and implementation. The information model and consistency rules are defined using the FUJABA tool as described in section 3 and section 4. The

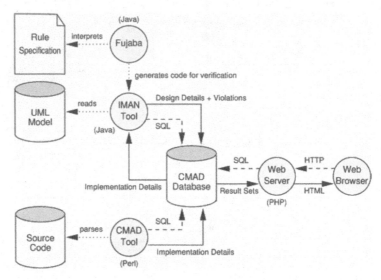

Fig. 9. Tool chain

tool uses the CMAD database to access the implementation artifacts and parses the UML model using the Rose Extensibility Interface [17]. The IMAN tool updates an extended CMAD database each day with changes found in the UML model. Finally, the consistency rules are checked and changes in the number of violations are propagated to the CMAD database as well. During office hours, software architects and developers can access the resulting statistics via a web browser as described in [19].

7   Related Work

When considering related work, different aspects can be addressed. First of all, our way of describing architectural concepts can be compared with architecture description languages. Next, we compare our work with other tools performing re-engineering using graph-transformations. Besides, there are alternative techniques to deal with re-engineering. Finally, though our approach is dedicated to a single system, it could be generalized to deal with other systems as well. So we compare our ideas for a more generic approach to evolve large software architectures with existing ones.

7.1   Architectural Concepts

When modeling the architectural concepts of the system, the goal was to find an accurate definition of the notion of architecture as it lives in the development department. From the software point of view, these concepts define a Module Interconnection Language (MIL) [16] for our system which does not fulfill all requirements of an Architecture Description Language (ADL) according to [12]. On the other hand, our concepts are closely related to the organization and the development process of our department, which is an aspect usually not considered by MILs or ADLs.

Having an explicit definition of these concepts enables us to communicate with software architects and developers. Thus, we are able to define consistency rules that are well understood within our department. We do not claim that, in general, our architectural concepts are better than others. They rather express exactly those concepts that we actually need.

7.2   Graph-Based Re-engineering Tools

Re-engineering for Distribution The REforDI project[1, 2] provides integrated tools for automated code analysis, interactive architecture recovery and re-design, and generated source code transformations based on the PROGRES environment [23]. The tools are applied on COBOL legacy systems which have to be migrated to a client/server-architecture.

In our approach, code modifications have to be performed manually according to established procedures, so we avoid automatic code transformation and restrict them to changes of the directory structure and include directives. Our system is not a legacy system, but evolving continually and both the existing and the desired module architecture are quite well understood. Rather than re-engineering the system in an isolated activity, we continuously provide feedback on the migration progress to integrate the transformations into the ongoing development process.

Ericsson Communication Architecture for Embedded Systems The E-CARES [11] project provides tools that perform structural and behavioral analyses of complex telecommunication systems written in PLEX. The structure of the system and related metrics are displayed in different graphs views. Metrics are displayed as node attributes or visible properties like size, color etc. Execution traces are used in combination with structural information are used to dynamically identify link chains and recover state diagrams. The tools are specified as graph transformations using the PROGRES environment.

Rather than visualizing information about our system as graphs, we generate navigable tables and focus on trends rather than absolute metrics in our approach. Our software architects and developers are more comfortable with this kind of representation. While absolute numbers are difficult to interpret, trends provide an understandable means to define and monitor improvement of the architecture.

Conditional Graph Rewriting In [13] a formal approach is described that allows to manage the evolution of software related documents such as source code, UML models and software architecture. Concurrent evolutions of the same software artifacts can be merged and both structural and behavioral inconsistencies can be detected. The formalism is based on graph transformation to describe atomic modifications of these documents. However, this approach is not related to any concrete re-engineering project.

Re-engineering of Information Systems In [7] an incremental consistency managed approach to understand the relational schema of an existing information system and transform it to a redesigned object-oriented conceptual model in an iterative process. This approach is based on triple graph grammars and uses the Varlet re-engineering environment. To propagate modifications, the Varlet system supports history graphs that do not only keep current model, but also manually applied re-engineering operations. In this way, changes from the analysis phase do not destroy manual operations of previous iterations.

This approach works well, as both models are initially consistent, though incomplete. In our case the models are inconsistent in the beginning, so we measure the gap between the existing and the desired situation and monitor the progress of narrowing this gap manually.

7.3   Alternative Re-engineering Techniques

Relation Partition Algebra In [15] a module architecture verification approach is described that applies Relation Partition Algebra [3] for consistency checks in the context of another medical imaging system. Compared to our approach, there is no emphasis on trends in violations an other metrics. As RPA expressions become very large for real systems and do not visualize structure as well as a graphic notation, UML class diagrams in combination with graph transformations are a better means to specify consistency rules for software architecture.

Constraining Software Evolution In [18] constraints are used to keep evolving source code and related documents consistent with each other. Modifications of either source code or documents trigger consistency checks. If inconsistencies are detected, either repair actions are performed or the developer is notified so he manually can fix the problem.

Graph Queries The GUPRO Repository Query Language (GReQL) [9] allows to formulate queries for a graph-based repository representing parsed C-Code. GReQL is particularly useful for software re-engineering, as queries may contain path expressions to deal with transitive closures and metrics. However, GUPRO uses a fixed graph scheme and does not support UML documents and site-specific architecture descriptions. Even if a language like GReQL would be more comfortable on the one hand, using a modified GUPRO project would bind additional resources on the other. So we decided to live with the limited querying capabilities of the relational DBMS which we needed anyway for the system under investigation.

Change Impact Analysis Zhao et al. propose a formal technique to analyze the impact of architectural changes a priori [25]. However, this would require a complete and formal architectural specification of the current and the desired architecture, and we do not have them at our disposal.

7.4 Generic Re-engineering Approaches

Re-engineering for C++ Columbus [4] is a re-engineering tool for C++ based on the GXL [6] reference schema for C++. This is a more generalized approach for analyzing C++ programs than we use in our approach, although it was not yet available when we started our research. Columbus performs a more detailed dependency analyses than just include dependencies. On the other hand Columbus does not deal explicitly with file systems artifacts which represent most of our architectural concepts.

Extensible Metrics Toolbench for Empirical Research The goal of the EMBER project [24] is to provide a multiple-language reverse engineering tool that allows to integrate complexity metrics into the development process. The tool uses a generic object-oriented programming language meta-model for its database schema. The tool does not support architecture concepts and is limited to source code metrics.

Combining the more detailed analyses of the presented source-centered tools with our architecture-centered and trend-oriented approach would result in an even better support for software evolution and integration in the development process.

8 Conclusions

This paper describes how graph transformations have been used to support the architectural reorganization of a large medical imaging system. UML-like graph schemata have been used to separate site-specific architectural concepts and generic design and implementation concepts. Graph rewriting rules describe the construction of correspondence relationships and consistency rules between these concepts following the idea of triple graph grammars.

In combination with database and web technology we are able to improve the system manually according to established procedures and monitor the progress of architectural improvement during the ongoing development process. Information about modifications and violations is updated daily and visualized by means of navigable tables. Hence, we can gradually align design documents and source code to each other in terms of an improved architectural meta model. This approach provides us with the necessary control over system modifications to meet our safety and quality requirements.

Our evolutionary, analytical approach has little impact on the ongoing development process and works well within the cultural and organizational constraints of an industrial setting. Here our work is different from other approaches where

re-engineering is considered an isolated transformation outside the ongoing development process. Although tailored towards the needs of a specific development
project, the approach could be generalized to work in similar settings as well.

References

[1] Katja Cremer. Graph-Based Reverse Engineering and Reengineering Tools. In
    AGTIVE, pages 95–109. Springer, 1999. 197
[2] Katja Cremer, André Marburger, and Bernhard Westfechtel. Graph-Based Tools
    for Reengineering. *Journal of Software Maintenance and Evolution: Research and
    Practice*, 14(4):257–292, 2002. 197
[3] L.M.G. Feijs and R.C. van Ommering. Relation Partition Algebra for Reverse
    Architecting - mathematical aspects of uses- and part-of-relations. *Science of
    Computer Programming*, 33:163–212, 1999. 198
[4] Rudolf Ferenc, Árpád Beszédes, Mikko Tarkiainen, and Tibor Gyimóthy. Colum-
    bus - Reverse Engineering Tool and Schema for C++. In *IEEE International
    Conference on Software Maintenance (ICSM)*, pages 172–181, October 2002. 199
[5] Thomas Fischer, Jörg Niere, Lars Torunski, and Albert Zündorf. Story Diagrams:
    A new Graph Grammar Language based on the Unified Modelling Language
    and Java. In *Workshop on Theory and Application of Graph Transformation
    (TAGT'98)*. University-GH Paderborn, November 1998. 191
[6] Ric Holt, Andreas Winter, and Andy Schürr. GXL: Towards a Standard Exchange
    Format. In *Proc. Working Conference on Reverse Enginering*, pages 162–171,
    November 2000. 199
[7] Jens Jahnke and Jø"rg Wadsack. Integration of Analysis and Redesign Activities
    in Information System Reengineering. In *Proc. of the 3rd European Conference
    on Software Maintenance and Reengineering (CSMR'99)*, pages 160–168, Ams-
    terdam, NL, 1999. IEEE Press. 198
[8] H.J. Köhler, U. Nickel, J. Niere, and A. Zündorf. Using UML as a Visual Pro-
    gramming Language. Technical Report tr-ri-99-205, University of Paderborn,
    Paderborn, Germany, August 1999. 191
[9] Berndt Kullbach and Andreas Winter. Querying as an Enabling Technology
    in Software Reengineering. In *IEEE Conference on Software Maintenance and
    Reengineering*, pages 42–50, 1999. 198
[10] Frank J. van der Linden and Jürgen K. Müller. Creating Architectures with
    Building Blocks. *IEEE Software*, pages 51–60, November 1995. 186
[11] André Marburger and Bernhard Westfechtel. Graph-Based Reengineering of
    Telecommunication Systems. In *Proc. of the 1st International Conference on
    Graph Transformation (ICGT 2002)*, volume 2505 of *LNCS*, pages 270–285.
    Springer, 2002. 197
[12] Nenad Medvidovic and Richard N. Taylor. A Classification and Comparision
    Framework for Software Architecture Description Languages. *IEEE Transactions
    on Software Engineering*, 1(26):70–93, January 2000. 197
[13] Tom Mens. Conditional Graph Rewriting as a Domain-Independent Formalism
    for Software Evolution. In *AGTIVE*, pages 127–143. Springer, 1999. 197
[14] David L. Parnas. A Technique for Software Module Specifications with Examples.
    Comunications of the ACM, 15:330–336, 1972. 187
[15] Andre Postma. A Method for Module Architecture Verification and its Applica-
    tion on a Large Component-Based System. *Information and Software Technology*,
    45:171–194, 2003. 198

[16] J.M. Neighbors R. Prieto-Diaz. Module interconnection languages. *Systems and Software*, 6(4):307–334, 1986. 196

[17] Rational Software Corporation. *Using the Rose Extensibility Interface*, 2001. http://www.rational.com/docs/v2002/Rose_REI_guide.pdf. 196

[18] Steven P. Reiss. Constraining Software Evolution. In *IEEE International Conference on Software Maintenance (ICSM)*, pages 162–171, October 2002. 198

[19] Tobias Rötschke and René Krikhaar. Architecture Analysis Tools to Support Evolution of Large Industrial Systems. In *IEEE International Conference on Software Maintenance (ICSM)*, pages 182–193, October 2002. 196

[20] Tobias Rötschke, René Krikhaar, and Danny Havenith. Multi-View Architecture Trend Analysis for Medical Imaging. In *IEEE International Conference on Software Maintenance (ICSM)*, page 107, November 2001. 195

[21] James Rumbaugh, Ivar Jacobson, and Grady Booch. *The Unified Modelling Language Reference Manual*. Addison-Wesley, 1999. 188

[22] Andy Schürr. Specification of Graph Translators with Triple Graph Grammars. In Ernst W. Mayr, Gunther Schmidt, and Gottfried Tinhofer, editors, *Graph-Theoretic Concepts in Computer Science, 20th International Workshop, WG '94*, volume 903 of *LNCS*, pages 151–163. Springer, 1995. 188, 189

[23] Andy Schürr, Andreas J. Winter, and Albert Zündorf. Developing Tools with the PROGRES Environment. In Manfred Nagl, editor, *Building Tightly Integrated Software Development Environments: The IPSEN Approach*, volume 1170 of *LNCS*, pages 356–369. Springer, 1996. 197

[24] F.G. Wilkie and T.J. Harmer. Tool Support for Measuring Complexity in Heterogeneous Object-Oriented Software. In *IEEE International Conference on Software Maintenance (ICSM)*, pages 152–161, October 2002. 199

[25] J. Zhao, H. Yang, L. Xiang, and B. Xu. Change Impact Analysis to Support Architectural Evolution. *Software Maintenance and Evolution: Research and Practice*, 14(5):317–333, 2002. 199

Behavioral Analysis of Telecommunication Systems by Graph Transformations*

André Marburger and Bernhard Westfechtel

RWTH Aachen University, Department of Computer Science III
Ahornstrasse 55, 52074 Aachen, Germany
{marand,westfechtel}@cs.rwth-aachen.de
http://www-i3.informatik.rwth-aachen.de

Abstract. The E-CARES project addresses the reengineering of large and complex telecommunication systems. Within this project, graph-based reengineering tools are being developed which support not only the understanding of the static structure of the software system under study. In addition, they support the analysis and visualization of its dynamic behavior. The E-CARES prototype is based on a programmed graph rewriting system from which the underlying application logic is generated. Furthermore, it makes use of a configurable framework for building the user interface. In this paper, we report on our findings regarding feasibility, complexity, and suitability of developing tool support for the behavioral analysis of telecommunication systems by means of graph rewriting systems.

1 Introduction

The E-CARES[1] research cooperation between Ericsson Eurolab Deutschland GmbH (EED) and Department of Computer Science III, RWTH Aachen, has been established to improve the reengineering of complex legacy telecommunication systems. It aims at developing methods, concepts, and tools to support the processes of understanding and restructuring this special class of embedded systems. The subject of study is Ericsson's Mobile-service Switching Center (MSC) for GSM-networks called AXE10. The AXE10 software system comprises approximately 10 million lines of code spread over circa 1,000 executable units.

Maintenance of long-lived, large, and complex telecommunication systems is a challenging task. In the first place, maintenance requires understanding the actual system. While design documents are available at Ericsson which do support understanding, these are informal descriptions which cannot be guaranteed to reflect the actual state of implementation. Therefore, reverse engineering and reengineering tools are urgently needed which make the design of the actual system available on-line and which support planning and performing changes to the system.

---

* Supported by Ericsson Eurolab Deutschland GmbH
[1] **E**ricsson **C**ommunication **AR**chitecture for **E**mbedded **S**ystems [10]

J.L. Pfaltz, M. Nagl, and B. Böhlen (Eds.): AGTIVE 2003, LNCS 3062, pp. 202–219, 2004.
© Springer-Verlag Berlin Heidelberg 2004

According to the "horseshoe model of reengineering" [7], reengineering is divided into three phases. Reverse engineering is concerned with step-wise abstraction from the source code and system comprehension. In the restructuring phase, changes are performed on different levels of abstraction. Finally, forward engineering introduces new parts of the system (from the requirements down to the source code level).

In E-CARES, a prototypical reengineering tool is being developed which addresses the needs of the telecommunication experts at Ericsson. Currently, it assumes that the systems under study are written in PLEX, a proprietary programming language that is extensively used at Ericsson. However, we intend to support other programming languages — e.g., C — as well, so that the prototype may handle multi-language systems. So far, tool support covers only reverse engineering, i.e., the first phase of reengineering. While structural analysis is covered as well, we put strong emphasis on behavioral analysis since the structure alone is not very expressive in the case of a telecommunication system [12]. These systems are process-centered rather than data-centered like legacy business applications. Therefore, tool support focuses particularly on understanding the behavior by visualizing traces, constructing state diagrams, etc.

Internally, telecommunication systems are represented by various kinds of graphs. The structure of these graphs and the effects of graph operations are formally defined in PROGRES [16], a specification language which is based on programmed graph transformations. From the specification, code is generated which constitutes the core part of the application logic of the reverse engineering tool. In addition, the E-CARES prototype includes various parsers and scripts to process textual information, e.g., source code. At the user interface, E-CARES offers different kinds of textual and graphical views which are realized with UPGRADE [5], a framework for building graph-based applications.

The general applicability of graph transformation systems in the reengineering domain and the advantages of using the PROGRES generator mechanism for tool development have already been discussed in [11]. A more detailed comparison of the different tools that contribute to the reengineering framework showed that there are major differences in the role of graph transformations in the different analyses implemented so far. Furthermore, there are significant differences concerning the proportion between graph transformation code and additional code (e.g., in Java) that is needed to implement these analyses. These differences will be discussed on the basis of the three examples for behavioral analysis of telecommunication systems that are introduced in this paper.

2   Background

The mobile-service switching centers are the heart of a GSM network (Figure 1). An MSC provides the services a person can request by using a mobile phone, e.g., a simple phone call, a phone conference, or a data call, as well as additional infrastructure like authentication. Each MSC is supported by several Base Station Controllers (BSC), each of which controls a set of Base Station

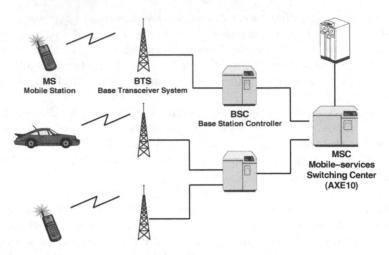

Fig. 1.   Simplified sketch of a GSM network

Transceivers (BTS). The interconnection of MSCs and the connection to other networks (e.g., public switched telecommunication networks) is provided by gateway MSCs (GMSC). In fact, the MSC is the most complex part of a GSM network. An MSC consists of a mixture of hardware (e.g., switching boards) and software units. In our research we focus on the software part of this embedded system.

Figure 2 illustrates how a mobile originating call is handled in the MSC. The figure displays logical rather than physical components according to the GSM standard; different logical components may be mapped onto the same physical component. The mobile originating MSC (MSC-MO) for the A side (1) passes an initial address message (IAM) to a GMSC which (2) sends a request for routing information to the home location register (HLR). The HLR looks up the mobile terminating MSC (MSC-MTE) and (3) sends a request for the roaming number. The MSC-MTE assigns a roaming number to be used for addressing during the call and stores it in its visitor location register (VLR, not shown). Then, it (4) passes the roaming number back to the HLR which (5) sends the requested routing information to the GMSC. After that, the GMSC (6) sends a call request to the MSC-MTE. The MSC (7) returns an address complete message (ACM) which (8) is forwarded to the MSC-MO. Now, user data may be transferred between A and B.

Ericsson's implementation of the MSC is called AXE10. Each MSC has a central processor which is connected to a set of regional processors for controlling various hardware devices by sensors and actors. The core of the processing is performed on the central processor. The AXE10 software is composed of blocks which constitute units of functionality and communicate by exchanging signals (see below). On each processor, a runtime system (called APZ) is installed which controls the execution of all blocks executing on this processor. An event raised

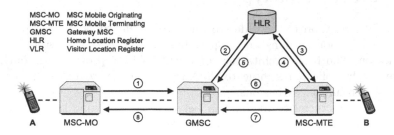

Fig. 2. Mobile originating call

by some hardware device is passed from the regional processor to the block handling this event on the central processor. In response to the event, an effect may be triggered on another hardware device.

The executable units of the AXE10 software system are implemented in Ericsson's in-house programming language PLEX (Programming Language for EX changes), which was developed in about 1970 and has been extended since then. PLEX is an asynchronous concurrent real-time language designed for programming of telecommunication systems. The programming language has a signaling paradigm as the top execution level. That is, only events can trigger code execution. Events are programmed as signals.

A PLEX program is composed of a set of blocks which are compiled independently. Each block consists of a number of sectors for data declarations, program statements, etc. Although PLEX does not support any further structuring within these sectors, we have identified some additional structuring through coding conventions in the program sector. At the beginning of the program sector, all signal reception statements (signal entries) of a block are coded. After these signal entry points, a number of labeled statement sequences follows. The bottom part of the program sector consists of subroutines.

The control flow inside a program sector is provided by goto and call statements. The goto statement is used to jump to a label of a labeled statement sequence. Subroutines are accessed by means of call statements. Both goto and call statements are parameter-less. That is, they affect only the control flow, but not the data flow.

Inter-block communication and data transport is provided by different kinds of signals. As every block has data encapsulation, signals are able to carry data. Therefore, signals may affect both the control flow and the data flow.

At runtime, every block can create several instances (processes). This again is not a feature of the PLEX programming language but achieved by means of implementation tricks and coding conventions. Therefore, these instances are managed by the block and not by the runtime environment.

3   E-CARES Prototype

In the E-CARES project, we design and implement tools for reengineering of telecommunication systems and apply them to the AXE10 system developed at Ericsson. The basic architecture of the E-CARES prototype is outlined in Figure 3. The solid parts indicate the current state of realization, the dashed parts refer to further extensions.

Below, it is crucial to distinguish between the following kinds of analyses: Structural analysis refers to the static system structure, while behavioral analysis is concerned with its dynamic behavior. Thus, the attributes "structural" and "behavioral" denote the outputs of analysis. In contrast, static analysis denotes any analysis which can be performed on the source code, while dynamic analysis requires information from program execution. Thus, "static" and "dynamic" refer to the inputs of analysis. In particular, behavior can be analyzed both statically and dynamically.

We obtained three sources of information for the static analysis of the structure of a PLEX system. The first one is the source code of the system. It is considered to be the core information as well as the most reliable one. Through code analysis (parsing) a number of structure documents are generated from the source code, one for each block. These structure documents form a kind of textual graph description. The second and the third source of information are miscellaneous documents (e.g., product hierarchy description) and the system documentation. As far as the information from these sources is computer processable, we use parsers and scripts to extract additional information, which is stored in structure documents, as well.

The static analysis tool processes the graph descriptions of individual blocks and creates corresponding subgraphs of the structure graph representing the overall application. The subgraphs are connected by performing global analyses in order to bind signal send statements to signal entry points. Moreover, the subgraphs for each block are reduced by performing simplifying graph transformations [10]. The static analysis tool also creates views of the system at different levels of abstraction. In addition to structure, static analysis is concerned with behavior (e.g., extraction of state machines or of potential link chains from the source code).

There are two possibilities to obtain dynamic information: using an emulator or querying a running AXE10. In both cases, the result is a list of events plus additional information in a temporal order. Such a list constitutes a trace which is fed into the dynamic analysis tool. Interleaved with trace simulation, dynamic analysis creates a graph of interconnected block instances that is connected to the static structure graph. This helps telecommunication experts to identify components of a system that take part in a certain traffic case. At the user interface, traces are visualized by collaboration and sequence diagrams.

The dashed parts of Figure 3 represent planned extensions of the current prototype. The re-design tool will be used to map structure graph elements to elements of a modeling language (e.g., ROOM [17] or SDL [4]). This will result in an architecture graph that can be used to perform architectural changes

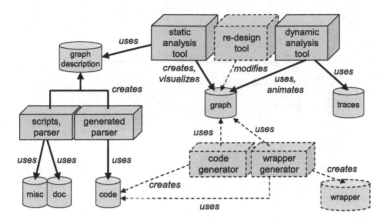

Fig. 3. Prototype architecture

to the AXE10 system. The code generator will generate PLEX code according to changes in the structure graph and/or the architecture graph. The wrapper generator will enable reuse of existing parts of the AXE10 system written in PLEX in a future switching system that is written in a different programming language, e.g., C++.

To reduce the effort of implementing the E-CARES prototype, we make extensive use of generators and reusable frameworks [11]. Scanners and parsers are generated with the help of JLex and jay, respectively. Graph algorithms are written in PROGRES [16], a specification language based on programmed graph transformations. From the specification, code is generated which constitutes the application logic of the E-CARES prototype. The user interface is implemented with the help of UPGRADE [5], a framework for building interactive tools for visual languages.

4 Structural Analysis

The static structure of a PLEX program is represented internally by a structure graph, a small (and simplified) example of which is shown in Figure 4[2]. This graph conforms with the graph scheme in Figure 5 which will be explained in Section 5. In the example, there is a subsystem which contains two blocks A and B. The subgraphs for these blocks are created by the PLEX parser. The subgraph for a block – the block structure graph – contains nodes for signal entry points, labels, (contiguous) statement sequences, subroutines, exit statements, etc. Thus, the block structure graph shows which signals may be processed by the block, which statement sequences are executed to process these signals, which subroutines are used for processing, etc. In addition, the block structure graph

---

[2] For the time being, please ignore all graph elements for link chains (lc), which will be explained in Section 5.

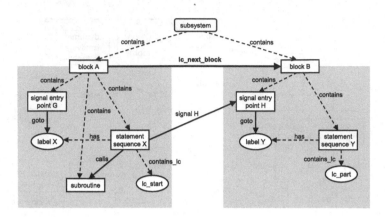

Fig. 4. Cut-out of a structure graph

initially contains nodes representing outgoing signals. Subsequently, a global analysis is carried out to bind outgoing signals to receiving blocks based on name identity. In our example, a signal H is sent in the statement sequence X of block A. This signal is bound to the entry point of block B. From the signal edges between statement sequences and signal entry points, more coarse-grained communication edges may be derived (between blocks and eventually between subsystems).

Externally (at the user interface), the structure graph is represented by multiple views [10]. The product hierarchy is displayed in a tree view. Furthermore, there is a variety of graphical views which display the structure graph at different levels of abstraction (internals of a block, block communication within a subsystem, communication between subsystems). The user may select among a set of different, customizable layout algorithms to arrange graphical representations in a meaningful way. He may also collapse and expand sets of nodes to adjust the level of detail. Graph elements representing code fragments are connected to the respective source code regions, which may be displayed on demand in text views.

5   Behavioral Analysis

The behavioral analysis of a software system either uses and extends the information gathered during structural analysis or processes graphs or graph like descriptions of the runtime systems structure, control flow, and data flow. The former is referred to as static analysis, the latter is called dynamic analysis. During the development of the different facilities for the behavioral analysis we found, that there are significant differences in the way graph transformations and the PROGRES system contribute to these facilities. In the sequel, we will discuss three example methods for behavioral analysis which show different complexity and different characteristics concerning the proportion of the graph transformations part and "externally" defined functionality.

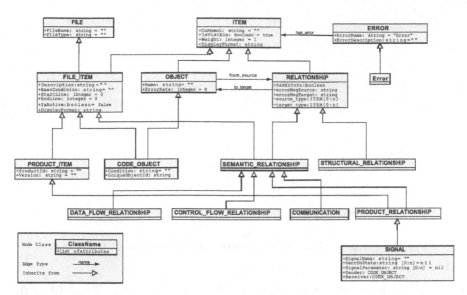

Fig. 5.   Graph scheme

5.1   Graph Scheme

In Figure 5, the basic graph scheme we set up for our reengineering tools is described in a notation similar to UML class diagrams. The node classes FILE, ITEM, and ERROR build the basis of the graph scheme. The node class ERROR has been introduced to be able to annotate errors with respect to structure or semantics of the graph or graph elements[3]. Nodes of type ERROR are connected to nodes of type ITEM via a **has_error** edge.

The two classes OBJECT and RELATIONSHIP, both derived from ITEM, divide the majority of nodes that occur in a structure graph into two groups. All node types derived from OBJECT are used to abstract from different kinds of fragments in the source code of an analyzed system or they describe runtime entities, respectively. Node types derived from RELATIONSHIP serve to describe relationships between code fragments/runtime entities. Starting from the source object, a relationship is represented by the concatenation of a **from_source** edge, a node of the corresponding subclass of RELATIONSHIP, and an edge of type **to_target**. This edge-node-edge construct allows to simulate attributed edges which we are lacking. In the remainder, we sometimes use the term edge if we refer to such a construct.

The node classes OBJECT and RELATIONSHIP are further refined to be able to differ between different kinds of code fragments (node class CODE_OBJECT), runtime entities (node class TRACE_OBJECT), and relationships. For example,

---

[3] Sometimes it is necessary to allow intermediate inconsistencies in a graph, e.g. during manual changes to the graph. In other cases, the correctness of a structure graph depends on a user's focus and interest.

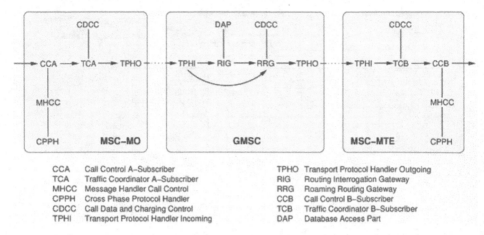

CCA	Call Control A–Subscriber		TPHO	Transport Protocol Handler Outgoing
TCA	Traffic Coordinator A–Subscriber		RIG	Routing Interrogation Gateway
MHCC	Message Handler Call Control		RRG	Roaming Routing Gateway
CPPH	Cross Phase Protocol Handler		CCB	Call Control B–Subscriber
CDCC	Call Data and Charging Control		TCB	Traffic Coordinator B–Subscriber
TPHI	Transport Protocol Handler Incoming		DAP	Database Access Part

Fig. 6. Simplified link chain for mobile originating call at GSM-layer 3

the structure graph we use combines structural information (STRUCTURAL_RELA-TIONSHIP), control flow information (CONTROL_FLOW_RELATIONSHIP), and data flow information (DATA_FLOW_RELATIONSHIP) in a single graph. This allows complex graph transformations and graph queries that utilize all three kinds of information at once. Some examples of these graph transformations are described in the following section.

5.2   Example 1: Static Link Chain Analysis

As stated in Section 1, we found that the static system structure is not very expressive in the case of telecommunication systems. These highly dynamic, flexible, and reactive systems handle thousands of different calls at the same time. The numerous services provided by a telecommunication system are realized by re-combining and re-connecting sets of small (stand alone) processes, block instances in our case, at runtime. Each of these block instances realizes a certain kind of (internal) mini-service. Some of the blocks can even create instances for different mini-services dependent on the role they have to play in a certain scenario.

Therefore, structural analysis as described in Section 4 is not sufficient to understand telecommunication systems. For example, the structure graph does not contain any information on how many instances of a single block are used to set up a simple phone call. In Figure 6, a so-called link chain for the GSM-layer 3 part of a simple mobile originating call is sketched. Link chains describe how block instances are combined at runtime to realize a certain service. Each node represents a block instance. An edge between two nodes indicates signal interchange between these blocks. Each link chain consists of a main part (directed edges) and side-links for supplementary services (authentication, charging, etc.). The directed edges between elements of the main link chain indicate its establishment from left to right; communication is bidirectional in all cases. In

st

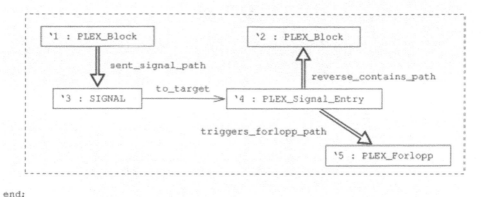

end;

Fig. 7.   Static path expression to establish link chains

correspondence to Figure 2, the link chain in Figure 6 is divided into the three parts MSC-MO, GMSC, and MSC-MTE.

This simple example shows that there are, e.g., three instances of the charging access (CDCC), two message handler instances (MHCC), and two instances of the CPPH protocol handler block needed to setup a mobile originating call. This kind of behavioral information is very important to telecommunication engineers. Therefore, we will now show how information on link chains can be derived via static analysis using a simple graph transformation, a (static) path expression.

Figure 7 shows this static path expression. Paths are simple graph queries that only consist of a left hand side. In case of static paths, there is a side effect that materializes all matches of the left hand side in the form of edges between the source node and the target node of a path. Materialized paths can be visualized in our graph visualization tool. Therefore, we use static paths, e.g., in the analysis part of the specification to gather additional information on a system that is not obtainable via code parsing.

The path expression in Figure 7 is used to detect blocks in the analyzed system that are inter-connected in a so-called forlopp link chain at runtime. Normally, the information on forlopp link chains is only available at runtime. But, using this path expression we are able to predict which blocks of the AXE10 software system are able to take part in a certain link chain. The path corresponding expression is defined to start and end at a node of type Block. Furthermore, it is defined that node 1 in the left hand side is the starting node of the path. Accordingly, node 2 is the target node of the path. Blocks 1 and 2 take part in the same link chain, if there is a path from block 1 via a signal to a forlopp action in block 2.

Though the task of extracting link chains from the system's source code seemed to be very difficult in the beginning, this simple static path expression

```
transformation Extract_StateMachine ( block : PLEX_Block)
   [0:1] =

   use readStateVariables, writtenStateVariables : DATA_ELEMENT [0:n];
       stateVariables : DATA_ELEMENT [0:n];
       exportInterface : PLEX_Signal_Entry [0:n];
       diagramRoot : ROOM_State_Diagram;
       fromState : string;
       toStates : string [0:n]
   do
     Determine_State_Variables ( block, out stateVariables )
     &
     choose
       when not empts ( stateVariables )
       then
            exportInterface := (block.=contains_path=>) : PLEX_Signal_Entry
          & New_State_Diagram ( block, block.Name, out diagramRoot )
          & for all signalEntry := elem ( exportInterface )
            do
               readStateVariables := (signalEntry.=reads_path=>)
             & readStateVariables := (readStateVariables and stateVariables)
             & for all stateVariable := elem ( readStateVariables )
               do
                  IdentifyFromState ( signalEntry, stateVariable, out fromState )
                & IdentifyToStates ( signalEntry, out toStates )
                & InsertTransitions ( diagramRoot, fromState, toStates )
               end
            end
     end
   end
end;
```

Fig. 8.   Transaction to extract state machines

is all we need to automatically perform a static link chain analysis on the structure graph. No additional functionality had to be defined and coded outside the PROGRES specification. Furthermore, no user interaction has to take place to start or end the analysis. Even if the structure graph is consecutively extended with new block subgraphs, these will be analyzed automatically as well. Therefore, this static path expression is a good example how graph transformations can help to reduce complex issues significantly.

5.3   Example 2: Static State Machine Extraction

Each block responsible for the instances in Figure 6 implements at least one state machine. As a result, each of the block instances has its own state machine. State machines are a common modeling means in the design of telecommunication systems. Therefore, telecommunication experts are interested in having a good knowledge about the state machines implemented in a block and their operation at runtime. Therefore, extraction of state machines from source code is another hot topic in behavioral analysis.

The basic ideas for the corresponding extraction algorithm have been defined in collaboration with telecommunication experts from Ericsson. The algorithm is based on special coding patterns within the source code. We have then transferred a first informal description of this algorithm into the graph transformation domain.

Figure 8 shows the main transaction for the extraction of a single block's state machines from the structure graph. Consequently, the corresponding block is supplied as input to this transaction. The transaction Extract_State_Machine is divided into three parts. First, the set of state variables associated with the given block is determined by calling the transaction Determine_State_Variables. The existence of at least one variable that meets the criteria for state variables is a necessary precondition to state machine extraction. If no state variable is found, that is, the set stateVariables is empty, then the current block does not implement a state machine.

If there is at least one state variable detected, the second part of the extraction transaction determines the export interface of the block, that is, all signal entries that are reachable from outside the block. Each of the signal entries contributing to the export interface represents a "service" offered to "clients" of a block. Next, a new state diagram is introduced by adding a corresponding root node to the structure graph, and connecting this node to current block node. This is achieved by calling the graph transformation New_State_Diagram.

In the third part, each of the signal entries in the block's export interface is processed separately to extract possible state transitions one by one. According to a design rule at Ericsson, every signal execution should at most trigger a single state change. Therefore, simply speaking, each state change within the statement sequences reachable from a signal entry point can be considered to represent a transition from the starting state in the signal entry to the state specified by the state change.

Consequently, the further procedure is as follows: The set of state variables read within the statement sequence represented by a node of type Signal_Entry is identified first. For each of these variables the expected starting state is determined by calling the transaction Identify_From_State. Next, the state changes of all control flow paths triggered by this starting state are returned by the transaction Identify_To_States. Finally, the transaction Insert_Transitions adds new state transitions into the state diagram – one for each pair of the state in fromState and one element of the set toStates.

Though the initial problem seemed to be only slightly more complex (on paper) as the link chain analysis discussed in Section 5.2, the resulting specification is much more complex. Instead of a single path definition, several transactions are necessary to implement the state machine extraction algorithm. Furthermore, the extraction has to be initiated by the user of the dynamic analysis tool (which is intentionally). Inspected in more detail, the specification of the state machine extraction algorithm cannot take too much advantage of special features of the PROGRES machinery like automatic path calculation or backtracking. Instead, the extraction algorithm utilizes normal control flow elements known from traditional programming languages. This is a matter of the characteristic of the extraction problem. Here, we are not interested in inserting

```
transaction + Trace_Step_Block( traceId : string ; senderBlockName : string ;
                    senderInstanceId : string ; signalName : string ;
                    executionNumber : integer ; receiverBlockName : string ;
                    receiverInstanceId : string )
   [0:1] =
   use
      senderBlock, receiverBlock : Block
      (* ... and other local declarations *)
   do
      (* ----------------------- changes to structure graph ----------------------- *)
      Add_or_Get_Node ( Block, senderBlockName & "UPROGRAM", out senderBlock )
      & Add_or_Get_Node ( Block, receiverBlockName & "UPROGRAM", out receiverBlock )
      & Add_or_Extend_Block_Communication ( senderBlock, receiverBlock, signalName,
                                 out communicationEdge )
      & Activate ( senderBlock, receiverBlock, communicationEdge )
      (* ----------------------- changes to trace graph ----------------------- *)
      & Suspend_Working_Instance ( traceId, signalName )
      & Add_or_Get_Block_Instance ( traceId, senderBlock, senderInstanceId,
                                 out senderInstance )
      & Switch_ExecStatus_Of_Instance ( senderInstance, "WaitingOrSuspended" )
      & Add_or_Get_Block_Instance ( traceId, receiverBlock, receiverInstanceId,
                                 out receiverInstance )
      & Switch_ExecStatus_Of_Instance ( receiverInstance, "Working" )
      & Insert_Block_Instance_Communication ( traceId, senderInstance, receiverInstance,
                                 signalName, out instanceCommunicationEdge )
      & executionId := ("[" & string ( executionNumber ) & "] " & signalName)
      & instanceCommunicationEdge.SignalExecutionOrder :=
                  (instanceCommunicationEdge.SignalExecutionOrder or executionId )
      (* ----------------------- changes to structure grap ----------------------- *)
      & currentSubsystem := senderBlock.=contained_in_subsystem_path=>
      & Propagate_Activation ( currentSubsystem )
      & receiverSubsystem := receiverBlock.=contained_in_subsystem_path=>
      & Propagate_Activation ( receiverSubsystem )
      & Propagate_Activation ( communicationEdge )
   end
end;
```

Fig. 9. Transaction to support animation of traces

a single edge but in creating a state machine consisting of named nodes and labeled edges. This cannot be achieved by using a static path expression. Instead, operations for finding state nodes, inserting state nodes and transitions, and the like are needed to perform the task. In addition, a number of non-graph-based operations like attribute value analysis at nodes in the structure graph or set operations are involved. As a result, the mixture of different operations — graph-based and non-graph-based — that contribute to the final goal and that have to be executed in a certain controlled order prevent from taking too much advantage of the PROGRES machinery.

5.4   Example 3: Dynamic Trace Analysis

The last example introduces another important issue in the behavioral analysis of telecommunication systems — trace analysis. As already described in Section 3, traces are lists of runtime events in temporal order. Traces are dynamic information, that is, this information cannot be obtained from the source code. Instead, a runtime inspection of the analyzed software system is necessary.

For trace analysis, a trace file is loaded into the trace simulator. This trace file is then analyzed stepwise — event by event. During the processing of the trace file, an instance graph is created that constitutes a new subgraph of the structure graph. This instance graph is used to present collaboration and sequence diagram like views to the user of the trace simulator tool.

In Figure 9, a transaction is presented that illustrates the connection of a trace to the structure graph as described above. It performs a single step of a trace file. That is, the transaction is used to transform a single signal action in a trace file into corresponding graph transformations. A signal action references a sending block instance, a receiving block instance, and a signal. Furthermore, each signal action is part of a specific trace. This information is again supplied as parameters of the transaction.

The transaction `Trace_Step_Block` is divided into three parts. The first part checks whether the counterparts of the two block instances referenced by a signal action are already elements of the current structure graph. If not, the transaction `Add_or_Get_Node` inserts appropriate block nodes into a special part of the structure graph and annotates them as formerly missing. This enables a user on the one hand to use the trace animation tool without having an appropriate structure graph. On the other hand, the separation of formerly missing parts in a special part of the structure graph gives quick access to this kind of inconsistencies and allows controlled corrections.

The second part of the transaction first switches the execution status of the block instance currently working to **suspended**. Next the node that corresponds to sending block instance is either added to or fetched from the current trace graph. The execution status of this block instance is either set to **waiting** or it is set to **suspended**. The value depends on the kind of signal action processed (combined signal or single signal). The corresponding information is obtained by querying the structure graph. This procedure is repeated for the receiving block instance node. But, its execution status is set to **working**. After the block instance communication edge between the sending and the receiving block instance has been inserted, an execution identifier for the current signal is calculated. This identifier is added to the attribute `SignalExecutionOrder` of the communication edge. Finally, in the third part of the transaction, the information about changes of the execution status of parts of the structure graph and trace graph is propagated from the block level to the more abstract levels (subsystems etc.).

Considering the trace simulation tool as a whole, this transaction represents just a small part of its implementation. The main control part does not even reside within the graph specification. Instead, the contribution of the graph transformation machinery is limited to "simple" database-like activities. The core of the trace processing and analysis is performed within hand-written Java code that is integrated as a module into the UPGRADE framework. One reason for this way of implementing the trace analysis is the difficulty in embedding non-graph-transformational functionality like parsers into PROGRES specifications. But, more striking is the necessity to create very complex temporary data structures while processing a trace file.

In general, it is possible to import additional functionality into PROGRES specifications through external functions. Furthermore, complex data structures (e.g., records) can be represented through equivalent graphs. But, it is more suitable to implement the corresponding functionality, which does not profit from the strengths of a graph machinery, in C, C++, or Java programs (to name just some programming languages). In the case of UPGRADE and PROGRES, the Java-part can access and manipulate the graph via a special interface, but not vice versa. All transformations specified in the specification can be accessed. That is, the PROGRES graph machinery is the passive part while the UPGRADE framework provides the active part.

Summarizing the findings of the three examples, there is a trade-off in the proportion of graph specification regarding the implementation of the different behavioral analysis algorithms. The more the solution of the initial problem is suited for the graph transformations domain, the higher is the share solved within the specification. But, the more the solution requires direct control, complex data structures, or an increasing amount of user interaction, the more of the realization will take place outside the graph transformation system. But, this is not a disadvantage in general. An appropriate combination of both worlds — graph transformation and traditionally programmed systems — resulted in flexible, manifold, and efficient solutions to our behavioral analysis problems.

6   Related Work

In contrast E-CARES, much work has been performed in business applications written in COLBOL. The corresponding tools such as e.g. Rigi [14] or GUPRO [9] primarily focus on the static system structure. Moreover, they are typically data-centered; consider e.g. [13, 18]. Here, recovery of units of data abstraction and migration to an object-oriented software architecture play a crucial role [1]. More recently, reengineering has also been studied for object-oriented programming languages such as C++ and Java. E.g., TogetherJ or Fujaba [19] generate class diagrams from source code.

Reengineering of telecommunication systems follows different goals. Telecommunication systems are designed in terms of layers, planes, services, protocols, etc. Behavior is described with the help of state machines, message sequence charts, link chains, etc. Thus, reengineering tools are required to provide views on the system which closely correspond to system descriptions given in standards, e.g., GSM. Telecommunication experts require views on the system which match their conceptual abstractions.

Graphs and graph rewriting systems play an essential role in E-CARES. In the following, we examine reengineering tools from a tool builder's perspective. We compare E-CARES to a set of other graph-based reengineering tools.

Rigi [14] is an interactive toolkit with a graphical workbench which can be used for reengineering. Internally, a software system is represented by a set of entities and relationships. Externally, program understanding is supported by graph visualization techniques. Rigi is also used in the Bauhaus project [8], whose aim is to develop methods and techniques for (semi-)automatic architecture recovery, and to explore languages to describe recovered architectures. In contrast to E-CARES, Rigi (and thus Bauhaus) is not based on a high-level specification language. Rather, graphs are accessed through a procedural interface, which makes coding of graph algorithms more painstaking.

The GUPRO project [9] is concerned with the development of a generic environment for program understanding. Internally, programs are represented as graphs. GUPRO offers parsers for several languages, including COBOL and C. Different kinds of analyzes may be specified with the help of a graph query language, but graph transformations cannot be specified in a declarative way. Moreover, GUPRO offers a textual user interface, while E-CARES provides graphical tools.

VARLET [6] addresses the problem of database reengineering. In VARLET, a relational schema is transformed into an object-oriented one; triple graph grammars [15] provide the underlying theoretical foundation. VARLET tools have been specified and implemented with the help of PROGRES. E-CARES addresses a different application domain. Up to now, we only address reverse engineering, i.e., the system under study is not modified. For the re-design of telecommunication systems, we intend to use triple graph grammars, as well, as we did in a previous project [2].

FUJABA [19] is a CASE tool for UML which supports round-trip engineering from UML to Java and back again. While reverse engineering of class diagrams is well understood and implemented in a couple of commercial tools (e.g., Together and Rational Rose), Fujaba also recovers collaboration diagrams from Java source code. In Fujaba, collaboration diagrams are not used as examples of computations; rather, they are executable and may be considered as generalizations of graph transformations. The E-CARES prototype is based on a different specification language (PROGRES), addresses PLEX rather than Java programs, and deals with the application domain of telecommunication systems, which has not been considered in Fujaba so far.

7   Conclusion

We have presented the E-CARES prototype for reengineering of telecommunication systems, which is based on a programmed graph rewriting system acting as an operational specification from which code is generated. In this paper, we have provided insight into the behavioral analysis part of this specification. Reengineering is an application domain which is well-suited for graph rewriting. But, only the combination of graph transformation and traditional programming resulted in an appropriate analysis tool. There are other reengineering tools which are also based on graphs, but most of them lack the support provided by a high-level specification language and force the tool developer to perform low-level programming.

So far, the implementation supports only reverse engineering, i.e., it aids in system understanding, but not yet in system restructuring. To date, approximately 2.5 million lines of PLEX code have successfully been parsed and analyzed by means of the E-CARES prototype. Current and future work addresses (among others) the following topics: reverse engineering of state diagrams (testing and improvement), application of metrics, multi-language support (e.g., C or SDL in addition to PLEX), improvement and extension of the ROOM architectural description language support, re-design, and source code transformation.

References

[1] G. Canfora, A. Cimitile, A. De Lucia, and G. A. Di Lucca. Decomposing legacy systems into objects: An eclectic approach. *Information and Software Technology*, 43:401–412, 2001. 216

[2] Katja Cremer, André Marburger, and Bernhard Westfechtel. Graph-based tools for re-engineering. *Journal of Software Maintenance and Evolution: Research and Practice*, 14(4):257–292, 2002. 217

[3] Hartmut Ehrig, Gregor Engels, Hans-Jörg Kreowski, and Grzegorz Rozenberg, editors. *Handbook on Graph Grammars and Computing by Graph Transformation: Applications, Languages, and Tools*, volume 2. World Scientific, 1999. 218, 219

[4] Jan Ellsberger, Dieter Hogrefe, and Amardeo Sarma. *SDL - Formal Object-oriented Language for Communicating Systems*. Prentice Hall, 1997. 206

[5] Dirk Jäger. Generating tools from graph-based specifications. *Information Software and Technology*, 42(2):129–140, 2000. 203, 207

[6] Jens Jahnke and Albert Zündorf. Applying graph transformations to database re-engineering. In Ehrig et al. [3], pages 267–286. 217

[7] Rick Kazman, Steven G. Woods, and Jeromy Carrière. Requirements for integrating software architecture and reengineering models: CORUM II. In *Working Conference on Reverse Engineering*, pages 154–163. IEEE Computer Society Press, 1998. 203

[8] Rainer Koschke. *Atomic Architectural Component Recovery for Program Understanding and Evolution*. PhD thesis, Institute of Computer Science, University of Stuttgart, 2000. 216

[9] Bernt Kullbach, Andreas Winter, Peter Dahm, and Jürgen Ebert. Program comprehension in multi-language systems. In *Proc. 4th Working Conference on Reverse Engineering*. IEEE Computer Society Press, 1998. 216, 217

[10] André Marburger and Dominikus Herzberg. E-CARES research project: Understanding complex legacy telecommunication systems. In *Proc. 5th European Conference on Software Maintenance and Reengineering*, pages 139–147. IEEE Computer Society Press, 2001. 202, 206, 208

[11] André Marburger and Bernhard Westfechtel. Graph-based reengineering of telecommunication systems. In *Proceedings international conference on graph transformations ICGT 2002*, LNCS 2505, pages 270–285. Springer, 2002. 203, 207

[12] André Marburger and Bernhard Westfechtel. Tools for understanding the behavior of telecommunication systems. In *Proceedings 25th International Conference on Software Engineering ICSE 2003*, pages 430–441. IEEE Computer Society Press, 2003. 203

[13] Lawrence Markosian, Philip Newcomb, Russell Brand, Scott Burson, and Ted Kitzmiller. Using an enabling technology to reengineer legacy systems. *Communications of the ACM*, 37(5):58–70, 1994. 216

[14] Hausi A. Müller, Kelly Wong, and Scott R. Tilley. Understanding software systems using reverse engineering technology. In *The 62nd Congress of L'Association Canadienne Francaise pour l'Avancement des Scienes (ACFAS)*, 1994. 216

[15] Andy Schürr. Specification of graph translators with triple graph grammars. In *Proceedings WG '94 Workshop on Graph-Theoretic Concepts in Computer Science*, LNCS 903, pages 151–163. Springer, 1994. 217

[16] Andy Schürr, Andreas Winter, and Albert Zündorf. The PROGRES approach: Language and environment. In Ehrig et al. [3], pages 487–550. 203, 207

[17] Bran Selic, Garth Gullekson, and Paul T. Ward. *Real-Time Object-Oriented Modeling*. Wiley & Sons, 1994. 206

[18] Henk J. van Zuylen, editor. *The REDO Compendium: Reverse Engineering for Software Maintenance*. John Wiley & Sons: Chichester, UK, 1993. 216

[19] Albert Zündorf. *Rigorous Object-Oriented Development*. PhD thesis, University of Paderborn, 2002. Habilitation thesis. 216, 217

Specifying Integrated Refactoring
with Distributed Graph Transformations*

Paolo Bottoni[1], Francesco Parisi Presicce[1,2], and Gabriele Taentzer[3]

[1] University of Rome "La Sapienza"
[2] George Mason University
[3] Technical University of Berlin

Abstract. With refactoring, the internal structure of a software system changes to support subsequent reuse and maintenance, while preserving the system behavior. To maintain consistency between the code (represented as a flow graph) and the model (given by several UML diagrams of different kinds), we propose a framework based on distributed graphs. Each refactoring is specified as a set of distributed graph transformations, structured and organized into transformation units. This formalism could be used as the basis for important extensions to current refactoring tools.

1   Introduction

Refactoring is the process of changing a software system so as to preserve its observable behavior while improving its readability, reusability, and flexibility. Refactoring has its origins in some circles of the Smalltalk community, though its principles can be traced back to the idea of writing subprograms to avoid repetitious code fragments. It is now a central practice of extreme programming [2], and can be carried out in a systematic way, as witnessed for instance in [4]. Although refactoring techniques can be applied in the context of any programming paradigm, refactoring is particularly effective with object-oriented languages and is usefully combined with the notion of design pattern.

The input/output view of the behavior of the system is not intended to change with refactoring. However, the changes can have several consequences for the computing process, as expressed for instance by the sequence of method calls, or by state changes of an object or an activity. Since refactoring is usually performed at the source code level, it becomes difficult to maintain consistency between the code and its model, expressed for example with UML diagrams, usually fitting to the original version of the code.

Several tools have been developed to assist refactoring: some are packaged as stand-alone executables, while others have integrated refactorings into a development environment. Many tools refer directly and exclusively to specific languages: **C# Refactory** (http://www.xtreme-simplicity.net/) supports 12 refactorings on pieces of C# code, including extraction of methods, superclasses and interfaces, and renaming of type, member, parameter or local value; **CoreGuide6.0** (http://www.omnicore.com) is based on

---

* Partially supported by the EC under Research and Training Network SeGraVis.

J.L. Pfaltz, M. Nagl, and B. Böhlen (Eds.): AGTIVE 2003, LNCS 3062, pp. 220–235, 2004.

Java and provides refactorings such as extract method and move/rename class/package. **Xrefactory** (http://www.xref-tech.com) manages refactoring in C and Java, including pushing down and pulling up of fields and methods, and insertion/deletion/shift/exchange of parameters, in addition to the usual method extraction and various renamings. None of these tools mentions diagrams or the effect that these refactorings have on other views of the system, including documentation.

To preserve consistency, one can either recover the specification after a chosen set of code changes, as in Fujaba [11], or define the effects of a refactoring on different model parts. The latter option is quite easily realized on structural models, where diagram transformations are notationally equivalent to lexical transformations on the source code, less so on behavioral specifications.

objectiF (http://www.microtool.de/objectiF) supports a wide variety of languages, and allows transformations both in the code and in the class model, expressed via class diagrams. Changes are propagated automatically to both views, but no mention is made of behavioral diagrams. System-wide changes of code are integrated in **Eclipse** (http://www.eclipse.org) with several refactoring actions (such as rename, move, push down, pull up, extract) into the Java Development Tools that automatically manages refactoring. Class diagrams are implicitly refactored. Finally, **JRefactory** supports 15 refactorings including pushing up/down methods/fields and extract method/interface (http://jrefactory.sourceforge.net). The only diagrams mentioned are class diagrams, which are reverse engineered from the .java files.

Class diagram editors do not extend changes to all related diagrams, limiting their "automation" to source code. Hence, direct intervention is needed to restore consistency among the different UML diagrams representing the same subsystem.

We discuss an approach to maintaining consistency between source code and both structural and behavioral diagrams, using the formal framework of graph transformations. In particular, the abstract syntax of UML diagrams is represented through graphs, and an abstract representation of the source code is expressed through suitable attributed graph structures. The UML diagrams and the code are seen as different views on a software system, so that the preservation of consistency between the views is accomplished by specifying refactoring as a graph transformation distributed on several graphs at once. Complex refactorings, as well as the checking of complex preconditions, are decomposed into collections of distributed transformations, whose application is managed by control expressions in appropriate transformation units. It is then possible to describe 'transactional' refactorings [17] as combinations of primitive transformations that, individually, may not be behavior-preserving, but that, in appropriate compositions under appropriate control, may globally preserve behavior.

Paper Outline. In Section 2, we review some approaches to refactoring and to software evolution using graph rewriting, and in Section 3 we present a motivating example. Background notions on distributed graph transformation are given in Section 4. In Section 5, we first reformulate the problem of maintaining consistency among different forms of specification and code as the specification of suitable distributed graph transformations, and then we illustrate our approach by means of two refactorings. Conclusions are given in Section 6.

2   Related Work

Since Opdyke's seminal thesis [13], where preconditions for behavior preservation are analysed, formal methods have been applied towards a clear definition of the conditions under which refactoring takes place. More recently, the approach has been complemented in Robert's thesis, where the effect of refactoring is formalized in terms of postconditions [14], so that it is possible to treat and verify composite refactorings, in which the preconditions for a primitive refactoring are guaranteed by the postconditions for a previous one. In general, it is very difficult to provide proofs of the fact that refactorings preserve the behavior of a program, due mostly to the lack of a formal semantics of the target language. The approach in [13] is based on proofs that the enabling conditions for each refactoring preserve certain invariants, without proving that the preservation of the invariants implies the preservation of the behavior.

Fanta and Rajlich proposed a collection of algorithms which implement a set of refactorings on C++ code, exploiting the abstract semantic tree for the code provided by the GEN++ tool [3]. These algorithms directly access the tree, but are not formalized in terms of tree modification.

Recent work in the UML community has dealt with the effects of refactoring on UML diagrams, mainly class or state diagrams [15]. As these works consider model refactoring prior to code generation, they do not take into account the possible associated modifications of code.

Mens [9] expresses code transformations in terms of graph transformations, by mapping diagrams onto type graphs. Such type graphs are comparable to the abstract syntax description of UML models via the UML metamodel. Mens et al. exploit several techniques also used here, such as control expressions, negative conditions, and parameterized rules [10]. As these studies focus on the effect of refactorings on the source code, they do not investigate the coordination of a change in different model views with that in the code.

Since some refactorings require modifications in several diagrams, we propose a constrained way of rewriting different graphs, as a model for a full formalization of the refactoring process, derived from distributed graph transformations. This is based on a hierarchical view of distributed systems, where high-level "network" graphs define the overall architecture of a distributed system, while low-level "specification" ones refer to the specific implementation of local systems [16]. Such an approach has also been applied to the specification of ViewPoints, a framework to describe complex systems where different views and plans must be coordinated [5]. In the ViewPoint approach, inconsistencies between different views can be tolerated [6], while in the approach proposed here different graphs have to be modified in a coordinated way so that the overall consistency of the specification is always maintained.

3   An Example

We introduce an example which represents in a synthetic way a situation often encountered when refactoring, i.e. classes have a similar structure, or follow common patterns of collaboration. In particular, we illustrate the problem of exposing a common client-server structure in classes otherwise unrelated.

Consider a set of client classes, indicated as Client⟨X⟩, and a set of server classes, indicated as Server⟨X⟩, providing services to the clients (the names are arbitrary and have been chosen only to facilitate reading).

```
class Client⟨X⟩ {
    protected Server⟨Y⟩ serv⟨Y⟩;
    protected Server⟨Y⟩ getServer⟨Y⟩() { // dynamic server lookup }
    protected void exploitService⟨X⟩() {
        serv⟨Y⟩ = getServer⟨Y⟩(); Object result = serv⟨Y⟩.service⟨Y⟩(this);
        // code using result
    }
}

class Server⟨Y⟩ {
    public Object service⟨Y⟩(Object requester) { // service implementation }
}
```

Several types of primitive refactorings can be combined to extract the client-server structure[1]. First of all, we rename all the variables serv⟨Y⟩ as serv using the rename_inst_variable refactoring. Similarly we rename all the methods of type getServer⟨Y⟩, exploitService⟨X⟩, and service⟨Y⟩ to their common prefixes using the rename_method refactoring. extract_code_as_method is then applied to all client classes to extract different versions of the method exploit, containing the specific code which uses the result provided by the call to service. We then add an abstract class, through the add_class refactoring, to which we pull up the instance variables serv and the now identical methods exploitService. The client classes are subclasses of this abstract class. We proceed in a similar manner to add an interface Server declaring a method service, of which the server classes are implementations (we can use interfaces here as we do not have structure or code to pull up).

The result of this sequence of refactoring is described by a new collection of classes, where the parts in bold indicate the modifications.

```
abstract class AbstractClient {
    Server serv;
    abstract Server getServer();
    void exploitService() {
        serv = getServer(); Object result = serv.service(this);
        exploit(result);
```

---

[1] We use the names of refactorings in the *Refactoring Browser* [14].

```
    }
    abstract void exploit(Object arg);
}

class Client⟨X⟩ extends AbstractClient {
    Server getServer() { //code from getServer⟨Y⟩ }
    void exploit(Object arg) { // previous code using result }
}

interface Server { Object service(Object requester); }

class Server⟨Y⟩ implements Server {
    public Object service(Object requester) {return service⟨Y⟩(requester);}
    Object service⟨Y⟩(Object requester) { // service implementation }
}
```

Each individual refactoring implies changes in class diagrams to modify the class structures and inheritance hierarchies. However, some refactorings also affect other forms of specification. For example, `extract_code_as_method` affects sequence diagrams, as it introduces a self call during the execution of `exploitService()`. State machine diagrams may be affected as well, if states such as `WaitingForService`, `ResultObtained` and `ResultExploited` are used to describe the process realized by `exploitService`. To reflect process unfolding, the latter state can be decomposed into `ExploitStarted` and `ExploitCompleted`.

4   The Formal Background

Distributed rule application follows the double-pushout approach to graph transformation as described in [16], using rules with negative application conditions. For further control on distributed transformations, transformation units [8] on distributed graph transformation are used.

4.1   Distributed Graph Transformation

We work with distributed graphs containing typed and attributed nodes and edges. Edge and node types for a given family of graphs $\mathcal{F}$ are defined in a type graph $T(\mathcal{F})$ and the typing of a graph $G \in \mathcal{F}$ consists of a set of injective mappings from edges and nodes in G to edges and nodes in $T(\mathcal{F})$. Distributed graph transformations are graph transformations structured at two abstraction levels: the network and the object level. The network level describes the system's architecture by a network graph, and allows its dynamic reconfiguration through network rules. At the object level, graph transformations are used to manipulate local object structures. To describe a synchronized activity on distributed object structures, a combination of graph transformations on both levels is needed. A distributed graph consists of a network graph where each network node is refined by a local object graph. Network edges are refined by graph morphisms on local object graphs, which describe how the object graphs are interconnected. A distributed graph morphism m is defined by a network morphism n – which is a normal graph morphism – together with a set S of local object morphisms,which are graph morphisms on local object graphs. Each node mapping in n is refined by a graph morphism of S on the corresponding local graphs. Each mapping of network edges guarantees a compatibility between the corresponding local object morphisms. The morphisms must also be consistent with the attribute values.

Following the double-pushout approach, a distributed graph rule $p : L \xleftarrow{l} I \xrightarrow{r} R$ is given by two distributed graph morphisms l and r. It transforms a distributed host graph G, into a target graph G' if an injective match $m : L \rightarrow G$ exists, which is a distributed graph morphism. A comatch $m' : R \rightarrow G'$ specifies the embedding of R in the target graph. The elements in the interface graph I must be preserved in the transformation. In our case, I can be understood to simply be the intersection of L and R. A rule may also contain negative application conditions (NAC) to express that something must not exist for the rule to be applicable. These are a finite set of distributed graph morphisms $NAC = \{L \xrightarrow{n_i} N_i\}$ and can refer to values of attributes [16]. Several morphisms $L \xrightarrow{n_i} N_i$ can become necessary in an NAC to express the conjunction of basic conditions. For the rule to be applicable, no graph present in a NAC must be matched in the host graph in a way compatible with $m : L \rightarrow G$. We also allow the use of set nodes, which can be mapped to any number of nodes in the host graph, including zero. The matching of a set node is exhaustive of all the nodes in the host graph satisfying the condition indicated by the rule.

The application of distributed graph transformations is synchronized via subrules. A subrule of an object level rule identifies that portion of the rule which modifies nodes or edges shared among different local graphs, so that modifications have to be synchronized over all the involved rules. Hence, synchronous application is achieved by rule amalgamation over subrules. For details, see [16]. In this paper, we use dotted lines to denote NACs and grey shading to indicate subrules. Non-connected NACs denote different negative application conditions (see Figure 3). The rules in the following figures are only the local components of distributed transformations. For the case discussed in the paper, the network components of the rules are usually identical rules. We will discuss later the few cases in which non-identical rules are needed at the network level.

4.2   Transformation Units

Transformation units are used to control rule application, with the control condition specified by expressions over rules [8]. The concept of a transformation unit is defined independently from any given approach to graph transformation.A graph transformation approach $\mathcal{A}$ consists of a class of graphs $\mathcal{G}$, a class of rules $\mathcal{R}$, a rule application operator $\Longrightarrow$ yielding a binary relation on graphs for every rule of $\mathcal{R}$, a class $\mathcal{E}$ of graph class expressions, and a class $\mathcal{C}$ of control conditions. Given an approach $\mathcal{A}$, a transformation unit consists of an initial and a terminal graph class expression in $\mathcal{E}$ (defining which graphs serve as valid input and output graphs), a set of rules in $\mathcal{R}$ and a set of references to other transformation units, whose rules can be used in the current one, together with a control condition over $\mathcal{C}$ describing how rules can be applied. Typically, $\mathcal{C}$ contains expressions on sequential application of rules and units as well conditions or application loops, e.g. applying a rule as long as possible.

Applying transformation units to distributed graph transformation, $\mathcal{A}$ is thus defined: $\mathcal{G}$ is the class of distributed graphs, $\mathcal{R}$ the class of distributed rules, and $\Longrightarrow$ the DPO way of rule application, as defined in [16]. The control expressions in $\mathcal{C}$ are of the type mentioned above and described in [7], while the class $\mathcal{E}$ can trivially be left empty, as no special initial and terminal graph classes need be specified.

We relate rule expressions to graph rules by naming rules and passing parameters to them, to be matched with specific attributes of some node. By this mechanism, we can restrict the application of rules to those elements which carry an actual reference to the code to be refactored. To this end, the rules presented in the transformation units are meant as rule schemes to be instantiated to actual rules, assigning the parameters as values of the indicated attributes.

5   Refactoring by Graph Transformation

In this section, we analyse examples of refactoring involving transformations in more than one view, i.e., in the code and at least one UML diagram. Following [14], refactorings are expressed by pre- and post-conditions. A typical interaction with a refactoring tool can be modelled by the following list of events: (1) The user selects a segment of code and (2) selects one from a list of available (possible composite) refactorings. (3) The tool checks the preconditions for refactoring. (4) If the preconditions are satisfied, refactoring takes place, with effects as described in the postconditions. Otherwise a message is issued.

The choice to perform a specific refactoring is usually left to the software designer, and we ignore here the relative process. We consider the effect of refactorings by defining graph transformation rules, possibly distributed over different view graphs, and managed in transformation units. Complex refactoring are modeled as sequences of individual refactoring steps. The composition of transformation units can give rise to complex refactorings. The effect of refactoring on different views is expressed through schemes of graph rewriting rules, which have to be instantiated with the proper names for, say, classes and methods, as indicated in the code transformation.

Preconditions, even if checked on the textual code, involve the analysis of structural properties, such as the visibility of variables in specific portions of code, or the existence of calls to some methods, which may be distributed over diagrams of different types.

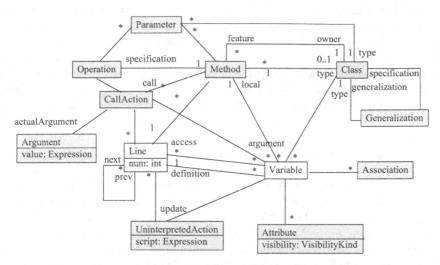

Fig. 1. The type graph for code representation

5.1   Graph Representation of Diagrams and Code

In the line of [9], we consider type graphs for defining the abstract syntax for concrete visual languages, such as those defined in UML. In particular, we refer to UML class, sequence, and state diagrams, with type graphs given by the metamodel definitions in [12].

We adopt a representation of the source code in the form of a flow graph for a method, as is typical in compiler construction [1]. This is a directed graph where nodes are lines of code and a prev/next relation exists between two nodes if they are consecutive lines, or the line represented by the source node can branch execution to the variable represented by the target node. Moreover, each Line node is attached to a set of nodes describing the variables mentioned in the line (for definition or usage), and the methods, if any, it calls. Some of these variables can be local to the Method environment, while those used in the method, but not local to it (and not defined in the class) are passed to the method. Local variables are defined in specific lines. Finally, Parameter nodes describe the types of arguments to a method, and Classifier nodes, describe types of variables, parameters and methods. Figure 1 describes the resulting type graph. This is simpler than the one in [10] for the representation of relations among software entities, which also considers inheritance among classes, and the presence of subexpressions in method bodies. Here, we deal with inheritance in the UML class diagram and, since we are not interested in representing the whole body of a method, we only keep trace of references to variables and methods and not of complete expressions. On the other hand, we maintain a representation of code lines and of reachability relations among them, which allows us to have a notion of block that will be used in Section 5.2.

Distributed graphs are suited to describe relationships between diagrams and code fragments. Following the approach of [5], a network graph NG describes the type graph for the specification of the whole software system. In NG, a network node is either associated with one local object graph – representing either a UML diagram or the code flow graph (we call such nodes content nodes) – or it is an interface node. Here, we consider only the Class, Sequence, and StateMachines families of diagrams discussed in the text, and the Code Flowgraph. For each pair of diagram nodes, a common interface node exists. Interface nodes are refined at the local level by the common graph parts of two diagrams in the current state. Network edges connect diagram nodes and interface nodes and are refined at the local level by defining how common interface parts are embedded in diagrams. Hence, an interface graph is related to its parent graphs by two graph embeddings (being injective graph morphisms). For example, an interface between Class diagrams and Flow graphs will present Method, Variable, and Class nodes, an interface between State Machine diagrams and Sequence diagrams may have nodes for states and transitions dealing with method calls[2] or other events, and an interface between Code and Sequence Diagrams contain nodes for call actions. Several network nodes of the same type can be used in the specification of a software system. For instance, different sequence diagrams are used to depict different scenarios, or a class can be replicated in different class diagrams to show its relationships with different sets of other classes.

5.2   Code Extraction

As a first example, consider the `extract_code_as_method` refactoring by which a segment of code is isolated, given a name, and replaced in the original code by a call to a newly formed method.

A precondition for such a refactoring is that the code to be extracted constitute a block, i.e., it has only one entry point and one point of exit, although it does not have to be maximal, i.e., it can be immersed in a fragment of code which has the block property itself. Moreover, the name to be given to the method must not exist in the class hierarchy to which the affected class belongs. The post-conditions for this refactoring assert that: 1) a new method is created whose body is the extracted code; 2) such a method receives as parameters all the variables used in the code but not visible to the class (they had to be passed or be local to the original method); 3) the code in the original method is replaced by a call to the new method.

Figure 2 contains the local rule for the code graph. It is part of a distributed rule, named ecamAll after the initials of the refactoring. In Figure 2, a set node indicates the lines of code, in the original version of the method, which lie between the first and last lines of the block to be moved. Another set node indicates the variables to be passed as parameters to the new methods. These are variables referenced in the block, but not defined within it, nor in the class. The rest of the method is left untouched. However, one has to check that no branches exist

---

[2] We could as well refer to Activity Diagrams in a similar way.

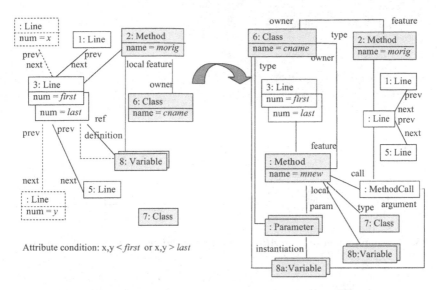

Fig. 2. The component of *ecamAll(cname, morig, mnew, first, last)* for the code flow graph

from lines in the block to other parts of the method, nor must branches exist from other parts of the method to the middle of the block. This is checked in the negative application conditions. The values of ■rst and last, as well as the names m orig and m new of the original method and of the new one, are provided by the rule expression activating the application of the transformation. Nodes of type Param eter, C lass and M ethod are shaded, as they and their interrelations are shared with other diagrams, thus identifying the subrules.

Figure 3 describes the local rule of ecam A ll acting on class diagrams. At the structural level, only the existence of a new method in the class shows up. The effects on the referred variables and the existence of a call for this method, observable in the textual description, are not reflected here. The two negative application conditions state that a method with the same signature as the new one must not appear in any class higher or lower in the inheritance hierarchy of the modified class.

These conditions use additional gen edges, produced by the rules of Figure 4, constructing the transitive closure of the G eneralization relation in the inheritance hierarchy. The negative application condition, indicated by the dashed lines, states that a gen arrow is inserted between two classes if they are not already related. The set of rules to compute the closure is completed by a rule insert down_gen, similar to insert gen, creating gen edges betwen a class and its descendants. An inverse sequence of operations eliminating the gen edges must be performed at the end of the refactoring process.

The newly created call is also observed at the behavioral level, as shown in Figure 5. It inserts a CallAction to the new Method, named m new, and the

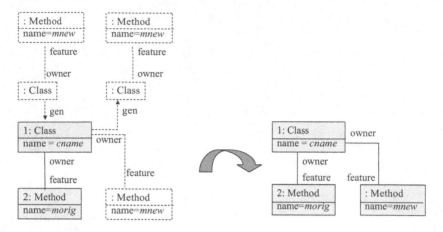

Fig. 3. The component of *ecamAll(cname, morig, mnew, first, last)* for class diagrams

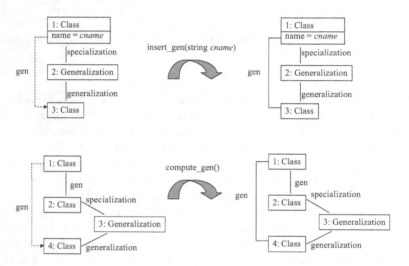

Fig. 4. The rule scheme for computing the transitive closure of the *Generalization* relation

corresponding activation of the `Operation` spawning from an activation of the old `Method`, m orig, in the `Class` cnam e. As the subrule identifies these elements for amalgamation, this transformation occurs in a synchronized way on all the sequence diagrams presenting an activation of m orig. However, as several activations of the same method can occur in one sequence diagram, this local rule has to be applied as long as possible on all sequence diagrams. The transformation of the sequence diagrams should be completed by transferring all the `CallActions`, originating from m orig and performed in lines that have been extracted to m new,

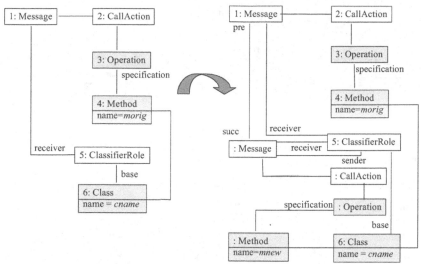

Fig. 5. The component of *ecamAll(cname, morig, mnew, first, last)* for sequence diagrams

to the activation of m new. A rule completeEcam Sequence, not presented here for space reasons, is responsible for that.

Finally, the component of ecam All which operates on state machine diagrams inserts a new state and a couple of transitions for all the states to which a **Transition** labelled with a **CallEvent** for m orig exists. The new state is reached with a **Transition** labelled with a **CallEvent** for m new. Return to the previous state occurs by a "completed" **Event**, indicating the completion of the operation.

All the transformations above have to be applied in a synchronized way to maintain consistency of diagrams and code. The network level transformations simply rewrites nodes into themselves. With each such rewriting, a transformation of the associated local graph occurs. An overall transformation unit describes this refactoring. This can be expressed as:

ExtrCodeAsMthd(String cname, String morig, String mnew, int first, int last) =
    asLongAsPossible insert_gen() end;
    asLongAsPossible insert_down_gen() end;
    asLongAsPossible compute_gen() end;
    if applicable(ecamAll(cname, morig, mnew, first, last)) then
        asLongAsPossible ecamAll(cname, morig, mnew, first, last) end;
        asLongAsPossible completeEcamSequence(cname, morig, mnew) end;
    else null end;
    asLongAsPossible remove_gen() end;

As transformations occur both at the network and local level (but at the network level they are identical), applying them as long as possible transforms

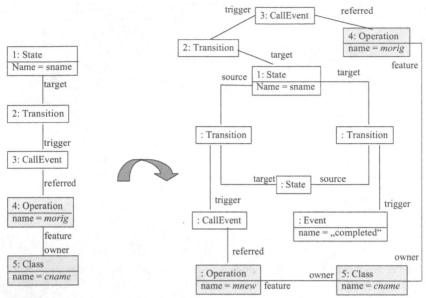

Fig. 6. The component of *ecamAll(cname, morig, mnew, first, last)* for state machine diagrams

all local graphs affected by refactoring, while at each time the associated network node is rewritten into itself. The rules presented above should be complemented so that the elements to which a transformation has already been applied are tagged and the presence of the tag must be checked to prevent the process from applying a rule again to the same element. The transformation unit is then completed by removing the tags.

5.3   Method Movement

As shown in Figure 7, the code of a Method m orig can be moved from its defining source Class to a different target Class in which it takes a new name m new. A Method with the name m new must not already appear (higher or lower) in the hierarchy of the target Class, as indicated by the NACs. Since the original method could refer to members of its original class, the signature for the method is enriched with a reference to the original class, as indicated by the Parameter node whose attribute nam e takes the value "orig" in Figure 7, illustrating the component of the distributed rule m m A ll(m orig, source, m new , target) for class diagrams. (Again, the name m m A ll derives from the initials of the refactoring.) The case where the method has to be moved to the superclass would be expressed by a rule scheme similar to the one in Figure 7, but which would require the existence of an inheritance relation between the nodes labelled ClassX and ClassY and where one of the NACs would be omitted. In the code flow graph,

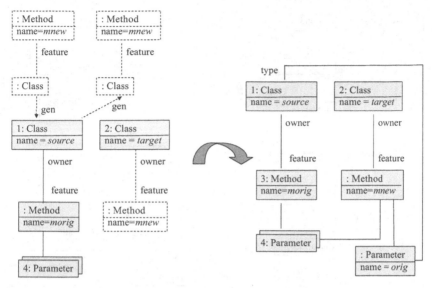

Fig. 7. The component of *mmAll(morig, source, mnew, target)* for class diagrams

the code of m orig has to be replaced with a forwarding method that simply calls m new in the target class. We do not show this transformation here.

Also sequence diagrams have to be modified, according to the transformation scheme depicted in Figure 8. Hence, a `CallAction` for m new towards the target `Class` must be inserted. While the call to the forwarding method does not modify the behavior of the class, it has to be reflected in this diagram to prevent subsequent refinements of the diagram from violating the correct sequence of calls. Again, the transformation of diagrams should be completed by a transfer of all calls originating from the m orig method to the activation of the m new method. The overall transformation unit is controlled by the rule expression:

MoveMethod(String morig, String source, String mnew, String target) =
    asLongAsPossible insert_gen() end;
    asLongAsPossible insert_down_gen() end;
    asLongAsPossible do compute_gen() end;
    asLongAsPossible mmAll(morig, source, mnew, target) end;
    asLongAsPossible completeEcamSequence(morig, source, mnew, target) end;
    asLongAsPossible remove_gen() end;

which causes the flow graph for the old method to be affected by the transformation, together with all the network nodes of type C lassD iagram and Sequence-D iagram whose local nodes contain references to the moved method.

In Figures 7 and 8 we consider the case for instance methods, with the code being moved between unrelated classes. The case for static methods, or for calls to methods in classes in the same hierarchy, requires some obvious modifications.

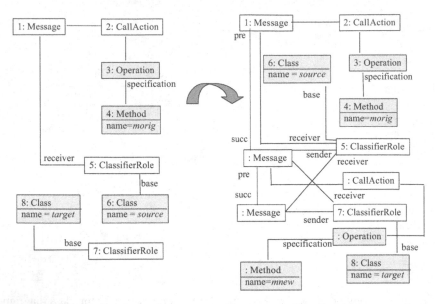

Fig. 8. The rule scheme for sequence diagram modification in move_method

6   Conclusions

We have presented an approach to maintaining consistency between code and model diagrams in the presence of refactorings. Each refactoring is described by a transformation unit, with parameters depending on the specific code modification, applied to the diagrams affected by the change. This framework avoids the need for reverse engineering, which reconstructs the models from the modified code, by propagating only the incremental changes to the original model. The model proposed can also be seen as a way to maintain consistency in the different diagrams through re-engineering steps, before proceeding to the actual code modification. Since there is no 'preferred' object in a distributed graph, the model can be used to propagate changes made on any diagram to the remaining ones, and not necessarily from the code to the documentation, as in refactoring. A more thorough study of existing refactorings, and experimentation on actual code, is needed to produce a library of distributed transformations which can be used in practical cases. The formalism can then be exploited to analyze possible conflicts or dependencies between different types of refactorings. We are now investigating the use of abstract syntax as a replacement for representation of concrete code, which would also favor the integration of the approach in existing refactoring tools based on abstract syntax representations.

References

[1] A.W. Appel. *Modern Compiler Implementation in Java*. Cambridge University Press, 1998. 227

[2] K. Beck and M. Fowler. *Planning Extreme Programming*. Addison Wesley, 2001. 220

[3] R. Fanta and V. Rajlich. Reengineering object-oriented code. In *Proceedings of ICSM 1998*, pages 238–246. IEEE Computer Society Press, 1998. 222

[4] M. Fowler, K. Beck, J. Brant, W. Opdyke, and D. Roberts. *Refactoring: Improving the Design of Existing Code*. Addison Wesley, 1999. 220

[5] M. Goedicke, B. Enders, T. Meyer, and G. Taentzer. Towards integration of multiple perspectives by distributed graph transformation. In M. Nagl, A. Schürr, and M. Münch, editors, *Proc. AGTIVE 1999*, pages 369–377, 2000. 222, 228

[6] M. Goedicke, T. Meyer, and G. Taentzer. Viewpoint-oriented software development by distributed graph transformation: Towards a basis for living with inconsistencies. In *Proc. 4th IEEE Int. Symp. on Requirements Engineering*, pages 92–99, 1999. 222

[7] M. Koch and F. Parisi Presicce. Describing policies with graph constraints and rules. In A. Corradini, H. Ehrig, H.-J. Kreowski, and G. Rozenberg, editors, *Proc. ICGT02*, pages 223–238, 2002. 226

[8] H.-J. Kreowski, S. Kuske, and A. Schürr. Nested graph transformation units. *Int. J. on SEKE*, 7(4):479–502, 1997. 224, 225

[9] T. Mens. Conditional graph rewriting as a domain-independent formalism for software evolution. In M. Nagl, A. Schuerr, and M. Muench, editors, *Proc. AGTIVE 1999*, pages 127–143, 1999. 222, 227

[10] T. Mens, S. Demeyer, and D. Janssens. Formalising behaviour preserving program transformations. In A. Corradini, H. Ehrig, H.-J. Kreowski, and G. Rozenberg, editors, *Proc. ICGT02*, pages 286–301, 2002. 222, 227

[11] J. Niere, J.P. Wadsack, and A. Zündorf. Recovering UML Diagrams from Java Code using Patterns. In J. H. Jahnke and C. Ryan, editors, *Proc. of the 2nd Workshop on Soft Computing Applied to Software Engineering*. Centre for Telematics and Information Technology, University of Twende, The Netherlands, February 2001. 221

[12] OMG. UML specification version 1.4. `http://www.omg.org/technology/documents/formal/uml.htm`, 2001. 227

[13] W.F. Opdyke. *Refactoring Object-Oriented Frameworks*. PhD thesis, University of Illinois at Urbana-Champaign, 1992. 222

[14] D.B. Roberts. *Practical Analysis for Refactoring*. PhD thesis, University of Illinois, 1999. 222, 223, 226

[15] G. Sunyé, D. Pollet, Y. Le Traon, and J.-M. Jézéquel. Refactoring UML models. In M. Gogolla and C. Kobryn, editors, *Proc. UML 2001*, pages 134–148, 2001. 222

[16] G. Taentzer, I. Fischer, M. Koch, and V. Volle. Visual Design of Distributed Systems by Graph Transformation. In H. Ehrig, H.-J. Kreowski, U. Montanari, and G. Rozenberg, editors, *Handbook of Graph Grammars and Computing by Graph Transformation, Volume 3: Concurrency, Parallelism, and Distribution*, pages 269–340. World Scientific, 1999. 222, 224, 225, 226

[17] L. Tokuda and D. Batory. Evolving object-oriented designs with refactorings. *Automated Software Engineering*, 8:89–120, 2001. 221

A Domain Specific Architecture Tool: Rapid Prototyping with Graph Grammars*

Thomas Haase, Oliver Meyer, Boris Böhlen, and Felix Gatzemeier

RWTH Aachen University
Department of Computer Science III
Ahornstraße 55, 52074 Aachen, GERMANY
{thaase,omeyer,boris,fxg}@i3.informatik.rwth-aachen.de

Abstract. Rapid prototyping is a feasible approach to determine the requirements and the functionalities of a software system. One prerequisite for successful prototyping is the existence of suitable tools to quickly implement the prototype. In this article we report about experiences by using the PROGRES system for rapid prototyping and whether we met the goal. Therefore, we will take a look at the development process of the prototype and its resulting specification. The development of a specific architecture design tool is used as an example.

1   Introduction

In [3] we present the outside behaviour of the Friendly Integration Refinement Environment (Fire3), a specific architecture design tool that addresses the problem of a-posteriori application integration. The prototype covers multiple refinement steps of architectural design on two levels: (1) An initial and coarse-grained view on an integration scenario is refined to a logical architecture denoting components and strategies to integrate them. (2) This architecture, in turn, is refined to a "concrete" one distributing the components on various processes.

Following the definition of a software architecture as "the structure of the components of a program/system [and] their interrelationships" [2] (see also [7]), graphs are a natural way to model software architectures. Consequently, the rules and constraints for the dynamic evolution of an architecture, e. g. adding or removing components and links between them, i. e. the architectural style, can be defined by graph transformations [6, 4].

In this article we report on experiences using the PROGRES language and programming environment [8, 9] and the UPGRADE framework [1, 5] to realize this problem-specific design tool.

2   Developing the Prototype

The task to implement the prototype mentioned in section 1 was split into (a) creating the executable specification in PROGRES containing the logic of the in-

---

* Financial support is given by Deutsche Forschungsgemeinschaft, Collaborative Research Centre 476 IMPROVE

J.L. Pfaltz, M. Nagl, and B. Böhlen (Eds.): AGTIVE 2003, LNCS 3062, pp. 236–242, 2004.

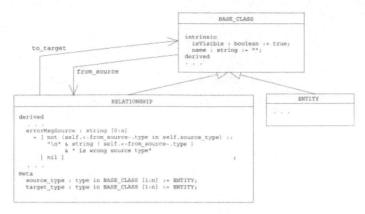

Fig. 1. Basic schema

tegration design tool (one person), and (b) adapting the UPGRADE framework to provide the user interface and a suitable presentation (two persons).

We followed a waterfall model of development, starting with defining strict requirements. First of all, we drew sketches of typical graphs we wanted to display in the demo of the tool. They showed what we expected to happen when refining the coarse-grained integration architecture.

Scripting the overall demo path helped in testing the feasibility of the prototype. We had to clearly define which input parameters are necessary for which graph transformations.

It showed that we need enhancements in the visualization. UPGRADE itself does not provide a gray-box visualization of nodes, needed for a view on the logical architecture (packages contain classes) as well as the concrete architecture (process nodes contain components).

The implementation, that is, the PROGRES specification (code can be directly generated from it) was created in three phases. They are described in the following. The overall structure of the specification is described in the closing subsection.

Generic Operations

The first step was to identify generic operations. To allow attributed relations, we model edges as nodes. The basic schema for this is shown in fig. 1. The basic node class **BASE_CLASS** also carries common attributes like a name. Operations of this simple PROGRES section include creation and destruction of nodes and relations. Using these basic operations makes sure that every node can be related with an attributed edge and that dangling edges are removed together with their connecting nodes. Writing this section is standard work for every PROGRES programmer.

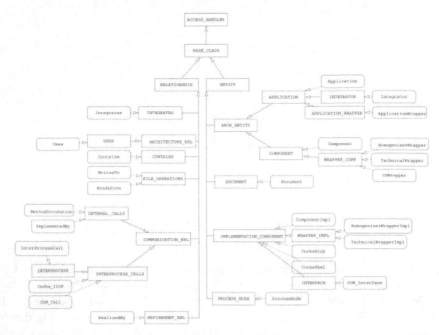

Fig. 2. Schema of the prototype (cutout)

Demo Graphs through Syntactical Transformations

The next step was to instantiate the graphs mentioned in the demo script. This gave the Java developers who extended the UPGRADE framework something to work with.

The corresponding operations transform one demo graph directly into the next one on a syntactical level. The first operation, for example, clears the graph and then creates an INTEGRATOR node labeled 'PFD_Ed_to_Ahead', two APPLICATION nodes labeled 'PFD_Editor' and 'Ahead', and two Integrates relations with the appropriate access rights. This is simply specified as a sequence of operation calls.

If a node has to be replaced from one demo graph to the next, it is removed from the graph and a new node is created. The operations of this phase do not make use of the left hand side of a graph replacement rule. All transformations are on the level of single nodes and not of complex graph patterns.

The schema, i. e. node and relation types, must be defined to create the demo graphs. A cutout of the schema is shown in fig. 2. It contains classes and types for the coarse-grained scenario, as APPLICATION and INTEGRATOR together with their Integrates relation. The logical architecture necessitates classes like COMPONENT and DOCUMENT with their Contains, Uses, and ReadsFrom/WritesTo relations. For the concrete architecture, analogous classes and relations were created. Also, further specializations like CorbaStub and CorbaSkel were added

```
1    transaction + IMP_MoveImplementationIntoProcessNode( component : IMPLEMENTATION_COMPONENT [1:1] ;
2                                              processNode : PROCESS_NODE [1:1]) [0:1] =
3      use wrongInternal : INTERNAL_CALLS [0:n];
4         wrongExternal : INTERPROCESS_CALLS [0:n];
5         oldProcessNode : PROCESS_NODE [1:1] := component.<=box_contains=.instance_of PROCESS_NODE;
6      do
7         choose
8            when ((component.<=box_contains=) = processNode)
9              then skip (* do nothing *)
10           else when  not empty (
11                         component.=relates_source ( MethodInvocation )=>.instance_of (CorbaStub or CorbaSkel)
12                      or component.=relates ( MethodInvocation )=>.instance_of (CorbaStub or CorbaSkel)
13                             (* calls CorbaStub or -Skel *)
14                      or component.=relates_source ( ImplementedBy )=>
15                             (* is called via COM *)
16                      or component.=relates ( COM_Call )=>
17                             (* makes a~COM_Call *) )
18                      or (component.type in (CorbaSkel or CorbaStub or COM_Interface)
19                             (* You may not move those components around *))
20              then
21                 fail (* Does work for InterProcess only. *)
22           else
23                 ARCH_MoveEntityIntoBox ( component, processNode )
24               & IMP_Priv_GetNowExternalCalls ( component, out wrongInternal )
25               & for_all internalCall : INTERNAL_CALLS [1:1] := wrongInternal do
26                    IMP_Priv_MakeExternal ( internalCall )
27                 end
28               & IMP_Priv_GetNowInternalCalls ( component, out wrongExternal )
29               & for_all externalCall := wrongExternal do
30                    IMP_Priv_MakeInternal ( externalCall )
31                 end
32               & choose when (oldProcessNode.isEmpty) then
33                       BAS_RemoveEntityWithRelations ( oldProcessNode )
34                 else
35                    skip
36                 end
37              end
38         end
39   end;
```

Fig. 3. Example transaction

and the unspecific Uses relation was refined to process INTERNAL_CALLS and
INTERPROCESS_CALLS.

From these demo graph creation operations, code was generated and the first
Java prototype was built. The prototype operations create the demo graphs, and
the unspecific basic operations allow to modify the graph instances in a powerful
but yet very inconvenient way.

In this state of development the Java programmers could start working on
representation and layout. The demo graphs allow to develop the gray-box visualization extensions and an automatic layouter for this special graph type and
corresponding element types for nodes and edges.

Semantical Transformation

Although a demo path can be presented, up to now we only have a prototype of
the prototype. It shows the different graphs of the prototype, but we still lack the
semantical transformation operations between them. Operations have names like
CreateDemoStep5From4 instead of RefineInterprocessCallToCorba. Also, we
are
stuck with a specific demo path we can not derivate from. In the following
development step we specify the transformations on a semantical level.

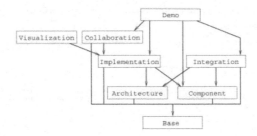

Fig. 4. Major dependencies between various sections of the specification

As an example, fig. 3 shows such a semantical operation. It moves an implementation component into another process[1] node. Former interprocess calls might internal methods calls and vice versa. As this transformation makes only use of the more basic operations, it is solely textual.

The prototype does not offer the full functionality needed to move implementation components between processes. Components that participate in a COM or CORBA call cannot be transfered between processes, as the transformation of the refined call is not implemented. On the main demo path components are moved betwwen processes before interprocess calls are refined. A user can, however, order the tool to move a component that uses e. g. a COM call. Lines 10–19 test for that condition and make the transaction fail. The failure mechanism of PROGRES provides perfect means to quickly handle these cases: No transformation is defined for them and the operation fails (line 21).

With these semantical operations, the syntactical operations from the previous phase are successively replaced. We also implemented further semantical transformations to allow for a more comprehensive final demo and to estimate the complexity of the final tool. On the user interface level, little changes were necessary to arrive at a complete prototype: Moving menu entries around and hiding additional technical edges.

The resulting prototype now truly transforms the architecture along the targeted refinement stages and even offers additional functionality like an animated collaboration diagram. The prototype, of course, still lacks the functionality of a complete tool. For example, the very complex back integration from concrete architecture to logical architecture and the propagation of changes of the logical architecture into the concrete architecture was not in the focus of the prototype and, therefore, not part of the specification.

3   Structure of the Specification

The resulting PROGRES specification is rather small, compared to other research projects at our department also realized by graph rewriting. In its current state it contains around 3.000 lines of PROGRES code.

---

[1] Here, in the operating system sense.

The specification contains eight sections as shown in fig. 4 together with their layering. There are generic operations in the base section. These are completely independent of the application domain. On top of these generic operations, the architecture and component sections allow to create and edit simple software architectures. As we handle specific architecture integration problems, the integration section allows creating the coarse-grained integration scenarios and deriving a logical architecture from it. From the created logical architecture the implementation section creates a concrete architecture. These five sections form the core of Fire3.

As an extension, the demo section realizes some predefined demo steps to quickly create demo scenarios. As such, it uses operations from multiple sections. The visualization section offers operations to hide the internals of interprocess communications by means of setting some attributes. The collaboration section realizes lists of communication calls, that are used to display collaboration diagrams. It defines its own graph structures and uses them to connect nodes defined in the implementation section.

4   Conclusion

PROGRES together with the UPGRADE framework are a perfect platform for rapid prototype development of visual tools. The prototype described in this paper was completed after only three weeks. The strict separation of logic, found in the graph specification, and visualization properties, found in the framework adaptions, allows to individually allocate development resources to specific tasks. There are very few dependencies between these two aspects. This supports incremental development.

If the demonstration prototype, for example, requires a better visualization, development resources can be allocated to that task, enhancing the prototype noticeably, without effecting the development of the logic part. If, on the other hand, additional functionality is needed, usually a single new operation suffices to offer an additional command.

Handling mainly special cases in the specification reduces the time between release states of the prototype. New perceivable functionality can be implemented very quickly. In the early phases of development, which we do not leave for this prototype, doubling the development effort leads to doubled functionality for the user.

With larger specifications the missing abstraction makes the graph rewriting specification much more difficult to handle. Packages do not offer the modularity needed here. Also their purely textual interfaces neglect the benefits of visual programming [10]. Graphical interfaces would make the PROGRES/UPGRADE couple more suitable for prototyping development. Even more, it is necessary for the unrestricted tool offering all integration and distribution commands.

References

[1] Boris Böhlen, Dirk Jäger, Ansgar Schleicher, and Bernhard Westfechtel. UP-GRADE: Building Interactive Tools for Visual Languages. In Nagib Callaos, Luis Hernandez-Encinas, and Fahri Yetim, editors, *Proc. of the 6th World Multiconference on Systemics, Cybernetics, and Informatics (SCI 2002)*, volume I (Information Systems Development I), pages 17–22, Orlando, Florida, USA, 2002. 236

[2] David Garlan and Dewayne E. Perry. Introduction to the special issue on software architecture. *IEEE Transactions on Software Engineering*, 21(4):269–274, 1995. 236

[3] Thomas Haase, Oliver Meyer, Boris Böhlen, and Felix Gatzemeier. Fire3 – architecture refinement for a-posteriori integration. this volume, 2003. 236

[4] Dan Hirsch, Paola Inverardi, and Ugo Montanari. Modeling software architectures and styles with graph grammars and constraint solving. In Patrick Donohoe, editor, *Software Architecture (TC2 1st Working IFIP Conf. on Software Architecture, WICSA1)*, pages 127–143, San Antonio, Texas, USA, 1999. Kluwer Acadamic Publishers. 236

[5] Dirk Jäger. Generating tools from graph-based specifications. *Information and Software Technology*, 42:129–139, 2000. 236

[6] Daniel Le Métayer. Describing software architecture styles using graph grammars. *IEEE Transactions on Software Engineering*, 27(7):521–533, 1998. 236

[7] Manfred Nagl. *Softwaretechnik: Methodisches Programmieren im Großen*. Springer, Berlin, Heidelberg, Germany, 1990. 236

[8] Andreas Schürr. *Operationelles Spezifizieren mit programmierten Graphersetzungssystemen*. Deutscher Universitätsverlag, Wiesbaden, Germany, 1991. PhD thesis, in German. 236

[9] Andreas Schürr, Andreas Joachim Winter, and Albert Zündorf. The PROGRES Approach: Language and Environment. In H. Ehrig, G. Engels, H.-J. Kreowski, and G. Rozenberg, editors, *Handbook of Graph Grammars and Computing by Graph Transformation*, volume 2: Applications, Languages and Tools, pages 487–550. World Scientific, Singapore, 1999. 236

[10] Andreas Joachim Winter. *Visuelles Programmieren mit Graph-Transformationen*, volume 27 of *Aachener Beiträge zur Informatik*. Wissenschaftsverlag Mainz, Aachen, Germany, 2000. PhD thesis, in German. 241

Graph Transformations in OMG's Model-Driven Architecture

(Invited Talk)

Gabor Karsai and Aditya Agrawal

Institute for Software Integrated Systems (ISIS)
Vanderbilt University, Nashville, TN, USA
{gabor.karsai,aditya.agrawal}@vanderbilt.edu
http://www.isis.vanderbilt.edu

Abstract. The Model-Driven Architecture (MDA) vision of the Object Management Group offers a unique opportunity for introducing Graph Transformation (GT) technology to the software industry. The paper proposes a domain-specific refinement of MDA, and describes a practical manifestation of MDA called Model-Integrated Computing (MIC). MIC extends MDA towards domain-specific modeling languages, and it is well supported by various generic tools that include model transformation tools based on graph transformations. The MIC tools are metaprogrammable, i.e. they can be tailored for specific domains using metamodels that include metamodels of transformations. The paper describes the development process and the supporting tools of MIC, and it raises a number of issues for future research on GT in MDA.

Keywords: Graph grammars, graph transformations, Model-Integrated Computing, domain-specific modeling languages, model-driven architecture, formal specifications

1 The MDA Vision

The Model-Driven Architecture initiative of OMG has put model-based approaches to software development into focus. The idea of creating models of software artifacts has been around for quite some time. However, this is the first time when mainstream software developers are willing to embrace the concept and demand tools that support this process. MDA is a "good thing" because it helps us develop software on a higher level of abstraction and - hopefully - will provide a "toolbox" that helps to keep the monster of complexity in check. This seems justified, as one can envision that multiple, yet interconnected models that represent requirements, design, etc. on different levels of abstraction, will offer a better way to work on complex software than today's UML models (used often only for documentation) and source code (spread across thousands of files).

Naturally, we need models to enact MDA; multiple, different kinds of models. Models capture our expectations ("requirements"), how the software is actually constructed ("design"), what kind of infrastructure the software will run on

J.L. Pfaltz, M. Nagl, and B. Böhlen (Eds.): AGTIVE 2003, LNCS 3062, pp. 243–259, 2004.
© Springer-Verlag Berlin Heidelberg 2004

("platform"), and other such details. There are - at least - two important observations that we can make about these models: (1) they are (or should be) linked to each other, (2) models can often be computed from each other via model transformation processes. Among the advocates of MDA an agreement seems to be forming that model transformations play a crucial role and tool support is needed, but this need is often understood in the context of PIM-to-PSM mappings only.

MDA introduces the concepts of the Platform-Independent Models (PIM), and Platform-Specific Models (PSM). The rationale for making this distinction can be found in the requirement that a model-driven development process must be platform-independent, and the resulting software artifacts must exist in a form that allows their specializations (perhaps optimization) for different kinds of software platforms (e.g. CORBA CCM, .NET and EJB). Hence, PIM is a representation of a software design, which captures the essence and salient properties of the design, without platform-specific details, while the PSM is an extended, specialized representation that does include all the platform-specific details. The two models are related through some transformation process that can convert a PIM to its semantically equivalent PSM.

2   Refining the MDA Vision

While the MDA vision provides a roadmap for model-based software development, we can and must step beyond the "canonical MDA", where PSM-s (which are closed, practical implementations on a particular platform) are automatically created from PIM-s (which are closer to abstract design) via transformations (which capture platform-specific details). We argue that this "one-shot" view of transformations is very limited, and there is a much broader role for transformations. We envision a model-driven process where engineers develop multiple, interlinked models that capture the various aspects of the software being produced, and, at the same time, they also develop transformations that relate models of different kind to each other, and apply these transformations whenever necessary. Transformations become "bridges" that link models of different nature, and maintain consistency between models: when an "upstream" model is changed (by the designer or by a tool), the "downstream" model is automatically updated.

But the question arises: Where are the models coming from and what do they exactly capture? We argue that models should capture the various aspects of software design in a domain-specific manner. Domain-specificity in development is being widely recognized as a potential way of increasing productivity in software engineering. The idea is simple: instead of having a programmer constantly translate domain-specific knowledge into low-level code, first a "language" is built for the domain, next a translator is created that maps the language into other, possibly executable artifacts, and then the software products are built in the form of domain-specific models, which are then often transformed into code. The goal is to raise the level of abstraction to a level such that programming happens closer to the application domain and away from the implementation domain. Arguably, the model-driven process offers a natural habitat for realizing domain-specific software development, and we can call this integration of concepts from MDA and domain-specific development as Model-Integrated Computing [1], or Domain-Specific Model-Driven Architecture.

Definition: MIC is a domain-specific, model-driven approach to software development that uses models and transformations on models as first class artifacts, where models are sentences of domain-specific modeling languages (DSML-s). MIC captures the invariants of the domain, the fixed constructs of the DSML-s (i.e. the "grammar"), and the variabilities of the domain in the models (i.e. the "sentences").

MIC advocates development in a linguistic framework: the developer should define a DSML (including a transformations that interpret its "sentences"), and then use it to construct the final product: the software. This approach to software development brings forward some questions such as (1) how to define new languages, and (2) how define transformation tools for those languages.

3   Tools for Domain-Specific Model-Driven Architecture: MIC

First, we need a language to formally and precisely define DSML-s. Formally, a DSML is a five-tuple of concrete syntax (C), abstract syntax (A), semantic domain (S) and semantic and syntactic mappings $(M_S,$ and $M_C)$ [18]:

$$L = \{C, A, S, M_S, M_C\} \tag{1}$$

The concrete syntax (C) defines the specific (textual or graphical) notation used to express models, which may be graphical, textual or mixed. The abstract syntax (A) defines the concepts, relationships, and integrity constraints available in the language. Thus, the abstract syntax determines all the (syntactically) correct "sentences" (in our case: models) that can be built. (It is important to note that the abstract syntax includes semantic elements as well. The integrity constraints, which define well-formedness rules for the models, are frequently called "static semantics".) The semantic domain (S) is usually defined by means of some mathematical formalism in terms of which the meaning of the models is explained. The mapping M_C: $A \rightarrow C$ assigns concrete syntactic constructs (graphical, textual or both) to the elements of the abstract syntax. The semantic mapping M_S: $A \rightarrow S$ relates syntactic constructs to those of the semantic domain. The definition of the (DSM) language proceeds by constructing metamodels of the language (to cover A and C), and by constructing a metamodel for the semantics (to cover M_C and M_S).

3.1   Defining the Syntax

A meta-programmable visual modeling environment, GME (see [2] for details)is available that provides a language called "MetaGME" for defining the abstract

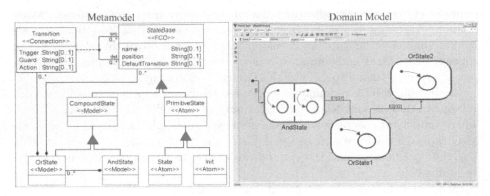

Fig. 1. Metamodel and model in MIC

and concrete syntax of DSML-s. The abstract syntax is specified using a UML class diagram [4] editor that captures the abstract syntax in the form of a diagram, and static semantics in the form of OCL [11] expressions. The metamodel of MetaGME (i.e. the meta-metamodel) is MOF compliant [3]. Note that the UML class diagram is used to represent a "grammar" whose sentences are the "object graphs" that conform to it. Concrete syntax is captured in MetaGME using idioms: patterns of classes and stereotypes, which have a have a specific meaning for the GME visualization and editing engine. Currently the GME editor supports a fixed set of visual modeling concepts. In the future, GME will be changed to enable adding modeling concepts and new visualization and manipulation techniques. Once the abstract and concrete syntax are defined, i.e. the metamodel of the language is built, a MetaGME "interpreter" translates this metamodel into a format that (the generic) GME uses to morph itself into a domain-specific GME that supports that (and only that) language which is defined by the metamodel. This GME instance strictly enforces the language "rules": only models that comply with the abstract syntax and the static semantics can be built. Figure 1 shows an illustrative metamodel and a compliant model.

3.2   Defining the Semantics

For mapping the domain specific models into a semantic domain we have chosen a pragmatic approach: we assume that there is always a "target platform" whose semantics is well-known. This approach defining semantic: "semantics via transformations," has been used in the past for the formal specification of semantics [8]. Note that the target platform also has an abstract syntax (with static semantics), and the transformation between the domain-specific models and target platform models establishes the semantics of the DSM-s in terms of the target models. One can observe that conceptually this is the same process employed in MDA's PIM to PSM transformation: the transformation provides semantics to platform-independent models via their mapping to platform-specific models. In MIC, just like in MDA, transformations play a crucial role: they specify the (dynamic) semantics of domain-specific models.

On a more general note, one can observe that in a model-based development process transformations appear in many, different situations. A few representative examples are as follows:

- Refining the design to implementation; this is the basic case in the PIM/PSM mapping.
- Pattern application; expressing design patterns as locally applied transformations on the software models [10, 13].
- Aspect weaving; the integration of aspect code into functional code is a transformation on the design [17].
- Analysis and verification; analysis algorithms can be expressed as transformations on the design [20].

One can conclude that transformations can and will play an essential role, in general, in model-based development, thus there is a need for highly reusable model transformation tools. These tools must be generic, in our terms: meta-programmable; i.e. their function should be determined by a "meta-program", which defines how models are transformed.

There exist well-known technologies today that seem to satisfy these requirements, and do not require sophisticated metamodeling: for instance XML and XSLT [12]. XML provides a structured way to organize data (essentially as tagged/typed data elements, organized in a hierarchy and untyped references that cut across the hierarchy), while XSLT provides a language to define transformations on XML trees. However, XSLT is not adequate for implementing sophisticated model transformations: (1) it lacks a type system, (2) it does not support reasoning about transformations, (3) and its performance is often not sufficient for practical development [19].

3.3   Defining Semantics via Transformations

We have created a model transformation system called "Graph Rewriting and Transformation" (GReAT). GReAT consists of a model transformation language called UML Model transformer (UMT), a virtual machine called Graph Rewrite Engine (GRE), a debugger for UMT called Graph Rewriting Debugger (GRD) and a Code Generator (CG) that converts the transformation models into efficient, executable code [6, 7].

In UMT transformations are specified on metamodel elements. This helps to strongly type the transformations and ensures the syntactic correctness of the result of the transformations, within the specification. The transformation rules consist of a graph pattern (which is matched against an input graph), a guard condition (a precondition, which is evaluated over the matched subgraph), a consequence pattern (which expresses the creation and deletion of target graph objects), and a set of attribute mapping actions that are used to modify the attributes of input or target objects). These transformations specify the mapping of input models (i.e. the input "graph") to target models (i.e. the "output graph"). For efficiency reasons rewriting rules accept "pivot" points: initial bindings for pattern variables. This reduces the search in the pattern matching process (effectively reducing it to matching in a rooted tree). One can also explicitly sequence the execution of the rules, and sequential, parallel, and conditional composition of the rules is also available.

GRE works as an interpreter: it executes transformation programs (which are expressed in the form of transformation metamodels) on domain-specific models to generate target models. GRE is slow compared to hand-written code but is still useful while creating and modifying the transformation. The GRD provides debugging capabilities on top of GRE such as setting break points, single step, step into, step out and step over functions. A visual front end to the debugger is also available.

After the transformations have been created, debugged and tested the Code Generator (CG) can be used to generate efficient code that executes the transformations. The generated code improves the performance of the transformations by at least two orders of magnitude, early experiments show. An in-depth coverage of the GReAT system is provided in the next section.

4   The GReAT Model Transformation System

This section provides a description of the details of the GReAT tool. First, the transformation language is described, followed by a description of the GRE.

4.1   The UML Model Transformer (UMT) Language

UMT consists of three sub languages: (1) the pattern specification language, (2) the graph rewriting language, and (3) the control flow language.

4.2   The Pattern Specification Language

At the heart of a graph transformation language is the pattern specification language and the related pattern matching algorithms. The pattern specifications found in graph grammars and transformation languages [6, 7, 22, 23] do not scale well, as the entire pattern to be matched has to be enumerated. The pattern matching language provides additional constructs for the concise yet precise description of patterns. String matching will be used to illustrate representative analogies.

Patterns in most graph transformation languages have a one-to-one correspondence with the host graph. Consider an example from the domain of textual languages where a string to match starts with an 's' and is followed by 5 'o'-s. To specify such a pattern, we could enumerate the 'o'-s and write "sooooo". Since this is not a scalable solution, a representation format is required to specify such strings in a concise and scalable manner. One can use regular expressions: for

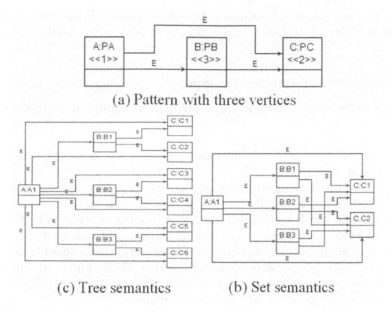

(a) Pattern with three vertices

(c) Tree semantics   (b) Set semantics

Fig. 2. Pattern with different semantic meanings

strings we could write it as "s5o" and use the semantic meaning that o needs to be repeated 5 times. The same argument holds for graphs, and a similar technique can be used. Cardinality can be specified for each pattern vertex with the semantic meaning that a pattern vertex must match n host graph vertices, where n is its cardinality. However, it is not obvious how the notion of cardinality truly extends to graphs. In text we have the advantage of a strict ordering from left to right, whereas graphs do not possess this property.

In figure 2(a) we see a pattern having three vertices. One possible meaning could be tree semantics, i.e., if a pattern vertex pv1 with cardinality c1 is adjacent to pattern vertex pv2 with cardinality c2, then the semantics is that each vertex bound to v1 will be adjacent to c2 vertices bound to v2. These semantics when applied to the pattern gives figure 2(b). The tree semantics is weak in the sense that it will yield different results for different traversals of the pattern vertices and edges and hence, it is not suitable for our purpose.

Another possible unambiguous meaning could use set semantics: consider each pattern vertex pv to match a set of host vertices equal to the cardinality of the vertex. Then an edge between two pattern vertices pv1 and pv2 implies that in a match each v1, v2 pair should be adjacent, where v1 is bound to pv1 and v2 is bound to pv2. This semantic when applied to the pattern in figure 2(a) gives the graph in figure 2(c). The set semantics will always return a match of the structure shown in figure 2(c), and it does not depend upon factors such as the starting point of the search and how the search is conducted.

Due to these reasons, we use set semantics in GReAT and have developed pattern-matching algorithms for both single cardinality and fixed cardinality of vertices.

4.3   Graph Transformation Language

Pattern specification is just one important part of a graph transformation language. Other important concerns include the specification of structural constraints in graphs and ensuring that these are maintained throughout the transformations [6]. These problems have been addressed in a number of other approaches, such as [22, 23].

In model-transformers, structural integrity is a primary concern. Model-to-model transformations usually transform models from one domain to models that conform to another domain making the problem two-fold. The first problem is to specify and maintain two different models conforming to two different metamodels, simultaneously. An even more relevant problem to address involves maintaining associations between the two models. For example, it is important to maintain some sort of references, links, and other intermediate values required to correlate graph objects across the two domains.

Our solution to these problems is to use the source and destination metamodels to explicitly specify the temporary vertices and edges. This approach creates a unified metamodel along with the temporary objects. The advantage of this approach is that we can then treat the source model, destination model, and temporary objects as a single graph. Standard graph grammar and transformation techniques can then be used to specify the transformation.

The rewriting language uses the pattern language described above. Each pattern object's type conforms to the unified metamodel and only transformations that do not violate the metamodel are allowed. At the end of the transformation, the temporary objects are removed and the two models conform exactly to their respective metamodels. Our transformation language is inspired by many previous efforts, such as [6, 7, 22, 23]

The graph transformation language of GReAT defines a production (also referred to as rule) as the basic transformation entity. A production contains a pattern graph (discussed above) that consists of pattern vertices and edges. Each object in the pattern graph conforms to a type from the metamodel. Each object in the production has another attribute that specifies the role it plays in the transformation. A pattern can play the following three, different roles:

1. Bind: Match object(s) in the graph.
2. Delete: Match object(s) in the graph, then remove the matched object(s) from the graph.
3. New: Create new object(s) (provided the pattern matched successfully).

The execution of a rule involves matching every pattern object marked either bind or delete. If the pattern matcher is successful in finding matches for the pattern, then for each match the pattern objects marked delete are deleted from the match and objects marked new are created.

Sometimes the pattern alone is not enough to specify the exact graph parts to match and we need other, non-structural constraints on the pattern. An example for such a constraint is: "the value of an (integer) attribute of a particular vertex should be within some limits." These constraints or pre-conditions are captured in a guard and are written using the Object Constraint Language (OCL) [11]. There is also a need to provide values to attributes of newly created objects and/or modify attributes of existing object. Attribute Mapping is another ingredient of the production: it describes how the attributes of the "new" objects should be computed from the attributes of the objects participating in the match. Attribute mapping is applied to each match after the structural changes are completed.

A production is thus a 4-tuple, containing: a pattern graph, mapping function that maps pattern objects to actions, a guard expression (in OCL), and the attribute mapping.

4.4   Controlled Graph Rewriting and Transformation

To increase the efficiency and effectiveness of a graph transformation tool, it is essential to have efficient implementations for the productions. Since the pattern matcher is the most time consuming operation, it needs to be optimized. One solution is to reduce the search space (and thus time) by starting the pattern-matching algorithm with an initial context. An initial context is a partial binding of pattern objects to input (host) graph objects. This approach significantly reduces the time complexity of the search by limiting the search space. In order to provide initial bindings, the production definition is expanded to include the concept of ports. Ports are elements of a production that are visible at a higher-level and can then be used to supply initial bindings. Ports are also used to retrieve result objects from the production (and pass them along to a downstream production).

An additional concern is the application order of the productions. In graph grammars there is no ordering imposed on productions. There, if the pattern to be matched exists in the host graph and if the pre-condition is met then the production will be executed. Although this technique is useful for generating and matching languages, it is not efficient for model-to-model transformations that are algorithmic in nature and require strict control over the execution sequence. Moreover, a well-defined execution sequence can be used to make the implementation more efficient.

There is a need for a high-level control flow language that can control the application of the productions and allow the user to manage the complexity of the transformation. The control flow language of GReAT supports the following features:

- Sequencing: rules can be sequenced to fire one after another.
- Non-determinism: rules can be specified to be executed "in parallel", where the order of firing of the parallel rules is unspecified.

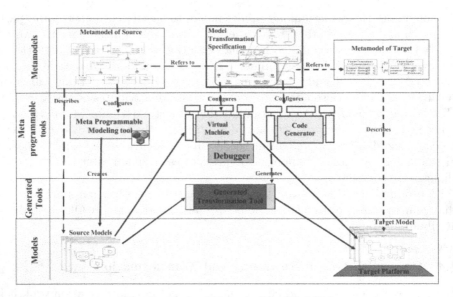

Fig. 3. Tools of Domain-Specific MDA

- H ierarchy: Compound rules can contain other compound rules or primitive rules.
- R ecursion: A rule can call itself.
- Test/Case: A conditional branching construct that can be used to choose between different control flow paths.

4.5   MIC Tools

Figure 3 shows the MIC tool suite included in GReAT, and how the tools rely on metamodels. In order to set up a specific MIC process one has to create metamodels for the (input) domain, the (output) target, and the transformations. One can then use the meta-level tools (such as the MetaGME interpreter and the Code Generator) to build a domain specific model editor and a model transformation tool. The model editor is then used to create and modify domain models, while the transformation tool is used to convert the models into target models.

We believe a crucial ingredient in the above scheme is the meta-programmable transformation tool: GRE that executes a transformation metamodel (as a "program") and facilitates the model transformation. Using the concepts and techniques of graph transformations allows not only the formal specification of transformations, but, arguably, reasoning about the properties of the transformations as well.

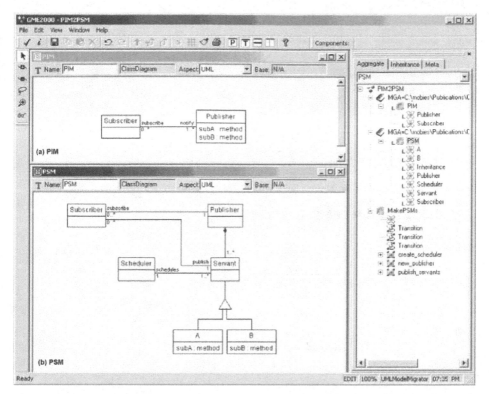

Fig. 4. Platform-Independent Model and Platform-Specific Model

5 Example

The tools described above can also be used to implement PIM to PSM transformations of MDA as illustrated through the following example. The example shows the transformation of software designs from a more abstract, generic
model to a model with more specialized components. Figure 4(a) shows the
platform-independent model: a UML class diagram. The model describes the
entities Publisher and Subscriber, and the relationship between them. In this
case the relationship specifies that multiple Subscribers can subscribe to one of
the multiple services provided by a Publisher.

Starting from the PIM, the transformer applies design patterns and adds
further implementation details to build a more detailed, platform-specific model
(PSM). Figure 4(b) shows the refined model where there is only one Publisher
(indicated by the cardinality on the subscribes association). This class could
be implemented using the Singleton design pattern. The Publisher class creates
a new, specific Servant for each Subscriber (the Servants could be created using
the "AbstractFactory" design pattern). The Publisher also hands over the Subscriber's location so that a Servant can notify its Subscriber directly. Moreover,

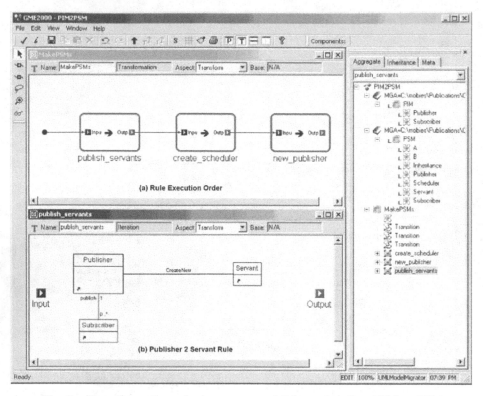

Fig. 5. Transformation rules to convert publisher subscriber PIM to PSM

in this implementation, only one Servant is assumed to be running at a time. Hence, for scheduling multiple servants the Scheduler class has been added to the PSM.

Transforming these models takes three steps. The first step is to transform all Publishers into Servants. After the appropriate publishers have been created, a Scheduler must be created. Finally, the new Publisher (which will be the only one in the new model) is created.

Figure 5 shows two levels of the transformation specification. Figure 5(a), "MakePSMs" specifies the order of execution of the transformation rules. Figure 5(b), the specification for the transformation that converts a publisher into a servant is shown. On the bottom left side of figure 5(b), the Publisher and Subscriber form the (PIM) pattern to be matched, and on the right side the Servant (PSM) denotes the new object to be created.

6   Extending MIC towards a Multi-model Process

In the previous discussions on MIC we have focused on a single DSML and a single target; similar to the simple PIM/PSM mapping. In a large-scale application of MIC, however, a multitude of metamodels and transformations are needed. As discussed in the introduction, models can capture requirements, designs, platforms (and many other subjects), as well as the transformations between them. We envision that the next generations of software development tools are going to support this multi-model development.

It is interesting to draw the parallel here with the design and development of very large-scale integrated circuits. VLSI circuits are design using a number of languages (VHDL being only one of them), and using transformation tools (datapath generators, etc.) that "link" the various design artifacts [9]. It usually takes a number of steps to go from a high-level representation of a design to the level of masks, and consistency and correctness must be maintained across the levels. We believe that with consistent and recursive application of MDA through the use of transformations, the software engineering processes will be able to achieve the kind of reliability VLSI design processes have achieved.

We envision that the developers who apply MIC develop a number of domain-specific modeling languages. Modeling languages are for capturing requirements, designs, but also platforms. It is conceivable that the engineer wants to maintain a hierarchy of design models that represent the system on different levels of abstractions. Lower-level design models can be computed from higher-level ones through a transformational process. Maintaining consistency across models is of utmost importance, but if a formal specification of the transformation is available then presumably this can be automated. Note that model transformations are also models. Thus they can be computed from higher-level models by yet another transformation, and, in fact, transformation specifications can be derived also through a transformation process.

Naturally, the toolset introduced above need to be extended to allow this multi-model MIC. Presumably models should be kept in a model database (or "warehouse"), with which the developers interact. Transformations can be applied automatically, or by the user. Transformation results may be retained, and/or directly manipulated by the developer. We believe that an environment that supports this process should allow multiple formalisms for visualizing and manipulating artifacts ("models"), and transformation objects should ensure the coherency across the objects.

7   Relationship to OMG's QVT

On can argue that the approach described above is MDA, instead of being an extension of MDA. Indeed, the approach described above has all the ingredients of the MDA vision as described in [3], and recent papers pointed towards using UML as a mechanism for defining a family of languages [21]. We fully agree with this notion, however we can envision applications where the full flexibility and power of MIC is not worth the cost. For example, "throw-away", "use-it-once" applications may not require all the features of MIC.

On the other hand, the MIC tools discussed above provide a way of implementing the MDA concepts, including QVT, although they do not support everything, at least not immediately. The reason is that we wanted to focus on the DSML-s and drive the entire development process using domain-specific models, instead of the modeling language(s) defined by UML. We were also concentrating on building configurable, meta-programmable tools that support arbitrary modeling approaches.

8   Challenges and Opportunities for Graph Transformations in the MDA

Graph Transformation (GT) is powerful technology that may lead to a solid, formal foundation for MDA. It offers a number of challenges and opportunities for further research, as listed below.

1. GT as the bridge between the programmer's intentions and their implementation. The vision of Intentional Programming (from Charles Simonyi) is that software should be built in the form of "intentions": high-level abstractions specify what the software needs to do, and explicit transformations that convert these intentions into executable code. We believe GT is the perfect vehicle to realize this vision.
2. "Provably correct" software via provably correct GT-s. In many critical applications software reliability is of utmost importance, yet reliability is assured using extensive testing and, to a very limited extent, through formal verification. Formal specification of the transformations can lead to formal verification of the software artifacts and may be able to provide some confidence in the generated software.
3. Non-hierarchical (de)composition supported by GT-s. Requirement analysis often leads to a non-hierarchical decomposition of the problem, and implementation could also lead to non-hierarchical composition of the system. Perhaps the best example for the latter is aspect-oriented programming. We argue that GT-s offer a uniform framework for describing, representing, implementing and analyzing these orthogonal (de)compositions.
4. Assurance of para-functional properties of the final SW. Especially in the field of embedded systems, it is often necessary to calculate para-functional properties (like schedulability, timeliness, performance, lack of deadlocks, etc.) of the system at design time. We conjecture that some of these properties can be formally defined and calculated using GT techniques.
5. Efficient implementations of GT. The usability of GT tools will determine success of the GT technology. Efficient implementation algorithms need to be developed such that the performance of GT based transformations is at acceptable levels and is comparable to that of the equivalent, hand-written code.

6. GT tools and support as an integral part of the software development process. A large number of IDE-s are used in software development today, some with extremely well-defined processes. These IDE-s (typically) do not handle GT-s (yet). Again, the industrial success of GT-s will depend on how well GT tools are integrated with existing tools and processes.

7. Design tool integration via GT. Many development processes require a large number of (possibly heterogeneous) tools. This implies a need for generic tool integration solutions [16]. We claim that GT-s offer an opportunity for implementing these solutions and provide a highly effective technology for the rapid creation of integrated tools.

8. Teaching software engineers about GT as a programming paradigm (design-time and run-time). Currently, GT technology is not a mainstream software engineering technology. One potential cause of this is the lack of trained software engineers who can use GT as an engineering tool. There is a need for courses and tutorials on this topic, and the training should cover the use of GT-s for both design-time (i.e. transformation on the software design artifacts), and run-time (i.e. GT on domain-specific data structures, to implement some application functionality).

9   Summary

We claim that a sophisticated model-driven software development process requires multiple, domain-specific models. Building transformation tools that link these models, by mapping them into each other (including mapping into executable) form a crucial tool component in the process, and graph transformations offer a fundamental technology for these transformations. We here briefly introduced a domain-specific model-driven process and its supporting tools, some of them based on graph transformations, and summarized the major concepts behind it.

The described toolset exists in a prototype implementation today and has been used in small-scale examples. However, it needs to be extended towards a multi-model process, where a wide variety of models (and modeling languages) can be used. This extension is the subject of ongoing research.

Acknowledgements

The NSF ITR on "Foundations of Hybrid and Embedded Software Systems" has supported, in part, the activities described in this paper. The effort was also sponsored by DARPA, AFRL, USAF, under agreement number F30602-00-1-0580.The US Government is authorized to reproduce and distribute reprints for Governmental purposes notwithstanding any copyright thereon. The views and conclusions contained therein are those of authors and should not be interpreted as necessarily representing the official policies and endorsements, either expressed or implied, of the DARPA, the AFRL or the US Government. Tihamer Levendovszky and Jonathan Sprinkle have contributed to the discussions and work that lead to GReAT, and Feng Shi has written the first implementation of GReAT-E. A shortened, preliminary version of this paper has appeared in the WISME workshop at the UML 2003 conference.

References

[1] J. Sztipanovits, and G. Karsai, "Model-Integrated Computing", IEEE Computer, Apr. 1997, pp. 110-112 245

[2] A. Ledeczi, et al., "Composing Domain-Specific Design Environments", IEEE Computer, Nov. 2001, pp. 44-51. 245

[3] "The Model-Driven Architecture", http://www.omg.org/mda/, OMG, Needham, MA, 2002. 246, 255

[4] J. Rumbaugh, I. Jacobson, and G. Booch, "The Unified Modeling Language Reference Manual", Addison-Wesley, 1998. 246

[5] A. Agrawal, T. Levendovszky, J. Sprinkle, F. Shi, G. Karsai, "Generative Programming via Graph Transformations in the Model-Driven Architecture", Workshop on Generative Techniques in the Context of Model Driven Architecture, OOPSLA , Nov. 5, 2002, Seattle, WA.

[6] Rozenberg G. (ed.), "Handbook on Graph Grammars and Computing by Graph Transformation: Foundations"; Vol.1-2. World Scientific, Singapore, 1997. 247, 248, 250

[7] Blostein D., Schürr A., "Computing with Graphs and Graph Transformations", Software - Practice and Experience 29(3): 197-217, 1999. 247, 248, 250

[8] Maggiolo-Schettini A., Peron A., "Semantics of Full Statecharts Based on Graph Rewriting", Springer LNCS 776, 1994, pp. 265–279. 246

[9] A. Bredenfeld, R. Camposano, "Tool integration and construction using generated graph-based design representations", Proceedings of the 32nd ACM/IEEE conference on Design automation conference, p.94-99, June 12-16, 1995, San Francisco, CA. 255

[10] A. Radermacher, "Support for Design Patterns through Graph Transformation Tools", Applications of Graph Transformation with Industrial Relevance, Monastery Rolduc, Kerkrade, The Netherlands, September 1999. 247

[11] Object Management Group, "Object Constraint Language Specification", OMG Document formal/01-9-77. September 2001. 246, 251

[12] XSL Transformations, www.w3.org/TR/xslt. 247

[13] Karsai G., "Tool Support for Design Patterns", NDIST 4 Workshop, December, 2001. (Available from: www.isis.vanderbilt.edu). 247

[14] U. Assmann, "How to Uniformly specify Program Analysis and Transformation", Proceedings of the 6 International Conference on Compiler Construction (CC) '96, LNCS 1060, Springer, 1996.

[15] J. Gray, G. Karsai, "An Examination of DSLs for Concisely Representing Model Traversals and Transformations", 36th Annual Hawaii International Conference on System Sciences (HICSS'03) - Track 9, p. 325a, January 06 - 09, 2003.

[16] Karsai G., Lang A., Neema S., "Tool Integration Patterns, Workshop on Tool Integration in System Developement", ESEC/FSE , pp 33-38., Helsinki, Finland, September, 2003. 257

[17] Uwe Assmann and A. Ludwig, "Aspect Weaving by Graph Rewriting", In U. Eisenecker and K. Czarnecki (ed.), Generative Component-based Software Engineering. Springer, 2000. 247

[18] T. Clark, A. Evans, S. Kent, P. Sammut, "The MMF Approach to Engineering Object-Oriented Design Languages", Workshop on Language Descriptions, Tools and Applications (LDTA2001), April, 2001. 245

[19] Karsai G., "Why is XML not suitable for Semantic Translation", Research Note, ISIS, Nashville, TN, April, 2000. (Available from: www.isis.vanderbilt.edu). 247

[20] U. Assmann, "How to Uniformly specify Program Analysis and Transformation", Proceedings of the 6 International Conference on Compiler Construction (CC) '96,LNCS 1060, Springer, 1996. 247

[21] Keith Duddy, "UML2 must enable a family of languages", CACM 45(11), p, 73-75, 2002. 255

[22] H. Fahmy, B. Blostein, "A Graph Grammar for Recognition of Music Notation", Machine Vision and Applications, Vol. 6, No. 2 (1993), 83-99. 248, 250

[23] G. Engels, H. Ehrig, G. Rozenberg (eds.), "Special Issue on Graph Transformation Systems", Fundamenta Informaticae, Vol. 26, No. 3/4 (1996), No. 1/2, IOS Press (1995). 248, 250

Computing Reading Trees
for Constraint Diagrams

Andrew Fish and John Howse

Visual Modelling Group, University of Brighton, Brighton, UK
{Andrew.Fish,John.Howse}@brighton.ac.uk
http://www.cmis.brighton.ac.uk/research/vmg/

Abstract. *Constraint diagrams* are a visual notation designed to complement the Unified Modeling Language in the development of software systems. They generalize Venn diagrams and Euler circles, and include facilities for quantification and navigation of relations. Their design emphasizes scalability and expressiveness while retaining intuitiveness. Due to subtleties concerned with the ordering of symbols in this visual language, the formalization of constraint diagrams is non-trivial; some constraint diagrams have more than one intuitive reading. A 'reading' algorithm, which associates a unique semantic interpretation to a constraint diagram, with respect to a reading tree, has been developed. A reading tree provides a partial ordering for syntactic elements of the diagram. Reading trees are obtainable from a partially directed graph, called the dependence graph of the diagram. In this paper we describe a 'tree-construction' algorithm, which utilizes graph transformations in order to produce all possible reading trees from a dependence graph. This work will aid the production of tools which will allow an advanced user to choose from a range of semantic interpretations of a diagram.

1   Introduction

The Unified Modeling Language (UML) [10] is the Object Management Group's industrial standard for software and system modelling. It has accelerated the uptake of diagrammatic notations for designing systems in the software industry.

In this paper, we are concerned with a diagrammatic notation, constraint diagrams, which may be used to express logical constraints, such as invariants and operation preconditions and postconditions, in object oriented modelling. It was introduced in [7] for use in conjunction with UML, and is a possible substitute for the Object Constraint Language (OCL) [13], which is essentially a textual, stylised form of first order predicate logic. An alternative approach was adopted in [1], where progress was made towards a more diagrammatic version of OCL.

Constraint diagrams were developed to enhance the visualization of object structures. Class diagrams show relationships between objects, as associations between classes, for example. Annotating, with cardinalities and aggregation, for example, enables one to exhibit some properties of these relationships between

J.L. Pfaltz, M. Nagl, and B. Böhlen (Eds.): AGTIVE 2003, LNCS 3062, pp. 260–274, 2004.
© Springer-Verlag Berlin Heidelberg 2004

objects. However, frequently one wishes to exhibit more subtle properties, such as those of composite relations. This is impossible using class diagrams, but the inherent visual structure of constraint diagrams makes this, and many other constructions, easy to express.

Constraint diagrams build on a long history of using diagrams to visualize logical or set-theoretical assertions. They generalize Venn diagrams [12] and Euler circles [2], which are currently rich research topics, particularly as the basis of visual formalisms and diagrammatic reasoning systems [11, 5, 6]. Constraint diagrams are considerably more expressive than these systems because they can express relations, whilst still retaining the elegance, simplicity and intuitiveness of the underlying diagrammatic systems. For constraint diagrams to be used effectively in software development, it is necessary to have strong tool support. Such tools are currently under development [9, 4].

In [3], we described a 'reading' algorithm which produces a unique semantic reading for a constraint diagram, provided we place a certain type of partial order (represented by a reading tree) on syntactic elements of the diagram. Briefly, this process involves constructing a unique, partially directed graph from the diagram. This dependence graph describes the dependences of certain syntactic elements of the diagram. Information from this graph allows one to construct reading trees. The constraint diagram, together with a single reading tree, allows one to provide a unique semantic interpretation (given by a logical formula). However, only the specification of the requirements of a reading tree are given in [3]. In this paper, we describe a 'tree-construction' algorithm which produces all possible reading trees, given a dependence graph. This work will aid the production of tools which allow an advanced user to choose from a range of semantic interpretations of a diagram.

In Sect. 2 we give a concise description of constraint diagrams. In Sect. 3 we define dependence between certain syntactic elements of the diagram. This enables us to associate a unique dependence graph to a diagram, in Sect. 4. Using this graph, we describe the reading trees in Sect. 5. The 'tree-construction' algorithm is described in Sect. 6 and a worked example, together with semantic interpretations, is given in Sect. 7. Finally, in Sect. 8, we highlight the usefulness of this work, especially with regard to tool creation, and indicate possibilities for further investigation. In particular we describe the potential usefulness of graph transformations when attempting to reason with the system by considering reasoning rules.

2   Constraint Diagrams

A contour is a simple closed curve in the plane. The area of the plane which constitutes the whole diagram is a basic region. Furthermore, the bounded area of the plane enclosed by a contour c is the basic region of c. A region is defined recursively: a basic region is a region, and any non-empty union, intersection, or difference of regions is a region. A zone is a region which contains no other

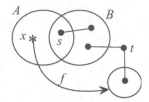

Fig. 1. A constraint diagram

region. A zone may or may not be shaded. A region is shaded if it is a union of shaded zones.

A spider is a tree with nodes (called feet) in distinct zones. It touches any region which contains (at least) one of its feet. The union of zones that a spider touches is called the spider's habitat. A spider is either an existential spider, whose feet are drawn as dots, or a universal spider, whose feet are drawn as asterisks.

The source of a labelled arrow may be a contour or a spider. The target of a labelled arrow may be a contour or a spider. A contour is either a given contour, which is labelled, or a derived contour, which is unlabelled and is the target of some arrow.

For example, in Fig. 1, there are two given contours, labelled by A and B, and one derived contour. These contours determine the five zones of the diagram. There is a universal spider, labelled by x, in the zone which is "inside A and outside B". This spider is the source of an arrow labelled f, which targets the derived contour. There are two existential spiders in the diagram, one of which has two feet and is labelled by s and the other has three feet and is labelled by t. The habitat of s is the region "inside B" and the habitat of t is the region "outside A".

Given contours represent sets, arrows represent relations and derived contours represent the image of a relation. Existential quantification is represented by existential spiders, and universal quantification is represented by universal spiders. Distinct spiders represent distinct elements. A shaded region with n existential spiders touching it represents a set with no more than n elements.

The diagram in Fig. 1 should mean "There are two sets A and B. There is an element, s, in B and an element t outside A, such that $s \neq t$. Every element x in $A - B$ is related, by f, to some set which is outside A and B (and may be different for each x)".

We will use the standard object-oriented notation $x.f$ to represent textually the relational image of element x under relation f, that is, $x.f = \{y : (x,y) \in f\}$. Thus $x.f$ is the set of all elements related to x under relation f. The expression $x.f$ is a navigation expression, so called because we can navigate from x along the arrow f to the set $x.f$. The relational image of a set S is then defined by

$$S.f = \bigcup_{x \in S} x.f.$$

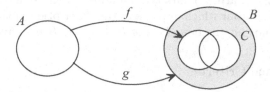

Fig. 2. An example of an arrow hitting the boundary of a region

3   Dependences

The informal semantics of a constraint diagram consists of a collection of pieces of information, some of which are related and need to be ordered. For the formal specification of the semantics, we will need to know precisely which diagrammatic elements are related to each other and need to be ordered. For example, if two spiders' habitats intersect (they have a foot in the same zone) then we wish to say that the elements, s and t, represented by the spiders, are not equal, forcing the scope of quantification of s to encompass t, or vice versa. The dependence criteria will encapsulate all of the necessary information of this type.

We informally describe the dependences of the relevant syntactic elements of the diagram. In Sect. 4, these are used to define the dependence graph of a diagram. The intuitive interpretation of these descriptions gives the general idea and, in fact, gives the correct dependence graph almost always. The precise, formal interpretations are given in [3].

Definition 1. A spider or an arrow in a constraint diagram is called an orderand.

These are the syntactic elements of the diagram which require ordering. In order to define the dependence criterion we first need a few definitions.

Definition 2. A minimal description of a region is a description that involves the minimal number of contour labels.

A minimal description of a region usually coincides with the labels or identifiers of contours appearing as part of the topological boundary of the region. It is not necessarily unique, but any regions appearing in this paper have a unique minimal description.

Example 1. The orderands in the diagram in Fig. 2 are f and g. Let $T(f)$ denote the derived contour which is the target of f. A minimal description of the shaded region is "inside B and outside $T(f)$ and C". This region represents the set $B \cap \overline{T(f)} \cap \overline{C}$, where $\overline{X}$ denotes the complement of X.

Definition 3. An arrow, f, hits the boundary of a region, r, if the target of f is a derived contour which is required in a minimal description of r.

Example 2. In Fig. 2, arrow f hits the boundary of the shaded region, because $T(f)$ is a derived contour which appears in the description of the shaded region. However, g does not hit the boundary of B, because B is a given contour.

3.1 Dependence Criteria

Dependence of orderands is defined by the six criteria listed below. The dependence relation is not transitive. This means that it is possible to have a dependence between s and t and a dependence between t and f without concluding that there is a dependence between s and f. These six dependence criteria are illustrated by reference to the diagrams in Figs. 3 - 8 respectively. These figures show both the constraint diagram and their dependence graphs (see Sect. 4). Note that the term spider refers to either existential or universal spiders.

1. If the habitats of spiders s and t have non-empty intersection, then there is a dependence between s and t.
2. If the source of arrow f is a derived contour c then f is dependent upon any arrow which hits c.
3. If the source or the target of an arrow f is a spider s then f is dependent upon s.
4. If arrow f hits the boundary of the habitat of spider s, then there is a dependence between s and f.
5. The shaded zones of a diagram can be collected into regions which are connected components. For each such shaded region r, construct the set of spiders which touch r. Add to this set the arrows which hit the boundary of the region r. There is pairwise dependence between the objects in this set.
6. Suppose that f and g are arrows whose targets are derived contours, denoted by $T(f)$ and $T(g)$ respectively. If $T(f) = T(g)$, or if any minimal description of the placement of $T(f)$ requires the use of $T(g)$, or vice versa, then there is dependence between f and g. A minimal description of the placement of a derived contour refers to a minimal description of the region inside the contour, without reference to the contour itself.

In Fig. 3, the habitats of the existential spiders s and t intersect in the zone which is "inside B and outside A" (representing the set $B \cap \overline{A}$). Therefore there is a dependence between s and t. The habitat of the existential spider u does not intersect either of the habitats of the spiders s or t.

In Fig. 4, the arrow f is dependent upon the arrow g because the source of f is the derived contour which is the target of g.

In Fig. 5, the source of the arrow f is the universal spider x and the target of f is the existential spider s, so f is dependent upon both x and s.

In Fig. 6, let $T(f)$ denote the derived contour which is the target of f. The region which is the habitat of s is described as "inside B and outside $T(f)$" (representing the set $B \cap \overline{T(f)}$). Thus f hits the boundary of the habitat of s and so there is dependence between f and s. However, g does not hit the boundary of the habitat of s because the target of g is a given contour, labelled by B,

Fig. 3. Dependence criterion 1

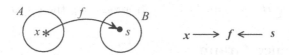

Fig. 4. Dependence criterion 2

Fig. 5. Dependence criterion 3

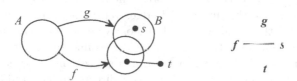

Fig. 6. Dependence criterion 4

and is not a derived contour. The spider t has habitat described as "outside A and B" (representing the set $\overline{A} \cap \overline{B}$). Neither f nor g hit the boundary of this region.

In Fig. 7, the shaded zone is described as "inside B and $T(f)$" (representing the set $B \cap T(f)$). Since the existential spider s touches this zone, and the arrow f hits the boundary of this zone, there is dependence between f and s.

In Fig. 8, a minimal description of the placement of the target of the arrow f is "inside B and outside $T(g)$" (representing the set $B \cap \overline{T(g)}$), so there is a

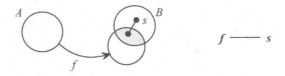

Fig. 7. Dependence criterion 5

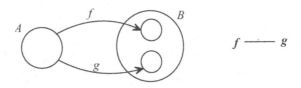

Fig. 8. Dependence criterion 6

dependence between f and g. Similarly, a minimal description of the placement of $T(g)$ is "inside B and outside $T(f)$" (representing the set $B \cap \overline{T(f)}$).

4   Dependence Graph

The dependences between orderands can be described collectively using a dependence graph, $G(d)$. This is a partially directed graph whose directed edges represent the dependence of one orderand upon another and whose undirected edges represent dependence between orderands. This is an unusual partially directed graph, because the undirected edges represent a disjunction of directed edges, rather than the usual conjunction. For example, if two nodes in the graph, corresponding to spiders in the diagram say, are connected by an undirected edge then this means that the quantification they represent must be nested (the scope of either one contains the other), whereas a directed edge would also force the order of the quantifiers.

Definition 4. Let d be a constraint diagram. The dependence graph of d, denoted by $G(d)$, is a partially directed graph with node set equal to the set of orderands of d. There is a directed edge from node o_2 to node o_1 if and only if orderand o_1 is dependent upon orderand o_2. There is an undirected edge between node o_2 and node o_1 if and only if:

– there is a dependence between orderands o_1 and o_2 and
– orderand o_1 is not dependent upon orderand o_2 and
– orderand o_2 is not dependent upon orderand o_1.

Definition 5. A diagram is unreadable if its dependence graph has a directed cycle, and readable otherwise.

Proposition 1. A diagram is unreadable if and only if there exists a sequence of arrows, $f_1, \ldots, f_n$, with $n \geq 1$, such that the target of f_i is a derived contour which is equal to the source of f_{i+1}, for each i, reading f_1 for f_{n+1}.

Proof. A directed edge in the dependence graph arises from the dependence upon criteria (Sect. 3.1, criteria 2 and 3). From these it can be seen that no spider can be dependent upon any orderand. Therefore any directed cycle in the dependence graph has only arrows as nodes. Furthermore, an arrow, f, in a diagram is dependent upon another arrow, g, if and only if the source of f is a derived contour which is the target of g.

The tree-construction algorithm, given in Sect. 6, will simply return no reading trees if there is directed cycle in the dependence graph.

5   Reading Trees

From the dependence graph of a diagram one can construct a reading tree, which is a rooted, directed tree. The semantic interpretation of the diagram depends upon the reading tree as well as the diagram. Note that the dependence graph is merely an internal artefact used to construct the reading trees. Essentially these trees are used to specify the order and scope of the quantifiers in the semantic interpretation of a diagram.

Let $G(d)$ be the dependence graph of a readable diagram d. Let $F(d)$ denote a directed forest satisfying the following conditions:

1. The node set of $F(d)$ equals the node set of $G(d)$.
2. No two directed edges in $F(d)$ end at the same node.
3. If there is a directed edge from node n_1 to n_2 in $G(d)$, then there is a directed path in $F(d)$ from n_1 to n_2.
4. If there is an undirected edge between nodes n_1 and n_2 in $G(d)$, then the nodes n_1 and n_2 must lie in a common directed path in $F(d)$.

As a consequence of these conditions, each component of $F(d)$ has a unique starting node, n, with no incoming edges.

The Plane Tiling Condition (PTC) is useful for defining the semantics of a diagram, and encapsulates certain topological properties of the given contours, such as disjointness. It states that the union of the sets assigned to the zones of the underlying Euler diagram consisting of the given contours is the universal set.

Definition 6. Let $G(d)$ be the dependence graph of a readable diagram d, and let $F(d)$ be any forest satisfying conditions 1–4 above. A rooted, directed tree, with root node labelled by PTC and directed edges from this node to each starting node in $F(d)$, is called a reading tree of d, denoted by $RT(d)$.

Example 3. The dependence graph in Fig. 5 gives rise to two reading trees:

$$PTC \to x \to s \to f \text{ and } PTC \to s \to x \to f.$$

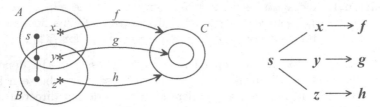

Fig. 9. A complicated constraint diagram and its dependence graph

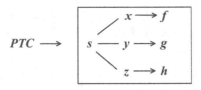

Fig. 10. The start graph for the dependence graph in Fig. 9

The semantics of the diagram with respect to the first tree is given by:

$$A \cap B = \emptyset \ \wedge \ \forall x \in A \ (\exists s \in B \ (x.f = s))$$

and with respect to the second tree:

$$A \cap B = \emptyset \ \wedge \ \exists s \in B \ (\forall x \in A \ (x.f = s)).$$

6   The Tree-Construction Algorithm

Given a partially directed graph, called a dependence graph, G, the following 'tree-construction' algorithm (Sect. 6.1) will construct the set of rooted, directed trees which are reading trees. The nodes of these reading trees are the nodes of G, together with one extra node, the root node, labelled by PTC. Essentially this algorithm involves constructing trees of graphs whose nodes contain subgraphs of the dependence graph. These nodes are deleted and replaced with trees of graphs whose nodes contain smaller subgraphs. The process terminates when each node in these trees of graphs contains only a single node. We use an algebraic approach to graph transformations, where matches satisfy the gluing condition (so that an application of a production deletes exactly what is specified by the production). We begin with some examples.

Example 4. Fig. 9 shows a complicated constraint diagram and its dependence graph, G. Fig. 10 shows the start graph for the dependence graph in Fig. 9. This is a tree of graphs with two nodes, one labelled by PTC and the other containing the dependence graph G. The leftmost tree in Fig. 11 is an example of replacing G (with respect to s) and the other two trees show examples of

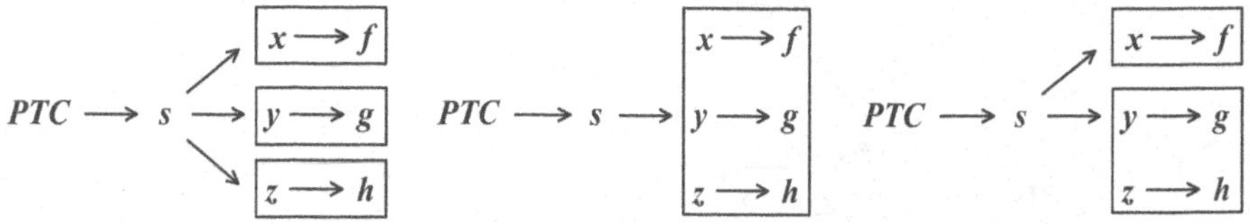

Fig. 11. Examples of replacements for the dependence graph in Fig. 9

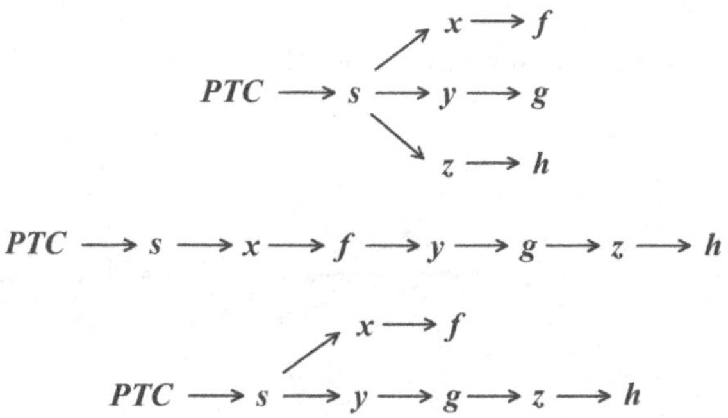

Fig. 12. Examples of reading trees obtainable using the replacements in Fig. 11

merging some of the branches of this tree after the replacement. Fig. 12 shows examples of reading trees obtainable from these trees.

Example 5. Fig. 5 shows an example of a constraint diagram and its dependence graph, G. In order to aid understanding of the following 'tree-construction' algorithm, we explain, using Fig. 13, the application of this algorithm to the dependence graph in Fig. 5. This graph, G, is connected, so the starting graph is the tree of graphs with two nodes shown at the top of Fig. 13. Looking inside the node containing G, we see that x and s are the only nodes with no incoming directed edges. Deleting x from G leaves one connected component and so there is only one replacement for G using x (there are no branches to merge). Similarly, there is only one replacement obtained by deleting s from G. This gives the two trees of graphs on the middle level of the figure. For each of these trees there is only one choice of replacement for the nodes containing the shown graphs, and the two reading trees are given at the bottom of the figure.

6.1   The 'Tree-Construction' Algorithm

Define the production, MERGE, as in Fig. 14, where n is a node of a tree of graphs which contains a singleton node as its graph and H_i and H_j are nodes containing graphs. The crossed out arrows depict a negative application condition.

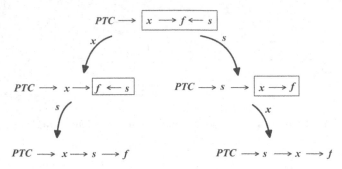

Fig. 13. Obtaining reading trees from the dependence graph in Fig. 5

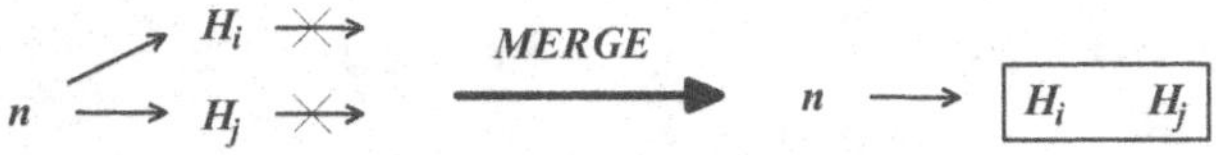

$$n \nearrow \begin{array}{c} H_i \not\longrightarrow \\ H_j \not\longrightarrow \end{array} \quad \text{MERGE} \quad \Longrightarrow \quad n \longrightarrow \boxed{H_i \quad H_j}$$

Fig. 14. The MERGE production

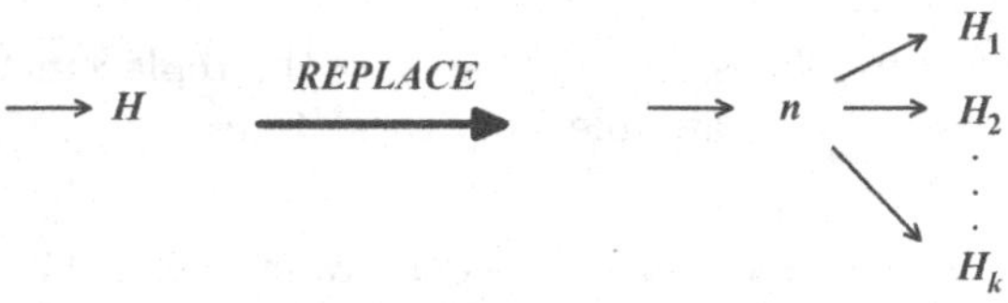

$$\longrightarrow H \quad \xrightarrow{\textbf{REPLACE}} \quad \longrightarrow n \xrightarrow{\nearrow \begin{array}{c} H_1 \\ \searrow H_2 \\ \vdots \\ H_k \end{array}}$$

Fig. 15. The REPLACE production

Define the production, REPLACE, as in Fig. 15, where H is a node containing a graph, $graph(H)$, with more than one node, n is a node of $graph(H)$ which has no incoming directed edges in $graph(H)$, and the H_i, $1 \leq i \leq k$, are nodes containing the connected components of $graph(H) - \{n\}$ (deleting dangling edges).

In order to simplify the description of the algorithm, define the productions, REPLACETHENMERGE to be the composites of REPLACE, followed by any number (including 0) of MERGE.

The initial start graph is the tree of graphs shown in Fig. 16, where PTC contains a single node, and the G_i, $1 \leq i \leq r$, are nodes containing the connected components of the dependence graph, G.

The 'tree-construction' algorithm is:

1. Construct the language corresponding to the graph grammar consisting of the initial start graph and the MERGE production.

Fig. 16. The start graph

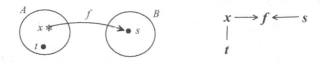

Fig. 17. A constraint diagram and its dependence graph

2. Use each of the words in this language as a start graph of a new graph grammar, where the productions are the REPLACETHENMERGE productions. Construct the languages corresponding to these graph grammars.
3. The words in these languages which admit no more matches, and have only singleton nodes as nodes of the tree are the reading trees required.

7   Worked Example

Applying the algorithm to the dependence graph in Fig. 17 yields eleven reading trees, four of which yield distinct semantic interpretations of the constraint diagram shown. Fig. 18 shows the effect of the 'tree-construction' algorithm. Nodes which contain graphs with more than one node are shown inside a rectangle. The labelled arrows between trees refer to applications of REPLACETHENMERGE, using the chosen nodes, which have no incoming edges. The start graph is shown at the top of the figure, where the node on the right contains G itself because G has only one connected component. In G, the nodes x, s and t have no incoming edges and so there are three matches for REPLACE in the tree of graphs (one for each of x, s and t). However, the REPLACETHENMERGE production, has four matches (one or no applications of MERGE may be used following the replacement involving x). Thus we produce four trees. Continuing the algorithm produces all reading trees, which are shown as the leaves of the structure in Fig. 18.

7.1   Semantics

The semantic interpretation of the diagram in Fig. 17, with respect to these four reading trees:

$$PTC \longrightarrow x \longrightarrow s \longrightarrow f \qquad PTC \longrightarrow s \longrightarrow x \longrightarrow f$$
$$\searrow t \qquad\qquad\qquad\qquad\qquad \searrow t$$

$$PTC \longrightarrow t \longrightarrow x \longrightarrow s \longrightarrow f \qquad PTC \longrightarrow t \longrightarrow s \longrightarrow x \longrightarrow f$$

gives, in order (reading from left to right and top to bottom), the following four distinct semantic interpretations:

$$A \cap B = \emptyset \ \wedge \ \forall x \in A \ (\exists s \in B \ (x.f = s) \ \wedge \ \exists t \in A \ - \ \{x\})$$

$$A \cap B = \emptyset \ \wedge \ \exists s \in B \ (\forall x \in A \ (x.f = s \ \wedge \ \exists t \in A \ - \ \{x\}))$$

$$A \cap B = \emptyset \ \wedge \ \exists t \in A \ (\forall x \in A - \{t\} \ (\exists s \in B \ (x.f = s)))$$

$$A \cap B = \emptyset \ \wedge \ \exists t \in A \ (\exists s \in B \ (\forall x \in A - \{t\} \ (x.f = s))).$$

Essentially the semantics of the diagram, with respect to a reading tree, are given by performing a depth first search of the tree, simultaneously building a copy of the diagram by adding the corresponding syntax each time a node is processed and reading the new semantic information at each stage (see [3] for more details).

8   Conclusion and Future Work

Formal semantics for constraint diagrams are a pre-requisite for their safe use in software specification. In [3] a 'reading' algorithm was presented, which constructs the semantics of a constraint diagram, with respect to a reading tree. In this paper, we develop a 'tree-construction' algorithm which generates all possible reading trees for a diagram. However, for complicated diagrams the number of reading trees is very large. ¿From a modelling perspective, assuming that disconnected components of the dependence graph remain disconnected in the construction of $F(d)$, seems very reasonable. The only change this makes to the 'tree-construction' algorithm is to ignore the initial merge (and is identical if the dependence graph has exactly one component). This simplifying assumption restricts the number of reading trees to sensible proportions for practical modelling purposes. Further work will be to investigate which reading trees yield equivalent semantic interpretations.

We envisage the development of an environment (a collection of software tools) which will produce the semantics of a diagram, automating the resolution of possible ambiguities. To maintain the inherent usability of the notation, the environment should provide a single default reading of a diagram, but allow more confident users to specify alternate readings. Such an environment is already under construction as part of [9].

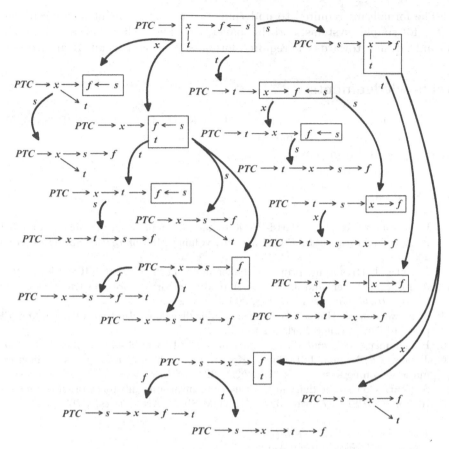

Fig. 18. Applying the algorithm to the dependence graph in Fig. 17

Unambiguous semantics are an essential foundation for the development of a set of diagrammatic reasoning rules. These rules can be applied to reason about system specifications. For example, one can check structural subtyping: that the invariants of a subclass are stricter than the invariants of a superclass. The environment should guide a modeller through a reasoning process. A similar environment is provided by the KeY project [8] using the non-diagrammatic language OCL. Reasoning rules for constraint diagrams can be described using graph transformations involving the abstract dual graph of a constraint diagram [4] (which provides abstract information about the underlying Euler diagram without reference to the topology of the plane), an abstract syntax graph (which provides abstract information about other important pieces of concrete syntax), and the dependence graph and reading trees (which provide information relating to the semantic interpretation of the diagram).

The formalisms required to underpin such an environment have been completed for simpler systems, and this paper, together with [3], constitutes a significant advance towards the required formalisms for constraint diagrams.

Acknowledgements

This research was partially supported by UK EPSRC grant GR/R63516. We thank Jean Flower, Gem Stapleton, John Taylor and the anonymous referees for their very helpful comments.

References

[1] P. Bottoni, M. Koch, F. Parisi-Presicce, and G. Taentzer. A visualisation of OCL using collaborations. In *Proc UML '01*, volume 2185 of *LNCS*. Springer-Verlag, 2001. 260

[2] L. Euler. Lettres a une princesse d'allemagne. *Letters*, (2):102–108, 1761. 261

[3] A. Fish, J. Flower, and J. Howse. A reading algorithm for constraint diagrams. In *Proc HCC '03*, pages 161–168, 2003. 261, 263, 272, 274

[4] J. Flower and J. Howse. Generating Euler diagrams. In *Proc of Diagrams '02*, pages 61–75. Springer-Verlag, 2002. 261, 273

[5] E. Hammer. *Logic and Visual Information*. CSLI Publications, 1995. 261

[6] J. Howse, F. Molina, J. Taylor, S. Kent, and J. Gil. Spider diagrams: A diagrammatic reasoning system. *JVLC*, 12:299–324, 2001. 261

[7] S. Kent. Constraint diagrams: Visualising invariants in object oriented models. In *Proc OOPSLA97*, pages 327–341. ACM SIGPLAN Notices, 1997. 260

[8] *The KeY Project.* 273

[9] *Kent Modelling Framework (KMF).* 261, 272

[10] *OMG: UML Specification, Version 1.4.* 260

[11] S.-J. Shin. *The Logical Status of Diagrams.* Cambridge University Press, 1994. 261

[12] J. Venn. On the diagrammatic and mechanical representation of propositions and reasonings. *Phil.Mag*, 1880. 261

[13] J. Warmer and A. Kleppe. *The Object Constraint Language.* Addison-Wesley, 1998. 260

UML Interaction Diagrams: Correct Translation of Sequence Diagrams into Collaboration Diagrams*

Björn Cordes, Karsten Hölscher, and Hans-Jörg Kreowski

University of Bremen, Department of Computer Science
P.O. Box 330440, D-28334 Bremen, Germany
{bjoernc,hoelsch,kreo}@informatik.uni-bremen.de

Abstract. In this paper, the two types of UML interaction diagrams are considered. A translation of sequence diagrams into collaboration diagrams is constructed by means of graph transformation and shown correct.

1   Introduction

The Unified Modeling Language (UML) is a graphical object-oriented modeling language used for the visualization, specification, construction, and documentation of software systems. It has been adopted by the Object Management Group (OMG) and is widely accepted as a standard in industry and research (cf., e.g., [2],[17]). The UML provides nine types of diagrams for different purposes. This paper focuses on sequence and collaboration diagrams collectively known as interaction diagrams. Both diagram forms concentrate on the presentation of dynamic aspects of a software system, each from a different perspective. Sequence diagrams stress time ordering while collaboration diagrams focus on organization. Despite their different emphases they share a common set of features. Booch, Rumbaugh, and Jacobson claim that they are semantically equivalent since they are both derived from the same submodel of the UML metamodel, which gives a systematic description of the syntax and semantics of the UML.

In this paper we provide some justification of this statement by presenting a correct translation of sequence diagrams into collaboration diagrams by means of graph transformation. As illustrated in Figure 1, the translation consists of two steps, each modeled as a graph transformation unit. In the first step a sequence diagram is translated into a metamodel object diagram. The metamodel object diagram is then translated into a corresponding collaboration diagram. In order to accomplish these tasks, the diagrams involved are considered and represented as labeled and directed graphs in a straightforward way where we use additional node and edge attributes during the translation. In this way, the two translations

---

* Research partially supported by the EC Research Training Network SegraVis and the DFG project UML-AID.

J.L. Pfaltz, M. Nagl, and B. Böhlen (Eds.): AGTIVE 2003, LNCS 3062, pp. 275–291, 2004.

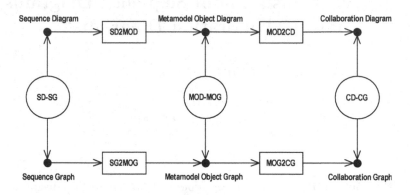

Fig. 1. Overview of the translation

on the level of diagrams are obtained by the two translations on the graph level and going back and forth between diagrams and corresponding graphs. The translation focuses on sequence diagrams on instance level with synchronous and nonconcurrent communication restricted to procedure calls. The diagrams may contain conditions, iteration conditions for procedure calls, and conditional branchings. A complete coverage of all features described in the UML is beyond the scope of this paper.

Our translation contributes to the on-going attempt to develop a formal semantics of UML based on graph transformation (cf. [6], [7], [8], [12], [13], [15], [16], [18]).

The paper is organized in the following way. In the next section, the basic notions of graph transformation are recalled as far as needed. Section 3 provides the translation of sequence graphs while in Section 4 sequence graphs are specified. A short example translation is presented in Section 5. In Section 6 we discuss how the translation can be proved correct, which is followed by some concluding remarks.

2   Graph Transformation

In this section, the basic notions and notations of graph transformation are recalled as far as they are needed in this paper.

In general, graphs consist of nodes and edges, where each edge has a source and a target node. Two different kinds of labels are assigned to the parts (i.e. nodes and edges) of a graph. The first kind of labels are fixed ones over a given alphabet, providing a basic type concept. The second kind are mutable labels, that are used as attributes. An attribute is regarded as a triple comprising the name of the attribute, its type and its current value. While the assignment of fixed labels to the parts is unique, there can be more than one attribute assigned to them.

The concept of graph transformation has been introduced about thirty years ago as a generalization of Chomsky grammars to graphs. Local changes in a graph are achieved by applying transformation rules. The graph transformation approach used in the context of this paper is based on the single pushout approach (cf., e.g., [5]). A formal embedding of the attribute concept into the single pushout approach can be found in [14]. A transformation rule consists of a left-hand side and a right-hand side, both of which are graphs. They share a common subgraph, which is called application context. In examples, the letter **L** indicates left-hand sides and **R** right-hand sides. The application context consists of all nodes of **L** and **R** with the same numbers. A rule can be applied to a given host graph if a match of the left-hand side of the rule in the host graph can be found. A match is a graph morphism which respects the structure of the involved graphs and the labels of their parts. The application of the rule to a host graph is achieved by adding those parts of the right-hand side that are not part of the application context to the host graph. Afterwards the parts of the left-hand side that are not part of the application context are deleted. The resulting structure is then made a graph by removing possible dangling edges. In this approach a rule can always be applied if a match is found, and in case of conflicting definitions deletion is stronger than preservation.

Note that it is possible to demand a concrete value of an attribute in the left-hand side of a rule. If the application of a rule is meant to change an attribute depending on its current value (e.g. increase a value by one), that value has to be identified as a variable in the left-hand side. The operation that is specified in the right-hand side of the rule then has to be performed on the value represented by that variable. If attributes are omitted in a rule, their values are of no interest. Thus attributes are only specified if one wants a concrete value to be present for the rule application or if that value has to be changed, either by calculating a new value or by simply overwriting the current value with a new one.

The left-hand side of a rule describes a positive application condition in the sense that its structure has to be found in the host graph. Sometimes it may be necessary to define a situation that is **not** wanted in the host graph. As proposed in [5], a situation that is not wanted in the host graph can be depicted as an additional graph called constraint. It shares a common subgraph with the left-hand side of the rule. If a match of the left-hand side into the host graph is found, and this match can be extended to map the constraint totally into the host graph, the rule must not be applied. If no such extension of the match can be found, the match satisfies the constraint and the rule can be applied. A rule can have a finite set of constraints, which is called negative application condition. In examples, the letter **N** indicates constraints (where indices distinguish different ones of the same rule). Nodes that are common in left-hand sides and constraints are numbered equally. A match satisfies a negative application condition if it satisfies all its constraints. Thus a rule with a negative application condition

NAC may only be applied if the match satisfies NAC. In [5], the left-hand side of a rule is required to be a subgraph of each constraint. This can always be met in our case by adding the missing parts to the constraints.

As a structuring principle, the notion of a graph transformation unit is employed (cf., e.g. [10],[11]). Such a transformation unit is a system tu $=$ (I, U, P, C, T) where I and T are graph class expressions specifying sets of initial and terminal graphs respectively, U is a set of (names of) imported transformation units, P is a set of local rules, and C is a control condition to regulate the use of the import and the application of the rules. In examples, we specify the five components after the respective keywords init, uses, rules, cond, and term. A keyword is omitted in case of a default component where all (graphs) is the default graph class expression, the empty set the default import and the default rule set, and true (allowing everything) the default control condition. The further graph class expressions and control conditions we use are explained when they occur the first time.

3   Translation

In this section, the translation of sequence graphs into collaboration graphs is modeled in a structured and top-down fashion by means of graph transformation units. It is beyond the scope of this paper to present the translation in all details. We introduce explicitly the structure of the translation and some significant transformation units, but we omit units which are defined analogously to given ones. Readers who want to see the complete specification are referred to [3].

The main unit splits the translation into two steps translating sequence graphs into metamodel object graphs first and then the latter ones into collaboration graphs. Moreover, it states that the translation is initially applied to sequence graphs generated by the transformation unit **GenerateSG**, which is specified in the next section.

SG2CG
init: **GenerateSG**
uses: **SG2MOG, MOG2CG**
cond: **SG2MOG; MOG2CG**

The control condition is a regular expression requiring that the two imported units are applied only once each in the given order.

Both the imported units enjoy essentially the same basic structure each except that all graphs are accepted as initial. In both cases, the collaboration parts and the interaction parts are handled separately followed by deleting graph components that are no longer needed. In the first unit, there is also an additional initial step.

SG2MOG
uses: **initMOG, SG2MOG-Collaboration, SG2MOG-Interaction, DeleteSG**
cond: **initMOG; SG2MOG_Collaboration; SG2MOG_Interaction; DeleteSG**

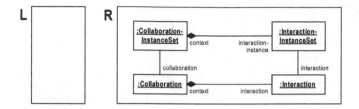

Fig. 2. Rule initMOG-1: Generating the initial metamodel object graph

MOG2CG
uses: **MOG2CG-Collaboration, MOG2CG_Interaction, DeleteMOG**
cond: **MOG2CG-Collaboration; MOG2CG_Interaction; DeleteMOG**

Altogether, the translation consists of seven sequential steps which are quite similarly structured to each other. None of them imports other units, but each consists of local rules only which are applied in a certain order given by the respective control conditions. Whenever we refer to UML features in the following units, their initial letters are printed in capitals as in the UML documents.

Let us start with the unit **initMOG**.

initMOG
rules: initMOG-1
cond: initMOG-1

It consists of a single rule which is applied only once. It adds an initial metamodel object graph disjointly to the given graph (see Figure 2).

Let us consider now the unit **SG2MOG-Collaboration**. It has got six rules, named sg2mog-C1a, -C1b, -C2a, -C2b, -C2c, and -C3. The rules are explicitly given in Figures 3 and 4. The control condition requires that either sg2mog-C1a or sg2mog-C1b is applied first, then either sg2mog-C2a or sg2mog-C2b is applied as long as possible, and afterwards sg2mog-C2c is applied once. This sequence is then repeated as long as possible, and finally sg2mog-C3 is applied once.

SG2MOG-Collaboration
rules: sg2mog-C1a, sg2mog-C1b, sg2mog-C2a, sg2mog-C2b, sg2mog-C2c, sg2mog-C3
cond:((sg2mog-C1a|sg2mog-C1b);(sg2mog-C2a|sg2mog-C2b)!;sg2mog-C2c)!;sg2mog-C3

The rules sg2mog-C1a and sg2mog-C1b add a new ClassifierRole, where sg2-mog-C1b adds a new confirming Object in addition. The regarded object is also marked with a writeClassifiers loop inidicating that all of its classifiers will be added next. The negative application conditions ensure, that neither the ClassifierRole nor the Object to be added already exists in the metamodel object graph. The rules sg2mog-C2a, sg2mog-C2b, and sg2mog-C2c concern base Classifiers. While sg2mog-C2a adds a new one, sg2mog-C2b links a role to an existing one. And sg2mog-C2c ends the addition. Finally, sg2mog-C3 inserts a first Message of the Interaction, with the ClassifierRole of the SGObject identified by the Current loop as both sender and receiver.

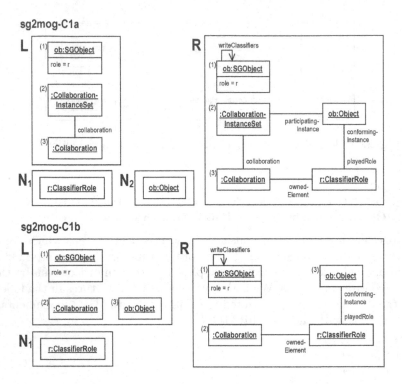

Fig. 3. Rules sg2mog-C1a and sg2mog-C1b

The control condition is formally described by a generalized regular expression. The symbol ';' denotes the sequential composition and the symbol '|' the alternative choice. The symbol '!' is similar to the regular Kleene star '*'. But while the latter one allows the repetition of the preceding expression as long as one likes, the exclamation mark requests repetition as long as possible. Therefore, the condition above allows to add as many ClassifierRoles and Objects as needed where each role is connected to all its base Classifiers and a first Message of the Interaction is inserted (which closes the role addition). The further transformation units used by **SG2MOG** are only sketched without the explicit rules.

The transformation unit **SG2MOG-Interaction** translates the interaction. The rules are designed to process the sequence graph in a certain order, thus a control condition is not necessary. The ordered translation is achieved by using a loop of type Current, that is moved from one node to the next node during the process. Terminal graphs contain a node that is incident with both an End and a Current loop. The class of all these graphs is denoted by **Current&End**.

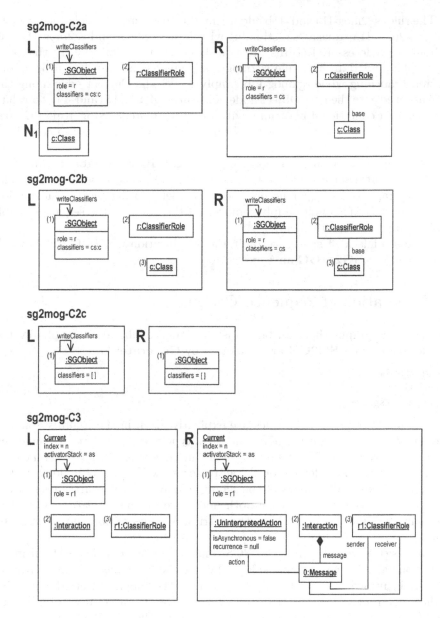

Fig. 4. Rules sg2mog-C2a, sg2mog-C2b, sg2mog-C2c, and sg2mog-C3

SG2MOG-Interaction
rules: sg2mog-I1a, sg2mog-I1b, sg2mog-I2a, sg2mog-I2b, sg2mog-I3a, sg2mog-I3b,
      sg2mog-I4a, sg2mog-I4b, sg2mog-I5a, sg2mog-I5b
term: **Current&End**

The rules sg2mog-I1a and -I1b add a stimulus to the metamodel object graph, where sg2mog-I1a translates a stimulus with a predecessor and sg2mog-I1b one without a predecessor. Returns are translated in a similar way by the rules sg2mog-I2a and -I2b. The rule sg2mog-I3a prepares the translation of a conditional branching while sg2mog-I3b simply moves the Current edge along the Activation edge to the next Object node. The rules sg2mog-I4a and -I4b translate a branch of a conditional branching with or without predecessor. Rule sg2mog-I5a completes the translation of one branch, while rule sg2mog-I5b completes the whole conditional branching.

Applying the transformation units **SG2MOG-Collaboration** and **SG2MOG-Interaction** yields the desired metamodel object graph. The sequence graph is no longer needed, so it is deleted using the transformation unit **DeleteSG**. This unit simply removes all the nodes of the sequence graph (and all of its edges with them).

Because of lack of space, we omit the specification of **MOG2CG**, which looks similar to **SG2MOG** anyway.

4   Generation of Sequence Graphs

The sequence graphs that can be translated into collaboration graphs by the transformation unit **SG2CG** are generated by **GenerateSG**.

GenerateSG
init: SG_0
rules: gsg1, gsg2, gsg3

The initial graph and the rules are explicitly given in the Figures 5 and 6. The initial graph SG_0 consists of two nodes of type SGObject that are connected by an Activation edge. The source node of this edge is marked as Current with index 1, and 0 the only element on the activatorStack. The target node is marked with an End edge. The values of the name, role and classifier attributes of the nodes can be freely chosen. Since both nodes are on the activation path they represent the same Instance. Thus the attribute values of both nodes must be equal.

Rule gsg1 is used to insert an SGstimulus with a matching Return in an activation path. In the left-hand side of the rule the two nodes of that activation path are identified. In order to ensure synchronous communication, the negative application graph N_1 secures that node 2 cannot be the target node of a Return edge. This means that the Instance represented by the identified nodes cannot send another procedure call while the current one is not yet completed.

The negative application graph N_2 prohibits node 1 from being the target node of a Return edge. This condition enforces that sequences are built from back to front.

The right-hand side of the rule adds two SGObject nodes to the activation path of the two identified nodes. These determine the attribute values of the newly added nodes. The first node (concerning the order in the activation path)

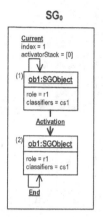

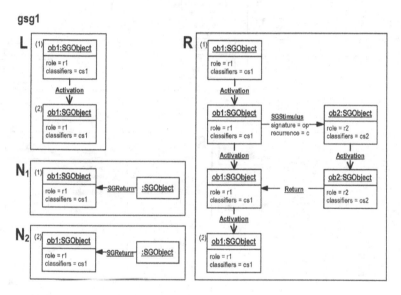

Fig. 5. Initial sequence graph and rule gsg1

is the source node of an SGStimulus edge and the second one is the target node of the matching Return edge. The attribute values of the SGStimulus edge can be freely chosen. The target node of the SGStimulus and the source node of the Return are added to the host graph as well. They form an activation path of their own. Their attribute values can be freely chosen and they might even equal those of the nodes of any other activation path. Note that the Instance they represent must possess the operation defined by the signature attribute of the SGStimulus edge. This rule enables one to model sequential and nested communication including recursive procedure calls of the same Instance.

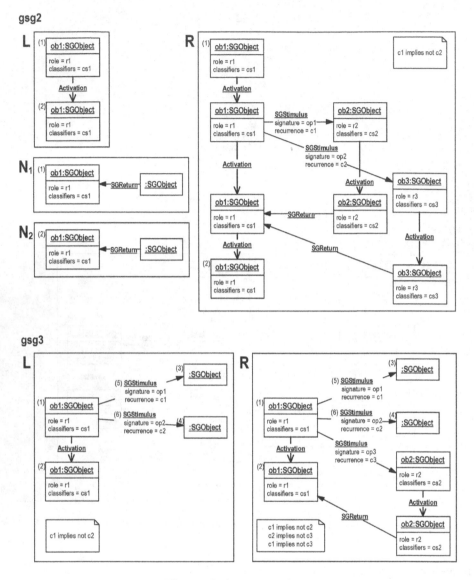

Fig. 6. Rules gsg2 and gsg3

Rule gsg2 is used to insert a conditional branching with two branches in an activation path. The left-hand side and the two negative application graphs N_1 and N_2 are identical to those of rule gsg1.

The right-hand side of this rule contains that of rule gsg1, i.e. the same changes are made. Additionally a second branch is inserted in the host graph starting at the source node of the newly added SGStimulus edge. It ends at the target node of the newly added Return. Note that both SGStimulus edges must have recurrence clauses and these must be mutually exclusive. This is necessary to ensure that the branching is not concurrent.

Rule gsg3 adds a new branch to an already existing conditional branching. Since an existing branching must always contain at least two branches and in order to avoid that a branching is created where none was before, the left-hand side must identify two different SGStimulus edges leaving the same source node. The end node of the conditional branching is also determined. It is the node directly succeeding the start node of the branching in the activation path. The application of the rule adds a new SGStimulus edge with a target node starting a new activation path to the host graph. A Return edge is also added which leaves the end of this activation path and points to the end node of the conditional branching. The recurrence clause of the newly added SGStimulus edge must be mutually exclusive with the ones of all the SGStimulus edges of the branching (which could in fact contain more than the ones identified by the left-hand side).

If a graph transformation unit is used as graph class expression, its semantics is the graph language consisting of all terminal graphs that are related to some initial graph. To make this notion feasible, we assume that there is only a single initial graph as in the case of **GenerateSG**. Formally, a graph is a graph class expression that specifies itself, i.e. $SEM(G) = \{G\}$ for all graphs G. Note that **GenerateSG** has no import, the rules are not regulated and all graphs are terminal. Therefore the language consists of all graphs that can be derived from the initial graph by the three rules. This is the set of sequence graphs that serve as inputs of the translation into collaboration graphs. Note that these graphs resemble sequence diagrams, they are not meant as abstract syntax description.

5 Example

The short example presented in this section is taken from the model of a computer chess game. In the considered situation player Amy moves the white queen to a target field. If this is a diagonal move it is validated as the move of a bishop, otherwise as the move of a rook.

Figure 7 shows the sequence diagram of this situation, which contains nested communication and a conditional branching.

The collaboration part of the metamodel object graph generated by the application of **SG2MOG_Collaboration** is depicted in Figure 8.

Applying **SG2MOG_Interaction** yields the intermediate graph, which is omitted due to its complexity. Figure 9 shows only the activator tree, which is the most important part with respect to the translation.

Finally, Figure 10 shows the resulting collaboration graph after applying **MOG2CG** and the corresponding collaboration diagram.

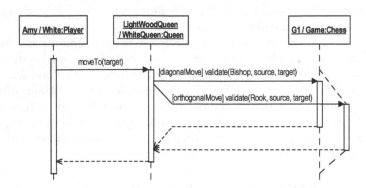

Fig. 7. Sequence diagram

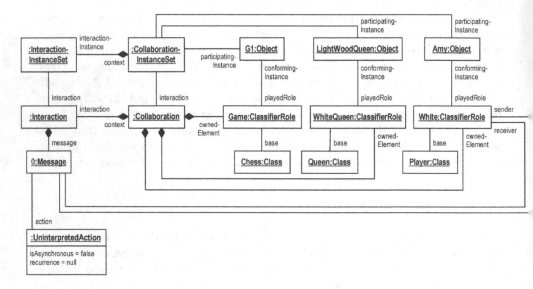

Fig. 8. Collaboration part of the metamodel object graph

6 Correctness

A translation can be called correct if the meaning of each input coincides with the meaning of the resulting output in a reasonable way. If both meanings are not equal, they should differ only in insignificant aspects (cf. [9]).

To consider the correctness of **SG2CG**, we must fix a meaning of sequential and collaboration diagrams. According to [2], the meaning of UML diagrams is given by their metamodel description besides the description in natural language. In other words, we may consider the translation **SG2MOG** of sequence graphs into metamodel object graphs (together with the one-to-one correspondence of these graphs with the respective types of diagrams) as the meaning of sequence

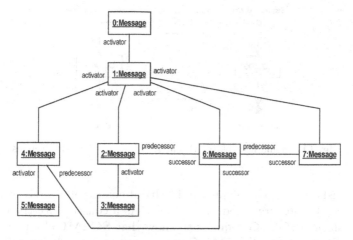

Fig. 9. Activator tree of the metamodel object graph

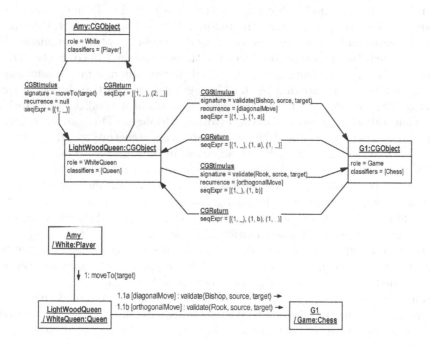

Fig. 10. Resulting collaboration graph and collaboration diagram

diagrams. Analogously, we consider a translation **CG2MOG** of collaboration diagrams onto the metamodel level to establish the meaning of collaboration diagrams.

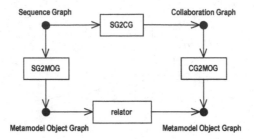

Fig. 11. Semantic relator discarding insignificant differences

This translation is given explicitly in the appendix of [3] and works analogously to translations presented in Section 3. As a consequence the correctness of our translation **SG2CG** requires to show that **SG2MOG** equals the composition of **SG2CG** and **CG2MOG** up to certain insignificant differences as illustrated in Figure 11.

Indeed, the metamodel object graph that is regarded as formal semantics of a sequence diagram is not necessarily unique. This is due to the fact, that in the case of conditional branchings the derivation process is not deterministic. The order in which the respective branches are translated is not fixed. So two translations of the same sequence graph may result in two different metamodel object graphs due to the numbering of the Messages during the translation process. Thus a different order in translating the branches of a conditional branching may yield different numbers for the involved Messages. This can be regarded as a minor aspect, since the decisive activator tree structure together with the predecessor-successor relationships is identical in both graphs. So in our case it makes sense to demand equality except for the numbering that is naming of the Messages. The numbering of the Messages has been established by us to aid in the translation. They are not demanded by the UML metamodel and must be omitted in the comparison of the metamodel object graphs. This can be realized by introducing a further transformation unit used as a semantic relator (cf. [9]), the purpose of which is to make the compared objects semantically equivalent by removing the insignificant differences. After applying this semantic relator to the regarded metamodel object graphs they must be equal.

This can be proved by complete induction on the lengths of derivations of the transformation unit **GenerateSG** that start with the initial sequence graph because, according to the construction in Section 4, the graphs derived in this way are the inputs of the translation.

The induction basis is a derivation length of zero. In this case, the transformation unit **GenerateSG** yields its initial graph as depicted in Figure 5. Now all the transformation units are applied, yielding the corresponding collaboration graph. This collaboration graph and the original sequence graph are then translated to their respective metamodel object graph by means of **CG2MOG** and **SG2MOG**. In this case the two resulting metamodel object graphs are already equal, since they both contain only one Message numbered 0. So the basis of the induction is proved.

The inductive step is set up by the statement, that we assume a correct translation for a derivation length of n and that we want to prove the correctness for a derivation length of $n + 1$. Since the transformation unit **GenerateSG** has three local rules, there are three cases that have to be distinguished in the inductive step. Each of these cases then has a number of subcases, depending on the concrete application situation. For example rule gsg1 can be used to either add a procedure call to a sequential communication or to insert it into a nested communication. For every case it has to be checked how the metamodel object graphs have to be changed during the translation of the generated sequence graphs in $n + 1$ derivation steps compared to the one generated in the first n identical derivation steps. So basically it suffices to compare the changes in the metamodel object graphs for every case after applying the semantic relators. This would prove that our approach yields a correct translation. But the explicit and detailed presentation of the various cases is beyond the scope of this paper.

7 Conclusion

We have presented a translation of UML sequence diagrams into UML collaboration diagrams in a structured way and sketched how the correctness can be proved. Due to Booch, Jacobson and Rumbaugh [2], the semantics of both types of diagrams is given in terms of metamodel object diagrams such that the correctness proof must involve this intermediate level.

We have constructed the translator in the framework of graph transformation units (cf. [10],[11]) because their interleaving semantics establishes translations between initial and terminal graphs by definition, supports correctness proofs by an underlying induction schema, and provides structuring concepts.

The translation of sequence diagrams into collaboration diagrams may also be seen as an example of the more recently introduced notion of a model transformation (see, e.g., [1], [4], [19]). But these approaches do not yet seem to support structuring and correctness proofs explicitly in the way they are used in this paper.

References

[1] D. Akehurst and S. Kent. A Relational Approach to Defining Transformations in a Metamodel. In J.-M. Jézéquel, H. Hussmann, and S. Cook, editors, *Proc. 5th Int. Conference on UML 2002—The Unified Modeling Language*, volume 2460 of *Lecture Notes in Computer Science*, pages 243–258. Springer, 2002. 289

[2] G. Booch, J. Rumbaugh, and I. Jacobson. *The Unified Modeling Language User Guide*. Addison-Wesley, 1998. 275, 286, 289

[3] B. Cordes and K. Hölscher. UML Interaction Diagrams: Correct Translation of Sequence Diagrams into Collaboration Diagrams. Diploma thesis, Department of Computer Science, University of Bremen, Bremen, Germany, 2003. 278, 288

[4] J. de Lara and H. Vangheluwe. AToM³: A Tool for Multi-formalism and Meta-modelling. In R.-D. Kutsche and H. Weber, editors, *Proc. 5th Int. Conference on Fundamental Approaches to Software Engineering*, volume 2306 of *Lecture Notes in Computer Science*, pages 174–188. Springer, 2002. 289

[5] H. Ehrig, R. Heckel, M. Korff, M. Löwe, L. Ribeiro, A. Wagner, and A. Corradini. Algebraic Approaches to Graph Transformation II: Single Pushout Approach and Comparison with Double Pushout Approach. In G. Rozenberg, editor, *The Handbook of Graph Grammars and Computing by Graph Transformation, Volume 1: Foundations*. World Scientific, 1997. 277, 278

[6] M. Gogolla. Graph Transformations on the UML Metamodel. In J.D.P. Rolim, A.Z. Broder, A. Corradini, R. Gorrieri, R. Heckel, J. Hromkovic, U. Vaccaro, and J.B. Wells, editors, *Proc. ICALP Workshop Graph Transformations and Visual Modeling Techniques (GVMT'2000)*, pages 359–371. Carleton Scientific, Waterloo, Ontario, Canada, 2000. 276

[7] M. Gogolla and F. Parisi-Presicce. State Diagrams in UML: A Formal Semantics using Graph Transformations. In M. Broy, D. Coleman, T.S.E. Maibaum, and B. Rumpe, editors, *Proc. ICSE'98 Workshop on Precise Semantics for Modeling Techniques*, pages 55–72. Technical Report TUM-I9803, 1998. 276

[8] M. Gogolla, P. Ziemann, and S. Kuske. Towards an Integrated Graph Based Semantics for UML. In P. Bottoni and M. Minas, editors, *Proc. ICGT Workshop Graph Transformation and Visual Modeling Techniques (GT-VMT'2002)*, volume 72(3) of *Electronic Notes in Theoretical Computer Science*. Springer, 2002. 276

[9] H.-J. Kreowski. Translations into the Graph Grammar Machine. In R. Sleep, R. Plasmeijer, and M. van Eekelen, editors, *Term Graph Rewriting: Theory and Practice*, pages 171–183. John Wiley, New York, 1993. 286, 288

[10] H.-J. Kreowski and S. Kuske. On the Interleaving Semantics of Transformation Units—A Step into GRACE. In J.E. Cuny, H. Ehrig, G. Engels, and G. Rozenberg, editors, *Proc. 5th Int. Workshop on Graph Grammars and their Application to Computer Science*, volume 1073 of *Lecture Notes in Computer Science*, pages 89–106. Springer, 1996. 278, 289

[11] H.-J. Kreowski and S. Kuske. Graph Transformation Units and Modules. In H. Ehrig, G. Engels, H.-J. Kreowski, and G. Rozenberg, editors, *The Handbook of Graph Grammars and Computing by Graph Transformation, Volume 2: Applications, Languages and Tools*. World Scientific, 1999. 278, 289

[12] S. Kuske. A Formal Semantics of UML State Machines Based on Structured Graph Transformation. In M. Gogolla and C. Kobryn, editors, *UML 2001 - The Unified Modeling Language. Modeling Languages, Concepts, and Tools*, volume 2185 of *Lecture Notes in Computer Science*, pages 241–256. Springer, 2001. 276

[13] S. Kuske, M. Gogolla, R. Kollmann, and H.-J. Kreowski. An Integrated Semantics for UML Class, Object and State Diagrams Based on Graph Transformation. In M. Butler and K. Sere, editors, *3rd Int. Conf. Integrated Formal Methods (IFM'02)*, volume 2335 of *Lecture Notes in Computer Science*. Springer, 2002. 276

[14] M. Löwe, M. Korff, and A. Wagner. An Algebraic Framework for the Transformation of Attributed Graphs. In R. Sleep, R. Plasmeijer, and M. van Eekelen, editors, *Term Graph Rewriting: Theory and Practice*, pages 185–199. John Wiley, New York, 1993. 277

[15] A. Maggiolo-Schettini and A. Peron. Semantics of Full Statecharts Based on Graph Rewriting. In H.-J. Schneider and H. Ehrig, editors, *Proc. Graph Transformation in Computer Science*, volume 776 of *Lecture Notes in Computer Science*, pages 265–279. Springer, 1994. 276

[16] A. Maggiolo-Schettini and A. Peron. A Graph Rewriting Framework for State-charts Semantics. In J. E. Cuny, H. Ehrig, G. Engels, and G. Rozenberg, editors, *Proc. 5th Int. Workshop on Graph Grammars and their Application to Computer Science*, volume 1073 of *Lecture Notes in Computer Science*, pages 107–121. Springer, 1996. 276

[17] OMG, editor. OMG Unified Modeling Language Specification, Version 1.4, September 2001. Technical report, Object Management Group, Inc., Framingham, MA, 2001. 275

[18] A. Tsiolakis and H. Ehrig. Consistency Analysis of UML Class and Sequence Diagrams using Attributed Graph Grammars. In H. Ehrig and G. Taentzer, editors, *Proc. of Joint APPLIGRAPH/GETGRATS Workshop on Graph Transformation Systems*, pages 77–86, 2000. Technical Report no. 2000/2, Technical University of Berlin. 276

[19] D. Varró. A Formal Semantics of UML Statecharts by Model Transition Systems. In A. Corradini, H. Ehrig, H. J. Kreowski, and G. Rozenberg, editors, *Proc. First Int. Conference on Graph Transformation*, volume 2505 of *Lecture Notes in Computer Science*, pages 378–392. Springer, 2002. 289

Meta-Modelling, Graph Transformation and Model Checking for the Analysis of Hybrid Systems

Juan de Lara[1], Esther Guerra[1], and Hans Vangheluwe[2]

[1] Escuela Politécnica Superior
Ingeniería Informática
Universidad Autónoma de Madrid
(Juan.Lara,Esther.Guerra_Sanchez)@ii.uam.es
[2] School of Computer Science
McGill University, Montréal
Québec, Canada
hv@cs.mcgill.ca

Abstract. This paper presents the role of meta-modelling and graph transformation in our approach for the modelling, analysis and simulation of complex systems. These are made of components that should be described using different formalisms. For the analysis (or simulation) of the system as a whole, each component is transformed into a single common formalism having an appropriate solution method. In our approach we make meta-models of the formalisms and express transformations between them as graph transformation. These concepts have been automated in the AToM^3 tool and as an example, we show the analysis of a hybrid system composed of a temperature controlled liquid in a vessel. The liquid is initially described using differential equations whose behaviour can be abstracted and represented as a Statechart. The controller is modelled by means of a Statechart and the temperature as a Petri net. The Statechart models are translated into Petri nets and joined with the temperature model to form a single Petri net, for which its reachability graph is calculated and Model-Checking techniques are used to verify its properties.

Keywords: Graph Rewriting, Meta-Modelling, Multi-Paradigm, Hybrid Systems, Model-Checking.

1   Introduction

Complex systems are characterized by interconnected components of very different nature. Some of these components may have continuous behaviour while the behaviour of other components may be discrete. Systems with both classes of components are called hybrid systems. There are several approaches to deal with the modelling, analysis and simulation of complex systems. Some approaches try to use a formalism general enough (a "super-formalism") to express the behaviour of all the components of the system. In general this is neither possible

J.L. Pfaltz, M. Nagl, and B. Böhlen (Eds.): AGTIVE 2003, LNCS 3062, pp. 292–298, 2004.

nor adequate. Other approaches let the user model each component of the system using the most appropriate formalism. While in co-simulation each component is simulated with a formalism-specific simulator; in multi-formalism modelling a single formalism is identified into which each component is symbolically transformed [10]. In co-simulation the simulator interaction due to component coupling is resolved at the trajectory (simulation data) level. With this approach it is no longer possible to answer questions in a symbolic way about the behaviour of the whole system.

In multi-formalism modelling however, we can verify properties of the whole system if we choose a formalism with appropriate analysis methods for the transformation. The Formalism Transformation Graph (FTG) [10] can help in identifying a common, appropriate formalism to transform each component. The FTG depicts a part of the "formalism space", in which formalisms are shown as nodes in the graph. The arrows between them denote a homomorphic relationship "can be mapped onto", using symbolic transformations between formalisms. Other arrows (vertical) denote the existence of a simulator for the formalism.

Multi-Paradigm Modelling [10] combines multi-formalism, meta-modelling, and multiple levels of abstraction for the modelling, analysis and simulation of complex systems. Meta-modelling is used to describe different formalisms of the FTG and generate tools for them. In our work, we propose to model both transformation and simulation arrows of the FTG as graph transformation, as models and meta-models can be represented as attributed, typed graphs. Other model manipulations, for optimisation and code generation can be expressed with graph transformation as well. We have implemented these concepts in the Multi-Paradigm tool AToM^3 [5], which is used in the following section to model and analyse a simple hybrid system.

2   Example: A Temperature-Controlled Liquid in a Vessel

As a very simple example to help clarifying our Multi-Paradigm modelling approach, we show the modelling and analysis of a temperature-controlled liquid in a vessel. The system is composed of a continuous part which represents the liquid behaviour and a discrete part which controls the temperature. The continuous part is described by the following equations:

$$\frac{dT}{dt} = \frac{1}{H}\left[\frac{W}{c\rho A}\right] \tag{1}$$

$$is\_cold = (T < T_{cold}) \tag{2}$$

$$is\_hot = (T > T_{hot}) \tag{3}$$

Where W is the rate at which heat is added or removed, A is the cross-section surface of the vessel, H is the level of the liquid, c is its specific heat and ρ its density. This system can be observed through two output sensors is_cold and is_hot, which are set in equations 2 and 3. These sensors allow us to discretize the

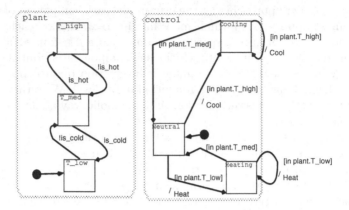

Fig. 1. Statechart representing the behaviour of a temperature-controlled liquid

state-space [10], in such a way that the system can be represented as a finite state automaton (shown to the left of Figure 1). The system's behaviour is governed by equation (1) in each automaton state, while transitions are fired by equations (2) and (3). Though at a much higher level of abstraction, the automaton alone (without the equation) captures the essence of the system's behaviour.

The controller can also be represented as a state automaton, and is shown to the right of Figure 1. We have combined both automata in a single Statechart with two orthogonal components, named as plant and control, using the meta-model for Statecharts available in AToM³. Equation 1 can be discretized and represented as a Petri net (it is shown together with the sensors in equations 2 and 3 to the right of Figure 2). This Petri net has a place Temp to represent the temperature value. The Heat and Cool methods invoked by the controller are modelled by transitions which add and remove a token from Temp. The number of tokens in Temp remains bounded (between 0 and 30 in this example) using a well-known technique for capacity constraining. This technique consists on creating an additional place Temp', in such a way that the number of tokens in both places is always equal to 30. Using the value of both places, we can model the events is_cold, !is_cold, is_hot and !is_hot (that can be considered as events produced by sensors). In this example we set the intervals [0-10] for cold and [21-30] for hot.

The Statechart can be automatically converted into Petri nets and joined with the component which models the temperature. The transformation from Statecharts into Petri nets was automated with AToM³ using the graph transformation described in [6]. States and events are converted into places, current states in orthogonal components are represented as tokens. Once the model is expressed in Petri nets, we can apply the available analysis techniques for this formalism. In AToM³ we have implemented transformations to obtain the reachability graph, code generation for a Petri net tool (PNS), simulation and simplification. The latter can be applied before calculating the reachability graph to help

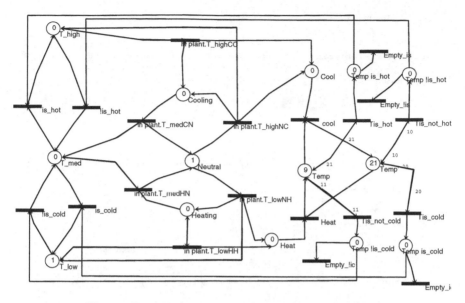

Fig. 2. Petri Net generated from the previous model

reducing the state-space. The Petri net model, after applying the simplification graph transformation is shown in Figure 2.

The reachability graph is shown in Figure 3 (labels depict tokens in places). Note how the reachability graph calculation is indeed another formalism transformation, from Petri nets into state automata. We have manually set priorities on the Petri net transitions to greatly reduce the state-space. We have set pr(T is_cold)>pr(T is_not_hot), pr(T is_hot)>pr(T is_not_cold), the priorities of Heat and Cool bigger than any other and the priorities of the plant transitions larger than the ones in the controller. Once we have the reachability graph, we can formalize the properties we want to check using Computational Tree Logic (CTL) and check the graph using a simple explicit Model-Checking algorithm [4] and a meta-model for CTL we have implemented in AToM³. The user can specify the CTL formula in two ways: graphically (drawing the graph of the CTL formula using the meta-model), or textually. In the latter case the formula is parsed and translated into an Abstract Syntax Graph (ASG). Note how if the user graphically specifies the formula, he is directly building the ASG. In the example, we may ask whether the plant reaches a certain temperature, or the T_med state (interval [11-20]), or whether the controller heats the liquid and eventually stops heating. With our initial conditions (liquid at 9 degrees), all these properties are true, given appropriate fairness constraints allowing the eventual firing of enabled plant and control transitions when the sensor transitions are also enabled.

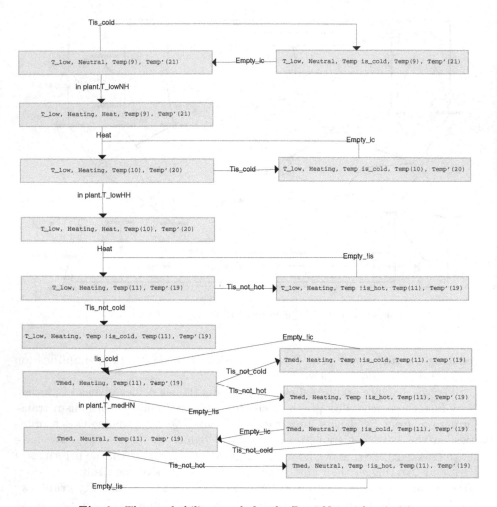

Fig. 3. The reachability graph for the Petri Net with priorities

3   Related Work

There are similar tools in the graph grammars community. For example,
GENGED [2] has a similar approach: the tool allows defining visual notations
and manipulate them by means of graph grammars (the graph rewriting engine
is based on AGG [1]). Whereas the visual language parsing in the GENGED
approach is based on graph grammars, in AToM³ we rely on constraints checking to verify that the model is correct. The mapping from abstract to concrete
syntax is very limited in AToM³, as there is a "one to one" relationship between
graphical icons and entities. On the contrary, this mapping is more flexible in
GENGED, which in addition uses a constraint language for the concrete syntax

layout. In AToM^3 this layout should be coded as Python expressions. This is lower-level, but usually more efficient.

Although other tools based on graph grammars (such as DiaGen [8]) use the concept of bootstraping, in AToM^3 there is no structural difference between the generated editors (which could be used to generate other ones!), and the editor which generated them. In fact, one of the main differences of the approach taken in AToM^3 with other similar tools, is the concept that (almost) everything in AToM^3 has been defined by a model (under the rules of some formalism, including graph grammars) and thus the user can change it, obtaining more flexibility.

With respect to the transformation into Petri nets, the approach of [3] is similar, but they create two places for each method in the Statechart (one for calling the method and the other for the return). Another approach for specifying the transformation is the use of triple graph grammars [9].

4   Conclusions

In this paper we have presented our approach (multi-paradigm modelling) for modelling and analysis of complex systems, based on meta-modelling and graph transformation. We make meta-models of the formalisms we want to work with, and use graph transformation to formally and visually define model manipulations. These include formalism transformation, simulation, optimisation and code generation. To analyse a complex system, we transform each component into a common formalism where the properties of interest can be checked. In the example, we have modelled a simple system, composed of a Statechart and a Petri net component into a single Petri net. Then we have calculated the reachability graph and verified some properties using Model-Checking. It must be noted however that we have made large simplifications for this problem. For example, for more complex systems, an explicit calculation of the reachability graph may not be possible, and other symbolic techniques should be used.

We are currently working in demonstrating properties of the transformations themselves, such as termination, confluence and behaviour preservation. Theory of graph transformation, such as critical pair analysis [7] can be useful for that purpose.

Acknowledgements

We would like to acknowledge the SEGRAVIS network and the Spanish Ministry of Science and Technology (project TIC2002-01948) for partially supporting this work, and the anonymous referees for their useful comments.

References

[1] AGG home page: http://tfs.cs.tu-berlin.de/agg/ 296

[2] Bardohl, R., Ermel, C., Weinhold, I. 2002 *AGG and GenGED: Graph Transformation-Based Specification and Analysis Techniques for Visual Languages* In Proc. GraBaTs 2002, Electronic Notes in Theoretical Computer Science 72(2) 296

[3] Baresi, L., Pezze, M.. *Improving UML with Petri nets.* Electronic Notes in Theoretical Computer Science 44 No. 4 (2001) 297

[4] Clarke, E. M., Grumberg, O., Peled, D. A. 1999. *Model Checking.* MIT Press 295

[5] de Lara, J., Vangheluwe, H. 2002 *AToM³: A Tool for Multi-Formalism Modelling and Meta-Modelling.* In ETAPS02/FASE. LNCS 2306, pp.: 174 - 188. See also the AToM³ home page: http://atom3.cs.mcgill.ca 293

[6] de Lara, J., Vangheluwe, H. 2002 *Computer Aided Multi-Paradigm Modelling to process Petri Nets and Statecharts.* ICGT'2002. LNCS 2505. Pp.: 239-253 294

[7] Heckel, R., Küster, J. M., Taentzer, G. 2002. *Confluence of Typed Attributed Graph Transformation Systems.* In ICGT'2002. LNCS 2505, pp.: 161-176. Springer 297

[8] Minas, M. 2003. *Bootstrapping Visual Components of the DiaGen Specification Tool with DiaGen* Proceedings of AGTIVE'03 (Applications of Graph Transformation with Industrial Relevance), Charlottesville, USA, pp.: 391-406. See also the DiaGen home page: http://www2.informatik.uni-erlangen.de/DiaGen/ 297

[9] Schürr, A. 1994. *Specification of Graph Translators with Triple Graph Grammars.* LNCS 903, pp.: 151-163. Springer 297

[10] Vangheluwe, H., de Lara, J., Mosterman, P. 2002. *An Introduction to Multi-Paradigm Modelling and Simulation.* Proc. AIS2002. Pp.: 9-20. SCS International 293, 294

Proper Down-Coloring Simple Acyclic Digraphs

Geir Agnarsson[1], Ágúst S. Egilsson[2], and Magnús M. Halldórsson[3]

[1] Department of Mathematical Sciences, George Mason University
MS 3F2, 4400 University Drive, Fairfax, VA 22030
geir@math.gmu.edu
[2] Department of Mathematics, University of California
Berkeley, CA 94720-3840
egilsson@Math.Berkeley.EDU
[3] Department of Computer Science, University of Iceland
Dunhaga 3, IS-107 Rvk, Iceland
mmh@hi.is

Abstract. We consider vertex coloring of a simple acyclic digraph $\overline{G}$ in such a way that two vertices which have a common ancestor in $\overline{G}$ receive distinct colors. Such colorings arise in a natural way when clustering, indexing and bounding space for various genetic data for efficient analysis. We discuss the corresponding chromatic number and derive an upper bound as a function of the maximum number of descendants of a given vertex and the inductiveness of the corresponding hypergraph, which is obtained from the original digraph.

Keywords: Relational databases, multidimensional clustering, indexing, bitmap indexing, vertex coloring, digraph, acyclic digraph, poset, hypergraph, ancestor, descendant, down-set.

1 Introduction

1.1 Purpose

The purpose of this article is to discuss a special kind of vertex coloring for acyclic digraphs in bioinformatics, where we insist upon two vertices, that have a common ancestor, receive distinct colors.

This article can be viewed as a continuation of [2], but we have attempted to make it as self-contained as possible. For a brief introduction and additional references to the ones mention here, we refer to [2].

Before setting forth our terminology and stating precise definitions we need to address the justification for such vertex colorings and why they are of interest, especially for various databases containing life science related data.

Digraphs representing various biological phenomena and knowledge are used all over in the life sciences and in drug discovery research, i.e. the gene ontology digraph maintained by the Gene Ontology Consortium, [11]. Many other open biological ontologies are available, and additionally medical or disease classifications in the form of acyclic digraphs are used throughout. In a typical setting

J.L. Pfaltz, M. Nagl, and B. Böhlen (Eds.): AGTIVE 2003, LNCS 3062, pp. 299–312, 2004.
© Springer-Verlag Berlin Heidelberg 2004

Table 1. This matrix indicates a data structure of size n^2, where in this case $n = 6$

	g_1	g_2	g_3	g_4	g_5	g_6
g_1	g_1	–	–	g_4	g_5	–
g_2	–	g_2	–	g_4	–	g_6
g_3	–	–	g_3	–	g_5	g_6
g_4	–	–	–	g_4	–	–
g_5	–	–	–	–	g_5	–
g_6	–	–	–	–	–	g_6

Table 2. This matrix indicates a data structure of size kn, where in this case $n = 6$ and $k = 3$

	1	2	3
g_1	g_1	g_4	g_5
g_2	g_6	g_4	g_2
g_3	g_6	g_3	g_5
g_4	–	g_4	–
g_5	–	–	g_5
g_6	g_6	–	–

the digraphs are referenced by numerous large tables containing genetic or other observed data and stored in relational data warehouses. An overview of several projects relating to indexing of semistructured data (i.e. graphs) can be found in [1].

The challenges addressed in this article include certain coloring analysis of an acyclic digraph $\overline{G}$, which provides an efficient structure for querying of relational tables referencing the digraph $\overline{G}$. This includes efficiently identifying and retrieving all rows in a given table, or matrix, that are conditioned, based on sets of ancestors of $\overline{G}$.

EXAMPLE: Consider the digraph $\overline{G}$, on $n = 6$ vertices representing genes, where a directed edge from one vertex to a second one indicates that the first gene is an ancestor of the second gene.

$$V(\overline{G}) = \{g_1, g_2, g_3, g_4, g_5, g_6\},$$
$$E(\overline{G}) = \{(g_1, g_4), (g_1, g_5), (g_2, g_4), (g_2, g_6), (g_3, g_5), (g_3, g_6)\}.$$

In the static data structure of a 6×6 matrix, we assign a column to each vertex g_i. Further we include the data of g_j in the (i, j)-th entry if, and only if, g_i is an ancestor of g_j, as shown in Table 1. As we see there, most entries are empty with no data in them, which is a waste of storing space. Relaxing the condition that each vertex has its own column, it suffices instead to insist that (i) each gene can only occur in one column, and to avoid clashing of data information, (ii) each column can only contain data of unrelated genes. In this case we obtain a compact form of Table 1 as shown in Table 2. There we have reduced the number of columns from $n = 6$ to $k = 3$. Note that k here is the maximum number of vertices which are descendants of a single vertex. All the vertices g_1, g_2 and g_3 have three descendants (including themselves), and all the vertices g_4, g_5 and g_6 have just one descendant, just themselves. Such sets will in this article be denoted by $D[g_i]$, so, for example $D[g_1] = \{g_1, g_4, g_5\}$. As we will see, it is, in general, is not always possible to reduce the the number of columns down to this maximum number of descendants of a single vertex. Here we can view the numbers 1,2 and 3 (the columns of Table 2) as distinct labels, data entries or colors assigned to each vertex. So g_1 and g_6 are assigned color 1, g_3 and g_4 are assigned color 2 and g_2 and g_5 are assigned color 3. We see that if two vertices g_i and g_j have the same color, then they must be unrelated in the sense that there is no vertex g_k such that $g_i, g_j \in D[g_k]$.

A more functional way of viewing these color labelings is given in the following subsection, where each color i can be viewed as a labeling function f_i.

1.2   A Functional Approach

Consistent with the notation introduced in the next section, we let $U[u] = \{u\} \cup \{x : x$ is an ancestor of $u\}$ for each vertex u of a given digraph $\overline{G}$. Also, assign a unique data entry $U[u]*$ to each of the ancestor sets considered. Since each ancestor set $U[u]$ is determined uniquely by the vertex u, then we can simply identify u and $U[u]*$, and assume that $u = U[u]*$.

In order to create an index on a table column referencing the vertices $V(\overline{G})$ of the digraph $\overline{G}$, it is necessary to develop a schema, such as a collection of functions, for assigning data entries to the search key values in the table. First though, we develop such a schema for the vertices $V(\overline{G})$ themselves. For a given vertex u there should be at least one function f, in the schema, defined on a subset of $V(\overline{G})$, satisfying $U[u] = f^{-1}(U[u]*)$, that is, f assigns the data entry $U[u]*$ only to search key elements from the set $U[u]$. Therefore, a complete schema is a collection $f_1, f_2, \ldots, f_k$ of functions so that for each vertex u there exists at least one integer $c(u)$ with

$$U[u] = f_{c(u)}^{-1}(U[u]*).$$

The vertex coloring $u \mapsto c(u)$ is what we will call a down-coloring of the digraph $\overline{G}$, as we will define in the next section, and it has the following property: If two distinct vertices u and v have a common ancestor, say w, then $f_{c(u)}(w) = U[u]*$ and $f_{c(v)}(w) = U[v]*$, so since $U[u]* \neq U[v]*$, we must have $c(u) \neq c(v)$, that is u and v receive distinct colors. This allows us to conclude that k is at least the down-chromatic number of $\overline{G}$, defined in the next section.

Conversely, given a down-coloring $c : V(\overline{G}) \to \{1, \ldots, k\}$ one can construct a complete schema for assigning data entries to the vertices by defining functions $f_1, \ldots, f_k$ as

$$f_i(u) = U[v]*, \text{ if } u \in U[v] \text{ and } c(v) = i.$$

The down-coloring condition ensures that the functions are well defined. The possible schemas of "functional indexes" are therefore in a one-to-one correspondence with the possible down-colorings.

One can realize the schema $f_1, \ldots, f_k$ in a relational database system in many different ways. In the relational database system, one may try additionally to devise a structure so that the set operations $(\cup, \cap, \backslash)$ can be optimally executed on elements from the collection $\{U[u] : u \in V(\overline{G})\}$. A straightforward way to do this is to have the functions physically share the domain $V(\overline{G})$ in the database, instead of using additional equijoins to implement the sharing of the domain. For digraphs with relatively small down-chromatic numbers we therefore materialize the relation

$$\{(u, f_1(u), \ldots, f_k(u)) : u \in V(\overline{G})\}$$

in the database system, in addition to the coloring map c. This requires that $f_j(u) = $ "NULL" if u is not in the domain of f_j, following standard convention. The table containing the above relation is referred to as Clique(U), it has a domain column "Vertex" and a column "CJ" for each of the colors $j \in \{1, 2, \ldots, k\}$. Of course, it is also the graph of the function $f_1 \times f_2 \times \cdots \times f_k$.

1.3   Concrete Databases

If a (large) table references the digraph $\overline{G}$ in one of its columns, then there are several possible methods to index the column using the data entries schema Clique(U) or, equivalently, the functions $f_1, \ldots, f_k$. Below we summarize two of the possible methods.

1. The Clique(U) relation may be joined (through the "Vertex" column) with any table that references the digraph $\overline{G}$ in one of its columns. In this way the Clique(U) table may be viewed as a dimension table, allowing queries conditioned on the ancestors sets and other conditioning to be evaluated as star queries. Additional search mechanisms are introduced, such as bitmap and other indexes and, most efficiently, bitmap join indexes may be used (e.g., one for each color column in Clique(U) when using the Oracle 9i system). Both the Oracle database system and DB2 from IBM are able to take advantage of this simple and space efficient setup for evaluating queries, see [4] and [6] for further information.
2. A table referencing the digraph $\overline{G}$ in one of its columns may be joined with, and clustered according to the schema Clique(U) using multidimensional clustering, if the down-chromatic number for the digraph is small. Ideally, the multidimensional clustering enables the referencing table to be physically clustered along all the color columns simultaneously. Commercially available relational database systems that implement multidimensional clustering, include IBM's DB2 Universal Database (version 8.1), see [5] for an introduction to the feature in DB2.

The first method is currently used by deCODE Genetics in Iceland. There the Clique(U) structure and bitmap and sometimes bitmap join indexes are combined to map a data entry $U[u]*$ to all the rows in a table referencing elements from $U[u]$. Comparisons, in this setting, favor greatly using the Clique(U) structure and star-query methods over other methods. We will demonstrate this with a concrete comparison.

EXAMPLE: A small table with about 1.5 million rows called "goTermFact" references the gene ontology digraph in a column called "acc". Additionally, the table has several other columns including a number column denoted by "m". The acyclic gene ontology digraph has 14,513 edges and 11,368 vertices, the edges are directed and converge at a root vertex, the root has three ancestors called "molecular function", "biological process" and "cellular component". One of the ancestors of "molecular function" in the graph is called "enzyme activity". We wish to summarize the column "m" for all the 394,702 rows that reference an ancestor of the vertex "enzyme activity". The digraph is down-colored using 36 colors and the vertex "enzyme activity" receives color "8", it is also assigned code "GO:0003824" by the Gene Ontology Consortium. The SQL query (Q1) is constructed as:

Q1 : select sum(f.m) from goTermFact f, clique_U d where
        f.acc = d.vertex and d.C8 = 'GO:0003824'

For the purpose of comparison, the referencing table can also be indexed using the digraph, in a more traditional way, by creating a lookup-table "Lookup" with columns "path" and "rid". The "path" column is a string path starting at the root vertex and the "rid" column contains row-ids from the referencing table. Since this particular digraph is not a tree structure, then one possibly needs to store several paths for each vertex (1.9 on the average in this case). The "Lookup" table is stored as an index-organized table with a composite primary key ("path" and "rid") in the Oracle 9.2i database, providing fast primary key based access to lookup data for range search queries. A path from the root vertex to the "enzyme activity" vertex, following the (reversed) digraph, is of the form: 'GO:0003673.GO:0003674.GO:0003824'. Therefore the above query (Q1) can now be rewritten using a range search of the "Lookup" table as follows:

Q2 : select sum(m) from goTermFact where
        rowid in (select rid from Lookup where
          path like 'GO:0003673.GO:0003674.GO:0003824%')

Executing the queries on a Windows XP based Pentium III, 1.2GHz, 512MB system using the Oracle 9.2i database reveals the following comparison:

Q1: Form Q1 executes up to 150 times faster than equivalent form Q2. The best performance is achieved using a bitmap join index to (pre-)resolve the join between the Clique(U) relation and the table. The query takes between 0.03 sec and 21 sec, depending on whether the data is located in memory or on disk. The query is also efficiently executed by using a (static) bitmap index on the "acc" column and the bitmap OR operator to dynamically construct, using Clique(U), a pointer to all the rows in the "goTermFact" table that satisfy the join.

Q2: In form Q2 the query takes between 4.02 sec and 104 sec, depending again on whether the table and index data is located in memory or on disk when the query is executed. Clearly, this form is much more space consuming and therefore less efficient than the previous one.

The above comparison, as well as many other results obtained using similar clique indexing schemas, demonstrate the power of the indexing, when combined with current relational database optimization techniques.

1.4   Related Work

Directed acyclic graphs are many times called DAG's by computer scientists, as is the case in [9, p. 194], but we will here call them acyclic digraphs.

Only few vertex coloring results rely on the structure that the directions of the edges in a digraph provide. In [7] the arc-chromatic number for a digraph $\overline{G}$ is investigated, and in [8] and [10] the dichromatic number of a digraph is studied.

Here in this article we define the down-chromatic number of an acyclic digraph, discuss some of its properties, similarity and differences with ordinary graph coloring, and derive an upper bound which, in addition, yields an efficient coloring procedure. We will give some different representations of our acyclic digraph, some equivalent and other more relaxed representations, which, from our vertex coloring point of view, will suffice to consider.

2   Definitions and Observations

We attempt to be consistent with standard graph theory notation in [14], and the notation in [12] when applicable.

For a natural number $n \in \mathbb{N}$ we let $[n] = \{1, \dots, n\}$. A simple digraph is a finite simple directed graph $\overline{G} = (V, E)$, where V is a finite set of vertices and $E \subseteq V \times V$ is a set of directed edges. For a digraph $\overline{G}$, the sets of its vertices and edges will many times be given by $V(\overline{G})$ and $E(\overline{G})$ respectively. The digraph $\overline{G}$ is said to be acyclic if $\overline{G}$ has no directed cycles. Henceforth $\overline{G}$ will denote an acyclic digraph in this section.

The binary relation $\leq$ on $V(\overline{G})$ defined by

$$u \leq v \Leftrightarrow u = v, \text{ or there is a directed path from } v \text{ to } u \text{ in } \overline{G}, \tag{1}$$

is a reflexive, antisymmetric and transitive binary relation, and therefore a partial order on $V(\overline{G})$. Hence, whenever we talk about $\overline{G}$ as a poset, the partial order will be the one defined by (1). Note that the acyclicity of $\overline{G}$ is essential in order to be able to view $\overline{G}$ as a poset. By the height of $\overline{G}$ as a poset, we mean the number of vertices in the longest directed path in $\overline{G}$. We denote by $\max\{\overline{G}\}$ the maximal vertices of $\overline{G}$ with respect to the partial order $\leq$.

For vertices $u, v \in V(\overline{G})$ with $u \leq v$, we say that u is a descendant of v, and v is an ancestor of u. The closed principal down-set $D[u]$ of a vertex $u \in V(\overline{G})$ is the set of all descendants of u in $\overline{G}$, that is, $D[u] = \{x \in V(\overline{G}) : x \leq u\}$. Similarly the open principal down-set $D(u)$ of a vertex $u \in V(\overline{G})$ is the set of all descendants of u in $\overline{G}$, excluding u itself, that is, $D(u) = D[u] \setminus \{u\}$.

Definition 1. A down-coloring of $\overline{G}$ is a map $c : V(\overline{G}) \to [k]$ satisfying

$$u, v \in D[w] \quad \text{for some } w \in V(\overline{G}) \Rightarrow c(u) \neq c(v)$$

for every $u, v \in V(\overline{G})$. The down-chromatic number of $\overline{G}$, denoted by $\chi_d(\overline{G})$, is the least k for which $\overline{G}$ has a proper down-coloring $c : V(\overline{G}) \to [k]$.

Just as in an undirected graph G, the vertices in a clique must all receive distinct colors in a proper vertex coloring of G. Therefore $\omega(G) \leq \chi(G) \leq |V(G)|$ where $\omega(G)$ denotes the clique number of G. In addition $\chi(G)$ can be larger than the clique number $\omega(G)$. Similarly for our acyclic digraph $\overline{G}$ we have $D(\overline{G}) \leq \chi_d(\overline{G}) \leq |V(\overline{G})|$, where

$$D(\overline{G}) = \max_{u \in V(\overline{G})} \{|D[u]|\},$$

and we will see below that the same holds for $\overline{G}$, that $\chi_d(\overline{G})$ can be much larger than $D(\overline{G})$.

A useful approach to consider down-colorings of a given acyclic digraph $\overline{G}$, is to construct the corresponding simple undirected down-graph G' on the same set of vertices as $\overline{G}$, but where we connect each pair of vertices that are contained in the same principal down-set

$$V(G') = V(\overline{G}),$$
$$E(G') = \{\{u, v\} : u, v \in D[w] \text{ for some } w \in V(\overline{G})\}.$$

In this way we have transformed the problem of down-coloring the digraph $\overline{G}$ to the problem of vertex coloring the simple undirected graph G' in the usual sense, and we have $\chi_d(\overline{G}) = \chi(G')$. Hence, from the point of down-colorings, both $\overline{G}$ and G' are equivalent, which is something we will discuss in the next section to come. However, some structure is lost. The fact that two vertices u and v are connected in G' could mean one of three possibilities, (i) $u < v$, (ii) $u > v$, or (iii) u and v are incomparable, but there is a vertex w with $u < w$ and $v < w$.

Although a down-set in $\overline{G}$ will become a clique in G', the converse is not true, as stated in Observation 1 below.

We conclude this section with a concrete example and some observations drawn from it. A special case of this following example can be found in [2].

EXAMPLE: Let $k \geq 2$ and $m \geq 1$ be natural numbers. Let $A_1, \ldots, A_k$ be disjoint sets, each A_i containing exactly m vertices. For each of the $\binom{k}{2}$ pairs $\{i, j\} \subseteq [k]$ define an additional vertex w_{ij}. Let $\overline{G}(k, m)$ be a digraph with vertex set and edge set given by

$$V(\overline{G}(k, m)) = \left(\bigcup_{i \in [k]} A_i \right) \cup \{w_{ij} : \{i, j\} \subseteq [k]\},$$

$$E(\overline{G}(k, m)) = \bigcup_{\{i,j\} \subseteq [k]} \{(w_{ij}, u) : u \in A_i \cup A_j\}.$$

Clearly $\overline{G}(k,m)$ is a simple acyclic digraph on $n = km + \binom{k}{2}$ vertices and with $\binom{k}{2} \cdot 2m = k(k-1)m$ directed edges. Each closed principal down-set is of the form $D[w_{ij}]$ for some $\{i,j\} \subseteq [k]$ and hence $D(\overline{G}(k,m)) = 2m + 1$. Note that in any proper down-coloring of $\overline{G}(k,m)$, every two vertices in $\bigcup_{i \in [k]} A_i$ are both contained in $D[w_{ij}]$ for some $\{i,j\} \subseteq [k]$, and hence $\bigcup_{i \in [k]} A_i$ forms a clique in $G'(k,m)$. From this we see that $\omega(G'(k,m)) = km$. In particular we have that

$$\chi_d(\overline{G}(k,m)) = \chi(G'(k,m)) \geq \omega(G'(k,m)) = km.$$

In fact, any coloring of $\bigcup_{i \in [k]} A_i$ with km colors can be extended in a greedy fashion to a proper down-coloring of $\overline{G}(k,m)$ with at most km colors. Therefore equality holds through the above display. In particular we have $D(\overline{G}(k,1)) = 3$ and $\omega(G'(k,1)) = k$, from which we deduce the following observation.

Observation 1 There is no function $f : \mathbb{N} \to \mathbb{N}$ with $\omega(G') \leq f(D(\overline{G}))$ for every simple acyclic digraph $\overline{G}$. In particular, there is no function f with $\chi_d(\overline{G}) \leq f(D(\overline{G}))$ for all simple acyclic digraphs $\overline{G}$.

Let $\alpha \in \mathbb{N}$ be a fixed and "large" natural number. Denoting by n the number of vertices of $\overline{G}(k,\alpha k)$, we clearly have $n = |V(\overline{G}(k,\alpha k))| = \alpha k^2 + \binom{k}{2} \sim k^2(\alpha + 1/2)$, where $f(k) \sim g(k)$ means $\lim_{k \to \infty} f(k)/g(k) = 1$. We now see that

$$D(\overline{G}(k,\alpha k)) = 2\alpha k + 1 \sim \left(\frac{2\alpha}{\sqrt{\alpha + 1/2}}\right)\sqrt{n},$$

$$\chi_d(\overline{G}(k,\alpha k)) = \alpha k^2 \sim \left(\frac{\alpha}{\alpha + 1/2}\right) n.$$

From this we have the following.

Observation 2 For every $\epsilon > 0$, there is an $n \in \mathbb{N}$ for which there is a simple acyclic digraph $\overline{G}$ on n vertices with $D(\overline{G}) = \Theta(\sqrt{n})$ and $\chi_d(\overline{G}) \geq (1 - \epsilon)n$.

REMARK: The above Observation 2 simply states that $\chi_d(\overline{G})$ can be an arbitrarily large fraction of $|V(\overline{G})|$ without making $D(\overline{G})$ too large. Hence, both the upper and lower bound in the inequality $D(\overline{G}) \leq \chi_d(\overline{G}) \leq |V(\overline{G})|$ are tight in this sense.

3   Various Representations

In this section we discuss some different representation of our digraph $\overline{G}$, and define some parameters which we will use to bound the down-chromatic number $\chi_d(\overline{G})$.

We first consider the issue of the height of digraphs. We say that two digraphs on the same set of vertices are equivalent if every down-coloring of one is also a valid down-coloring of the other, that is, if they induce the same undirected down-graph. We show that for any acyclic digraph $\overline{G}$ there is an equivalent acyclic digraph $\overline{G}_2$ of height two with $\chi_d(\overline{G}) = \chi_d(\overline{G}_2)$. However, the degrees of vertices in $\overline{G}_2$ may necessarily be larger than in $\overline{G}$.

Proposition 1. A down-graph G' of an acyclic digraph $\overline{G}$ is also a down-graph of an acyclic digraph $\overline{G}_2$ of height two.

Proof. The derived digraph $\overline{G}_2$ has the same vertex set as $\overline{G}$, while the edges all go from $\max\{\overline{G}\}$ to $V(\overline{G}) \setminus \max\{\overline{G}\}$, where $(u, v) \in E(\overline{G}_2)$ if, and only if, $v \in D(u)$. In this way we see that two vertices in $\overline{G}$ have a common ancestor if, and only if they have a common ancestor in $\overline{G}_2$. Hence, we have the proposition.

Therefore, when considering down-colorings of digraphs, we can by Proposition 1 assume them to be of height two. Moreover, there is a natural correspondence between acyclic digraphs and certain hypergraphs.

Definition 2. For a digraph $\overline{G}$, the related down-hypergraph $H_{\overline{G}}$ of $\overline{G}$ is given by:

$$V(H_{\overline{G}}) = V(\overline{G})$$
$$\mathcal{E}(H_{\overline{G}}) = \{D[u] : u \in \max\{\overline{G}\}\}.$$

Note that the down-graph G' is the clique graph of the down-hypergraph $H_{\overline{G}}$, that is, the simple undirected graph where every pair of vertices which are contained in a common hyperedge in $H_{\overline{G}}$ are connected by an edge in G'.

As we shall see, not every hypergraph is a down-hypergraph. There is a simple criterion for whether a hypergraph is a down-hypergraph or not. An edge in a hypergraph has a unique-element if it contains a vertex contained in no other edge. A hypergraph has the unique-element property if every edge has a unique element.

Observation 3 A hypergraph is a down-hypergraph if, and only if, it has the unique-element property.

Proof. A down-hypergraph is defined to contain the principal closed down-sets of a digraph $\overline{G}$ as edges. Each such edge contains a maximal element in $\overline{G}$, and this element is not contained in any other down-set. Hence, a down-hypergraph satisfies the unique-element property.

Suppose a hypergraph H satisfies the property. Then we can form a height-two acyclic digraph as follows: For each hyperedge, add a source vertex in the digraph corresponding to the representative unique element of the hyperedge. For the other hypervertices, add sinks to the digraph with edges from the sources to those sinks that correspond to vertices in the same hyperedge.

Note that a hypergraph with the unique element property is necessarily simple, in the sense that each hyperedge is uniquely determined by the vertices it contains.

We see that we can convert a proper down-hypergraph to a corresponding acyclic digraph of height two, and vice versa, in polynomial time.

Definition 3. A strong coloring of a hypergraph H, is a map $\Psi : V(H) \to [k]$ satisfying

$$u, v \in e \text{ for some } e \in \mathcal{E}(H) \Rightarrow \Psi(u) \neq \Psi(v).$$

The strong chromatic number $\chi_s(H)$ is the least number k of colors for which H has a proper strong coloring $\Psi : V(H) \to [k]$.

Note that a strong coloring of a down-hypergraph $H_{\overline{G}}$ is equivalent to a down-coloring of $\overline{G}$, and hence $\chi_s(H_{\overline{G}}) = \chi_d(\overline{G})$. Since we can convert to and from hypergraph and digraph representations, the two coloring problems are polynomial-time reducible to each other. Strong colorings of hypergraphs have been studied, but not to the extent of various other types of colorings of hypergraphs. In [13] a nice survey of various aspects of hypergraph coloring theory is found, containing almost all fundamental results in the past three decades.

In the next section we will bound the down-chromatic number of our acyclic digraph $\overline{G}$, partly by another parameter of the corresponding down-hypergraph $H_{\overline{G}}$.

4   Upper Bound in Terms of Degeneracy

As we saw in Observation 1, it is in general impossible to bound $\chi_d(\overline{G})$ from above solely in terms of $D(\overline{G})$, even if $\overline{G}$ is of height two. Therefore we need an additional parameter for that upper bound, but first we need to review some matters about a hypergraph $H = (V(H), \mathcal{E}(H))$.

Two vertices in $V(H)$ are neighbors in H if they are contained in the same edge in $\mathcal{E}(H)$. An edge in $\mathcal{E}(H)$ containing just one element is called trivial The largest cardinality of a hyperedge of H will be denoted by $\sigma(H)$. The degree $d_H(u)$, or just $d(u)$, of a vertex $u \in V(H)$ is the number of non-trivial edges containing u. Note that $d(u)$ is generally much smaller than the number of the neighbors of u. The minimum and maximum degree of H are given by

$$\delta(H) = \min_{u \in V(H)} \{d_H(u)\},$$

$$\Delta(H) = \max_{u \in V(H)} \{d_H(u)\}.$$

The subhypergraph $H[S]$ of H, induced by a set S of vertices, is given by

$$V(H[S]) = S.$$
$$\mathcal{E}(H[S]) = \{X \cap S : X \in \mathcal{E}(H) \text{ and } |X \cap S| \geq 2\}.$$

Definition 4. Let H be a simple hypergraph. The degeneracy or the inductiveness of H, denoted by $\mathrm{ind}(H)$, is given by

$$\mathrm{ind}(H) = \max_{S \subseteq V(H)} \{\delta(H[S])\}.$$

If $k \geq \mathrm{ind}(H)$, then we say that H is k-degenerate or k-inductive.

Note that Definition 4 is a natural generalization of the degeneracy or the inductiveness of a usual undirected graph G, given by $\mathrm{ind}(G) = \max_{H \subseteq G} \{\delta(H)\}$, where H runs through all the induced subgraphs of G. Note that the inductiveness of a (hyper)graph is always greater than or equal to the inductiveness of any of its sub(hyper)graphs.

To illustrate, let us for a brief moment discuss the degeneracy of an important class of simple graphs, namely that of simple planar graphs. Every subgraph of a simple planar graph is again planar. Since every planar graph has a vertex of degree five or less, the degeneracy of every planar graph is at most five. This is the best possible for planar graphs, since the graph of the icosahedron is planar and 5-regular. That a planar graph has degeneracy of five, implies that it can be vertex colored in a simple greedy fashion with at most six colors. The degeneracy has also been used to bound the chromatic number of the square G^2 of a planar graph G, where G^2 is a graph obtained from G by connecting two vertices of G if, and only if, they are connected in G or they have a common neighbor in G, [3].

In general, the degeneracy of a graph G yields an ordering $\{u_1, u_2, \ldots, u_n\}$ of $V(G)$, such that each vertex u_i has at most $\mathrm{ind}(G)$ neighbors among the previously listed vertices $u_1, \ldots, u_{i-1}$. Such an ordering provides a way to vertex color G with at most $\mathrm{ind}(G) + 1$ colors in an efficient greedy way, and hence we have in general that $\chi(G) \leq \mathrm{ind}(G) + 1$.

The inductiveness of a simple hypergraph is also connected to a greedy vertex coloring of it, but not in such a direct manner as for a regular undirected graph, since, as noted, the number of neighbors of a given vertex in a hypergraph is generally much larger than its degree.

Theorem 4. If the simple undirected graph G is the clique graph of the simple hypergraph H, then $\mathrm{ind}(G) \leq \mathrm{ind}(H)(\sigma(H) - 1)$.

Proof. For each $S \subseteq V(G) = V(H)$, let $G[S]$ and $H[S]$ be the subgraph of G and the subhypergraph of H induced by S, respectively. Note that for each $u \in S$, each hyperedge in $H[S]$ which contains u, has at most $\sigma(H[S]) - 1 \leq \sigma(H) - 1$ other vertices in addition to u. By definition of $d_{H[S]}(u)$, we therefore have that $d_{G[S]}(u) \leq d_{H[S]}(u)(\sigma(H) - 1)$, and hence

$$\delta(G[S]) \leq \delta(H[S])(\sigma(H) - 1). \tag{2}$$

Taking the maximum of (2) among all $S \subseteq V(G)$ yields the theorem.

Recall that the intersection graph of a collection $\{A_1, \ldots, A_n\}$ of sets, is the simple graph with vertices $\{u_1, \ldots, u_n\}$, where we connect u_i and u_j if, and only if, $A_i \cap A_j \neq \emptyset$.

Observation 5 For a simple connected hypergraph H, then $\mathrm{ind}(H) = 1$ if, and only if, the intersection graph of its hyperedges $\mathcal{E}(H)$ is a tree.

What Observation 5 implies, is that edges of H can be ordered $\mathcal{E}(H) = \{e_1, \ldots, e_m\}$, such that each e_i intersects exactly one edge from $\{e_1, \ldots, e_{i-1}\}$. If now G is the clique graph of H, this implies that $\mathrm{ind}(G) = \sigma(H) - 1$.

Also note that if H has the unique element property and $\sigma(H) = 2$, then clearly the clique graph G is a tree, and hence $\mathrm{ind}(G) = 1 = \sigma(H) - 1$. We summarize in the following.

Observation 6 Let H be a hypergraph that satisfies the unique element property. If either $\mathrm{ind}(H) = 1$ or $\sigma(H) = 2$, then the clique graph G of H satisfies $\mathrm{ind}(G) = \sigma(H) - 1$.

For a hypergraph H with the unique element property, we can obtain some slight improvements in the general case as well.

Theorem 7. Let H be a hypergraph with the unique element property. Assume further that $\mathrm{ind}(H) > 1$ and $\sigma(H) > 2$. Then the clique graph G of H satisfies

$$\mathrm{ind}(G) \leq \mathrm{ind}(H)(\sigma(H) - 2).$$

Proof. Since H has the unique element property, then by Observation 3 there is an acyclic digraph $\overline{G}$ such that $H = H_{\overline{G}}$. Let H'' be the hypergraph induced by $V(H) \setminus \max\{\overline{G}\}$ and G'' be the corresponding clique graph of H''. Since each $u \in \max\{\overline{G}\}$ is simplicial in H and in G, their removal will not affect the degeneracy of the remaining vertices, so $\mathrm{ind}(H'') = \mathrm{ind}(H)$ and $\mathrm{ind}(G'') = \mathrm{ind}(G)$. Also note that $\sigma(H'') = \sigma(H) - 1$. By Theorem 4 we get $\mathrm{ind}(G) = \mathrm{ind}(G'') \leq \mathrm{ind}(H'')(\sigma(H'') - 1) = \mathrm{ind}(H)(\sigma(H) - 2)$, thereby completing the proof.

REMARK: Recall that for any simple undirected graph G on n vertices, we have $\chi(G) \leq \mathrm{ind}(G) + 1$. In fact, the upper bound of $\mathrm{ind}(G)$ given in Theorem 7 yields an on-line down-coloring of $\overline{G}$ that uses at most $\mathrm{ind}(G) \log n$ colors, where $H = H_{\overline{G}}$ is the down-hypergraph of $\overline{G}$, as well.

Let $\overline{G}$ be an acyclic digraph. Since now $D(\overline{G}) = \sigma(H_{\overline{G}})$ and $\chi_d(\overline{G}) = \chi(G') \leq \mathrm{ind}(G') + 1$, we obtain the following summarizing corollary.

Corollary 1. If $\overline{G}$ is an acyclic digraph, then its down-chromatic number satisfies the following:

1. If $\mathrm{ind}(H_{\overline{G}}) = 1$ or $D(\overline{G}) = 2$, then $\chi_d(\overline{G}) = D(\overline{G})$.
2. If $\mathrm{ind}(H_{\overline{G}}) > 1$ and $D(\overline{G}) > 2$, then $\chi_d(\overline{G}) \leq \mathrm{ind}(H_{\overline{G}})(D(\overline{G}) - 2) + 1$.

Moreover, the mentioned upper bounds in both cases are sufficient for greedy down-coloring of $\overline{G}$.

EXAMPLE: Let $k, m \in \mathbb{N}$, assuming them to be "large" numbers. Consider the graph $\overline{G}(k, m))$ from Section 2. Here with $\mathrm{ind}(H_{\overline{G}(k,m)}) = k-1$ and $\sigma(H_{\overline{G}(k,m)}) = 2m + 1$. By Corollary 1 we have immediately that $\chi_d(\overline{G}(k, m)) \leq (k - 1)(2m - 1) + 1 = \Theta(km)$, which agrees with the asymptotic value of the actual down-chromatic number km (also a $\Theta(km)$ function,) which we computed in Section 2. Hence, Corollary 1 is asymptotically tight.

Moreover, if we were to just color each vertex with its own unique color, we compare $\chi_d(\overline{G}(k, m)) \leq (k - 1)(2m - 1) + 1$ with the actual number $km + \binom{k}{2}$ of vertices, and we see that for large k, this is a substantial reduction.

REMARK: Although we have assumed our digraphs to be acyclic, we note that the definition of down-coloring can be easily extended to a regular cyclic digraph $\overline{G}$ by interpreting the notion of descendants of a vertex u to mean the set of nodes reachable from u. In fact, if $\overline{G}$ is an arbitrary digraph, then there is an equivalent acyclic digraph $\overline{G}'$, on the same set of vertices, with an identical down-graph: First form the condensation $\hat{G}$ of $\overline{G}$ by shrinking each strongly connected component of $\overline{G}$ to a single vertex. Then form $\overline{G}'$ by replacing each node of $\hat{G}$ which represents a strongly connected component of $\overline{G}$ on a set $X \subseteq V(\overline{G})$ of vertices, with an arbitrary vertex $u \in X$, and then add a directed edge from u to each $v \in X \setminus \{u\}$. This completes the construction. – Observe that each node $v \in X$ has exactly the same neighbors in the down-graph of $\overline{G}'$ as u, as it is a descendant of u and u alone. Further, if node v was in a different strong component of $\overline{G}$ than u but was reachable from u, then it will continue to be a descendant of u in $\overline{G}'$. Hence, the down-graphs of $\overline{G}$ and $\overline{G}'$ are identical.

Acknowledgements

The authors are grateful to John L. Pfaltz for his interest and encouragements to submit this article.

References

[1] Serge Abiteboul, Peter Buneman, and Dan Suciu. *Data on the Web, From Relations to Semistructured Data and XML*. Morgan Kaufmann Publishers, 2000. 300

[2] Geir Agnarsson and Ágúst Egilsson. On vertex coloring simple genetic digraphs. *Congressus Numerantium*, 2004. to appear. 299, 305

[3] Geir Agnarsson and Magnús M. Halldórsson. Coloring powers of planar graphs. *SIAM Journal of Discrete Mathematics, No 4*, 16:651–662, 2003. 309

[4] An Oracle White Paper. Key Data Warehousing Features in Oracle9i: A Comparative Performance Analysis. Available on-line from Oracle at: http://otn.oracle.com/products/oracle9i/pdf/o9i_dwfc.pdf, September 2001. 302

[5] B. Bhattacharjee, L. Cranston, T. Malkemus, and S. Padmanabhan. Boosting Query Performance: Multidimensional Clustering. *DB2 Magazine, Quarter 2, Issue 2*, 8, 2003. Also, available on-line at: http://www.db2mag.com. 302

[6] Michael W. Cain and (iSeries Teraplex Integration Center). Star Schema Join Support within DB2 UDB for iSeries Version 2.1. Available on-line from IBM at: http://www-919.ibm.com/developer/db2/documents/star/, October 2002. 302

[7] C. C. Harner and R. C. Entringer. Arc colorings of digraphs. *Journal of Combinatorial Theory, Series B*, 13:219 – 225, 1972. 304

[8] H. Jacob and H. Meyniel. Extensions of Turán's Brooks' theorems and new notions of stability and colorings in digraphs. *Combinatorial Mathematics, North-Holland Math. Stud.*, 75:365 – 370, 1983. 304

[9] Clifford A. Shaffer. *A Practical Introduction to Data Structures and Algorithm Analysis*. Prentice Hall, java edition, 1998. 304

[10] Xiang Ying Su. Brooks' theorem on colorings of digraphs. *Fujian Shifan Daxue Xuebao Ziran Kexue Ban, No. 1*, 3:1 – 2, 1987. 304

[11] The Gene Ontology Consortium. Gene Ontology: Tool for the unification of biology. The Gene Ontology Consortium (2000). *Nature Genet*, 25:25 – 29, 2000. 299

[12] William T. Trotter. *Combinatorics and Partially Ordered Sets, Dimension Theory.* Johns Hopkins Series in the Mathematical Sciences. The Johns Hopkins University Press, 1992. 304

[13] Weifan Wang and Kemin Zhang. Colorings of hypergraphs. *Adv. Math. (China), No. 2,* 29:115–136, 2000. 308

[14] Douglas B. West. *Introduction to Graph Theory.* Prentice Hall, 2nd edition, 2001. 304

Local Specification
of Surface Subdivision Algorithms

Colin Smith, Przemyslaw Prusinkiewicz, and Faramarz Samavati

University of Calgary
Calgary, Alberta, Canada T2N 1N4
{smithco,pwp,samavati}@cpsc.ucalgary.ca

Abstract. Many polygon mesh algorithms operate in a local manner, yet are formally specified using global indexing schemes. This obscures the essence of these algorithms and makes their specification unnecessarily complex, especially if the mesh topology is modified dynamically. We address these problems by defining a set of local operations on polygon meshes represented by graph rotation systems. We also introduce the vv programming language, which makes it possible to express these operations in a machine-readable form. The usefulness of the vv language is illustrated by the application examples, in which we concentrate on subdivision algorithms for the geometric modeling of surfaces. The algorithms are specified as short, intuitive vv programs, directly executable by the corresponding modeling software.

1 Introduction

Locality is one of the most fundamental characteristics of structured dynamical systems. It means that a neighborhood relation is defined on the elements of the system, and that each element changes its state according to its own state and the state of its neighbors. The elements positioned farther away are not considered. A challenging problem is the characterization and modeling of dynamical systems with a dynamical structure [13], in which not only the state of the elements of the system, but also their number and configuration, change over time. In this paper, we consider a class of such systems pertinent to geometric modeling and represented by surface subdivision algorithms.

Subdivsion algorithms generate smooth (at the limit) surfaces by iteratively subdividing polygon meshes. This process involves the creation of new vertices, edges, and faces. The operations on the individual mesh elements are described in local terms using m asks [33], also referred to as stencils [29]. A mask is a graph that depicts a vertex of interest (either a newly created point or an old vertex being repositioned) and the neighbors that affect it. The new vertex position is determined as an affine combination[1] of the vertices identified by the mask.

---

[1] An affine combination of n points $P_1, P_2, \ldots, P_n$ is an expression of the form $\alpha_1 P_1 + \alpha_2 P_2 + \cdots + \alpha_n P_n$, where the scalar coefficients α_i add up to one: $\alpha_1 + \alpha_2 + \cdots + \alpha_n = 1$. The meaning of the affine combination is derived from its transformation to the form

J.L. Pfaltz, M. Nagl, and B. Böhlen (Eds.): AGTIVE 2003, LNCS 3062, pp. 313–327, 2004.
© Springer-Verlag Berlin Heidelberg 2004

Despite the locality and simplicity of the masks, formal descriptions of subdivision algorithms often rely on a global enumeration (indexing) of the polygon mesh elements [33]. Indices have the advantage of being the standard mathematical notation, conducive to stating and proving properties of subdivision algorithms. They are also closely related to the array data structures supported by most programming languages. On the other hand, indices provide only an indirect access to the neighbors of an element in a polygon mesh: all mesh elements must be first enumerated and then index arithmetic must be performed to select a specific neighbor. This arithmetic becomes cumbersome for irregular meshes. Moreover, for dynamically changing mesh structures, indices may have to be updated after each iteration of the algorithm. The index notation is also too powerful: providing a unique index to each vertex makes it possible to access vertices at random, thus violating the locality constraint. We seek a notation that would be as intuitive as masks, but also sufficiently precise to formally specify and execute subdivision algorithms.

Within the confines of linear and branching structures, an example of such a notation is provided by L-systems. In addition to the generation of fractals and the modeling of plants [26], L-systems have recently been applied to the geometric modeling of subdivision curves [27]. Unfortunately, known extensions of L-systems to polygonal structures, namely map L-systems [26, 22] and cell systems [7], lack features needed for the specification of subdivision algorithms for surfaces. They do not offer a flexible control over geometry, and do not provide a mechanism for accessing context information. Limited geometric interpretations and a focus on the context-free case are also prevalent in the broader scope of graph grammars (c.f. [28, 12]).

Our proposed solution preserves the purely local operation of L-systems and graph grammars, but departs from their declarative character. Instead, context information is accessed and modeled structures are modified by sequences of imperative operations.

The ease of performing local operations on polygon meshes depends on their representation. Known examples of such representations, conducive to both local information gathering and mesh transformations, include the winged-edge [4], and quad-edge [15] representations. Pursuing objectives closer to ours, Egli and Stewart [11] applied cellular complexes [25] to specify Catmull-Clark [5] subdivision in an index-free manner, and Lienhardt [21] showed that local operations involved in subdivision algorithms can be defined using G-maps [19, 20]. More recently, Velho [31] proposed a method for describing subdivision algorithms using stellar operators [18] that act on a half-edge structure [24].

We have chosen yet another representation, based on the mathematical notion of graph rotation systems [10, 32]. A graph rotation system associates each vertex of a polygon mesh with an oriented circular list of its neighboring vertices. A set of these lists, defined for each vertex, completely represents the topology of a 2-manifold mesh [32]. Graph rotation systems have been introduced to com-

---

$P = P_1 + \alpha_2(P_2 - P_1) + \cdots + \alpha_n(P_n - P_1)$, which is a well-defined expression of vector algebra [8, 14].

puter graphics by Akleman, Chen and Srinivasan [1, 2, 3] as a formal basis for
the doubly linked face list representation of 2-manifold meshes. Akleman et al.
have also defined a set of operations on this representation, which they used to
implement interactive polygon mesh modeling tools.

We introduce vertex-vertex systems, related to the adjacency-list graph representation [6], as an alternative data structure based on the graph rotation
systems. We also define the vertex-vertex algebra for describing local operations
on the vertex-vertex systems, and vv, an extension of the C++ programming
language, for expressing these operations in a machine-readable form. This leads
us to the language + engine modeling paradigm, which simplifies the implementation of individual algorithms by treating them as input to a multi-purpose
modeling program. We illustrate the usefulness of this paradigm by presenting
concise vv specifications of several subdivision algorithms.

2   Vertex-Vertex Systems

2.1   Definitions

Let U be an enumerable set, or the universe, of elements called abstract vertices.
We assume that U is ordered by a relation $<$; this assumption simplifies the
implementation of many algorithms (Section 3). Next, let $N : U \mapsto 2^U$ be
a function that takes every vertex $v \in U$ to a finite subset $v^\star \subset U$ of other
vertices ($v \notin v^\star$). We call the set $v^\star$ the neighborhood, and its elements the
neighbors[2] of v. Finally, let the vertex set $S \subset U$ be a finite subset of the
universe U, and N_S be the restriction of the neighborhood function N to the
domain S; thus $N_S(v) = v^\star$ if $N(v) = v^\star$ and $v \in S$ (the elements of $v^\star$ may
lie outside S). We call the pair $\langle S, N_S \rangle$ a vertex-vertex structure over the set S
with neighborhood N_S.

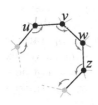

An undirected graph over a vertex set S is a vertex-vertex
structure over S, in which: (a) all neighborhoods are included
in S (the vertex set S is closed with respect to the function N),
and (b) vertex u is in the neighborhood of v if and only if
vertex v is in the neighborhood of u ($u \in v^\star$ if and only if
$u \in v^\star$, the symmetry condition). The pairs (u, v) of vertices
that are in the neighborhood of each other are called edges of
the graph. An edge is oriented if the pair (u, v) is considered
different from (v, u).

Fig. 1.   A
polygon iden-
tification   in
a graph rotation
system

A vertex-vertex rotation system, or vertex-vertex system for
short, is a vertex-vertex structure in which the vertices in each
neighborhood form a cyclic permutation (i.e., are arranged into
a circular list). A graph rotation system is a vertex-vertex system that is both a graph and a vertex-vertex rotation system.

---

[2] Our terminology is motivated by the practice of referring to adjacent cells in a grid
   as neighbors.

A polygon mesh is a collection of vertices, edges bound by vertex pairs, and polygons bound by sequences of edges and vertices. A mesh is a closed 2-manifold if it is everywhere locally homeomorphic to an open disk [32].

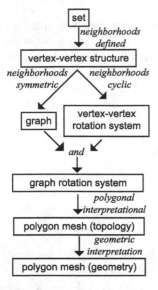

A polygonal interpretation of a vertex-vertex system maps it into a polygon mesh. The interpretations that we consider in this paper are variants of the Edmonds' permutation technique [10, 32, 2], which is defined for connected graph rotation systems. It defines polygons of the mesh using the following algorithm (Fig. 1). Given an oriented edge (u, v) in S, we find the oriented edge (v, w) such that w immediately follows u in the cyclic neighborhood of v. Next, we find the oriented edge (w, z) such that z immediately follows v in the neighborhood of w. We continue this process until we return to the starting point u. The resulting orbit (cyclic permutation) of vertices $u, v, w, z, \ldots$ and the edges that connect them are the boundaries of a polygon. By considering all such orbits in S, we obtain a polygon mesh with polygons on both sides of each (unoriented) edge. From this construction it immediately follows that the resulting mesh is a uniquely defined, orientable, closed 2-manifold (see [32] for a proof).

Fig. 2. Relations between notions pertinent to vertex-vertex systems

Vertex positions are a crucial aspect of the geometric interpretation of vertex-vertex systems. We will consider geometric interpretations in which edges are drawn as straight lines between vertices, and polygons are properly defined if their vertices and edges are coplanar.

The above progression of notions is summarized in Fig. 2. It suggests that polygon meshes can be manipulated using three types of operations: set-theoretic, topological, and geometric operations. The most difficult problem is the manipulation of topology. We address it by introducing a set of operations that modify at most one neighborhood at a time, and transform a vertex-vertex system into another vertex-vertex system. The individual operations do not necessarily transform graphs into graphs, because they may create incomplete neighbors that violate the symmetry condition ($u \in v^\star$ but $v \notin u^\star$).

2.2 The Vertex-Vertex Algebra

The vertex-vertex algebra consists of the class of vertex-vertex rotation systems with a set of operations defined on them. We introduce these operations using a mathematical notation that combines standard and new mathematical symbols. We also present the equivalent expressions and statements of the vv language. A further description of this language and its implementation is given in Section 2.3.

Table 1. Set-theoretic operations supported by the vv language

Name	Math. notation	vv statement
set creation	let $S \subset U$	mesh S
assignment	$S = T$	$S = T$
union	$S = S \cup T$	merge S with T
addition of an element	$S = S \cup \{v\}$	add v to S
removal of an element	$S = S - \{v\}$	remove v from S
iteration over a set	$\forall v \in S$	forall v in S
iteration over neighbors	$\forall x \in v^{*}$	forall x in v

In the vv language, vertex sets are a predefined data type. A set S is created using the declaration mesh S, and is in existence according to the standard scoping rules of C++. The vv language supports a subset of the standard set operations, listed in Table 1. In addition to operations that return a set as the result, vv includes iteration operators for flow control in vv programs.

Topological operations are the core of the vertex-vertex algebra. They are divided into three groups: query, selection, and editing operations. Query operations return information about vertices. Selection operations return an element of a vertex neighborhood. Editing operations locally modify a vertex-vertex system. Definitions of these operations are given in Table 2. The last column points to the illustrations below the table.

We use the standard functional notation $f(v)$ or vv expression $v\$f$ to associate property f with a vertex v. A special case is the position of a vertex, denoted $\bar{v}$ or $v\$pos$. Positions can be assigned explicitly, by referring to an underlying coordinate system, or result from affine geometry combinations and vector operations applied to the previously defined points. We use the standard C++ operator overloading mechanism to extend arithmetic operators to positions and vectors.

Operations of the vertex-vertex algebra are commonly iterated over vertex sets. This raises important questions concerning the sequencing of these individual operations. For example, if the same operation is to be performed on a pair of neighboring vertices u and v, the results may be different depending on whether u is modified first, v is modified first, or both vertices are modified simultaneously. To eliminate the unwanted dependence on the execution sequence, we introduce the coordination operation synchronize S, which creates a copy $`v$ of each vertex $v$ in the set $S$. All subsequent operations on the vertices $v \in S$ (until the next synchronize statement) do not affect the vertices $`v$, which continue to store the "old" values of vertex attributes. For example, $`v\$pos$ denotes the position of vertex $v$ at the time when the synchronize statement was last issued, whereas $v\$pos$ denotes the current position of $v$. Similarly, $`v^{*}$ and v^{*} denote the old and current neighborhoods of v. The use of old attributes instead of the current ones makes it possible to iterate over the elements of a set in any order without affecting the iteration results.

Table 2. Top: definition of the topological operations of the vertex-vertex algebra. Bottom: graphical interpretation of the selection and editing operations. a) Setting the initial neighborhood of vertex v. b-g) The results of various operations applied to v

Name	Math. notation	vv statement	Description	Note	Fig
		Query operations			
membership	$x \in v^*$	is x in v	true iff vertex x is in the neighborhood of v		
order	$x < v$	$x < v$	true iff vertex x precedes vertex v in the universe U		
valence	$|v^*|$	valence v	returns the number of neighbors of vertex v		
		Selection operations			
any	let $x \in v^*$	any in v	returns a neighbor of v	1	
next	$v^* \uparrow x$	nextto x in v	returns vertex that follows x in the neighborhood of v	2	b
previous	$v^* \downarrow x$	prevto x in v	returns vertex that precedes x in the neighborhood of v	2	c
		Editing operations			
create	let $v \in U$	vertex v	create a vertex		
set neighborhood	$v^* = \{a, b, c\}$	make $\{a, b, c\}$ nb_of v	set the neighborhood of v to the given circular list	3	a
erase	$v^* = v^* - x$	erase x from v	remove x from the neighborhood of v if $x \in v^*$	4	d
replace	$v^* = v^* - a + x$	replace a with x in v	substitute x for a in the neighborhood of v	5	e
splice after	$v^* + x \succ a$	splice x after a in v	insert x after a in the neighborhood of v	5	f
splice before	$v^* + x \prec a$	splice x before a in v	insert x before a in the neighborhood of v	5	g

1) Returns the null vertex if v^* is empty. 2) Returns the null vertex if $x \notin v^*$.
3) Not defined (error reported) if v appears in the list, or the same vertex is listed twice.
4) No effect if $x \notin v^*$. 5) No effect if $a \notin v^*$; not defined (error reported) if $x = v$ or $x \in v^*$.

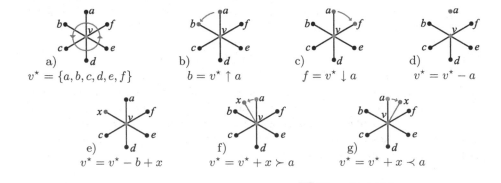

a) $v^* = \{a, b, c, d, e, f\}$

b) $b = v^* \uparrow a$

c) $f = v^* \downarrow a$

d) $v^* = v^* - a$

e) $v^* = v^* - b + x$

f) $v^* = v^* + x \succ a$

g) $v^* = v^* + x \prec a$

```
1  vertex insert(vertex p, vertex q) {
2     vertex x;
3     make {p, q} nb_of x;
4     replace p with x in q;
5     replace q with x in p;
6     return x;
7  }
```

Fig. 3. The vv code and illustration of the insertion of a vertex x between vertices p and q. Vertex x replaces p as the neighbor of q and q as the neighbor of p; vertices p and q become neighbors of x

2.3 Implementation of Vertex-Vertex systems

We have implemented vertex-vertex systems as a set of programs and libraries collectively called the vv environment. The central component of this environment is vvlib, a C++ library containing data structures and functions implementing the vertex-vertex polygon mesh representation and algebra. The user can refer to these structures and functions directly from a program written in C++, or from a program written in the vv language.

The vv language extends C++ with keywords and expressions implementing the vertex-vertex algebra. They are listed under the column 'vv statement' in Tables 1 and 2. All of the examples presented in this paper are actual code written in the vv language. To enhance code legibility, we set variable names in italics.

In order to be executed, a vv program is first translated to a C++ program, with the keywords and expressions specific to vv translated into calls to the vvlib library. This C++ program is then compiled into a dynamically linked library (DLL). The modeling program, called vvinterpreter, loads this DLL, runs, and produces the graphical output. This whole processing sequence is automated: from the user's perspective, the vvinterpreter treats the vv program as an input and runs accordingly. This approach is based on that introduced by Karwowski and Prusinkiewicz to translate and execute L-system-based programs in [16].

3 Subdivision Algorithms

To illustrate the usefulness of the vertex-vertex algebra, we now provide compact descriptions of several subdivision algorithms. These descriptions can be directly executed by vvinterpreter.

3.1 Insertion of a Vertex

A simple routine that is of much use in writing subdivision algorithms is a function that creates a new vertex x and inserts it between two given vertices p and q (Fig. 3).

```
 1  void polyhedral(mesh& S) {
 2    synchronize S;
 3    mesh NV;
 4
 5    forall p in S {
 6      forall q in 'p {
 7        if (p < q) continue;
 8        vertex x = insert(p, q);
 9        x$pos = (p$pos + q$pos) / 2.0;
10        add x to NV;
11      }
12    }
13    forall x in NV {
14      vertex p = any in x;
15      vertex q = nextto p in x;
16      make {nextto x in q, q, prevto x in q,
17        nextto x in p, p, prevto x in p} nb_of x;
18    }
19    merge S with NV;
20  }
```

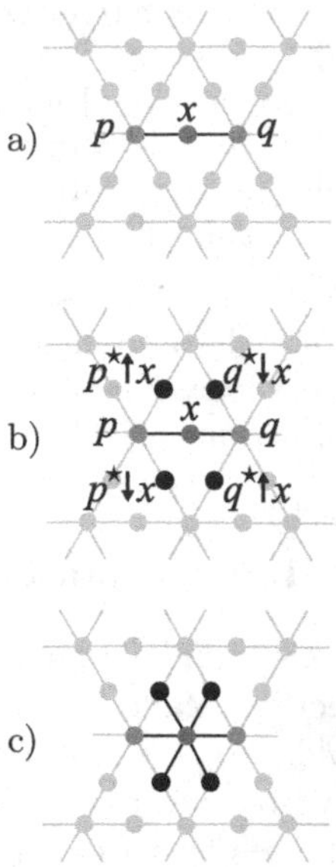

Fig. 4. Left: the polyhedral subdivision algorithm specified using vertex-vertex systems. Right: the vv identification of points involved in the creation of a new vertex x (a), the vv identification of vertices that will become neighbors of x (b), and the updated neighborhood of the new vertex x (c)

3.2   Polyhedral Subdivision

One of the simplest subdivision algorithms is the polyhedral subdivision of triangular meshes [30]. The algorithm inserts a new vertex at the midpoint of each edge, and divides each triangle of the mesh into four co-planar triangles. While the overall shape of the initial polyhedron does not change, the faces are subdivided.

The vv program that implements one step of the polyhedral subdivision consists of two loops (Fig. 4). The first loop (lines 5 to 12) iterates over pairs of neighboring vertices in the old vertex set S. The condition $p < q$ in line 7 assures that each vertex pair (i.e. edge of the polygon mesh) will be considered only once. New vertices are inserted at the midpoint of each edge (line 9) and added to the set NV (line 10). The second loop (lines 13 to 18) inserts new edges by redefining the neighborhoods of the new points. The intervening neighborhoods and the result of insertion are shown on the right side of Figure 4. An example of a polygon mesh and the results of its polyhedral subdivision are shown in Figure 5.

3.3   Loop Algorithm

The Loop subdivision scheme [23] is topologically equivalent to the polyhedral subdivision scheme, in the sense that both operate on triangular meshes and sub-

Fig. 5. From left to right: The vv specification of an initial polygon mesh topology, a sample polyhedron with that topology, and three steps of its polyhedral subdivision (with hidden lines eliminated)

divide a triangular face into four triangles in every iteration step. The vertex-vertex implementations of both schemes have, therefore, a similar structure. The difference is in the positioning of vertices: the Loop case aims at constructing a smooth surface with a general shape controlled by the initial polyhedron (Fig. 6). To this end, the Loop algorithm places new vertices using a mask involving four old vertices, and repositions old vertices using another mask that incorporates all of their immediate neighbors. A vv implementation of the Loop subdivision algorithm for closed surfaces and the corresponding masks are given in Figure 7. The derivation of the coefficients of the masks is presented in [23].

3.4 Butterfly Algorithm

The butterfly subdivision algorithm [9], like that for Loop subdivision, is topologically equivalent to the polyhedral subdivision. In contrast to the Loop subdivision, however, which approximates the shape of the initial polyhedron, the butterfly algorithm is an interpolating scheme. Consequently, the old vertex positions are not adjusted in the course of the algorithm. In order to produce a smooth limit surface, the butterfly algorithm uses a more extensive mask for the new vertices, which includes points outside the immediate neighborhood of the subdivided edge. This mask and the complete vv implementation of the butterfly algorithm for closed surfaces are presented in Figure 8. An example application of the algorithm is illustrated in Figure 9.

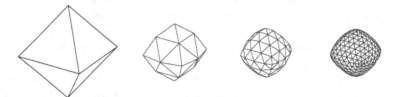

Fig. 6. An initial polygon mesh and the results of three iterations of Loop subdivision

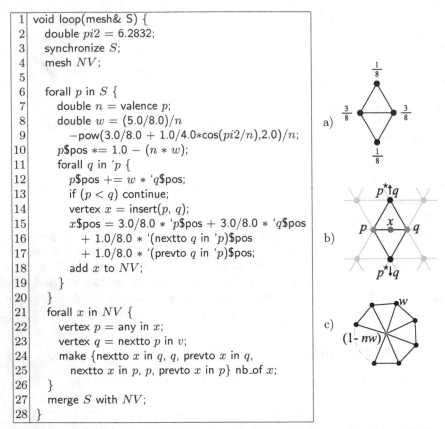

```
 1  void loop(mesh& S) {
 2      double pi2 = 6.2832;
 3      synchronize S;
 4      mesh NV;
 5
 6      forall p in S {
 7          double n = valence p;
 8          double w = (5.0/8.0)/n
 9              −pow(3.0/8.0 + 1.0/4.0*cos(pi2/n),2.0)/n;
10          p$pos *= 1.0 − (n * w);
11          forall q in 'p {
12              p$pos += w * 'q$pos;
13              if (p < q) continue;
14              vertex x = insert(p, q);
15              x$pos = 3.0/8.0 * 'p$pos + 3.0/8.0 * 'q$pos
16                  + 1.0/8.0 * '(nextto q in 'p)$pos
17                  + 1.0/8.0 * '(prevto q in 'p)$pos;
18              add x to NV;
19          }
20      }
21      forall x in NV {
22          vertex p = any in x;
23          vertex q = nextto p in v;
24          make {nextto x in q, q, prevto x in q,
25              nextto x in p, p, prevto x in p} nb_of x;
26      }
27      merge S with NV;
28  }
```

Fig. 7. Left: the vv implementation of the Loop subdivision algorithm. Right: illustration of the algorithm. a) The Loop mask for a new vertex. Vertex labels are the weights used in the affine combinations of vertex positions. b) The vv identification of vertices involved in the application of the mask to a new vertex x. c) The Loop mask for old vertices

3.5 $\sqrt{3}$ Algorithm

Kobbelt's $\sqrt{3}$-subdivision algorithm [17] is an example of a scheme that changes the topology of a triangular mesh in a manner different from the polyhedral subdivision. The vv specification of the $\sqrt{3}$-subdivision algorithm is given by Figure 10. In the first loop (lines 11 to 15), a new vertex c is created at the centroid of each triangle. The neighborhoods are then updated such that each triangle is divided into three, that is each vertex v, x, y of the original triangle is connected to c, and the vertices v, x, y form the neighborhood of c (lines 16 to 19). In the second loop (lines 23 to 31), the topology is updated by "flipping" all the edges between pairs of old vertices. An example of the operation of the algorithm is shown in Figure 11.

```
 1  void butterfly(mesh& S) {
 2      double k = 1.0/16.0, l = 1.0/8.0, m = 1.0/2.0;
 3      synchronize S;
 4      mesh NV;
 5
 6      forall p in S {
 7          forall q in 'p {
 8              if (p < q) continue;
 9              vertex x = insert(p, q);
10              x$pos = m * 'p$pos + m * 'q$pos
11                  + l * '(prevto q in 'p)$pos
12                  + l * '(nextto q in 'p)$pos
13                  − k * '(nextto (nextto q in 'p) in 'p)$pos
14                  − k * '(nextto (nextto p in 'q) in 'q)$pos
15                  − k * '(prevto (prevto q in 'p) in 'p)$pos
16                  − k * '(prevto (prevto p in 'q) in 'q)$pos;
17              add x to NV;
18          }
19      }
20      forall x in NV {
21          vertex p = any in x;
22          vertex q = nextto p in x;
23          make {nextto x in q, q, prevto x in q,
24              nextto x in p, p, prevto x in p} nb_of x;
25      }
26      merge S with NV;
27  }
```

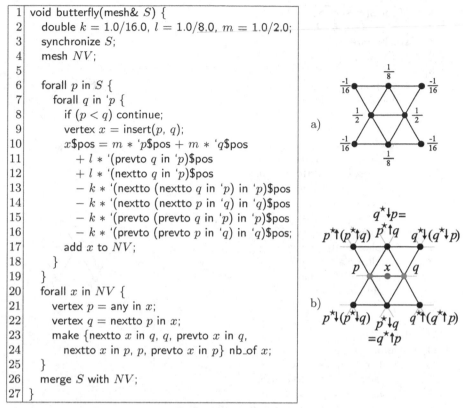

a)

b)

Fig. 8. Left: the vv implementation of the butterfly algorithm. Right: the mask (a) and vv identification (b) of points involved in its application to a new vertex x

4   Conclusions

We have addressed the problem of specifying polygon mesh algorithms in a concise and intuitive manner. To this end, we introduced a set of operations for locally changing the topology of a mesh, and we defined these operations in local terms. We have focused on subdivision algorithms as an application area, and

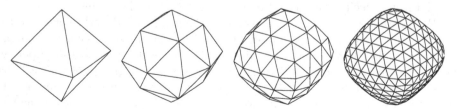

Fig. 9. An initial polygon mesh and the results of three iterations of butterfly subdivision

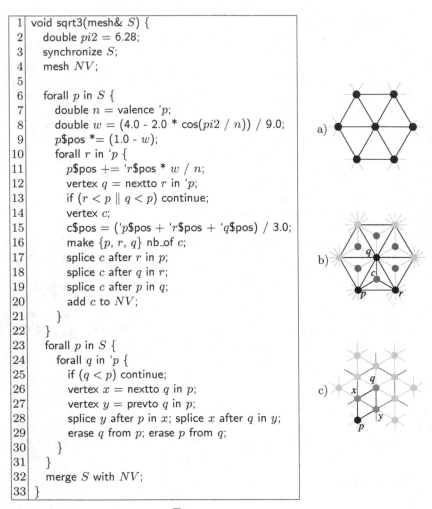

```
 1  void sqrt3(mesh& S) {
 2     double pi2 = 6.28;
 3     synchronize S;
 4     mesh NV;
 5
 6     forall p in S {
 7        double n = valence 'p;
 8        double w = (4.0 - 2.0 * cos(pi2 / n)) / 9.0;
 9        p$pos *= (1.0 - w);
10        forall r in 'p {
11           p$pos += 'r$pos * w / n;
12           vertex q = nextto r in 'p;
13           if (r < p || q < p) continue;
14           vertex c;
15           c$pos = ('p$pos + 'r$pos + 'q$pos) / 3.0;
16           make {p, r, q} nb_of c;
17           splice c after r in p;
18           splice c after q in r;
19           splice c after p in q;
20           add c to NV;
21        }
22     }
23     forall p in S {
24        forall q in 'p {
25           if (q < p) continue;
26           vertex x = nextto q in p;
27           vertex y = prevto q in p;
28           splice y after p in x; splice x after q in y;
29           erase q from p; erase p from q;
30        }
31     }
32     merge S with NV;
33  }
```

Fig. 10. Left: the algorithm for $\sqrt{3}$-subdivision. Right: a portion of the original mesh (a) after the insertion of central points and subdivision of triangles (b) and after the flip operation (c)

we have shown that the resulting vertex-vertex algebra leads to very compact and intuitive specifications of some of the best known algorithms.

We have also designed vv, a programming language based on the vertex-vertex algebra, and we implemented a modeling environment in which vv programs can be executed. In addition to the subdivision algorithms described in this paper, we used vv to generate fractals and aperiodic tilings, simulate growth of multicellular biological structures, and create procedural textures on non-regular meshes. In these tests, we found vv programs extremely conducive to rapid prototyping and experimentation with polygon mesh algorithms.

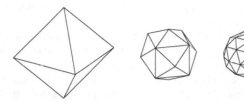

Fig. 11. Example of subdivision using the $\sqrt{3}$ algorithm

Our implementation of the vertex-vertex algebra was guided by the elegance of programming constructs, rather than performance. For example, profiling of vv programs showed that approximately 50% of the algorithm execution time is spent on dynamic memory management. It is an interesting open question whether vertex-vertex systems could reach the speed of the fastest implementations of polygon mesh algorithms.

Another interesting class of problems is related to the temporal coordination of vertex-vertex operations. The synchronization mechanism introduced in Section 2.2 is in fact a method for simulating parallelism on a sequential machine. This suggests that it may be useful to extend vv with constructs for explicitly specifying parallel rather than sequential execution of operations. Such an extension could further clarify vv programs and lead to their effective implementation on parallel processors with a suitable architecture. Finally, the problem of providing a declarative, grammar-like method for specifying subdivision algorithms remains open. Such specification, if possible, may provide the ultimately concise and clear specification of these algorithms.

References

[1] E. Akleman and J. Chen. Guaranteeing the 2-manifold property for meshes with doubly linked face list. *International Journal of Shape Modeling*, 5(2):149–177, 2000. 315

[2] E. Akleman, J. Chen, and V. Srinivasan. A new paradigm for changing topology during subdivision modeling. In *Proceedings of Pacific Graphics*, pages 192–201, October 2000. 315, 316

[3] E. Akleman, J. Chen, and V. Srinivasan. A prototype system for robust, interactive and user-friendly modeling of orientable 2-manifold meshes. In *Shape Modeling and Applications – Proceedings of Shape Modeling International*, pages 43–50, May 2002. 315

[4] B. Baumgart. Winged-edge polyhedron representation. Technical Report STAN-CS-320, Stanford University, 1972. 314

[5] E. Catmull and J. Clark. Recursively generated B-spline surfaces on arbitrary topological meshes. *Computer Aided Design*, 10(6):350–355, 1978. 314

[6] T. Cormen, C. Leiserson, R. Rivest, and C. Stein. *Introduction to algorithms. Second Edition*. MIT Press, Cambridge, MA, 2001. 315

[7] M. de Boer, F. Fracchia, and P. Prusinkiewicz. A model for cellular development in morphogenetic fields. In G. Rozenberg and A. Salomaa, editors, *Lindenmayer*

systems: Impacts on theoretical computer science, computer graphics, and developmental biology, pages 351–370. Springer, Berlin, 1992. 314

[8]   T. DeRose. A coordinate-free approach to geomeric programming. In W. Strasser and H.-P. Seidel, editors, *Theory and practice of geometric modeling*, pages 291–305. Springer, Berlin, 1989. 314

[9]   N. Dyn, D. Levin, and J. Gregory. A butterfly subdivision scheme for surface interpolation with tension control. *ACM Transactions on Graphics*, 9(2):160–169, 1990. 321

[10] J. Edmonds. A combinatorial representation of polyhedral surfaces (abstract). *Notices of the American Mathematical Society*, 7:646, 1960. 314, 316

[11] R. Egli and N. F. Stewart. A framework for system specification using chains on cell complexes. *Computer-Aided Design*, 32:447–459, 2000. 314

[12] H. Ehrig, G. Engles, H.-J. Kreowski, and G. Rozenberg, editors. *Handbook of Graph Grammars and Computing by Graph Transformation, Vol. 2: Applications, Languages and Tools*. World Scientific, Singapore, 1999. 314

[13] J.-L. Giavitto and O. Michel. MGS: A programming language for the transformation of topological collections. Research Report 61-2001, CNRS - Université d'Evry Val d'Esonne, 2001. 313

[14] Ron Goldman. On the algebraic and geometric foundations of computer graphics. *ACM Transactions on Graphics*, 21(1):52–86, January 2002. 314

[15] Leonidas Guibas and Jorge Stolfi. Primitives for the manipulation of general subdivisions and the computation of Voronoi diagrams. *ACM Transactions on Graphics*, 4(2):74–123, April 1985. 314

[16] R. Karwowski and P. Prusinkiewicz. Design and implementation of the L+C modeling language. *Electronic Notes in Theoretical Computer Science*, 86.2:19pp., 2003. 319

[17] Leif Kobbelt. $\sqrt{3}$-subdivision. In *Proceedings of SIGGRAPH*, pages 103–112. ACM, 2000. 322

[18] W. B. R. Lickorish. Simplicial moves on complexes and manifolds. *Geometry and Topology Monographs*, 2:299–320, 1999. 314

[19] P. Lienhardt. Subdivisions de surfaces et cartes généralisées de dimension 2. *Informatique Théorique et Applications*, 25(2):171–202, 1991. 314

[20] P. Lienhardt. Topological models for boundary representation: a comparison with *n*-dimensional generalized maps. *Computer-Aided Design*, 23(1):59–82, 1991. 314

[21] P. Lienhardt. Subdivision par opérations locales. Manuscript, Université de Poitiers, 2001. 314

[22] A. Lindenmayer and G. Rozenberg. Parallel generation of maps: Developmental systems for cell layers. In V. Claus, H. Ehrig, and G. Rozenberg, editors, *Graph grammars and their application to computer science; First International Workshop*, Lecture Notes in Computer Science 73, pages 301–316. Springer, Berlin, 1978. 314

[23] C. Loop. Smooth subdivision surfaces based on triangles. Master's thesis, The University of Utah, August 1987. 320, 321

[24] Martti Mäntylä. *An introduction to solid modeling*. Computer Science Press, Rockville, 1988. 314

[25] R. Palmer and V. Shapiro. Chain models of physical behavior for engineering analysis and design. *Research in Engineering Design*, 5:161–184, 1993. 314

[26] P. Prusinkiewicz and A. Lindenmayer. *The algorithmic beauty of plants*. Springer-Verlag, New York, 1990. 314

[27] P. Prusinkiewicz, F. Samavati, C. Smith, and R. Karwowski. L-system description of subdivision curves. *International Journal on Shape Modeling*, 9(1):41–59, June 2003. 314

[28] G. Rozenberg, editor. *Handbook of graph grammars and computing by graph transformation.* World Scientific, Singapore, 1997. 314

[29] M. Sabin. Subdivision surfaces. Shape Modeling International tutorial notes, 2002. 25pp. 313

[30] E. Stollnitz, T. DeRose, and D. Salesin. *Wavelets for computer graphics.* Morgan Kaufman, San Francisco, 1996. 320

[31] Luiz Velho. Stellar subdivision grammars. In *Proceedings of Eurographics Symposium on Geometry Processing.* Eurographics Association, 2003. 12pp. 314

[32] A. White. *Graphs, groups and surfaces.* North-Holland, Amsterdam, 1973. 314, 316

[33] Denis Zorin, Peter Schröder, Tony DeRose, Leif Kobbelt, Adi Levin, and Wim Sweldens. Subdivision for modeling and animation. In *SIGGRAPH Course Notes*, New York, 2000. ACM. 313, 314

Transforming Toric Digraphs

Robert E. Jamison

Department of Mathematical Sciences
Clemson University
Clemson, SC 29634-0975
rejam@clemson.edu

Abstract. A digraph embedded on a torus can be flattened out into the plane to form a 2-dimensional partially ordered set (poset) by cutting a pair of orthogonal fundamental cycles. The family of 2-dimensional posets arising from a single toric digraph is called the *web* of the digraph. In 1994 Halitsky noted that tree families important in linguistics had additional symmetry if embedded on a torus and then transformed into another member of the web. Halitsky has attempted to use this "hidden symmetry" idea to predict tertiary structure of proteins from their primary structure.

1   Background

The work to be presented here grew out of some theoretical models proposed by David Halitsky in an attempt to describe structural redundancies in DNA and proteins. Halitsky is a linguist by training but has spent most of his career in computing. Underlying his ideas are two main principles. First, natural language, DNA, amino acid sequences in proteins, and even computer databases are all methods for storing and communicating information, and he posits that there should be certain basic structural similarities among them. Second, he believes that seemingly complex low dimensional phenomena are sometimes the result of a projection of very regular higher dimensional structures, as is the case, say, in quasi-crystals. Symmetry is, of course, a way of measuring structure and redundancy, and it is Halitsky's recognition of previously unnoted symmetries in language-theoretic structures which differentiates his bio-molecular application of these constructs from their well-known application to bio-informatics by David Searls [8, 9, 10].

The philosophy underlying this note is as follows. The world is basically holistic and very interconnected. In ancient times, the connections were made automatically, subconsciously. At some point (around 600 BC perhaps with Thales introduction of deductive geometry) people began to sequentialize, think linearly, and use deductive reasoning. The trend has continued with a vengeance, so that now we must linearize things before we can understand them. There is a certain truth to this because nature does contain linear objects like the DNA code and time.

J.L. Pfaltz, M. Nagl, and B. Böhlen (Eds.): AGTIVE 2003, LNCS 3062, pp. 328–333, 2004.
© Springer-Verlag Berlin Heidelberg 2004

This shows itself in language, where highly inflected languages naturally express cross-relations with ease in total disregard for linear structure. The more linear the thinking, the more positional the language, the fewer inflections. For example, it is now unfortunately commonplace to hear barbarisms like "just between you and I" and "construct an altitude through one vertice of this triangle" which attack the last vestiges of inflection in English. We linearize and find inflections hard, just as the ancients would have found the linear reasoning hard and the inflections second nature.

There is a basically nonlinear order to things but to express relations, understand things, do things in time, we are forced to linearize. That is we must break cycles. This is what finding spanning subtrees and Feedback Arc Sets (FAS) is all about. The cycles can be broken in many different ways, resulting in many possible linearizations of the original nonlinear phenomenon. In the linearization some of the symmetry of the original object might be lost. For example, linguistically, there is an underlying digraph, full of cycles, expressing linguistic connections. To form sentences, to parse expressions, to write computer code, to search for data in a database, we somehow have to linearize, which is done by deleting part of the relations — cutting a FAS.

In all of these contexts, there are trees that must be updated or modified to create other trees representing the same or similar information. A motivating example for Halitsky is the case of natural language in which the "rules of grammar" allow transformations to sentence structure, permitting different sentences (syntax) with the same meaning (semantics) cf. [3, 6]. Halitsky's goal was to find a structured set of transformations rather than an ad hoc set which had been used previously in the theory.

Halitsky's approach arose as follows. Every tree is a 2-dimensional partially ordered set (poset) and hence realizable by two linear orders, one accounting basically for the up-down structure of the tree, the other giving the left-right relationship among children [1]. The two linear orders together can be viewed as a permutation. Halitsky's novel idea was to represent this permutation by points in the plane and then use wrap around to put it on a discrete torus. Changing the way the torus is projected onto the plane changes the poset that emerges – in particular, it is not always a tree. Halitsky conjectured that tree structures important in linguistics had a toriodally transformed form which had more symmetry. I was able to prove this and in fact characterize [4] all trees which can be transformed on the torus into objects processing half-turn (180°) symmetry.

Halitsky used certain standard binary oppositions on amino acids — for example, hydrophobic versus hydrophilic — to locate an amino acid sequence on a planar grid. The grid could then be wrapped up (conceptually) to form a torus and cut in a new place to yield a different planar pattern, which using the standard order on the grid coordinates could be regarded as a digraph and in fact a 2-dimensional partial order.

The primary structure of a protein is just its sequence of amino acids, which is determined in a well-known way from the DNA or RNA which builds the protein. The secondary structure has to do with certain local sheets and folds. The tertiary structure is how the protein molecule configures itself in space. The chemically active sites are determined by the 3D geometry, i.e., tertiary structure, of the protein. Most of the amino acids in the protein chain play only a structural role, such as serving as filler to keep active sites at the right distance from each other. Bio-engineering of a protein focuses on the active sites where its function is carried out. It is thus important to know where the active sites are in order determine where to modify the DNA which produces the protein. The process of sequencing the DNA or the protein is a separate and simpler process than determining the 3D structure which involves sterio-chemistry and X-ray diffraction. Thus prediction of the active sites from just the amino acid sequence is a very important unsolved problem.

Halitsky's hypothesis is that configurations which posess hidden symmetry — that is, transform on the torus into a configurations with greater symmetry — should have greater structural significance [2]. Halitsky has run many tests on proteins whose primary and tertiary structures are known, trying to predict the chemically active sites and test his hypothesis. Although the tests have not been conclusively positive, they do indicate that further work is warranted.

In the next section, we will show how some of the rudiments of the theory can be raised to higher dimensions. The main concern is with graph transformations which arise by removing different feedback arc sets. In particular, we will prove that a large, natural family of axial cuts for minimal feedback arc sets.

2   Feedback Arc Sets in Toric Digraphs

Let $G = (V, A)$ be a directed graph with vertex set V and arc set A. G is strongly connected iff there is a directed path between any two vertices of G. A digraph is acyclic iff it contains no directed cycles. A feedback arc set (FAS) in a digraph is a set of arcs whose removal leaves an acyclic digraph.

Let Z_n denote the integers $\{0, 1, 2, \ldots, n-1\}$ endowed with two structures:

1. the ring structure of arithmetic modulo n, and
2. the digraph structure as a directed cycle — i.e., there is an arc from g to $g + 1 \pmod{n}$ for each g in Z_n.

For any two points a and b in Z_n, define the cyclic interval $ci(a, b)$ to be the directed path in Z_n from a to b. The vertex set $V(ci(a, b))$ is given by the following conjunction/disjunction rules:

1. $V(ci(a, b)) = \{x \in Z_n : a \leq x \text{ AND } x \leq b\}$ if $a < b$
2. $V(ci(a, b)) = \{x \in Z_n : a \leq x \text{ OR } x \leq b\}$ if $a > b$
3. $V(ci(a, a)) = \{a\}$

Notice that if g lies in $V(ci(a, b))$, then

$$V(ci(a, b)) = V(ci(a, g)) \cup V(ci(g, b)) \tag{1}$$

It is crucial that the cyclic intervals be understood as directed paths, having both a vertex set and an arc set, and NOT just as a set of points. Thus, for example, the vertex set of ci$(1,0)$ is all of $\{0, 1, 2, \ldots, n-1\}$ but its arc set misses the arc $\overrightarrow{01}$.

Let $d > 1$ be a fixed integer. For p and q in $(Z_n)^d$, define the patch $P[p, q]$ to consist of all points x in $(Z_n)^d$ such that x_i is in the (vertex set of the) cyclic interval ci(p_i, q_i) for all coordinates i. Thus the patch $P[p, q]$ is just the Cartesian product of the (vertex sets of the) cyclic intervals ci(p_i, q_i). If x lies in $P[p, q]$, then it follows at once from Eq. 1 that $P[p, x]$ and $P[x, q]$ are both subsets of $P[p, q]$.

The 2-dimensional objects built by Halitsky's procedure [5] have the property that the projection onto each of the coordinate axes is a bijection. They can thus be viewed as graphs of permutations, considered as functions. It is desirable to extend this condition to higher dimensions. A set S in $(Z_n)^d$ is nondegenerate provided $p_i \neq q_i$ for any two points p and q in S and for each coordinate $1 \leq i \leq d$. That is, S is nondegenerate iff its projection onto each coordinate axis is one-to-one. If in addition, S has n points, so the coordinate projections are actually bijections, then S is a transversal. Transversals are the appropriate analogues of Halitsky's 2-dimensional permutations. However, for the purposes here, the weaker assumption of nondegeneracy suffices.

Let S be a nondegenerate set of points. Place a directed edge (arc) from u in S to v in S iff the patch $P[u, v]$ contains no points of S other than u and v. The result is the toric digraph $TD(S)$ on S. Let g be any element of Z_n. The ith axial cut $AC_i(g, S)$ consists of all arcs $\overrightarrow{pq}$ in the toric digraph $TD(S)$ on S such that the arc from g to $g+1$ (addition mod n) lies in the cyclic interval ci(p_i, q_i).

The first goal is to show that every toric digraph is strongly connected and the axial cuts are minimal feedback arc sets. For this it is convenient to introduce some more structure. Given any arc $\overrightarrow{pq}$ in $TD(S)$, the ith axial projection of $\overrightarrow{pq}$ is the cyclic interval ci(p_i, q_i). Given any path in $TD(S)$, its ith projection is obtained by sequentially gluing together the ith projections of its constituent arcs. The result is a directed walk, that is a sequence $p_1, p_2, p_3, \ldots, p_\ell$ of consecutively adjacent vertices joined by arcs $p_i p_{i+1}$, where repetition of vertices and edges is allowed. In Z_n, as a directed cycle, the winding number of a directed walk is the number of times it revisits its starting vertex. For any directed path in $TD(S)$, its ith winding number is the winding number of its ith axial projection. A directed path is principal provided all of its winding numbers are zero.

Lemma 21 Let S be a nondegenerate set of points. Let P be a directed path in $TD(S)$ from p to q. For each coordinate index i, the following are equivalent:

  i) the ith axial projection of the path is contained in ci(p_i, q_i),
 ii) the ith axial projection of the path equals ci(p_i, q_i),
iii) the ith winding number of the path is zero.

Proof. The ith projection of P is a walk around Z_n as a directed cycle, starting at p_i and ending at q_i. The cyclic interval is the shortest such walk and the only one that does not go all the way around the cycle at least once.   □

Lemma 22 Let S be a nondegenerate set of points. For any $p \neq q$ in a set S, there is a principal path from p to q in $TD(S)$.

Proof. We proceed by induction on the number of points of S in the patch $P[p,q]$. The least $|S \cap P[p,q]|$ can be is 2, and that occurs when p and q are the only points of S in their patch. In this case, there is, by definition, a directed arc from p to q. This is clearly a principal path from p to q.

If $|S \cap P[p,q]|$ is more than 2, let x be any point of S other than p or q in the patch $P[p,q]$. Then $P[p,x]$ is a subset of $P[p,q]$. Moreover, $P[p,x]$ does not contain q. Thus $|S \cap P[p,x]| < |S \cap P[p,q]|$, so by induction there is a principal path R from p to x. Similarly, there is a principal path Q from x to q. Thus RQ is a directed path from p to q. It remains to verify that it is principal. Fix a coordinate i. Then the ith projection of RQ is by definition just the ith projection of R followed by the ith projection of Q.

Since R is principal, Lemma 21 implies that the ith projection of R is just $\mathrm{ci}(p_i, x_i)$; likewise, the ith projection of Q is $\mathrm{ci}(x_i, q_i)$. The ith projection of RQ is thus

$$\mathrm{ci}(p_i, x_i) \ \cup \ \mathrm{ci}(x_i, q_i) \ = \ \mathrm{ci}(p_i, q_i)$$

by Eq. 1, since x is in the patch $P[p,q]$. Since this is true for all i, Lemma 21 says that RQ is principal as desired.   □

Theorem 23   i) Every toric digraph is strongly connected.
ii) Every arc of a toric digraph lies in a directed cycle with winding number one.
iii) Every axial cut is a minimal feedback arc set.

Proof. i) follows at once from Lemma 22.

ii) Given an arc $\overrightarrow{pq}$, by Lemma 22 there is a principal path from q to p in $TD(S)$. Adding the arc $\overrightarrow{pq}$ to this produces the desired cycle.

iii) Consider an ith axial cut $AC_i(g, S)$. The number of arcs of a directed cycle in $TD(S)$ which lie in this cut is precisely the winding number of the cycle. Since a cycle has winding number at least one, any cycle will contain at least one edge of $AC_i(g, S)$. Hence $AC_i(g, S)$ is an FAS. Now consider any arc $\overrightarrow{pq}$ of $AC_i(g, S)$. By ii), $\overrightarrow{pq}$ lies in a cycle C with winding number one. Thus $\overrightarrow{pq}$ is the unique arc of C in the cut $AC_i(g, S)$. Hence if $\overrightarrow{pq}$ is omitted from $AC_i(g, S)$, the result is no longer an FAS. Hence $AC_i(g, S)$ is minimal as an FAS.   □

3   Conclusion

I would like to conclude with some open questions concerning the higher dimensional ideas just introduced.

- In addition to the axial cuts, it is natural to ask what other minimal FAS are there?
- Halitsky's orginal ideas in the plane were concerned with the structure of permutations which produced trees. How can transversals that produce a rooted tree be characterized?
- Determine all transversals which contain a minimal FAS whose removal leaves a tree.
- The graph transformations suggested here arise from changing the axial cut used to destroy the feedback and produce an acyclic digraph. The axial cuts are especially nice since they partition the arcs of the full toric digraph. Are there any other such sets of minimal feedback arc sets?

References

[1] A. Aho, J. Hopcroft, G. Ullman, *Introduction to Automata Theory, Languages, and Computation* Addison-Wesley, 1979. 329

[2] R. C. Dougherty, David Halitsky, R. E. Jamison, and W. F. Mann, A principled modification of the Chomsky language hierarchy: languages whose derivation trees are decomposable into ordered trees and forests invariant under a half-turn, submitted for presentation at IEEE FOCS (Foundations of Computer Science), October 2003. 330

[3] Seymour Ginsburg, *The Mathematical Theory of Context-Free Languages*, McGraw Hill, NY 1966. 329

[4] Robert E. Jamison, Trees invariant under a half-turn, manuscript, 1995. 329

[5] Robert E. Jamison and William F. Mann, Symmetries of Halitsky building sequences and the geometric objects they produce, in preparation. 331

[6] Arto Salomaa, *Formal Languages*, Academic Press, 1973. 329

[7] D. B. Searls, Representing genetic information with formal grammars, Proc. of AAAI, (1988) 386–391.

[8] D. B. Searls, The linguistics of DNA, American Scientist, **80**(6):579–591,1992. 328

[9] D. B. Searls, The computational linguistics of biological sequences, In *Artificial Intelligence and Molecular Biology* (L. Hunter ed.), AAAI Press, The MIT Press, 1993, 47–120. 328

[10] D. B. Searls, Formal language theory and biological macromolecules, Series in Discrete Mathematics and Theoretical Computer Science **47**(1999), 117-140. 328

Graph-Based Specification of a Management System for Evolving Development Processes*

Markus Heller[1], Ansgar Schleicher[2], and Bernhard Westfechtel[1]

[1] RWTH Aachen University, Department of Computer Science III
Ahornstrasse 55, D-52074 Aachen
{bernhard,heller}@i3.informatik.rwth-aachen.de
[2] DSA Daten- und Systemtechnik GmbH
Pascalstr. 28, D-52076 Aachen
Ansgar.Schleicher@dsa.de

Abstract. Development processes are inherently difficult to manage. Tools for managing development processes have to cope with continuous process evolution. The management system AHEAD is based on long-term experience gathered in different disciplines (software, mechanical, or chemical engineering). AHEAD provides an integrated set of tools for evolving both process definitions and their instances. AHEAD is based on graphs which are formally specified and manipulated by programmed graph transformations.

1 Introduction

Development processes in different disciplines such as software, mechanical, or chemical engineering share many features. Unfortunately, one of these common features is that they are hard to manage. Development processes are highly creative and therefore can be planned only to a limited extent. The tasks to be performed depend on the product to be developed, which is not known in advance. Alternative designs (variants) are explored to arrive at an optimal solution. Feedback may occur frequently — including not only spontaneous feedback raised by design errors in earlier steps, but also anticipated feedback which may be used to improve the design or to select among variants of the design. Finally, development methods such as concurrent or simultaneous engineering require sophisticated coordination between inter-dependent design activities.

In order to build effective tools for managing development processes, one must face the challenge of process evolution. While this has been recognized widely, current management systems can cope with process evolution only to a limited extent. In particular, this applies to workflow management systems [9] which were designed for repetitive business processes, e.g., by automating routine work in banks, insurance companies, administrations, etc. In such systems

---

* The work presented in this paper was carried out in the Collaborative Research Center IMPROVE, which is supported by the Deutsche Forschungsgemeinschaft.

J.L. Pfaltz, M. Nagl, and B. Böhlen (Eds.): AGTIVE 2003, LNCS 3062, pp. 334–351, 2004.
© Springer-Verlag Berlin Heidelberg 2004

a high number of workflows are executed according to a common definition, ensuring that work is performed following a pre-defined procedure. This approach cannot be transferred to development processes because it does not take process evolution into account: Developers would perceive themselves being tied in a straight-jacket so that they cannot perform their creative work as desired.

In this paper, we present the comprehensive evolution support [14] offered by AHEAD [4], an Adaptable and Human-Centered Environment for the MAnagement of Development Processes. AHEAD is based on nearly 10 years of work on development processes in different engineering disciplines. So far, we have applied the concepts underlying the AHEAD system in mechanical, chemical and software engineering.

The AHEAD system is based on a formal specification for two reasons. First, the specification clearly defines the management data and the effects of commands provided by the system. Second, code is generated from a high-level specification, resulting in significant savings of implementation effort. As specification language, we selected PROGRES[15], a high-level language for specifying programmed graph rewriting systems. Development processes can be represented in a natural way as graphs, and their evolution can be specified by graph transformations.

2   Example

This section demonstrates how the AHEAD system supports the management of evolving development processes. For this purpose, AHEAD offers dynamic task nets [2]. Within dynamic task nets, tasks describe units of work to be done. Tasks are organized hierarchically, i.e., a task may be decomposed into a set of subtasks. Control flow relationships define the order of subtask enactment. Data flow relationships connect input and output parameters of tasks and allow to exchange documents between them. Feedback relationships model cycles within the process which may stem from planned iterations or occurring exceptional situations. Software processes evolve continuously and it is usually impossible to completely plan a process before its enaction. Therefore, dynamic task nets are designed to be editable and enactable in an interleaved fashion.

With respect to dynamic task nets, we have to distinguish between process definitions and process instances. A process instance represents a specific development process being executed. In contrast, a process definition describes a whole class of processes. It describes the structures of processes of this class, and the constraints that have to be met. Process instances refer to definitions; e.g., the task ImplementUI is an instance of the task type Implement.

2.1   Wide Spectrum Approach

Development processes span a wide spectrum, ranging from highly structured, repetitive, and well-known ones to ad hoc, one-shot, and unknown ones. In this

Table 1. Parametric definition of behavior

model element	property	values
control flow	enactment order	standard, sequential, simultaneous
parameter	versioned	true, false
input	available	on start, later
input	consumption mode	automatic, manual

subsection, we describe how AHEAD covers the whole spectrum of processes by offering flexible means for process definitions.

Figure 1a shows a type-level definition for an idealized software process. It consists of task types for system design, component implementation, and test. These task types are connected through control flow and feedback flow types. The task types own input and output parameter types which are connected by data flow types defining the potential channels for data exchange. The process definition is vague, which is modeled through the partial-flag. This means that the process modeler decides to permit unconstrained types: In addition to explicitly defined types, the process manager may instantiate pre-defined unconstrained types for those tasks which have not been anticipated by the process modeler. Permitting unconstrained types potentially leads to partially typed task nets but enhances flexibility.

Besides these structural constraints, the behavior of elements may be defined using behavioral patterns (some of which are shown in Table 1). For each structural element, a number of behavioral patterns are predefined. It is e.g. stated that a control flow may have standard, simultaneous or sequential semantics. Standard semantics means that the flow's source is to be terminated before its target. Simultaneous semantics additionally requires the source to be activated before the target. With sequential semantics the flow's source has to be terminated before the target is activated. For parameters behavioral patterns regarding the versioning of documents exchanged via a particular parameter, the time of availability of a document at a particular input parameter or the automation degree of input consumption are specified. One set of behavioral patterns is defined as the default behavior. E.g. for input parameters the default behavior is defined as (versioned:=true; available:=later; consumption-mode:=manual). The most unrestricted behavioral patterns are used to define the default semantics which are e.g. used for unconstrained types.

Assigning one or multiple of these behavioral patterns to particular types of the process definition results in parameterization of an enacted dynamic task net's behavior. The enactment order of related tasks may be adapted using different control flow semantics; the activation of a task can be delayed by requiring on-start availability of its inputs, etc. In the example of Figure 1a we have assigned sequential and simultaneous behavioral patterns to the control flow types.

Figure 1a defines a partially typed subprocess. In the current version of the process definition, we assume there is no refining definition for the design task type (untyped, i.e., unconstrained subprocess) and a type-level definition for the

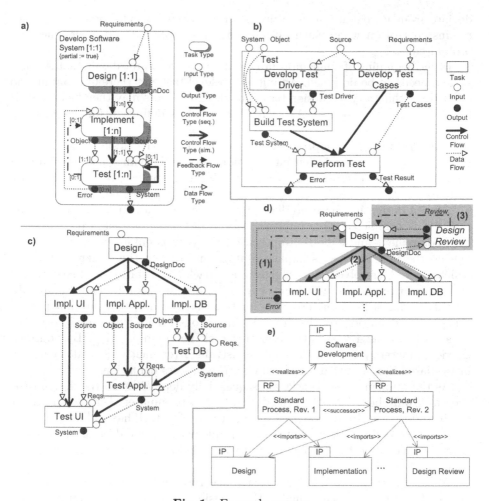

Fig. 1. Example process

implementation task type whose description we omit. Testing is well understood, and a stable process has been defined as an instance-level process definition. The corresponding instance pattern is displayed in Figure 1b. It consists of tasks for test driver and test case development, test system building and test execution. This instance pattern can be directly implanted into a task net without interactive planning steps. To summarize, the definition of our sample process covers the whole spectrum from unconstrained processes on one end to instance patterns on the other end.

2.2   Consistency Control

We now turn to the description of flexible process instance support. Our main focus lies on consistency control To enhance flexibility, the process instance may deviate from its definition, resulting in inconsistencies. The process manager retains control on where such inconsistencies are allowed and where they are prohibited.

In the beginning of our sample process it can be determined only that the new system needs to be designed. An initial top-level task net therefore contains no other tasks than a typed design task. The design process may then be described by a subordinate task net. As there is no process definition for design tasks available, the process manager creates an unconstrained ad-hoc task net (not shown). This task net comprises tasks for creating a coarse design to identify the system's components and their interrelations and for providing a detailed design for each of these components (e.g. user interface, application logic, database system).

After a design has been produced by the complex design task, the top-level task net is completed. One implementation and one test task are created for each component and a bottom-up testing strategy is enforced (cf. Figure 1c). During the user interface's implementation an error in the design is detected which has to be resolved. Accordingly, a feedback flow is introduced into the net and an error report is created and sent to the design task (cf. (1) in Figure 1d). As this feedback flow, the error parameters and the data flow are not part of the process definition (cf. Figure 1a) but unconstrained types are allowed by it, a weak consistency results from this task net manipulation. By contrast, strong consistency implies that only constrained (explicitly defined) types are instantiated in a way that conforms with the process definition.

As the implementation tasks are sequentially dependent on the design task, they may only be active when the design task is terminated. But, as a result of the introduced feedback flow, the terminated design task has to be reactivated for error correction. This results in the implementation and design task being active at the same time which leads to a behavioral inconsistency (cf. (2) in Figure 1d).

As design errors may become costly the further the software process proceeds, the process manager makes the decision to let the next design version be reviewed. Accordingly, a design review task is created in the context of the design task. A feedback flow is used to model the planned iteration between the design and review tasks (cf. (3) in Figure 1d). As the design review task is typed but the type is not part of this process definition, a structural inconsistency emerges.

Let us summarize the general features of consistency control: By default, a process instance must be consistent with the process definition. However, the process manager may allow inconsistencies selectively by means of local "switches" which are provided for each subnet in a task hierarchy. By default, consistency enforcement is switched on, but the process manager may switch it off deliberately to tolerate temporary or even permanent inconsistencies. When consistency enforcement is switched off, the process manager may manipulate the respect subnet (being part of the process instance) such that it deviates from the process definition. Consistency enforcement may be switched on again only after all inconsistencies have been removed (either by operations applying only to the process instance or by migrating to a revised process definition, see below).

2.3 Definition-Level Evolution and Migration

In some cases, deviating process instances contain new and valuable process knowledge which should be made available to other process instances. This leads to an extension of the process definition, which is structured in a modular way through packages. Interface packages contain the task and parameter types of a task, while realization packages contain the refining definitions of one interface. In the context of our example, the process modeler decides to revise the top level process definition. The new package version contains a feedback flow type between implementation and design task types together with the necessary parameter and data flow types. The design review type is embedded into the definition and the behavioral description of control flow types between design and implementation tasks is switched from sequential to simultaneous. The resulting package structure is shown in Figure 1e. The successor dependency between packages denotes the history of process definition extensions.

As mentioned in the beginning of this section, there might be multiple running instances of one process definition. The manager may select the instances to migrate on-demand and determine the time of migration. M igration means that an instance is upgraded to the new definition. In our example, the process manager decides to migrate the process instance of Figure 1d such that it is consistent with the new definition. Other running instances may also be migrated. Their consistency with respect to the new process definition version is very unlikely. Therefore, the inconsistencies induced by migration are again signaled to the manager, but migration can always be performed.

If the manager of another instance which is consistent with the original process definition decides to migrate to the new version, parts of the instance are inconsistent or incomplete. The missing design review task denotes an incompleteness of the task net, while the sequential control flows between design and implementation tasks are a behavioral inconsistency. The manager may selectively decide which incompletenesses and inconsistencies to remove from the net. E.g., he may create and embed a design review task into the net, removing the incompleteness.

Thus, instances may be migrated to new (versions of) definitions, but that does not necessarily imply that a consistent state will be reached after migration. Both temporary and persistent inconsistencies may be tolerated by turning off consistency enforcement.

3 Formal Specification

3.1 The AHEAD System

We have realized the process evolution approach described above in the AHEAD system (Figure 2). A process modeler creates an external process definition in

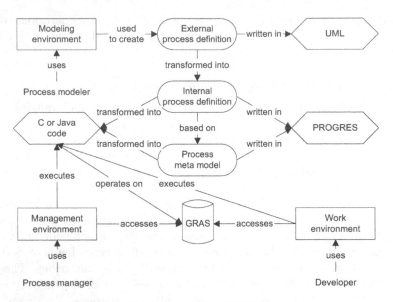

Fig. 2. Overview of the AHEAD system

a m odeling environm ent for UML (in this paper, we used a different, more concise notation for demonstration purposes). The model packages are transformed automatically into an internal process definition which is hidden from the external user. Internally, processes are defined in PROGRES [15], a specification language for programmed graph transformations. Internal process definitions are based on a process m eta m odel which has been defined once beforehand (again in PROGRES). The specifications for internal process definitions and the process meta model are translated by the PROGRES compiler into equivalent C (or Java) code which operates on a graph-based DBMS (GRAS). The generated code is executed by a m anagem ent environm ent supporting process managers in planning, analyzing, and monitoring software processes, and a work environm ent which provides software engineers with tools managing agendas and workspaces.

Wide spectrum processes are defined in the modeling environment, which is also used to evolve process definitions by versioning model packages. Consistency control is handled by the management environment, which offers commands for enabling and disabling inconsistencies and informs process managers about deviations from the process definition. In addition, the management environment supports migration of process instances to a modified process definition. In the following, we constrain the presentation to the level of PROGRES specifications since our approach to process evolution is formally defined at that level. For the UML modeling environment and the transformation from UML to PROGRES, see [5]. Since the overall specification covers far more than 100 pages, we can present only small fragments to illustrate the underlying principles. Furthermore,

```
node class ENTITY is a ITEM;
  intrinsic Name : string;
end;
node class TASK is a ENTITY
  intrinsic
    State : {InDefinition, ..., Failed} := InDefinition;
end;
node class TASK_RELATION is a ITEM end;
edge type fromSourceT : TASK -> TASK_RELATION;
edge type toTargetT : TASK_RELATION -> TASK;
node class CONTROL_FLOW is a TASK_RELATION end;
node class FEEDBACK_FLOW is a TASK_RELATION end;
node class REALIZATION is a ENTITY, TASK_RELATION end;
edge type assigned : REALIZATION -> TASK;

production BaseCreateSubTask
  ( Parent : TASK; TaskName : string;
    TaskType : type in TASK; out NewTask : TASK ) =
```

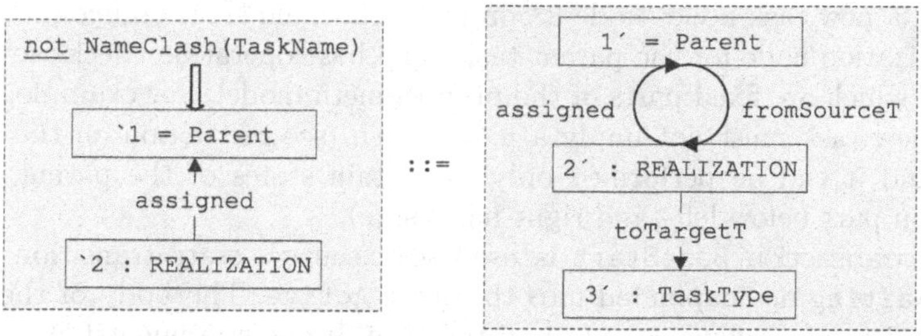

```
    condition `1.State in {InDefinition, Active};
    transfer 3´.Name := TaskName;
    return NewTask := 3´;
end;

transaction BaseStart( Task : TASK ) =
    (Task.State in {Waiting, Suspended}
      and (Task.IsRoot or Task.=ToParent=>.State = Active))
   & Task.State := Active
end;
```

Fig. 3. Specification the process meta model

we can give only rather cursory explanations; for further technical details, the reader is referred to [14].

3.2 Process Meta Model

In the sequel, we sketch how the process meta model is defined in PROGRES (see [2] for a more comprehensive description). We start out in this subsection by presenting the meta model elements and their interrelations. In the next subsections we explain enhancements of this process meta model needed for wide spectrum process model definition, consistency control and process evolution, respectively.

The upper part of Figure 3 shows a cutout of the graph schema for dynamic task nets. The graph schema defines the components of dynamic task nets in terms of node classes, attributes, and edge types. Tasks are modeled by the node class TASK, which is a subclass of ENTITY. A task has an intrinsic Name attribute — inherited from ENTITY — as well as an additional State attribute. The process meta model defines a common state diagram which applies to all types of tasks, but which can further be adapted at the process definition level. Task relations are represented by nodes of class TASK_RELATION and edges of type fromSourceT and toTargetT, respectively. Finally, a node of class REALIZATION is used to represent the realization of a task being assigned to its interface (represented by the task node itself).

The lower part of Figure 3 gives two examples of base operations defined in the process meta model. With the help of the graph rewrite rule (production) BaseCreateSubTask, a subtask is inserted into a task net. This is achieved by creating a new task node (node 3' on the right-hand side) and connecting it to the realization node for the parent task. The base operation checks several constraints which are fixed parts of the process meta model. For example, insertion of the new task must not imply a name clash (see restriction on the left-hand side), and it can be performed only in certain states of the parent task (see condition part below left- and right-hand side).

The transaction BaseStart is used to execute a state transition from the states Waiting or Suspended into the state Active. The body of the transaction consists of a sequence (&) of statements. If one statement fails, the whole transaction fails and leaves the task net unaffected. First, it is checked that the current task resides in a valid source state for the Start transition. Furthermore, the task must either be located at the root of the task hierarchy, or its parent must be Active. If these conditions hold, the State attribute of the current task is assigned the value Active.

Please note that all base operations check and enforce both structural and behavioral constraints. Since planning and enactment may be interleaved, a structural operation such as BaseCreateSubTask has to check behavioral constraints with respect to the state of the parent task. Moreover, we would like to emphasize that only minimal behavioral constraints are enforced in the process meta model. For example, BaseStart merely checks the parent task's state. All constraints checked at this level are hard constraints built into the process meta model.

3.3   Wide Spectrum Approach

In the following, we describe how the wide spectrum of process definitions is represented in the PROGRES specification. First, we go into type-level process definitions; the representation of instance patterns will be discussed briefly at the end of this subsection.

The production BaseCreateTask presented above instantiates a domain-specific task type, but it does not check domain-specific constraints (e.g., whether an instance of that type is permitted in the realization of the parent task). In

```
node class TASK is a ENTITY
  meta
     DeclaredParameters   :
         type in PARAMETER [0:n] := PARAMETER;
     DeclaredRealizations :
         type in REALIZATION [0:n] := REALIZATION;
  intrinsic
     State : {InDefinition, ..., Failed} := InDefinition;
end;
node class TASK_RELATION;
  meta
     SourceTypes : type in TASK [0:n] := TASK;
     TargetTypes : type in TASK [0:n] := TASK;
end;
node class CONTROL_FLOW is a TASK_RELATION
  meta
     EnactmentOrder :
         {standard, simultaneous, sequential} := standard;
end;
```

```
node type Task : TASK end;
node type Realization : REALIZATION end;
node type ControlFlow : CONTROL_FLOW end;
node type DevelopSoftwareSystem : TASK
  redef meta
     DeclaredParameters = {Requirements, System};
     DeclaredRealizations = {PhaseDevelopment};
end;
node type DesignToImplement : CONTROL_FLOW
  redef meta
     SourceTypes := {Design};
     TargetTypes := {Implement};
     EnactmentOrder := sequential;
end;
```

Fig. 4. Declaration of node types and meta attributes

Subsection 3.4, we will describe how base operations are extended with these checks. Before that, we explain how the information used by these checks is represented in the graph schema.

Domain-specific types introduced in a process definition are represented by node types, which are instances of node classes. Node types may be passed as parameters (e.g., the type of a task to be created, see Figure 3), and they may be stored as attribute values (see below). To express domain-specific constraints, we make use of meta attributes, i.e., attributes which are attached to classes or types rather than node instances.

The upper part of Figure 4 shows a cutout of the graph schema for the process meta model, extended with meta attributes. At the level of the meta model, these attributes are initialized in the most "liberal" way. For example, the meta attribute **DeclaredParameters** of class **TASK** is a set-valued attribute (cardinality [0:n]) which contains all **PARAMETER** types, i.e., parameters of any type are allowed. The lower part of Figure 4 shows a cutout of a process definition, which introduces node types and redefines the values of meta attributes to express domain-specific constraints. For example, nodes of type DevelopSoftwareSystem may have only parameters of type Requirements (input) and System (output). Please note that both structural and behavioral constraints are represented by meta attributes (see e.g. the type DesignTo-Implement, which is used to represent sequential control flows from Design to Implement tasks).

Instance patterns such as given in Figure 1b are not mapped onto the graph schema. Rather, a transaction is generated which contains a sequence of calls to base operations creating the instance pattern's elements (see [5], which also explains why we do not generate a production instead).

3.4 Consistency Control

Now we can establish the connection between the base operations of Section 3.2 and the domain-specific constraints of Section 3.3. We use the term soft constraints to emphasize that they can be enforced selectively (in contrast to the process meta model's hard constraints). For checking soft constraints, we introduce derived attributes which signal inconsistencies. While the values of intrinsic attributes are assigned explicitly, the values of derived attributes are defined by evaluation rules and are updated automatically.

For consistency checking, we attach Boolean attributes to the nodes of task nets (Figure 5). Consistency calculation is performed in an object-oriented manner. Every language element can calculate its own consistency within its context. In the node class ITEM, which serves as the root of the class hierarchy, attributes are defined for checking structural and behavioral consistency. As default, these attributes evaluate to true; the evaluation rules are redefined in subclasses of ITEM. These evaluation rules refer to the values of meta attributes. For example, in the class CONTROL_FLOW structural consistency is determined using the derived attribute TypesConsistent, which checks whether the type of the source and target tasks are contained in the sets SourceTypes and TargetTypes, respectively. Similarly, the behavioral consistency depends on the derived attribute StatesConsistent, which checks whether the states of source and target are compatible with the EnactmentOrder of the control flow type. For example, in the case of a sequential control flow the target may only be active, suspended, failed, or done if the source is terminated. Thus, there are two derived attributes signaling the overall behavioral and structural consistency of an element and an arbitrary number of additional derived attributes performing fine-grained checks. By retrieving the values of these latter attributes the cause of a particular inconsistency can be identified.

In TASK, an intrinsic attribute AllowInconsistencies is defined whose default value is false. To allow for inconsistencies in a certain task net, the process manager changes its value to true. After all inconsistencies have been removed, AllowInconsistencies may be switched off again.

```
node class ITEM
  derived
    StructurallyConsistent : boolean = true;
    BehaviorallyConsistent : boolean = true;
end;
node class ENTITY is a ITEM ... end;
node class TASK is a ENTITY
  intrinsic
    AllowInconsistencies : boolean := false;
  ...
end;
node class TASK_RELATION is a ITEM ... end;
node class CONTROLFLOW is a TASK_RELATION
  derived
    TypesConsistent  : boolean =
      self.<-fromSourceT-.type in self.SourceTypes and
      self.-toTargetT->.type in self.TargetTypes;
    StatesConsistent : boolean =
      [self.EnactmentOrder = sequential ::
         self.-toTargetT->.State in
           {Active, Suspended, Done, Failed} =>
         self.<-toSourceT-.State in {Done, Failed}
        | ... ];
  redef derived
    StructurallyConsistent = self.TypesConsistent and ...;
    BehaviorallyConsistent = self.StatesConsistent and ...;
  ...
end;

transaction CreateSubTask
  ( Parent : TASK; TaskName : string;
    TaskType : type in TASK; out NewTask : TASK ) =
    BaseCreateSubTask(Parent, TaskName, TaskType, out newTask)
  & [ not Parent.AllowInconsistencies ::
        NewTask.StructurallyConsistent and ...
      | true ]
end;

transaction Start( Task : TASK ) =
    BaseStart(Task)
  & [ not Task.ToParent.AllowInconsistencies ::
        (for all cf in AdjacentControlFlows(Task)
           cf.BehaviorallyConsistent
         end) and ...
      | true ]
end;
```

Fig. 5. Specification of consistency control

The base operations of the process meta model are embedded into wrappers
which can check soft constraints. The base operation is called first. Afterwards,
the AllowInconsistencies flag is checked which is attached to the surrounding
task net's root node. If inconsistencies are not allowed, but have been introduced
by the base operation, the operation is rolled back, and the wrapper transaction
fails. Figure 5 demonstrates this by the wrappers for creating a subtask and
starting a task.

```
production BaseMigrateRealization
( Realization : REALIZATION;
  RealizationType : type in REALIZATION;
  out NewRealization : REALIZATION ) =
```

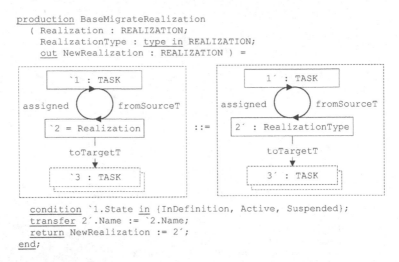

```
condition `1.State in {InDefinition, Active, Suspended};
transfer 2´.Name := `2.Name;
return NewRealization := 2´;
end;
```

Fig. 6. Base operation for migrating a realization

In the example of Figure 1d, the process manager may create an unconstrained feedback flow from implementation to design when inconsistencies are still disabled (weak consistencies are allowed). After that, the `AllowInconsistencies` flag has to be set to true. Only then may the `Design` task be reactivated, resulting in a behavioral inconsistency with respect to the sequential control flow to the source of the feedback flow. Moreover, insertion of a (typed) `DesignReview` task results in a structural inconsistency. These inconsistencies are signaled by the respective derived attributes.

3.5   Definition-Level Evolution and Migration

Due to the lack of space, we cannot elaborate on the technical details of definition-level evolution and migration. Rather, we briefly describe the basic ideas.

Evolution of process definitions is supported at the level of packages. PROGRES does provide packages, but it supports neither package versions nor type versions. Thus, package versions as shown in Figure 1e are simulated by appending version numbers to package names (likewise for node types). From the perspective of PROGRES, definition-level evolution implies an extension of the graph schema and (for instance patterns) addition of new operations (i.e., no loss of information). To make definition-level evolution effective, the extended external process definition is transformed in a batch process into an extended internal process definition in PROGRES, which is compiled into C in turn. The source code has to be compiled and linked to produce an AHEAD system for the extended process definition. Subsequently, migration be initiated.

Migrating a subnet to a new definition may involve insertion of new task net elements as well as deletion or conversion of old task net elements. Correspond-

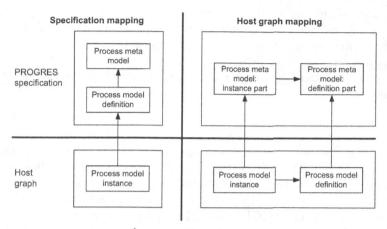

Fig. 7. Realization alternatives

ing base operations for insertion and deletion are available anyhow. The base operations for conversion upgrade task net elements to new types. Since PRO-GRES does not offer type changes of nodes, conversion has to be simulated by deleting the old node and inserting a new one, preserving the embedding. As an example, Figure 6 shows the graph rewrite rule for migrating a task realization. Please note that the condition part excludes realization of terminated tasks from migration because historical data should not be changed post mortem.

In addition, complex operations are provided to (partially) automate the migration process. A lot of user interactions are required if the process manager has to call the base operations on each individual element. Please note that the base operations require that the target type be passed explicitly, which makes the migration even more tedious. Therefore, AHEAD assists the process manager by a set of transactions which select the elements to be migrated and try to infer the respective target type. Unfortunately, it may happen that the target type may not be inferred in a unique way, in particular in the case of unconstrained types. Thus, currently migration normally requires user interactions.

3.6   Discussion

To conclude this section, we discuss a technical issue concerning the way we have applied the PROGRES language to process modeling. Our overall approach is illustrated on the left-hand side of Figure 7. The process meta model is encoded in PROGRES at the level of node classes, the process definition adds domain-specific parts of the specification at the level of node types, and process instances are represented as node instances (of node types) in the host graph at run time. This solution is called specification mapping (i.e., mapping of the process definition into the specification).

The right-hand side of Figure 7 shows an alternative realization: the host graph mapping (i.e., mapping of the process definition into the host graph).

Following this alternative, the host graph would be composed of two parts representing process definitions and instances, respectively. Accordingly, the PROGRES specification would consist of two analogous parts dealing with graphs and operations for process definitions and instances, respectively.

Specification and host graph mapping have complementary advantages and drawbacks:

Uniformity In the case of the specification mapping, process definitions and instances are represented in different ways. In contrast, the host graph mapping would handle process definitions and their instances in a uniform way.

Modeling Effort The specification mapping reduces the modeling effort considerably because all features of PROGRES may be exploited for process definitions. In contrast, for the host graph mapping features which are already available in PROGRES would have to be simulated.

Efficiency In the case of the specification mapping, the process definition is compiled into C code, which can be executed efficiently. In contrast, the host graph mapping would require time-consuming interpretation of the process definition.

Flexibility In the case of the specification mapping, process evolution at the definition level requires two batch compilation steps (UML $\rightarrow$ PROGRES $\rightarrow$ C). In contrast, the host graph mapping would allow for changes to the process definition on the fly, without any disruption.

This discussion shows that selecting one of these alternatives is a classical engineering design decision: In neither case, a solution is obtained which is optimal with respect to all aspects discussed above. We have favored specification mapping because it is more efficient and because it reduces the modeling effort. On the other hand, the host graph mapping would handle process definitions and instances uniformly, and it would be more flexible. We consider host graph mapping a viable alternative in the case of rather simple process definitions; otherwise, the modeling effort would become too high.

4   Related Work

AHEAD differs from commercial workflow management systems [9] considerably. Essentially, in a workflow management system a process instance is created by (conceptually) copying the definition (a "process program") and executing it. In contrast, in AHEAD a task net is built up only at runtime; planning and enactment may be interleaved seamlessly. The standards of the Workflow Management Coalition (see www.wfmc.org), having partially been adopted by the OMG [11], do not address process evolution.

AHEAD is based on a process meta model which bears some similarities to other meta models, e.g., the SPEM metamodel [12] defined by the OMG as an extension of the UML meta model. In SPEM, activities are modeled as operations, and software processes are primarily modeled with the help of activity diagrams. AHEAD follows a different approach and makes use of class diagrams to support process evolution through dynamic instantiation of task classes and associations [5].

The need for a wide spectrum approach to process management was recognized as a research challenge in [16]. It is addressed in some research prototypes of workflow management systems which provide a wide range of control flow types (e.g., Mobile [3] and FLOW.NET [7]). However, the main focus still lies on highly or medium-structured processes.

There are only a few other approaches which are capable of managing inconsistencies. In PROSYT [1], users may deviate from the process definition by enforcing operations violating preconditions and state invariants. However, eventually consistency has to be re-established; otherwise enactment has to be aborted. In contrast, the approach described in [10] does allow for potentially persistent inconsistencies, which raise exceptions that are handled manually or automatically. However, all of the approaches we are aware of do not address weak consistency in the presence of partial process knowledge.

A key and unique feature of our approach consists in its support for round-trip process evolution. In contrast, most other approaches are confined to top-down evolution. In [3, 6, 8], the process definition has to be created beforehand, while we allow for enacting partially known process definitions. Furthermore, both structural and behavioral consistency must be maintained during migration. This is not required in our approach, which is more flexible.

Finally, there are a few approaches which are confined to instance-level evolution (e.g., [13]). A specific process instance is modified, taking the current enactment state into account. However, the process definition remains unaffected. Our approach is more general since it covers both instance- and definition-level evolution (the latter of which may be used to propagate changes to more than just one instance).

5   Conclusion

In this paper, we have demonstrated the use of graph technology for the specification of process evolution support in the AHEAD management system. Development processes can be represented in a natural way as graphs, and their evolution can be specified by graph transformations. In our specification, we have exploited virtually all features offered by the specification language PROGRES, including the stratified type system, meta attributes, derived attributes, graph rewrite rules, transactions, and backtracking.

Finally, we have compared rather briefly two ways of using PROGRES for process modeling, namely specification mapping (i.e., mapping of the process definition into the PROGRES specification) and host graph mapping (i.e., mapping of the process definition into the host graph). We have deliberately pursued the specification mapping alternative to reduce modeling effort and to increase efficiency. However, we still consider comparison and evaluation of different styles of specification an important research topic to be addressed in the future.

References

[1] Gianpaolo Cugola. Tolerating deviations in process support systems via flexible enactment of process models. *IEEE Transactions on Software Engineering*, 24(11):982–1001, November 1998. 349

[2] Peter Heimann, Carl-Arndt Krapp, Bernhard Westfechtel, and Gregor Joeris. Graph-based software process management. *Intern. Journal of Software Eng. and Knowledge Eng.*, 7(4):431–455, December 1997. 335, 341

[3] Petra Heinl, Stefan Horn, Stefan Jablonski, Jens Neeb, Katrin Stein, and Michael Teschke. A comprehensive approach to flexibility in workflow management systems. In Dimitrios Georgakopoulos, Wolfgang, and Alexander L. Wolf, editors, *Proceedings of the International Joint Conference on Work Activities Coordination and Collaboration (WACC '99)*, volume 24-2 of *ACM SIGSOFT Software Engineering Notes*, pages 79–88, March 1999. 349

[4] Dirk Jäger, Ansgar Schleicher, and Bernhard Westfechtel. AHEAD: A graph-based system for modeling and managing development processes. In Manfred Nagl, Andy Schürr, and Manfred Münch, editors, *Proc. of the Intl. Workshop on Applications of Graph Transformations with Industrial Relevance (AGTIVE)*, LNCS 1779, pages 325–339. Springer, September 1999. 335

[5] Dirk Jäger, Ansgar Schleicher, and Bernhard Westfechtel. Using UML for software process modeling. In Oscar Nierstrasz and Michel Lemoine, editors, *Software Engineering — ESEC/FSE '99*, LNCS 1687, pages 91–108. Springer, September 1999. 340, 344, 349

[6] Gregor Joeris and Otthein Herzog. Managing evolving workflow specifications. In *Proc. of the Intl. Conf. on Cooperative Information Systems (CoopIS'98)*, pages 310–321. IEEE Comp. Soc. Press, 1998. 349

[7] Gregor Joeris and Otthein Herzog. Towards flexible and high-level modeling and enacting of processes. In Matthias Jarke and Andreas Oberweis, editors, *Proc. of the Intl. Conf. on Advanced Information Systems Engineering (CAiSE'99)*, LNCS 1626, pages 88–102. Springer, June 1999. 349

[8] Markus Kradolfer and Andreas Geppert. Dynamic workflow schema evolution based on workflow type versioning and workflow migration. In *Proc. of the Intl. Conf. on Cooperative Information Systems (CoopIS'99)*, pages 104–114. IEEE Comp. Soc. Press, September 1999. 349

[9] Peter Lawrence, editor. *Workflow Handbook*. John Wiley, Chichester, UK, 1997. 334, 348

[10] Takahiro Murata and Alex Borgida. Handling of irregularities in human centered systems: A unified framework for data and processes. *IEEE Transactions on Software Engineering*, 26(10):959–977, October 2000. 349

[11] Object Management Group, Needham, Massachusetts. *Workflow Management Facility Specification*, version 1.2 edition, April 2000. http://www.omg.org. 348

[12] Object Management Group, Needham, Massachusetts. *Software Process Engineering Metamodel Specification*, version 1.0 edition, November 2002. http://www.omg.org. 348

[13] Manfred Reichert and Peter Dadam. ADEPT$_{flex}$ — supporting dynamic changes without loosing control. *Journal of Intelligent Information Systems*, 10(2):93–129, March 1998. 349

[14] Ansgar Schleicher. *Management of Development Processes — An Evolutionary Approach*. Dissertation RWTH Aachen, Deutscher Universitäts-Verlag, Wiesbaden, Germany, 2002. 335, 341

[15] Andy Schürr, Andreas Winter, and Albert Zündorf. Graph grammar engineering with PROGRES. In W. Schäfer and P. Botella, editors, *Proc. of the European Software Engineering Conference (ESEC '95)*, LNCS 989, pages 219–234. Springer, September 1995. 335, 340

[16] Amit Sheth et al. NSF workshop on workflow and process automation. *ACM Software Engineering Notes*, 22(1):28–38, January 1997. 349

Graph-Based Tools for Distributed Cooperation in Dynamic Development Processes*

Markus Heller[1] and Dirk Jäger[2]

[1] RWTH Aachen University, Department of Computer Science III
Ahornstrasse 55, 52074 Aachen, Germany
heller@cs.rwth-aachen.de
[2] Bayer Business Services GmbH
51368 Leverkusen, Germany
dirk.jaeger@bayerbbs.com

Abstract. The highly dynamic character of development processes makes it a challenging task to provide support for the management of such processes within an organization. The process management system AHEAD addresses the specific problems related to the management of development processes in engineering disciplines. The system stores all management data as graphs. The application logic is specified in a formal specification based on a programmed graph rewriting system. From this specification several management tools of the AHEAD system are generated. Recently, the AHEAD system has been extended to support distributed development processes. Two or more organizations use their own instances of AHEAD and these instances are coupled at run-time. The coupling logic is specified by graph transformations and the executable code for the coupling can be automatically generated from this specification. Furthermore, the precise notation of the coupling by a formal specification makes it easy to enhance or extend the coupling mechanism. This paper describes how graph transformations are used to realize the demanded functionality.

1 Introduction

The study of development processes in our group focuses on managing the development of a complex end-product in engineering disciplines like software or chemical engineering. Development processes tend to be highly creative and dynamic. For example, it may be difficult to predict all activities, their order, and their duration. Changes in the product specifications may induce variations of planning the development activities. Moreover, development processes tend to be highly unique. As a result of this, the management of activities, performers and resulting products is rather complex.

Today the development of an end-product is not always accomplished by one organization alone, e.g. a company or department of a company. For example, the

---

* Financial support is given by Deutsche Forschungsgemeinschaft, Collaborative Research Center 476.

J.L. Pfaltz, M. Nagl, and B. Böhlen (Eds.): AGTIVE 2003, LNCS 3062, pp. 352–368, 2004.
© Springer-Verlag Berlin Heidelberg 2004

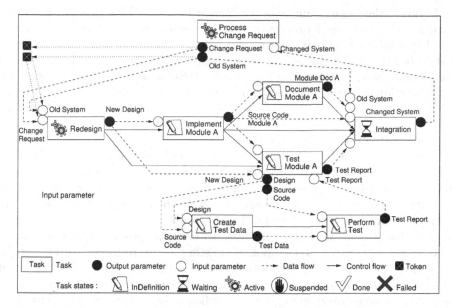

Fig. 1. Dynamic task net for a change request process in software engineering

know-how needed to perform some development activity may only be available in another organization. In a distributed development process the development activities are performed by employees of different organizations working together while development products are exchanged across organizational boundaries.

The process management system AHEAD addresses these problems. AHEAD has been developed within a long-term running Collaborative Research Center IMPROVE (Information Technology Support for Collaborative and Distributed Design Processes in Chemical Engineering)[9]. In the supported scenario it is assumed that all participating organizations use the AHEAD system for their process management. The developed concepts and supporting tools are described in detail in [1, 5]. A demonstration scenario is described in [4].

2 The Process Management System AHEAD

2.1 Graph-Based Integrated Management Model

The management of a development process to develop a certain end product comprises the coordination of all development activities, the management of all related product data, e.g. technical documents and plans, and the management of the related resources.

The activities of a development process and the relationships between these activities can be modeled by a dynam ic task net Dynamic task nets are defined by the underlying model DYNAMITE[17]. They can be planned, executed, and analyzed in an integrated way.

Figure 1 shows an example of a dynamic task net. This task net resembles a specific change request process in software engineering. Such a process is executed in a software company during the development process for a new software system (and even after the release of the system to customers). A change request describes a change of the existing software system in order to fix a bug or to add some functionality to the system. The first activity of this process can be a redesign of the current system, followed by the implementation, documentation, and test of all affected software modules. Finally, the changed modules have to be integrated into the software system.

The dynamic task net for this example is built of tasks representing activities, e.g. Redesign, Implement Module A. Each task has an execution state as 'InDefinition', 'Active', 'Suspended', or 'Finished'. Tasks can be connected with each other by directed edges representing controlflow relationships defining the temporal execution order of tasks. Tasks are characterized by an interface of all available output and input products. To limit the possible types of input or output products, input and output parameters are introduced. An output parameter of a task is linked to a corresponding input parameter of another task by an edge denoting a data flow relationship. Tokens representing products can be passed between tasks along these data flows. Tasks can be either atomic or complex. Complex tasks can be further refined by task nets, e.g. the task Test Module A is refined by a task net with tasks Create TestData and Perform Test. Thus dynamic task nets are hierarchical.

The product model COMA [15, 16] defines the representation of products (or documents) and the relationships between documents. Documents can be versioned. Configurations containing a set of product versions can be defined. The management of human and non-human resources is defined in the resource model RESMOD[8]. Plan resources, e.g. project teams and roles, and actual resources, e.g. organizational team units and team members, as well as a mapping between them can be represented.

The management models of AHEAD rely on graphs as the fundamental data structure. For example, a dynamic task net can naturally be represented as a graph of task nodes connected by edges denoting control or data flow relationships between different tasks. Operations on task nets (e.g. the insertion or deletion of tasks) are realized by graph transformations.

An example of such a graph structure is shown in figure 2. Different instances of the node classes TASK and PARAMETER as well as different instances of edge types for relationships between nodes are shown. Node classes are connected via has-edges with instances of the node types Input or Output.

The graph schema for dynamic task nets contains the common superclass of all model elements ITEM as root of the class hierarchy from which the node classes ENTITY and RELATION are derived. While ENTITY is used to represent entities like TASK and PARAMETER, the node class RELATION is introduced to model relationships between entities as, for example has.

In a similar way, the human and non-human resources and the product model are modeled using graphs and all operations on these models are realized as graph

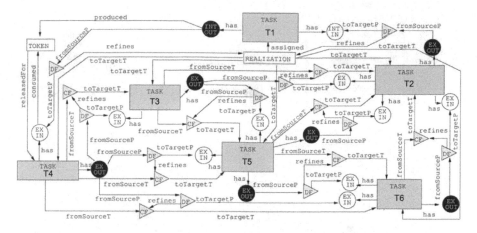

Fig. 2. Internal representation of dynamic task nets

transformations. All management data for an instance of a development process
are stored within a single graph containing all information about development
tasks, product data, and resource data.

The models for dynamic task nets, resources, and products form the base
model layer of the integrated management model of AHEAD. These base models
are then integrated with each other pair-wise in a higher layer. The full integration of all three models is realized in a third layer above. A detailed discussion
of this integrated management model can be found in [17].

The base models and the integration of these models are all combined in one
single specification written in the language PROGRES[11]. PROGRES uses directed, attributed graphs with labeled nodes and edges as its fundamental data
model. A PROGRES specification defines a graph schema and graph transformations. In a graph schema different node classes, node types and edge types can
be defined. A node type is an instance of a node class. Attributes can be defined
for node classes or types (not for edge types) to store additional information.
Graph transformations are written in the form of graph rewrite rules. Each graph
rewrite rule consists of a left-hand side, describing a matching graph pattern,
and a right-hand side, with a subgraph for the replacement of the pattern on the
left-hand side for each matching part of the graph. Further details can be found
in [11].

2.2   System Architecture of the AHEAD System

Figure 3 gives an overview of the AHEAD system. Users of the AHEAD system
are usually either process managers or developers. Process managers use the
management tool to create and manipulate instances of dynamic task nets which
represent development processes. The states of all activities of the process can
be controlled, e.g. by starting or suspending activities. The managers may also

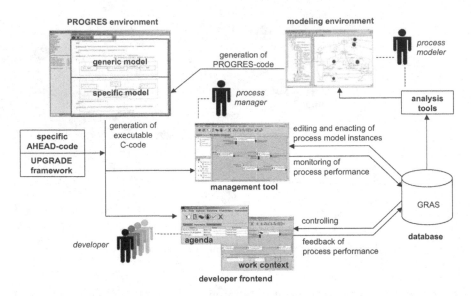

Fig. 3. Overview of the AHEAD system

define project teams (from human or non-human resources), manipulate the status of product data and assign team members to tasks. Developers use the developer frontend to get an agenda containing all tasks that are assigned to them. For every task, a work context can be opened showing detailed information about this task together with all associated input or output documents. There, developers can activate specific development tools working on the documents, too.

So all management data (task net, resource, and product data) are represented as graphs. Management tools and developer frontends have access to the management data, which are stored in the graph-based database GRAS[7]. The underlying base models are combined with each other in a PROGRES specification.

The PROGRES environment provides a graphical editor for specifications and an interpreter to execute the graph code, generated from specifications as executable C-code. With the help of the framework UPGRADE[2] the specific AHEAD-code can be embedded in graphical user interfaces of the AHEAD system, e.g. the management tool or the developer frontend. With our graph-based tool machinery (GRAS, PROGRES, UPGRADE) tools can be generated by rapid prototyping, starting with a formal specification. Changes of the functionality of the generated prototype are done by changing the related parts of the graph schema and graph transformations in the specification, followed by an almost mechanical procedure to generate the new version of the prototype.

The AHEAD system can be adapted for different applications. Therefore, the specification is split into two parts. The first part of the specification, the

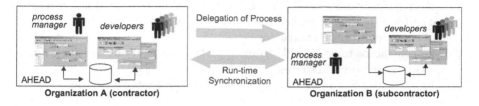

Fig. 4. Scenario of a distributed development process

generic m odel, contains all concepts which are independent of a special domain. This generic part is coded only once. The second part of the specification, the specific m odel, comprises all concepts (e.g. special types of tasks or special relationships between tasks) valid only in a specific application domain. For every new application domain a corresponding specific model can be defined. A process m odeler does not need to use PROGRES for the definition of the specific part of the model directly, but he can rely on a m odeling environm ent which allows to model the specific parts using the UML (Unified Modeling Language)[6]. The modeling environment is based on the commercial tool R ationalR ose. From the model in UML a PROGRES specification can be generated. Additional analysis tools help to evaluate the management data and aid process modelers in defining domain-specific models [10].

3 Concept and Tool Support of Distributed Development Processes

Often two organizations cooperate in delegation relationship where one organization (contractor) delegates a set of activities to another organization (subcontractor). The delegated activities have to be carried out by the subcontractor and some resulting products have to be returned to the contractor.

If each participating organization uses a process management system (each with its own database) to manage its development process, the question arises, how these management systems can be coupled to support a distributed development process. Up to now, it is assumed that both organizations use the AHEAD system with separate databases. Both AHEAD systems use the same graph-based PROGRES specification. This scenario is shown in figure 4 where organization A acts as a contractor and delegates a part of the overall process to organization B. Both organizations have to establish a run-time synchronization of their process management systems in order to inform each other about changes of those process parts which are executed on either side.

3.1 Requirements for an Efficient Support of Distribution

In the AHEAD-project several important requirem ents for an efficient support of distributed development processes have been identified[5]:

1. The contractor must be able to delegate one or more related activities to be carried out by the subcontractor. The results of the delegated activities have to be returned to the contractor.
2. The contractor must be able to monitor the execution of the delegated activities (based on m ilestones).
3. The subcontractor must be able to refine the delegated activities.
4. Both sides must be able to work independently with their own management systems.
5. Both parties must be able to hide process details from each other.
6. Both parties understand every delegation of activities as a contract. The violation of this contract by the contractor or the subcontractor may have legal consequences.
7. The execution semantics of the delegated activities have to be preserved by the management systems used by both sides.

These requirements are chosen in order to balance the interests of organizations taking part in a distributed development process. For example, contractors are interested in gaining extensive control of the progress of delegated activities, while subcontractors aim to shield their internal process structure. Both interests tend to conflict with each other. A detailed discussion of these requirements is given in [1].

One extreme case in the bandwidth of possible concepts for a proper support of distributed development processes is the so called white-box approach where the contractor has full access to all data maintained by the subcontractor. This is typically the case when the subcontractor is forced to work on the contractor's database. In the other extreme, the subcontractor uses an independent instance of a database to which the contractor has no access (black-box approach). The proposed concept is right-hand in the middle of these extremes and, therefore, called a grey-box approach.

3.2 Strategy for the Coupling of Process Management Systems

Figure 5 shows a part of a development process which is distributed over two different process management systems A and B. The process consists of four tasks T1 - T4 which are linked by control flows. As T2 and T3 reside in different

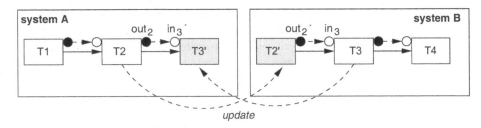

Fig. 5. Strategy for the distribution of a task net

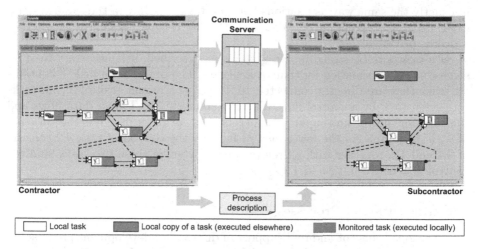

Fig. 6. Coupling of two instances of the AHEAD system

systems, the control flow between these tasks has to be realized in a different way. It is assumed that a command to start T3 was activated in system B. Due to the control flow between T2 and T3, the state of T2 has to be tested before T3 can start (for example, if the control flow is sequential, T2 has to be finished before activation of T3).

A possible strategy is to store duplicates of all tasks of system A, which have to be accessed from system B. For instance, in figure 5 a new task T2' (shown as a grey box) is created in system B as a local copy of the master copy T2 of A. From the perspective of system A, the task T2 is called a monitored task, as a local copy of it is stored in another system. A private task is neither monitored nor a (local) copy of a monitored task. In the example, T1 and T4 are private tasks. All changes to the state of the monitored task T2 are passed by change messages to system B, and system B updates its local copy of T2. Thus, both tasks T2 and T2' are kept consistent with respect to their state. Of course, the same strategy can be applied to duplicate copies of tasks of system B and maintaining them in system A, for instance, a duplicate task T3' is created for task T3 in figure 5. The definition of private, monitored tasks and local copies is always seen from the perspective of one of the two systems, which is called a local system. The other system then is a coupled system.

It can be decided within the system boundaries of system B, if a certain command, where a task of system A is involved, will be applicable or not, because all relevant information is stored within system B. Both systems do not need to be coupled when such a command is triggered. It is sufficient to ensure that every change message, sent by one of the systems, is delivered to the other system. If one of the two systems becomes unavailable, all change messages are temporarily stored in queues. Thus, both systems can work autonomously. With this idea of duplicate tasks, the relations between tasks in different systems can be handled in exactly the same way as relations between tasks residing in the same system. Concerning the execution of commands it now becomes transparent whether a task is private or not.

If a task is to be stored in two systems as described, it is necessary to store some context information for that task as well. The context of a task T contains all tasks that are directly connected to T, e.g. the parent task of T, all tasks connected to T by a control flow, all tasks connected to T by a data flows and the relevant output and input parameters, and all tasks in a refining task net of T. For example, in the lower part of figure 5 the context of task T2 consists of T1 and T3, together with all relevant parameters of these two tasks, and the control flow between T2 and T3.

3.3   Delegation of Processes

The developed concept and tool support of distributed development processes in the AHEAD system is based on the delegation of process fragments[5]. Figure 6 provides an overview of the coupling of two instances of AHEAD:

1. Preparation Step: The contractor selects a part of the task net for delegation within his own instance of the AHEAD system and exports this subprocess into a file. This file is sent to the subcontractor. The subcontractor imports the file into his own instance of AHEAD. A task net is set up there which corresponds to the delegated subprocess in the AHEAD system on the contractor side. The subcontractor's task net contains all delegated tasks together with their context elements. Copies thereof remain in the system of the contractor and they are updated every time the corresponding tasks within the system of the subcontractor change their state.

2. Run-Time Coupling: The management systems of contractor and subcontractor are connected via a communication server at run-time to exchange change messages. With these messages the two management systems inform each other about changes of the state of the tasks maintained on either side. The contractor is informed about changes at delegated tasks, which are executed by the subcontractor, and vice versa. Each task can only be manipulated by exactly one of the two coupled AHEAD systems. For example, the subcontractor is responsible for the delegated tasks. He may also refine the delegated tasks with private task nets, which are hidden from the contractor. The contractor cannot manipulate the delegated tasks and can only monitor changes performed by the subcontractor. Conversely, the responsibility for the tasks of the context tasks remains with the contractor. These tasks can only be changed in their state by the contractor and these changes can only be monitored on the subcontractor side. If one of the two management systems is disconnected for a while and can no longer process incoming events, messages are stored in queues maintained by the communication server between the two AHEAD systems. After the system is connected again, queued events are processed.

3. Manipulation of the Delegated Sub-Process: When the two systems are coupled at run-time, the delegated sub-process can be altered: for example, new tasks may be added to the delegated task net, input and output parameters may be attached to tasks, control flow and data flow relationships between tasks can be created, between local tasks as well as non-local tasks. Private tasks can be made visible for the partner by including them into the corresponding context. As contractor and subcontractor have to agree about these structural changes of the delegated process fragment, a change protocol is enforced by the AHEAD system. All structural changes have to be carried out by the contractor and are propagated to the subcontractor. The subcontractor can reject any of the propagated changes.

The AHEAD system offers a set of remote link commands for each of these phases. For example, there are commands for the export and import of process description files and for the refinement of delegated tasks. Contractor and subcontractor use special commands to register with the communication server. The largest set of commands deals with the structural manipulation of the delegated subprocess.

The described delegation model satisfies all of the requirements for a proper support of distributed development processes mentioned above. The idea to duplicate tasks in both management systems allows for the monitoring of duplicated tasks, thereby covering, together with the developed delegation model, requirements 1 to 4. The duplication of tasks for use in other systems can be forbidden (requirement 5). Management systems can work independent of each other, when messages are queued, e.g. in a communication server (also requirement 4). Requirement 6 is met by the change protocol for the manipulation of the delegated parts of a task net. Finally, both organizations use AHEAD for their process management and thus the execution semantics for tasks is the same in both systems (requirement 7).

4 Realization Based on Graph Concepts

The graph-based realization of the proposed delegation concept is explained in this section. A couple of technical problems had to be solved: First, it had to be made clear how a delegated process fragment (a portion of a task net) is transferred between two coupled systems. Second, a mechanism to synchronize two coupled AHEAD systems had to be developed. Third, the remote link commands had to be specified.

4.1 Export and Import of Delegated Task Nets

After a part of a task net has been selected for delegation in the AHEAD system of a contractor, it has to be transferred into the AHEAD system of the subcontractor. For this purpose an asynchronous method has been chosen: The delegated subnet and its context are exported to a file and then passed to the subcontractor, where it is being imported into his system. The XML-based language

GXL (Graph eXchange Language)[18] has been selected for the implementation. A synchronous transfer, for example using a network connection between both systems, cannot be used here, because both systems have to work independently.

All information about delegation relationships is maintained within the graph structures of AHEAD. The existing AHEAD specification had to be extended to designate whether elements of the task net are "private", "remote" or "monitored", as well as the unique id's of remote elements. These id's refer to elements stored in a coupled system, and the local system receives them from the coupled system. They are used within changes messages to denote the corresponding remote elements.

4.2   Run-Time Synchronization of a Delegated Distributed Task Net

AHEAD stores all graph related data in the graph database GRAS. Unfortunately, GRAS has not been designed to access graphs in another instance of a GRAS system. Therefore, a JAVA-based coupling module has been realized to handle the technical details of the coupling, such as the connection of two instances of the AHEAD system and the execution of the remote link commands.

As GRAS is an active database, it generates a database event of a certain type when the graph is altered and propagates this event to all interested event listeners. For example, when a graph node of type t is inserted into the graph, a database event of type NodeCreated(t) is sent to all interested listeners. Similarly, changing the value of an attribute of a graph node leads to the generation of a corresponding modification event.

This kind of event handling is used to realize the delegation of tasks between two management systems. If a task T1, stored in process management system (PMS) A, is delegated to system B for execution, then a new copy T1' of T1 is created in PMS B. Only for monitoring purposes the original task T1 remains in PMS A. When the state of T1' in PMS B changes, a change message is sent back to PMS A to trigger appropriate steps to update the local copy T1 to the new state. The same applies for all task net elements, e.g. control flows, parameters, and data flows. Every instance of the process management system can at the same time act as a producer of events regarding all elements which are monitored elsewhere and as a consumer of events regarding all elements which are executed elsewhere and only monitored locally.

This idea is now illustrated in the following example (see figure 7):

1. A user triggers a command "Start(T1)" at the graphical user interface of the (local) PMS A. Then a corresponding PROGRES transaction is executed in PMS A.
2. In this transaction it is tested whether the command is applicable in the current state of T1 or not. If so, the state of task T1 is changed to Active. If T1 is a task which is monitored elsewhere, a node of type EventStart is inserted into the graph of PMS A. This new event node is connected with T1 by an edge and T1 contains all data needed in a coupled system to update the maintained copy of T1 there (in the example, the identification of T1 is stored in changedTask as the only necessary information).

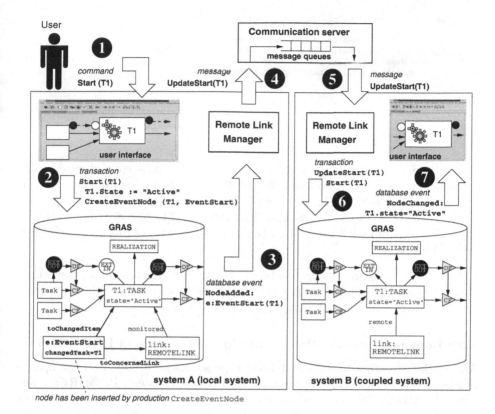

Fig. 7. Delegation of process fragments between two instances of the AHEAD system

3. The insertion of the new event node raises a new database event NodeAdded generated by GRAS. This database event is propagated to all registered listeners of this event type.

4. The Remote Link Manager of PMS, listening to all events of such type, receives this event. The Remote Link Manager sends (via a communication server between the two Remote Link Managers) a message UpdateStart together with the stored data of the event node to the Remote Link Manager of the coupled system. As stated before, the communication server can store such messages in case that one of the two systems is temporarily unavailable. The stored messages are processed when the system is available again. In the current prototypical implementation it is assumed that no change message is lost during transport from one system to another. This standard problem in distributed systems can be addressed by using a more reliable transport protocol. The messages can be numbered and the receipt of a message with a certain number can be signaled to the sender.

```
node type EventAddConFlowNeutralDelegated : RELATION_EVENT
  derived
      taskID    = getNetworkID ( self.-toChangedItem-> );
      parentID  = getNetworkID ( self.-toChangedItem-> : TASK.=toParent=> );
      taskName  = self.-toChangedItem-> : TASK.Name );
      taskState = self.-toChangedItem-> : TASK.State );
      [...]
  end;
```

Fig. 8. Event node type `EventAddConFlowNeutralDelegated`

5. In PMS B the Remote Link Manager receives the message and executes an appropriate graph transaction `UpdateStart(T1)`. The parameters for this transaction are taken from the received message.
6. The execution of the transaction `UpdateStart(T1)` first determines the ID of the copy of task T1 in system B. Second, the state of this task is changed to `Active`.

The data exchange mechanism between the systems using the Remote Link Manager and the communication server is realized in JAVA, while the underlying logic of this coupling mechanism is specified in PROGRES.

4.3   Implementation of the *remote link commands* in **PROGRES**

As mentioned earlier, every event node type stores all needed data to update the state of the local copy of the task in a coupled system. PROGRES offers derived attributes for these node types. These attributes are not set to a specific value. Rather, it is specified which graph data have to be evaluated to determine the correct value of the attribute. By this mechanism it is possible to calculate the data of the event node types automatically in the instance the attribute is queried.

Figure 8 shows the corresponding specification of an event node type `EventAddConFlowNeutralDelegated`. This event is raised, when a control flow between two tasks is inserted while one of the two connected tasks, which has not been part of the delegated task net before, becomes a context task. As this task has not been instantiated in the coupled system before, a new context task has to be created in the coupled system, before the new control flow can be inserted between the two tasks. The corresponding event node type aggregates all data needed to perform the two described steps. To obtain the `taskID`, the evaluation starts at the event node (`self`) and if possible, navigates over an edge of type `toChangedItem` to the changed model entity. Next, the function `getNetworkID` is called with this entity as an argument to obtain a network-wide ID that can be used in a coupled system to access this entity in the local system. The name of the changed task is retrieved in a similar way. Here the changed entity, which is expected to have type `TASK`, is reached during the graph navigation and its attribute `Name` is retrieved.

In the local system the insertion of a control flow between two tasks is done within a specific graph transformation. If both connected tasks are located in the local system, only the control flow is inserted. But if one of the connected tasks is located in a coupled system, it is decided during a case analysis, which type of event node is inserted into the local system.

The insertion of such event nodes is detected by the Remote Link Manager of the local system and a corresponding change message is sent to a coupled system. The Remote Link Manager of the coupled system then executes an update transaction. Every update transaction corresponds to exactly one event type, e. g. EventAddConFlowNeutralDelegated, and specifies all necessary steps on the side of the coupled system to reach a synchronized state with the local system concerning this event.

This mechanism serves for the run-time synchronization of two AHEAD systems. In general, this mechanism can be used for the coupling of two graph rewriting systems operating on different (graph) databases. By this mechanism it is possible to execute remote transactions in one of the two rewriting systems triggered by the other system. For instance, within a graph transaction gt, executed in rewriting system A, the execution of a second graph transaction gt' in rewriting system B has to be triggered.

To achieve this, a new node of a special event node type et is created in system A. This event node can carry in its attributes some information of interest to system B in order to execute the remote transaction gt'. The new event node is inserted into the graph of system A. This leads to the generation of an event node of type NodeAdded(et). A coupling module listens to all events regarding such event nodes. If the insertion of an event node of a specific type is detected, a corresponding message is sent to the coupling module of the other system B. The coupling module there by executing a predefined graph transaction gt', which can manipulate the graph of system B in a deliberate way.

5   Related Work

Related work is shortly discussed on a conceptual level (delegation of processes) and on a technical level (coupling of two graph rewriting systems).

Van der Aalst gives an overview about concepts for distributed cooperation in [14]. Although the focus is on workflow management, the classification can also be used for the management of dynamic development processes. Subcontracting is similar to the delegation of process fragments proposed in this paper, but only allows to assign single activities of a process to another organization for execution. Delegation allows to assign whole task nets to another organization.

In [3] Dowson describes the use of contracts in the system IStar. There, the structure of a software development process is hierarchical and the responsibilities for all process parts are described by contracts. However, these contracts do not define how the process is structured (black-box approach). Instead, every subcontractor is free to refine the assigned activities individually.

The Contract Net Protocol (CNP)[12] also deals with relationships between a contractor and a subcontractor. This work focuses on the fully automated negotiation of buying and selling of goods between systems, e.g. on electronic markets. A communication protocol with standardized messages is followed between an initiating agent and one or more participating agents who compete for an offered contract. The main focus lies on the automatic calling for and receiving of bids and awarding the contract to the most suitable bidding agent. This negotiation phase is not addressed in the AHEAD system. Here, both organizations have to come to a mutual agreement about details of the delegation relationship outside of AHEAD. After both organizations have come to an agreement, the AHEAD system handles the technical realization of the delegation of process parts and the coupling of both process management systems.

Taentzer et. al[13] use distributed graph transformations for the specification of static and dynamic aspects of distributed systems. There, a network structure consisting of local systems is modeled by a network graph. The internal state of each local system can also be described as a graph where nodes are data objects and edges between nodes express relationships between these data objects. Local views containing export and import interfaces are used to define the coupling of local systems within a network. The interfaces of a local system store a subgraph with all accessible nodes and edges. Coupling means connecting an export interface of one local system with an import interface of another. In contrast to this general specification approach, the coupling mechanism of this paper is developed with respect to the special case of coupled instances of the AHEAD system (based on PROGRES). The coupling of two AHEAD systems is the basis for the proposed concept of delegation of processes in the research area of process management. By a research prototype this delegation concept and the coupling of two AHEAD systems have been implemented.

6   Summary

In this paper a delegation-based concept supporting distributed development processes and its realization by the graph-based AHEAD system has been presented. In the scope of the IMPROVE project[9] this approach has been applied to a case study in chemical engineering, where the development of a chemical plant for polyamide is carried out in more than one organization.

The decision to use graph technology for the realization of the AHEAD system has proven to be a good basis for future extensions. The formal specification of the application logic of AHEAD gives a precise definition of the semantics of commands operating on the management data with AHEAD. The generation of executable code from this specification avoids the programming effort of implementing the commands in a traditional way. The application logic of AHEAD and the coupling logic are specified in PROGRES. As the coupling logic is based on the application logic of AHEAD, the same specification language can be used for both parts. Furthermore, changes in the logic are easy to implement by changing the specification and generating new executable code.

References

[1] Simon Becker, Dirk Jäger, Ansgar Schleicher, and Bernhard Westfechtel. A delegation-based model for distributed software process management. In Vincenzo Ambriola, editor, *Proc. 8th Europ. Workshop on Software Process Technology (EWSPT 2001)*, LNCS 2077, pages 130–144. Springer, June 2001. 353, 358

[2] Boris Böhlen, Dirk Jäger, Ansgar Schleicher, and Bernhard Westfechtel. UP-GRADE: Building interactive tools for visual languages. In Nagib Callaos, Luis Hernandez-Encinas, and Fahri Yetim, editors, *Proc. of the 6th World Multiconference on Systemics, Cybernetics, and Informatics (SCI 2002)*, volume I (Information Systems Development I), pages 17–22, July 2002. 356

[3] Mark Dowson. Integrated project support with ISTAR. *IEEE Software*, 4(6):6–15, November 1987. 365

[4] Markus Heller and Dirk Jäger. Interorganizational management of development processes. In Manfred Nagl and John Pfaltz, editors, *Proc. AGTIVE 2003*, LNCS. Springer, 2004. In this volume. 353

[5] Dirk Jäger. *Unterstützung übergreifender Kooperation in komplexen Entwicklungsprozessen*, volume 34 of *Aachener Beiträge zur Informatik*. Wissenschaftsverlag Mainz, Aachen, 2003. 353, 357, 360

[6] Dirk Jäger, Ansgar Schleicher, and Bernhard Westfechtel. Using UML for software process modeling. In Oscar Nierstrasz and Michel Lemoine, editors, *Software Engineering — ESEC/FSE '99*, LNCS 1687, pages 91–108. Springer, September 1999. 357

[7] Norbert Kiesel, Andy Schürr, and Bernhard Westfechtel. GRAS, a graph-oriented software engineering database system. *Information Systems*, 20(1):21–51, January 1995. 356

[8] Sven Krüppel-Berndt and Bernhard Westfechtel. RESMOD: A resource management model for development processes. In Gregor Engels and Gregorz Rozenberg, editors, *TAGT '98 — 6th Intern. Workshop on Theory and Application of Graph Transformation*, LNCS 1764, pages 390–397. Springer, November 1998. 354

[9] Manfred Nagl and Bernhard Westfechtel, editors. *Integration von Entwicklungssystemen in Ingenieuranwendungen*. Springer, Heidelberg, 1998. 353, 366

[10] Ansgar Schleicher. *Management of Development Processes – An Evolutionary Approach*. Deutscher Universitäts-Verlag, Wiesbaden, 2003. 357

[11] Andy Schürr, Andreas Winter, and Albert Zündorf. The PROGRES approach: Language and environment. In Hartmut Ehrig, Gregor Engels, Hans-Jörg Kreowski, and Grzegorz Rozenberg, editors, *Handbook on Graph Grammars and Computing by Graph Transformation: Applications, Languages, and Tools*, volume 2, pages 487–550. World Scientific, Singapore, 1999. 355

[12] R. G. Smith. The contract net protocol: High level communication and control in a distributed problem solver. *IEEE Transactions in Computers*, 29(12):1104–1113, 1980. 365

[13] G. Taentzer, I. Fischer, M. Koch, and V. Volle. Distributed graph transformation with application to visual design of distributed systems. In H. Ehrig, H.-J. Kreowski, U. Montanari, and G. Rozenberg, editors, *Handbook on Graph Grammars and Computing by Graph Transformation: Parallelism, Concurrency and Distribution*, volume 3. World Scientific, Singapore, 1999. 366

[14] W. M. P. van der Aalst. Process-oriented architectures for electronic commerce and interorganizational workflows. *Information Systems*, 24(8):639–671, 1999. 365

[15] Bernhard Westfechtel. A graph-based system for managing configurations of engineering design documents. *Intern. Journal of Softw. Eng. and Knowledge Eng.*, 6(4):549–583, December 1996. 354

[16] Bernhard Westfechtel. Integrated product and process management for engineering design applications. *Integrated Computer-Aided Engineering*, 3(1):20–35, January 1996. 354

[17] Bernhard Westfechtel. *Models and Tools for Managing Development Processes.* LNCS 1646. Springer, Heidelberg, 1999. 353, 355

[18] A. Winter, B. Kullbach, and V. Riediger. An overview of the GXL graph exchange language. In S. Diehl, editor, *Software Visualization*, LNCS 2269, pages 324–336, Heidelberg, 2002. Springer. 362

MPEG-7 Semantic Descriptions:
Graph Transformations, Graph Grammars, and the Description of Multimedia
(Invited Talk)

Hawley K. Rising III

Sony PTCA-MPD, San Jose, California, U.S.A.
hawley.rising@am.sony.com

1 An Overview of MPEG-7

MPEG-7 is the name given to the international standard created by ISO/IEC for description of multimedia. The formal name is ISO/IEC/JTC1/SC29/WG11 15938, parts 1 through 8 [1]. These eight parts are an attempt to address, in a comprehensive manner, everything from how to describe multimedia, to how to transmit or record those descriptions. The eight parts are:

Part 1: Systems. This part deals with transmission issues. It allows that multimedia descriptions are to be written either in XML in readable form, or in a compressed form called BIM. It details the construction of this binary form, which is a lossless compression based on Huffman coding. Part I also deals with such issues as what the units of transmission are to be, and how they are to be combined to form descriptions.

Part 2: DDL. This part details the data description language or DDL. As mentioned, this is a minor variant of XML, which adds certain features to XML via the automation fields, and this section also details what key words and XML expressions are normative for use in MPEG-7. It lays out the two main methods of description, which are the atomic units or descriptors, and the more complex units or description schemes. What descriptors and description schemes are used are detailed mostly in other parts of the standard.

Part 3: Visual. This part of the standard deals with descriptions of the visual aspects of the media. It standardizes certain methods and descriptors for describing color histograms, texture descriptions and shape descriptions. It also specifies highly optimized methods for writing these descriptions, and specifies the normative color spaces in which they may be written.

J.L. Pfaltz, M. Nagl, and B. Böhlen (Eds.): AGTIVE 2003, LNCS 3062, pp. 369–382, 2004.

Part 4: Audio. This part details audio descriptors and description schemes. The descriptions include descriptions of sound effects, the descriptions of speech, and methods for "query-by-humming" or identifying multimedia from scraps of recorded music. This part also standardizes transliterations used for describing speech.

Part 5: Multimedia Description Schemes. By far the largest part of the standard, this part details basic data types, descriptors and description schemes that are used to make up complex descriptions of the media. The description schemes are used to describe aspects of production, transmission, and use, together with descriptions of the content of the media itself.

Part 6: Reference Software. This portion of the standard provides a set of software that is to be used as a reference in generating applications and devices that conform to the MPEG-7 standard. There are example pieces of software for each of the descriptors and description schemes.

Part 7: Conformance. This part, written after the completion of the rest of the MPEG-7 standard, details methods for determining whether an application or device is in compliance with the standard, and what methods and descriptions are to be considered normative.

Part 8: This portion of the standard was announced when the standard was being completed, and it intended as a repository for tools that are useful, in conjunction with the reference software in Part 6 for understanding and building applications and devices that conform to the MPEG-7 standard.

Two parts are under consideration to be appended to these, these two parts deal with profiles and levels. They limit the scope of the descriptors and description schemes in use for specialized applications and for hardware with limited capabilities.

2   MPEG-7 Multimedia Description Schemes

The fifth part of the MPEG-7 standard, the multimedia description schemes, is a complex and interconnected set of description schemes for describing many aspects of multimedia. It has five main parts:

Structure of Content. This set of description schemes is used to break a piece of multimedia content down in a hierarchical manner, either temporally or spatially. Temporally, the intent is to break video content into scenes, shots and rushes, while spatially, the idea is segmentation into spatio-temporal regions. This set of description schemes is intended to interact with the semantic description schemes to provide a description of content that is accessible by temporal or spatial entities or by logical or semantic entities. In the original conception, when the standard was being formulated, the structural description was seen as the description that could be machine derived directly from the input bitstream, that is, without reference to the meaning of any entity in the sequence. Since that description is potentially more complex than a stream of bits, especially when the content described is not digital video, the standard formulation evolved significantly from this interpretation.

Semantics of the Content. This set of description schemes describes the multimedia from the point of view of the objects, events, concepts, roles and themes that make up the meaning of the content. In contrast to the structural description, this description is not limited to a hierarchical tree or forest of trees description. It allows for complex descriptions of places and times, and allows for storage and distribution of descriptions, as well as transforming the descriptions that are written.

Metadata. This part of the description deals with data about the multimedia rather than a description of the content portrayed. Included in the metadata description
schemes are methods for describing the content production, and ways to describe how the content should be classified (for instance, the genre). This information is important to multimedia description the way that titles and credits are important to a cinema film. Classification of the content is important for certain applications that archive content in library fashion.

Summaries and User Preferences. This set of description schemes allows two types of summaries of (principally video) content to be generated. A sequential summary is a summary that consists of thumbnail stills and corequisite descriptions that is temporally in sequence with the original multimedia. A hierarchical summary breaks down the multimedia content into ever increasing temporal detail, providing, again, thumbnails and corequisite description. Finally, this section of description schemes provides ways to describe user preferences and usage histories that enable some applications to tailor their behavior to individual users. Some of the material in the summaries portion overlaps in function other description schemes in the structural and semantic portions of the standard, but is specifically used in this section to create consise summaries, the application in mind for this part of the standard having been a video on demand or recorded video television application.

Transfer Information. This set of description schemes deals with information about the transfer of multimedia content. It describes the available formats for distribution of content, where the content can be found, how it can be served, and methods for optimizing the transfer to the display and server devices. Some of the parts of these descriptions will probably be superceded by the upcoming

MPEG-21 standard which goes into more detail on these matters, but this part of the MPEG-7 standard provides the basis for such description with in the context of multimedia description.

3   Important Parts

We will restrict our attention to those parts of the multimedia description schemes that deal with writing complex semantic descriptions of multimedia. These parts are:

Semantic Description Schemes. These description schemes comprise Objects, Events, Concepts, and their supportive description schemes, Semantic Place, Semantic Time, Agent Objects, States. They also include the relations and packaging to put these together into descriptions. Most of these description schemes are derived from Semantic Entity, and therefore share the ability to act as containers, to contain graphs and relations, and to contain nested semantic descriptions.

Classification Schemes. These are tree-structured ontologies, consisting of lists of classifying terms, together with links to other terms, as well as sublists, and the ability to import other classification schemes to create large lists. The items in these schemes are refered to by URN, so that they may be implied: it is possible in some application settings to assume the existence of the list at both ends of the transmission.

Graphical Classification Schemes. This variant of the classification schemes is designed expressly for generic semantic descriptions, templates, and graph transformation production rules allowing descriptions to be built or reused. They can be used for validation, or for elaboration, propagation and creation of descriptions.

4   Semantic Descriptions

Semantic Descriptions in MPEG-7 are used for high-level descriptions necessary to describe multimedia. They contain three basic entities: Objects, Events, and Concepts, and many smaller description schemes used to flesh out the description. The description that they encode would be best characterized as the kind of description that a friend might give when asked what a movie was about. This is an important point, since the way in which such a description would be structured by the friend, and even what it was that the friend was describing are essential to structuring a semantic description properly. On the one hand, it could be argued that this could best be captured by free text description, but free text does not give up its meaning to computers readily, and MPEG-7 is about nothing if not the use of automatic processing of descriptions. On the other hand, the descriptions most compatible with automatic processing, exhaustive lists of characteristics, or statically constructed sets of fields, are not well suited to the human description from which semantic descriptions originate.

Consequently, semantic descriptions are built of many complex and simple parts, including special semantic description versions of location and time elements, simple and complex methods for delineating properties, a way to monitor state for descriptions that must change over time, and attributes that are specialized to denote how much abstraction, and therefore how many choices or uncertainties exist in a description. Semantic descriptions are meant for both distribution and reuse. The mechanisms for this are contained in the graphs, graph transformations, and the graphical classification schemes described below.

Building Semantic Descriptions

At the very least, a semantic description usually has an object, and an event, and a relation between the two. It need not have even this much, since the event can hold the necessary objects within it, with the relations to the event being implicit. This dual way of constructing descriptions is characteristic of semantic descriptions. For most descriptions, the implicit description, in which all relations are implied, and relationships are, to the highest degree possible, delineated by inclusion, is the briefest way to describe the scene. All the description schemes, including objects and events, contain possible property lists, and the properties of any element of the description can be listed there. Times and places can be nested inside events and objects, and any relations that need to be expressed can be done with implied sources. No graphs need to be built, and no states or abstractions need to be employed.

At the other extreme, all of the description may be explicitly delineated in a graph, with minimal implicit or default behavior. Objects, events, and Agent Objects (people) may be given as subentities to the Semantic description scheme (semantic world) to which they belong directly, all relations can be specified, and, if necessary given weights, sources and targets may be named, and the graph object may be employed to consolidate them into a list, Semantic Places, and Semantic Times may be at the object level, their precise relations to the Objects and Events given in the graph, States may be created to monitor the changes in parameters, and Concepts, either simple or complex, may be used to elucidate properties, and connect these properties to the appropriate Semantic Entity.

These descriptions can be considered interchangeable up to a point: Any implicit description can be "unfolded" into the equivalent explicit description, hence there is an embedding of the implicit descriptions in the explicit descriptions. Such a description could be rewritten again, without loss, as an implicit description, and the original description recovered. However, this is an embedding, not an isomorphism: Explicit descriptions carry information that cannot be implied, since they can carry relations that are not the default, and since they can carry complex concepts that would need to be collapsed to a list of properties. When we talk of operations on descriptions, below, especially those that correspond to complex mental space operations like blending, we will always assume that the descriptions involved are or have been rendered into their explicit formulations.

5   Semantic Relations and Graphs

Semantic relations are used to express the relationships between semantic entities. They are allowed to be N-ary, but many are used as binary relations. Although the standard permits any relation to be written, there is a listing of those relations that are considered normative.

Every relationship in a semantic description can be written in explicit and implicit form. This is to allow notational shorthands to exist for some of the most common description occurences, while allowing great flexibility when needed. Relations, too exist in multiple forms. A completely implicit relation exists in the form of containment. This relation makes the contained description scheme a sub-description of its container, as when a sub-Object resides inside an Object. There is as well the relation that is specified with its source implicit, naming only the target.

Graphs in the Semantic Description Schemes are created by using the graph properties of all Semantic Entities. This is a field in the Semantic Entity that comprises a list of edges, together with their sources and targets. This is not a mandatory element of a Semantic Entity, and frequently will only be used in the container entity holding the description of a "semantic world." There are two important reasons to use them: First, when a description gets complex, they can consolidate the relations in the description in a single place to allow some ease of processing. Second, they are instrumental to doing validation, transformation and other operations using the Graphical Classification Schemes (GCS).

6   Graphical Classification Schemes

In the metadata portion of the multimedia description schemes, the standard outlines Classification Schemes (CS) as a description scheme for building ontologies of multimedia metadata. Several of these classification schemes are normative in the standard, principally a classification scheme for genre, and a classification scheme giving the normative relations for the various description schemes, including the semantic relations mentioned above.

Graphical Classification Schemes (GCS) are put together with units derived from the units in the Classification Schemes. As such, they are ostensibly tree structured ontologies, which allow importation, and some degree of cross linkage. The entries in the GCS are graphs which are semantic description with some degree of generality, and/or they are graph transformation specifications that specify a particular operation on a semantic description. These entries may be used to validate semantic descriptions (for instance, to restrict subject matter to a particular framework for a particular application domain), to reuse description (by populating a frame that is stored as a GCS entry), or to transform a description either to propagate it, to generalize or restrict it, or to solicit more information by prompting the user.

The GCS allow four types of production mechanisms, all taken from categorical graph transformation. These four are:

- Single Push Out (SPO). This production method is in the GCS standard to allow simple glueing operations.
- Double Push Out (DPO). This production method is in the GCS standard as a cut and paste operation. In its simplest form, it specifies the parts of the original graph to be cut out, those to be kept, and what is to be added.
- Single Pull Back (SPB). This production method is in the GCS standard as a means to replace a node, or nodes by (identical copies of) a graph.
- Double Pull Back (DPB). This production method is in the GCS standard as a means for replacing all occurences of a particular subgraph by a new subgraph.

It is important to note that the range of use and application of these methods is far more powerful than what has been presented in the text of the standard. The section of the standard is written to be accessible, and hopefully to promote the use of these classification schemes to validate and propagate descriptions, and to build ontologies of useful descriptions for well delineated application fields. Propagation and reuse are major problems for human-generated descriptions (since humans do not like to generate them), and these production rules are intended to provide methods to do that efficiently. All four of these methods have been put in the standard simply because there are well defined applications for which each is the most efficient and compact way of accomplishing a task.

It is also important to note, that for experts in graph transformations and graph grammars, there have been no limitations written into the standard on how these methods are used. The most limiting part of the whole graphical representation of semantic descriptions is the standardization of a set of "normative" relations between semantic entities. This was done principally so that an "MPEG-7 compliant" receiver of information could have coded behavior in place to interpret any relation that came through the transmission stream. For circumscribed application domains, it is allowable (but not "normative") to extend these relations with relations that are suitable to the application. However, no graph transformations, uses of these transformations, or methods of using the GCS to build grammars or any other graph transformation related entity is proscribed in the standard. Consequently, these GCS represent an opportunity to encode semantic descriptions using the full force of the work of the graph transformation community.

7   Current Status of MPEG-7

MPEG-7 became a standard in late 2001. Since the standard was published, there have been several proposed corregenda, and proposed amendments to the eight parts delineated above. There have also been new parts proposed, and worked on, to integrate MPEG-7 with the other MPEG standards, and to delineate profiles and levels. Profiles are subsets of the standard that are self-contained, and are used to allow simpler subsets of the standard for certain application domains or on certain hardware where either much of the standard will not be used, or it may not fit. Levels are a substratification of profiles. The amendments have added features to the audio, video, multimedia description schemes, and systems parts of the standard. the corregenda have corrected errors found subsequent to releasing the standard.

7.1   Industry Application

Industry application of MPEG-7 has been mixed. There are some companies that have been basing their future multimedia archiving strategies on MPEG-7, and there has been at least one company that has released a library for the entire standard set of descriptors and description schemes. Generally available tools also exist at IBM and Ricoh. Many companies have been adopting a "go slow" strategy, trying to figure out if other companies will adopt the standard before they commit themselves. There have been two motivating factors here, one is that there are other standardization needs that are being addressed, for instance, Digital Cinema. Once these are brought in line, so that widespread dissemination of multimedia is a large enough commercial interest, there will possibly be competition on the metadata and descriptive entries to a medium. The other is that there has been a general drawing back from multimedia retrieval and archiving following the "dot com bust". As companies find there feet in the aftermath of this decline, the interest in multimedia will again rise.

8   Description is Inherently Complex

Having surveyed those parts of MPEG-7 that are relevant to the relation between multimedia description and graph transformation, we now turn to the question of descriptions of multimedia themselves. The first thing to note about multimedia description, or about any description, is that it can be far from trivial to encode human or human-like description. It is, as well, difficult to devise a scheme that can accomodate human description, and will need few extensions, because standards are supposed to be at least a little immutable, and human description can be so varied as to be frustrating.

8.1   Multimedia as Pre-sheaves

MPEG public document N3414[6] attempts to find mathematical descriptions for multimedia entities and processes. These processes seem simple, and the categories meaningful. It is only when they are examined in greater scrutiny that it becomes obvious that there are inherent difficulties. The most difficult of these

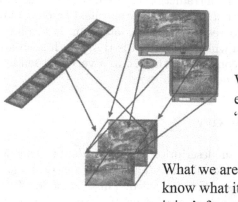

Media differentiated
for creation metadata

Well defined media are
equated in the concept of
"movie"

What we are describing. We
know what it is by its models,
it isn't frames, sets, bits...

Fig. 1. Multimedia and multimedia description are properly modeled as sheaves over an entity that we would describe as the "movie" or "picture" or "song". They are valid when there is no contradiction between the model and the underlying entity, usually measured by comparison to another model

is to specify what it is that we are describing when we give a semantic description. Take the example of a video scene. We are not describing the sequence of frames, since these frames can be changed for format changes. For instance, a movie is shot in 70 mm film, then transcribed to video tape to be shown on TV (say in the 1960s). It is subsequently "digitally remastered" and released in widescreen digital format on DVD. There is sufficient demand, so a version for standard digital TV is created by the well known technique of "pan and scan". Clearly our semantic description does not describe any of these completely different, but we would all agree our description of the "movie" should remain unchanged. Consequently, the description is not describing a sequence of frames, or a digital stream at all. Does it describe the underlying reality captured by the camera? Hardly. The movie is spliced together from short rushes, many filmed out of sequence, using props and costumes that we in no way wish to realistically describe. Add to this animation and special effects dubbing, and we don't even necessarily have an "underlying reality that the camera captured" for all of the movie to be described. To say we are describing the world and events portrayed in the movie would be correct, however, this world does not and never has existed, in general.

We can however leave the underlying "reality" of the movie undefined, and describe the process of modeling it. We suppose that the underlying "reality" has one or several spatiotemporal continua, that we can essentially make a representation of these by making a set of frames, or by generating a bit stream, or by generating any form of description of these continua. We will test the validity of this model by determining whether there are discrepancies between the model and a given continuum. If there is a contradiction between the two, the model is deemed to be not valid. A model that can go to any specified amount of detail, while maintaining its validity with respect to this continuum, is a valid model (to this level of detail). The model which is the best is that which generates the least entropy with respect to the continuum, that is, in a very real sense, is the most parsimonious, while not sacrificing validity.

9   Accomodating Complexity

The inherent complexity of human description is accomodated using Mental Spaces. Descriptions consist of temporally transient mental spaces, built to get a specific piece of information across, then discarded. This is a model of the way humans communicate. There is no exhaustive listing of features and attributes, as is common in descriptions that are machine based, rather subject matter is alluded to, and descriptions contain agreed upon analogy, form and elements. Mental spaces that are set up to do this perform two operations that allow them to build sufficient richness: They recruit frames, that is, frameworks of descriptions or situations, like, for instance the rules and typical behavior for a sports event, and they can recruit other mental spaces as needed, providing the ability to quickly generate complex description by recursion and transitivity.

The second action that mental spaces do is called blending. This allows two mental spaces with common items to be glued on these items. The resulting space is then pruned for contradictions, and a recruitment occurs to add applicable frames and elements. The blend is then "run" to allow the consequences implicit in the new mental space to be constructed. This process is a creative one, that allows new description to be created. It is inherently a complex process.

9.1   Parts of Mental Spaces

When people communidcate, they neither transmit images nor replicate sounds. Excluding creating media themselves, they speak in compact descriptions that rely on the building of internal models of the world that describe the same scene each in an idiosyncratic waw, touched off by the language of the conversation. Following Fauconnier, we will call these temporary, idiosyncratic models of the world that arise in large numbers and form the basis of conversation m ental spaces[2].

Description in human form is not a list of attributes or features, and certainly not a lossless or exhaustive transfer of media between people. This compact communication is desirable as a way to do description: it is parsimonious, it is very understandable to humans, and it is extensible as we shall see, to things which have not been previously described.

Unfortunately, this leaves the machine at an insurmountable disadvantage, without its own ability to construct or describe these mental spaces, it is stuck with free text comparison, which may work or not, as anyone who has searched

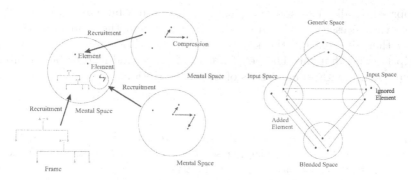

Fig. 2. Elements of a mental space. In Fauconnier's delineation, the elements are points in the space, together with frames, and subspaces. In other descriptions, these are connected by relations. The prototype for a blend is shown at right

the internet is aware. Another approach is to construct the mental spaces themselves as a method of storing description.

Mental spaces consist of short scenes that "run". They are populated with actors and actions which create this scene, together with anything necessary to complete the description. The semantic description schemes were influenced by this model. Usually, a mental space can be scene as being populated with three types of entities:

- Elements. This would include all of the semantic entities that were delineated in the description above of semantic descriptions. People, objects, events, and so forth.
- Subspaces. This allows the inclusion of another mental space to create the current mental space model. Usually, in the work by Fauconnier, this inclusion is not transitive, but there are reduction operations that abbreviate this subspace, and the resultant descriptive fields are then sometimes added to the parent. This would qualify as a transitive subspace, but would probably not be so described.
- Frames. Frames are important paradigms and sets of rules. They are recruited into the mental space to give it form and structure. At a much more basic level, they can be seen as giving the mental space an interpretation, in that many of the actors and events are to some extent defined by the frame in some mental spaces. This would be especially true for a very highly structured event like a rite of passage, or a sports event.

In the semantic description model, the elements and subspaces may or may not be generated specifically for a semantic description. The entities that correspond to the frames would necessarily occur in correspondence to a GCS. Consequently, when using mental spaces as a model for description, especially multimedia description, which often describes something like a film, that would appropriately fall into a mental space description if a human were to do it, one can see the necessity of the graphical ontologies and the transformational nature of them in creating the kind of description that closely patterns human description and thought. The belief was that this would lead to better interface correspondence with human users of these descriptions.

9.2   Blending

Blending at its very simplest consists of taking a pair of mental spaces and gluing them together by identifying some or all of their elements. Turner and Fauconnier describe a process in several parts:

- Identification – elements of two mental spaces are identified with each other.
- Generic space – a space called a generic space is created, from just the common elements.
- Blend space – the input spaces are glued together to form the blend space.
- Pruning – there is a pruning operation, during which inherent contradictions are resolved in the blend space by ignoring or deleting the offending parts.
- Recruiting – the new space recruits additional appropriate material, new elements or subspaces that complete descriptions, or new frames that interpret the new construct.
- Running – the new space is run to identify the new consequences of the blend.

This operation is especially useful in creating analogies and metaphoric speech. Such speech is used in small amounts nearly constantly by people. It shortens description and borrows or propagates descriptive information from topics that have an abundance of terminology to use in less well described subject matter.

9.3   Residue Productions

Because blending creates both the composite blend space and the generic space from the two input spaces, and the generic space is represented in the blend space, and in each of the input spaces, one can create residue spaces from the input spaces, by deleting the generic spaces from the input spaces. The reverse process must include the knowledge of how to glue the generic space and the residue space together. This may be done in several ways, in the GCS system, the necessary templates to ensure that the gluing is done properly could be stored with the residue or with the generic space. Alternatively, the definition of the residue space may take a form that makes the gluing unambiguous.

9.4   Image Analysis Style Pyramids

Generating Gaussian and Laplacian Pyramids

Once we have the creation of residues and generic spaces, we may create pyramids from the successive application of the blending process. This creates two pyramids, as usual, the pyramid of generic spaces, and the pyramid of residue spaces. These are in direct correspondence to the Gaussian and Laplacian pyramids used in image processing. We may refer to them as generalized Gaussian and Laplacian pyramids. They represent, for the Gaussian pyramid, levels of increasing "genericity" which corresponds to generality. The elements of the mental spaces that represent the generic spaces at the high levels of the pyramid are either elements held in common with many more specific mental spaces, or they are generic elements, categories that encompass increasingly more parts of the descriptions below. Notice that, in essence, an iteration process needs to happen with respect to the generic space and the blend. The blend is originally created by generating the generic space and pushing out the blend as the direct sum modulo the generic space image. The blend runs, and this changes the mapping between the input spaces and the blend. A generic space can now be generated as a pullback from this changed map. In many cases, there may not be any change between this space and the original generic space, but there are cases in which it would be a richer space, and encode more of the blend process than the original.

Building A Wavelet Structure

Consider now the special case where a blend consists precisely of the identification of the elements in the generic space. The elements in either input space which are not in the generic space are assumed to not correspond between input spaces, and there are no elements discarded and added as a result of running the blend. Under these circumstances, the residue created by deleting the generic space from the blend space contains precisely two components, and these components are the residues of the input spaces. For a single step of pyramid building, namely one blend, two residues and a generic space, the entire operation and structure can be represented without loss by the generic space and the blend. By blending the generic spaces, under the same rules, a pyramid structure can be set up that losslessly represents all of the original input spaces, and has the same number of spaces, all of which will now be blend spaces, with the exception of the top of the pyramid. The top will, of course, be a generic space. This situation is built in a manner reserved for wavelet transformations, from two constituent parts, initially, the common space of the inputs, and their blend. It is efficient, having only as many elements as the number of original input spaces, as opposed to the less frugal Gaussian and Laplacian pyramids. It is generated by the equivalent of the lifting scheme [5] method of creating wavelet pyramids, and creates only as many spaces as the original set of inputs, as opposed to the Laplacian pyramid structure which has 4/3 the number. It also allows hierarchy to be created by combining descriptions, an important method to create propagation and reuse.

Expanding this result, we may lay down a set of rules under which such a pyramid can be created. They are that the generic space must be represented in the finished blend, that the removal of the generic space from the blend space (and any subsequent processing in the general case) must produce two distinct residual mental spaces, and that the blend of the generic space with either residual space must produce the one of the two original constituent input spaces. The rule that requires the formation of two distinct residual spaces is hard, in that without this requirement, there is no regeneration of the proper number of input spaces. The others can be relaxed, however, and the degree to which the result differs from the original input spaces be considered the lossiness of the transformation. When this lossiness does not exist, however that is accomplished, then we can say the process permits perfect reconstruction.

10   Conclusions and Applications

We have shown that the process of forming descriptions from semantic descriptions modeled on mental spaces provides a flexible and comprehensive method of trying to model the way humans view multimedia content. We have provided extensibility through blending, as well as portability and replication through the Graphical classification schemes. Finally, we have demonstrated the ability to generate hierarchical structures of differing detail that correspond well to the types of hierarchical structures used in signal and image processing. Given these methods, and the flexibility of the descriptive tools, the topic of semantic description can be extended and refined, until it becomes an easy and common way to annotate multimedia. This is the promise of MPEG-7, and an opportunity for people in the graph transformation community to contribute to the future of multimedia.

References

[1] ISO/IEC JTC1/SC29/WG11 International Standard 15938, parts 1-8, Sydney, Australia, August, 2001. 369

[2] Fauconnier, G. Mental Spaces. New York: Cambridge University Press, 1994. 378

[3] Fauconnier, G., and M. Turner. The Way We Think. New York: Basic Books, 2002.

[4] Goguen, J. An Introduction to Algebraic Semiotics, with Application to User Interface Design. Lecture Notes in Computer Science, 1562, pp. 242-291, 1999.

[5] Sweldens, W. Building Your Own Wavelets at Home. ACM SIGGRAPH Course notes, pp. 15-87, 1996. 381

[6] Rising, H.K., and Y. Choi. Mathematical Models for Description Scheme Concepts, Update. ISO/IEC JTC1/SC29/WG11 public document N3414, Geneva, Switzerland, May 2000. 376

Collage Grammars for Collision-Free Growing of Objects in 3D Scenes[*]

Renate Klempien-Hinrichs, Thomas Meyer, and Carolina von Totth

Department of Computer Science, University of Bremen
P.O. Box 33 04 40, D-28334 Bremen, Germany
{rena,mclee,ariel}@informatik.uni-bremen.de

Abstract. In the real world, solid objects rarely collide without serious damage. We study collision-freeness in the context of collage grammars, which are picture- and scene-generating devices based on hyperedge replacement. We identify a type of collage grammars that generate only collision-free collages, and prove the undecidability of the problem whether an arbitrary collage grammar generates any collision-free collage. Moreover, we discuss collision avoidance in the derivation process as an appropriate tool for practical modelling purposes.

1 Introduction

Three-dimensional scenes occur in many application areas, such as landscape gardening, architecture, and interior design; transportation; biology, chemistry, and other sciences; or the entertainment industry. Computer-generated three-dimensional models of such scenes can be used in particular to visualise a design, to substantiate or disprove a scientific hypothesis, or to provide a virtual setting for a computer game, a commercial or a movie. Especially in highly complex scenes, it is advantageous to use rule-based modelling techniques rather than assemble the components of a scene by hand.

The familiar principle of putting together simple pieces like building blocks is used for the rules of collage grammars, which are picture- and scene-generating devices based on hyperedge replacement [8, 9, 2, 4, 5, 6]. A collage consists of a set of geometrical volumes called parts, a sequence of pin points, and a set of hyperedges each coming with a nonterminal label and a sequence of attachment points. A hyperedge can be replaced by a collage if its pin points meet the attachment points. This defines a direct derivation in a collage grammar provided that the label of the replaced hyperedge and an affine transformation of the replacing collage form a rule of the grammar. The generated language comprises all collages without hyperedges that are derived in this way from the start collage of the grammar. In other words, a collage grammar has parts as terminal items, hyperedges as nonterminal items, and comprises all necessary spatial information

---

[*] This research was partially supported by the EC Research Training Network SegraVis (Syntactic and Semantic Integration of Visual Modelling Techniques).

J.L. Pfaltz, M. Nagl, and B. Böhlen (Eds.): AGTIVE 2003, LNCS 3062, pp. 383–397, 2004.

Fig. 1. Collision of objects: a tree growing through a wall

so that it may be seen as a non-deterministic 'program' that generates a variety of similar objects or scenes.

If one computes a scene consisting of a collection of solid objects, and funny effects are not intended, one would like to avoid collision. For example, two trees should not share any branches, a plant should not grow through a wall without causing damage (see Figure 1), and a moving object should avoid obstacles. However, it is often desirable to position many objects closely together, for instance the twigs and leaves of a tree, the buildings in a city, or the aminoacids of a protein in its folded structure.

Two volumes collide if they have any point in common. That means that only completely distinct volumes do not collide. In contrast, volumes sharing at most boundary points are considered non-overlapping in, e.g., [4]. Yet, our notion of collision is appropriate in particular for two reasons: First, humans tend to distinguish two objects in the real world, for instance two leaves on a twig, if they do not share any cell or even an atom. Secondly, on a computer screen separate objects are usually represented in such a way that any one pixel belongs to at most one object.

Where specific knowledge on a particular modelling domain is present, it is possible to develop specific methods to deal with unwanted collision. Successful approaches have been proposed especially in the area of plant modelling, of which we name a few: For instance, a model for phyllotaxis, i.e. a regular arrangement of organs on some form such as seeds on a cone, is presented in [7]. Another example is the study of environmental influences on plant growing e.g. in [13, 12].

In this paper, collage grammars are studied as an application-independent method for generating collision-free scenes. Ideally we are looking for methods to avoid collision, already in the rules or at the latest during the derivation process. The obvious approach would be to use backtracking as soon as a collision occurs (algorithms for the detection of a collision are surveyed in [11, 10]). However, it may not be possible to generate a collision-free collage at all, or backtracking may lead to the loss of interesting intermediate scenes.

The results presented in this paper are of two kinds. On the one hand, we identify a type of collage grammars that generate only collision-free scenes. Since we have to use a rather strong restriction here, we are also interested in the collision-free section of the collage languages generated by general collage grammars. However, it turns out that one cannot decide whether an arbitrary collage grammar generates any collision-free collages. On the other hand, we study ways to obtain collision-free results from unrestricted collage grammars. This approach of collision avoidance provides first ideas well suited for practical modelling purposes.

The paper is organised as follows. In Section 2, we recall the basic notions and notations of context-free collage grammars. The concept of collision-freeness is introduced in Section 3, where we also give syntactic restrictions to collage grammars ensuring the collision-freeness of the derived collages. Section 4 contains a reduction of the membership problem for two-counter machines to the problem whether an arbitrary collage grammar generates a collision-free collage, implying the undecidability of the latter problem. In Section 5, we discuss the generation of collision-free collages from a perspective of practical modelling. The conclusion contains a number of topics for further research.

2   Collage Grammars

In this section, the basic notions and notations concerning collages and collage grammars (see [8, 9]) are recalled and illustrated by an example.

The sets of natural numbers and real numbers are denoted by $\mathbb{N}$ and $\mathbb{R}$, respectively. For a set A, A^* denotes the set of all finite sequences over A including the empty sequence λ, $\wp(A)$ denotes its powerset, and $A \smallsetminus B$ denotes the complement of a set B in A. For a function $f \colon A \to B$, the canonical extensions of f to $\wp(A)$ and to A^* are denoted by f as well, i.e., $f(S) = \{f(a) \mid a \in S\}$ for $S \subseteq A$, and $f(a_1 \cdots a_n) = f(a_1) \cdots f(a_n)$ for $a_1 \cdots a_n \in A^*$.

Familiarity with the basic notions of Euclidean geometry is assumed (see, e.g., Coxeter [1]). $\mathbb{R}^d$ denotes the Euclidean space of dimension d for some $d \geq 1$; our examples will be of dimension 2 or 3. For the purposes of this paper, the term 'transformation' refers to affine mappings $a \colon \mathbb{R}^d \to \mathbb{R}^d$.

A collage consists of a set of volumes called parts and a (possibly empty) sequence of so-called pin points; it specifies a scene by the overlay of all its parts. To generate sets of collages, collages are decorated with hyperedges. A hyperedge has a label and an ordered finite set of tentacles, each of which is attached to

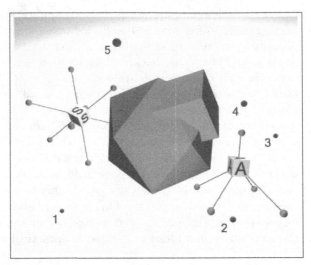

Fig. 2. A (decorated) collage

a point. A handle—a collage consisting of a single hyperedge—is used as an initial collage. All these notions are defined more precisely as follows.

A collage (in $\mathbb{R}^d$) is a pair (PART, pin) where PART $\subseteq \wp(\mathbb{R}^d)$ is a finite set of so-called parts (thus a part is any subset of $\mathbb{R}^d$) and pin $\in (\mathbb{R}^d)^*$ is the sequence of pin points. $\mathcal{C}$ denotes the class of all collages.

Let N be a set of labels and let, for each $A \in N$, $\text{pin}_A \in (\mathbb{R}^d)^*$ be a fixed sequence of pin points associated with A. A (hyperedge-)decorated collage (over N) is a construct $C = (\text{PART}, E, \text{att}, \text{lab}, \text{pin})$ where (PART, pin) is a collage, E is a finite set of hyperedges, att: $E \to (\mathbb{R}^d)^*$ is a mapping, called the attachment, and lab: $E \to N$ is a mapping, called the labelling, such that for each hyperedge $e \in E$ there is a unique (affine) transformation $a(e): \mathbb{R}^d \to \mathbb{R}^d$ satisfying $\text{att}(e) = a(e)(\text{pin}_{lab(e)})$. For a label A in N, the handle induced by A is the decorated collage $A^\bullet = (\emptyset, \{e\}, \text{att}, \text{lab}, \text{pin}_A)$ with $\text{lab}(e) = A$ and $\text{att}(e) = \text{pin}_A$. The class of all decorated collages over N is denoted by $\mathcal{C}(N)$, and PART $_C$, E_C, att_C, lab_C, and pin_C denote the components of $C \in \mathcal{C}(N)$.

A collage can be seen as a decorated collage C where $E_C = \emptyset$. In this sense, $\mathcal{C} \subseteq \mathcal{C}(N)$. Thus, we may drop the components E_C, att_C, and lab_C if $E_C = \emptyset$. Furthermore, we will use the term collage also in the case of decorated collages as long as this is not a source of misunderstandings.

Example 1 (collage). In Figure 2 one view of a three-dimensional collage decorated with two hyperedges is given. The grey polygon in the middle is the overlay of all parts. There are five pin points, which are depicted as small spheres and numbered according to their position in the sequence of pin points. The relative size of the spheres indicates approximately the distance to the viewer. Hyperedges are denoted as labelled cubes, together with numbered lines indicating the attachment to points which are again represented by small spheres. Since numbering attachment lines in three-dimensional collages readily becomes confusing and is often impossible to read, it is left out whenever convenient. In Figure 2 the attachment of the S-labelled hyperedge spans a cube, and the attachment of the A-labelled hyperedge spans a pyramid.

As we can present only two-dimensional projections on paper, we will sometimes show the same three-dimensional scene from different perspectives.   △

Removing a hyperedge $e \in E_C$ from a collage C yields the collage $C - e = (\text{PART }_C, E_C \setminus \{e\}, \text{att}, \text{lab}, \text{pin}_C)$ where att and lab are the respective restrictions of att_C and lab_C to $E_C \setminus \{e\}$.

The addition of a collage C' to a collage C is defined by $C + C' = (\text{PART }_C \cup \text{PART }_{C'}, E_C \uplus E_{C'}, \text{att}, \text{lab}, \text{pin}_C)$ where $\uplus$ denotes the disjoint union of sets,

$$\text{att}(e) = \begin{cases} \text{att}_C(e) & \text{for } e \in E_C \\ \text{att}_{C'}(e) & \text{for } e \in E_{C'}, \end{cases} \quad \text{and} \quad \text{lab}(e) = \begin{cases} \text{lab}_C(e) & \text{for } e \in E_C \\ \text{lab}_{C'}(e) & \text{for } e \in E_{C'}. \end{cases}$$

The transformation of a collage C by a transformation $a: \mathbb{R}^d \to \mathbb{R}^d$ is given by $a(C) = (a(\text{PART }_C), E_C, \text{att}, \text{lab}_C, a(\text{pin}_C))$ where $\text{att}(e) = a(\text{att}_C(e))$ for all $e \in E_C$.

Hyperedges in decorated collages serve as place holders for (decorated) collages. Hence, the key construction is the replacement of a hyperedge in a decorated collage with a collage. While a hyperedge is attached to some points, a collage is equipped with a number of pin points. If there is an affine transformation that maps the pin points to the attached points of the hyperedge, the transformed collage may replace the hyperedge.

Let $C \in \mathcal{C}(N)$ be a collage and let $e \in E_C$. Furthermore, let $\text{repl}(e) \in \mathcal{C}(N)$ be such that there is a unique (affine) transformation $a(e)$ that satisfies $\text{att}_C(e) = a(e)(\text{pin}_{repl(e)})$. Then the replacement of e in C by $\text{repl}(e)$ yields the decorated collage $C[\text{repl}]$ constructed by

1. removing the hyperedge e from C,
2. transforming $\text{repl}(e)$ by $a(e)$, and
3. adding the transformed collage $a(e)(repl(e))$ to $C - e$.

Thus, to be more precise, $C[\text{repl}] = (C - e) + a(e)(\text{repl}(e))$.

Whenever a notion of replacement like the one above is given, one may easily define productions, grammars, and derivations. In our particular case this leads to the notions of collage grammars and languages.

Let N be a set of labels. A production (over N) is a pair $p = (A, R)$ with $A \in N$ and $R \in \mathcal{C}(N)$ such that pin_A is the pin point sequence of R. A is called the left-hand side of p and R is its right-hand side, denoted by $\text{lhs}(p)$ and $\text{rhs}(p)$, respectively. A production $p = (A, R)$ is also denoted by $A ::=_p R$ or simply $A ::= R$.

Let $C \in \mathcal{C}(N)$, $e \in E_C$, and let P be a set of productions over N. We call $b: \{e\} \to P$ a base on e in C if $\text{lab}_C(e) = \text{lhs}(b(e))$. As, moreover, $\text{pin}_{lhs(b(e))} = \text{pin}_{rhs(b(e))}$, there is a unique affine transformation $a(e)$ that satisfies the equation $\text{att}_C(e) = a(e)(\text{pin}_{rhs(b(e))})$. Thus we can say that C directly

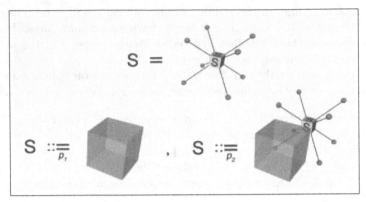

Fig. 3. Axiom and productions of the collage grammar G_{Cubes}

derives $C' \in \mathcal{C}(N)$ through b if $C' = C[\text{repl}]$, where $\text{repl}(e) = \text{rhs}(b(e))$. A direct derivation is denoted by $C \Longrightarrow_P C'$, or $C \Longrightarrow C'$ if P is clear from the context. A derivation from C to C' of length n is a sequence of direct derivations $C = C_0 \Longrightarrow_P C_1 \Longrightarrow_P \dots \Longrightarrow_P C_n = C'$; it may be abbreviated by $C \Longrightarrow_P^n C'$, or $C \Longrightarrow_P^* C'$ if its length is not important.

Definition 1 (*Collage Grammar*). A (context-free) collage grammar is a system $G = (N, P, Z)$ where N is a finite set of nonterminal labels (or nonterminals, for short), P is a finite set of productions (over N), and $Z = S^\bullet$ for some $S \in N$ is the axiom.

The collage language generated by G consists of all collages that can be derived from Z by applying productions of P:

$$L(G) = \{C \in \mathcal{C} \mid Z \xRightarrow[P]{*} C\}. \qquad \diamond$$

Example 2. Consider the collage grammar $G_{Cubes} = (\{S\}, \{p_1, p_2\}, S^\bullet)$ with axiom and productions as shown in Figure 3; the pin points of the right-hand sides are the corners of the respective cubes and do not appear explicitly. For better visibility the parts are rendered semi-transparent. The size of the cubes produced by repeated applications of production p_2 grows with the factor 1.5 for each new cube. Figure 4 contains the first three collages of the generated language. $\qquad \triangle$

3   Collision and Collision-Freeness

In this section, we introduce the notion of collision and discuss collision-freeness for collage grammars and their generated collage languages. Since our interest focusses on collision-free collages, we develop a type of grammar that generates only collision-free collages.

Definition 2 (*Collision*). Two volumes $v, v' \subseteq \mathbb{R}^d$ collide if they intersect, i.e., if $v \cap v' \neq \emptyset$. $\qquad \diamond$

Fig. 4. Some collages generated by the collage grammar G_{Cubes}

In order to extend the notion of collision to hyperedges we associate with every hyperedge a canonical part that depends on the volume spanned by the attached points of the hyperedge.

For a sequence $w \in (\mathbb{R}^d)^*$, $[w]$ denotes the convex hull spanned by the points in w and $]w[$ denotes the interior of $[w]$. Furthermore, the volume spanned by a hyperedge $e \in E_C$ in a collage C is denoted by $\mathrm{vol}_C(e) =]\mathrm{att}(e)[$, and the volumes in C form the set $\mathrm{VOL}_C = \mathrm{PART}_C \cup \{\mathrm{vol}_C(e) \mid e \in E_C\}$ and can be composed to $\mathrm{composite}(C) = \bigcup_{X \in VOL_C} X$.

Definition 3 (*Collision-Freeness*). A set S of volumes is collision-free if no two distinct volumes $v, v' \in S$ collide. A collage $C \in \mathcal{C}(N)$ is collision-free if VOL_C is collision-free. A production $A ::= R$ is collision-free if its right-hand side R is collision-free. A collage grammar is collision-free if all its productions are collision-free. A collage language $L \subseteq \mathcal{C}$ is collision-free if all collages in L are collision-free. ◇

Example 3.

1. The collage grammar G_{Cubes} presented in Example 2 is collision-free and generates a collision-free collage language.
2. Figure 5 contains three collages generated by a collage grammar G_{Cantor}, which produces a three-dimensional version of approximations of the Cantor tree. For the construction, copies of the part displayed in the first generated collage—basically a prism fused with a tetrahedron on top—are attached to the tetrahedron at the end of each branch.

   The collage grammar, which we leave out here for reasons of space, is collision-free. However, it can generate collages where collisions occur. Such a collage is shown in Figure 6 from two different perspectives (on the left the same perspective as in Figure 5 is used). The colliding parts are distinguished by black and white colouring, and the other parts are rendered semi-transparent for a better view on the collisions. △

A stronger restriction than collision-freeness of the grammars is needed to ensure collision-freeness of their generated languages. For this purpose, we introduce a variant of the pin-boundedness property used in [4]. The idea of pin-boundedness is to subsequently subdivide the space spanned by a hyperedge, rather than growing beyond it.

Fig. 5.   A series of collision-free collages generated by the grammar G_{Cantor} from Example 3

Fig. 6.   A collage in $L(G_{Cantor})$ where collisions occur

Definition 4 (*Strict Pin-Boundedness*). A collage $C \in \mathcal{C}(N)$ is strictly pin-bounded if $X \subseteq]pin_C[$ for all $X \in \text{VOL}_C$. A production $A ::= R$ is strictly pin-bounded if its right-hand side R is strictly pin-bounded. A collage grammar G is strictly pin-bounded if all productions in G are strictly pin-bounded.   ◇

Observation 5. Let G be a strictly pin-bounded collage grammar.

1. For every production p in G we have $\text{composite}(rhs(p)) \subseteq \text{composite}(lhs(p)^\bullet)$.
2. For all collages C derived from the axiom Z we have $\text{composite}(C) \subseteq \text{composite}(Z)$.   □

Theorem 6. Strictly pin-bounded collision-free collage grammars generate collision-free collage languages.

Proof (sketch). One can show for every derivation $Z \overset{*}{\underset{P}{\Longrightarrow}} C$ by induction on its length that C is strictly pin-bounded and collision-free.

In each derivation step, the collage replacing a hyperedge e is collision-free and transformed before its insertion in such a way that the result lies completely within $vol(e)$. Consequently, the collision-freeness of the generated collages results directly from the pin-boundedness of the grammar coupled with the definition of collision.   □

It should be noted that (strict) pin-boundedness is a strong restriction for collage grammars since the axiom determines a boundary beyond which no parts or hyperedges may be added in a derivation. There are, however, many interesting collision-free collage grammars that are not strictly pin-bounded and yet generate collision-free languages (for instance G_{Cubes} in Example 2).

4   Undecidability of Collision-Freeness

In this section, we show that the problem whether an arbitrary collage grammar generates a collision-free collage is undecidable. This result is shown by simulating a two-counter machine by a two-dimensional collage grammar in such a way that an accepting computation corresponds to the generation of a collision-free collage. The simulation used here is related to earlier ones used to obtain undecidability results in [4, 3].

Theorem 7. For a given collage grammar G, it is undecidable whether $L(G)$ contains a collision-free collage.

Informally, a two-counter machine is a Turing machine with a read-only input tape and two counters instead of a work tape. A computation step depends on the current state and the input symbol read. In every step, the state can be changed, the input head can stay or move one cell to the left or right, and one of the counters can stay or be incremented or decremented by 1 (the other remains unchanged). The computation step is sound if a counter is decremented only if its value is at least 1, and stays at its value only if that value is 0 (which means that the counter is testet for 0).

For a set A of symbols with $\triangleright, \triangleleft \notin A$, let $A_+ = A \cup \{\triangleright, \triangleleft\}$. The symbols $\triangleright$ and $\triangleleft$ are added as end markers to a word $w \in A^*$, yielding $\triangleright w \triangleleft$. For $1 \leq i \leq |w|$, let w_i denote the ith symbol in w, and let $w_0 = \triangleright$, $w_{|w|+1} = \triangleleft$.

Let $D = \{-1, 0, 1\}$ be the set of tape directions and $CA = \{+_1, 0_1, -_1, +_2, 0_2, -_2\}$ the set of counter actions. A two-counter machine is a tuple $M = (S, A, d, s_0, s_f)$ where S is a finite set of states, A is a finite set of input symbols, $s_0 \in S$ is the initial state, $s_f \in S$ is the final state, and $T \subseteq S \times A_+ \times S \times D \times CA$ is the transition table such that $(s, a, s', d, c) \in T$ implies $d \neq -1$ if $a = \triangleright$ and $d \neq 1$ if $a = \triangleleft$.

For an input word $w \in A^*$, a sequence $W \in T^*$ of transitions is a w-computation of M with input position $pos(W) \in \{0, \ldots, |w| + 1\}$, state $state(W) \in S$, and counter state $count(W) \in \mathbb{N} \times \mathbb{N}$ if

- $W = \lambda$, $pos(W) = 1$, $state(W) = s_0$, and $count(W) = (0,0)$; or
- $W = Vt$ for $V \in T^*$ and $t = (s, a, s', d, c) \in T$, where

$$pos(W) = pos(V) + d, \quad state(W) = s', \quad \text{and} \quad count(W) = count(V) + act(c)$$

and act is defined as follows: $act(+_1) = (1,0)$, $act(0_1) = (0,0)$, $act(-_1) = (-1,0)$, $act(+_2) = (0,1)$, $act(0_2) = (0,0)$, and $act(-_2) = (0,-1)$.

A w-computation W is sound if W is empty; or $W = Vt$ for a sound w-computation V and a transition $t = (s, a, s', d, c)$ such that $c = -_k$ implies $\operatorname{count}(V)_k > 0$ and $c = 0_k$ implies $\operatorname{count}(V)_k = 0$ $(k = 1, 2)$.

M accepts w if there is an accepting w-computation, i.e., a sound w-computation W with $\operatorname{state}(W) = s_f$.

Fact 8. For a two-counter machine M and a word $w \in A^*$ as input, it is undecidable whether M accepts w.

In the following, let M be an arbitrary two-counter machine and w an input word for M. We will construct a (linear) collage grammar $G_{M,w}$ whose derivations correspond to the w-computations of M in such a way that the resulting collage is collision-free if and only if the w-computation is sound. For this, the label of the hyperedge is used to keep track of the current state s of M and position i on the input word. The first production introduces a part start with which a part added later on collides if and only if in the w-computation, a counter action $-_k$ is performed while the value of counter k is 0, or a counter action 0_k is performed while the value of counter k is not 0. The hyperedge can be deleted if the current state is s_f.

Let $\lfloor (x_1, y_1), (x_2, y_2) \rceil = [(x_1, y_1)(x_2, y_1)(x_2, y_2)(x_1, y_2)]$ denote the rectangle with lower left corner (x_1, y_1) and upper right corner (x_2, y_2) (for $x_1 = x_2$ or $y_1 = y_2$, it is a vertical or horizontal line).

Define $G_{M,w} = (N, P, X^\bullet)$ with $N = \{X\} \cup (S \times \{0, \ldots, |w| + 1\})$ and $P = \{p_{start}\} \cup \{p_{i,t} \mid 0 \le i \le |w| + 1, \ t = (s, w_i, s', d, c) \in T\} \cup \{p_i \mid 0 \le i \le |w| + 1\}$ containing productions as follows, whose right-hand sides all have the same sequence of pin points $\operatorname{pin} = (0, 1)(1, 0)(0, 0)(1, 1)$:

- $p_{start} : X ::= (\{start\}, \{e\}, att, lab, pin)$ with $att(e) = pin$, $lab(e) = (s_0, 1)$, and

$$start = (\lfloor (0,0), (1,1) \rceil \cup \bigcup_{n \in \mathbb{N}} \lfloor (-1, \tfrac{1}{6^n}), (0, \tfrac{1}{6^n}) \rceil \cup \bigcup_{n \in \mathbb{N}} \lfloor (\tfrac{1}{6^n}, -1), (\tfrac{1}{6^n}, 0) \rceil)$$
$$\diagdown (\bigcup_{n \in \mathbb{N}} \lfloor (0, \tfrac{1}{6^n}), (1, \tfrac{1}{6^n}) \rceil \cup \bigcup_{n \in \mathbb{N}} \lfloor (\tfrac{1}{6^n}, 0), (\tfrac{1}{6^n}, 1) \rceil),$$

- $p_{i,t} : (s, i) ::= (part_c, \{e\}, att_c, lab, pin)$ with $lab(e) = (s', i + d)$ and
  - for $c = +_1$: $part_c = \emptyset$ and $att_c(e) = (0, \tfrac{1}{2})(1, 0)(0, 0)(1, \tfrac{1}{2})$,
  - for $c = -_1$: $part_c = \lfloor (-1, 1), (0, 1) \rceil$ and $att_c(e) = (0, \tfrac{1}{3})(1, 0)(0, 0)(1, \tfrac{1}{3})$,
  - for $c = 0_1$: $part_c = \lfloor (0, 1), (1, 1) \rceil$ and $att_c(e) = (0, \tfrac{1}{6})(1, 0)(0, 0)(1, \tfrac{1}{6})$,
  - for $c = +_2$: $part_c = \emptyset$ and $att_c(e) = (0, 1)(\tfrac{1}{2}, 0)(0, 0)(\tfrac{1}{2}, 1)$,
  - for $c = -_2$: $part_c = \lfloor (1, -1), (1, 0) \rceil$ and $att_c(e) = (0, 1)(\tfrac{1}{3}, 0)(0, 0)(\tfrac{1}{3}, 1)$,
  - for $c = 0_2$: $part_c = \lfloor (1, 0), (1, 1) \rceil$ and $att_c(e) = (0, 1)(\tfrac{1}{6}, 0)(0, 0)(\tfrac{1}{6}, 1)$,

  and
- $p_i : (s_f, i) ::= (\emptyset, pin)$ for all $0 \le i \le |w| + 1$.

Now the following claims can be proved easily by induction on n.

Claim. 1. $W = t_1 t_2 \ldots t_n$ is a w-computation of M with α_k $+_k$-actions, β_k $-_k$-actions, and γ_k 0_k-actions ($k = 1, 2$) if and only if

$$X^\bullet \underset{p_{start}}{\Longrightarrow} C_0 \underset{p_{i_1}, t_1}{\Longrightarrow} C_1 \Longrightarrow \ldots \underset{p_{i_n}, t_n}{\Longrightarrow} C_n$$

is a derivation in $G_{M,w}$, with the hyperedge in C_n labelled $(\mathsf{state}(W), \mathsf{pos}(W))$ and attached to the points $(0, y_n)$, $(x_n, 0)$, $(0, 0)$, (x_n, y_n) where $x_n = \frac{1}{2^{\alpha_2}} \cdot \frac{1}{3^{\beta_2}} \cdot \frac{1}{6^{\gamma_2}}$ and $y_n = \frac{1}{2^{\alpha_1}} \cdot \frac{1}{3^{\beta_1}} \cdot \frac{1}{6^{\gamma_1}}$.

2. W is sound if and only if PART_{C_n} is collision-free.

Since the terminating productions p_i require the label of the replaced hyperedge to be (s_f, i), we have the following theorem as a consequence, which in conjunction with Fact 8 implies Theorem 7.

Theorem 9. M accepts w if and only if $G_{M,w}$ generates a collision-free collage.

It may be noted that there is a strictly pin-bounded variant of our construction of $G_{M,w}$, so that Theorem 7 holds already for this class of collage grammars. Moreover, we expect that similar properties of collage grammars—for instance, whether all generated collages are collision-free, or whether finitely many generated collages are collision-free—are undecidable, too.

5 Dynamic Collision Avoidance in Collage Grammars

From a modelling point of view, one can deal with the undecidability of collision-freeness by selecting the collision-free yield of a collage grammar algorithmically and with reasonable complexity.

Let $L(G)$ be the language generated by a collage grammar G. By $\mathsf{cfs}(L(G))$ we denote the maximal collision-free subset of $L(G)$, i.e.,

$$\mathsf{cfs}(L(G)) = \{C \in L(G) \mid C \text{ is collision-free}\}.$$

Since a collage containing colliding parts cannot be derived into a collision-free collage, derivations of this type must be avoided.

Moreover, testing the volumes of hyperedges in a collage C for collision does not make sense for generating grammars that are not pin-bounded, since the hyperedge volume no longer serves as an upper bound for the volume of the replacing collage. Therefore, for general purposes it is enough to limit the collision testing to PART_C.

Definition 10 (*Part Collision-Freeness*). A collage $C \in \mathcal{C}(N)$ is part collision-free if PART_C is collision-free. A derivation $C_0 \Longrightarrow C_1 \Longrightarrow \ldots \Longrightarrow C_n$ is part collision-free if for all $0 \leq i \leq n$, C_i is part collision-free. For a collage grammar G,

$$L_{pcf}(G) = \{C \in \mathcal{C} \mid Z \xrightarrow[P]{*} C \text{ is part collision-free}\}$$

denotes the collages generated with a part collision-free derivation.   ◇

Fig. 7. Nearly the same collage as in Figure 6, this time collision-freely terminated in $L(G_{Cantor_\emptyset})$

Observation 11. Let G be a collage grammar.

1. A collage $C \in L(G)$ is collision-free if and only if there exists a part collision-free derivation $Z \overset{*}{\underset{P}{\Longrightarrow}} C$.
2. cfs$(L(G)) = L_{pcf}(G)$. ☐

Whenever collision occurs during the derivation process, the most obvious way of achieving a collision-free result would be employing a backtracking algorithm. However, there are two main arguments against the sole use of backtracking: First, there is no guarantee that the grammar can generate any collision-free result, and by Theorem 7 it is not possible to check the grammar for this property. Secondly, even if there are collision-free results, they might not be visually interesting. So, especially with grammars that are as complex as is needed for practical modelling purposes, backtracking means in general a long computation possibly without useful result.

On the other hand, it may happen that a grammar generates a collision-free intermediate collage that would make an interesting result from a modelling point of view but cannot be terminated collision-freely. In order to resolve such a derivation deadlock, we propose to modify collage grammars in such a way that hyperedges can be erased.

For a collage grammar $G = (N, P, Z)$, let $P_\emptyset = P \cup \{A ::= (\emptyset, pin_A) \mid A \in N\}$ and $G_\emptyset = (N, P_\emptyset, Z)$.

Example 4. By modifying the grammar G_{Cantor} to $G_{Cantor_\emptyset}$ one can derive the collage shown in Figure 7 instead of the collage with collisions shown in Figure 6.

△

Remark 1. Let G be a collage grammar.

1. $L(G) \subseteq L(G_\emptyset)$, since $P \subseteq P_\emptyset$ and otherwise G is the same as $G_\emptyset$.
2. This implies cfs$(L(G)) \subseteq$ cfs$(L(G_\emptyset))$.

3. In general, the converse inclusion is not true. For instance, if G does not have any productions, then $L(G)$ is empty, and $L(G_\emptyset)$ consists of the empty collage.

A less radical way than erasing is to require for every nonterminal at least one terminating production where the volume of the right-hand side is very small. An example for this is adding a bud instead of the basis of a branch. Such a decision lies of course in the modeller's responsibility. However, our considerations are not restricted to just one application domain like plant modelling, and in some domains there may not be any 'small' alternative for a collision-producing part. In any case this idea does not help with the principal problem whether a collision-free collage can be derived.

To sum up, a practical approach to derive collision-free collages in a system could look as follows:

Algorithm 12. Let G be a collage grammar where the right-hand sides of all productions are part collision-free. Transform G into $G_\emptyset$. A derivation in $G_\emptyset$ develops as follows: In the current collage, select the next hyperedge to be replaced. Select the next production that can be applied to the hyperedge and does not produce a collision. Apply the production to the hyperedge, and repeat until a collage in the language is reached.

The selection algorithms need to be specified in more detail. For instance, one can impose an order on the hyperedges by storing them in a queue, or one can assign priorities to the productions. While these restrictions will in general not yield all collages in the language, they may produce sufficiently interesting results for modelling purposes. Moreover, it may be possible to specify non-context-free collage languages. This depends solely on the selection mechanism, since adding erasing productions does not change the context-freeness of a collage grammar.

6   Conclusion

This paper is a first step towards studying collision-freeness in the context of collage grammars. We have determined a type of collage grammars that generates collision-free collages only. Moreover, we have shown that the existence of a collision-free collage in the language of an arbitrary collage grammar cannot be decided algorithmically. For practical applications we have studied a method for collision avoidance in the derivation of a collage.

A number of directions for further work have revealed themselves, in particular the investigation of the following questions:

- Is there a collision-free normal form for collage grammars whose generated language is collision-free? It may be possible to find a counterexample based on the well-known dragon curve.
- Is the collision-free sublanguage of a collage language context-free? The criteria to disprove context-freeness of collage languages developed in [6] may help with this question.

Among others, the following decidability problems may be investigated:

- Given a collage grammar G as input, is $L(G)$ collision-free?
- Given a collage grammar G as input, does $L(G)$ contain finitely many collision-free collages?
- Given a collage grammar G as input, does $L(G)$ contain finitely many collages with collision?

The implementation of our approach in a collage-generating system is another topic for future work. First of all, the derivation algorithm needs to be worked out in more detail. Secondly, it would be nice to provide a modeller with a number of significantly different collages in a short span of time. Finally we need a method to control collision and collision avoidance within one grammar and derivation. An example for this is the modelling of a tree, where leaves should not collide, but twigs have to be firmly attached into branches, and branches into the tree trunk.

Acknowledgements

We are grateful to Hans-Jörg Kreowski and Frank Drewes for stimulating discussions on the subject of this paper, and we thank the referees for their interesting remarks.

References

[1] H. S. M. Coxeter. *Introduction to Geometry*. John Wiley & Sons, New York, Wiley Classics Library, second edition, 1989. 385

[2] Frank Drewes. Language theoretic and algorithmic properties of d-dimensional collages and patterns in a grid. *Journal of Computer and System Sciences*, 53:33–60, 1996. 383

[3] Frank Drewes, Renate Klempien-Hinrichs, and Hans-Jörg Kreowski. Table-driven and context-sensitive collage languages. *Journal of Automata, Languages and Combinatorics*, 8:5–24, 2003. 391

[4] Frank Drewes and Hans-Jörg Kreowski. (Un-)decidability of geometric properties of pictures generated by collage grammars. *Fundamenta Informaticae*, 25:295–325, 1996. 383, 384, 389, 391

[5] Frank Drewes and Hans-Jörg Kreowski. Picture generation by collage grammars. In H. Ehrig, G. Engels, H.-J. Kreowski, and G. Rozenberg, editors, *Handbook of Graph Grammars and Computing by Graph Transformation*, volume 2, chapter 11, pages 397–457. World Scientific, 1999. 383

[6] Frank Drewes, Hans-Jörg Kreowski, and Denis Lapoire. Criteria to disprove context freeness of collage languages. *Theoretical Computer Science*, 290:1445–1458, 2003. 383, 395

[7] Deborah R. Fowler, Przemyslaw Prusinkiewicz, and Johannes Battjes. A collision-based model of spiral phyllotaxis. In *Computer Graphics*, volume 26,2, pages 361–368. ACM SIGGRAPH, 1992. 384

[8] Annegret Habel and Hans-Jörg Kreowski. Collage grammars. In H. Ehrig, H.-J. Kreowski, and G. Rozenberg, editors, *Proc. Fourth Intl. Workshop on Graph Grammars and Their Application to Computer Science*, volume 532 of *Lecture Notes in Computer Science*, pages 411–429, 1991. 383, 385

[9] Annegret Habel, Hans-Jörg Kreowski, and Stefan Taubenberger. Collages and patterns generated by hyperedge replacement. *Languages of Design*, 1:125–145, 1993. 383, 385

[10] Ming C. Lin and Stefan Gottschalk. Collision detection between geometric models: a survey. In *Proc. IMA Conference on Mathematics of Surfaces*. 1998. 385

[11] Ming C. Lin, Dinesh Manocha, Jon Cohen, and Stefan Gottschalk. Collision detection: Algorithms and applications (invited submission). In J.-P. Laumond and M. Overmars, editors, *Algorithms for Robot Motion and Manipulation*, pages 129–142. A. K. Peters, 1996. 385

[12] Radomír Měch and Przemyslaw Prusinkiewicz. Visual models of plants interacting with their environment. In Holly Rushmeier, editor, *Computer Graphics Proceedings*, Annual Conference Series, pages 397–410. ACM SIGGRAPH, 1996. 384

[13] Przemyslaw Prusinkiewicz, Mark James, and Radomír Měch. Synthetic topiary. In *Computer Graphics Proceedings*, Annual Conference Series, pages 351–358. ACM SIGGRAPH, 1994. 384

VisualDiaGen – A Tool for Visually Specifying and Generating Visual Editors

Mark Minas

Institute for Software Technology, Department of Computer Science
University of the Federal Armed Forces, Munich
85577 Neubiberg, Germany
minas@acm.org

Abstract. VISUALDIAGEN is a tool for visually specifying visual languages and generating graphical editors from such specifications that are mainly based on graph transformation and graph grammars. VISUAL-DIAGEN is an extension of DIAGEN that has already allowed for specification and generation of visual editors; however, DIAGEN's specifications have been based on a textual and, therefore, a less user-friendly representation. This paper describes how VISUALDIAGEN has been built on top of DIAGEN and by using DIAGEN as well.
VISUALDIAGEN reuses DIAGEN's specification tool. However, components that have still used a textual notation instead of the "naturally" visual one have been replaced by visual editors which have been specified and generated with DIAGEN.

1 Introduction

Diagram editors, i.e., graphical editors for visual languages like UML class diagrams, Petri nets, Statecharts etc., are cumbersome to implement. Therefore, various generators for diagram editors have been proposed and developed that create diagram editors from specifications. GENGED [1] and VISPRO [11] are two of these tools. This paper reports on continued work on the diagram editor generator DIAGEN [8].

DIAGEN provides an environment for the rapid development of diagram editors based on hypergraph grammars and hypergraph transformation. Such editors support free editing as well as structured editing: Free editors allow the user to create and modify diagrams like in drawing tools without any restrictions. The editor analyzes such diagrams, reports errors to the user, and makes diagrams accessible to other programs. Structured editors, however, provide a limited set of editing operations which are the only means for creating and modifying a diagram. DIAGEN editors (i.e., visual editors that have been specified and generated with DIAGEN) allow combining both editing modes in a single editor [8].

DIAGEN has been used to create editors and environments for a wide variety of diagram languages, e.g., process description languages like Statecharts and Petri nets [6, 10]. Actually we are not aware of a diagram language that cannot be specified so that it can be processed with DIAGEN. However, the process of

J.L. Pfaltz, M. Nagl, and B. Böhlen (Eds.): AGTIVE 2003, LNCS 3062, pp. 398–412, 2004.

generating a diagram editor required the editor developer to write a textual specification together with some additional Java classes (cf. Section 2), and, hence, one of the more urgent user requests was to provide a more user-friendly way to specify and generate diagram editors with DIAGEN. The first step in that direction was a graphical specification tool for DIAGEN, called DIAGEN designer [9]. This tool already offered support for the complete editor specification and generation process. It consisted of a graphical user interface that has greatly simplified the process of specifying and generating diagram editors with DIAGEN. However, hypergraph grammar productions etc. – although they are inherently visual constructs – still had to be specified textually. This paper describes how the DIAGEN system has been improved in this aspect. It shows how DIAGEN together the DIAGEN designer – a tool with a graphical user interface – have been extended to VISUALDIAGEN, a completely visual tool that provides visual editors for its inherently visual constructs of the diagram language specification. Actually, VISUALDIAGEN has been built on top the DIAGEN designer in a bootstrapping process: the DIAGEN designer has been used for textual specifying and generating each of those visual languages resp. their corresponding graphical editors, that have then been plugged into the DIAGEN designer, replacing textual by visual editing. Finally, VISUALDIAGEN has been used for visual specifying and generating those editors that are now components of VISUALDIAGEN. This paper describes that bootstrapping process by means of the visual language of hypergraph grammars.

There are many systems for creating visual editors from a specification. VIS-PRO [11], GENGED [1], VLCC [3], and PENGUINS [2] are only some of them, and all of them, except PENGUINS, allow visual editing specifications. All of those systems appear to allow specifying and generating at least parts of their own specification language and tool. However, VISPRO is – as far as we know – the only visual specification tool that has actually been bootstrapped by itself, i.e., the tool has been specified by the tool and generated from this specification. Unfortunately, there are no detailed reports on this bootstrapping process.

The rest of this paper is structured as follows: The next section introduces into DIAGEN and discusses the DIAGEN designer that still uses textual specifications inside of a graphical user interface. Section 3 then describes how VISUAL-DIAGEN has been created by extending the DIAGEN designer by visual editors that have been specified and generated with DIAGEN. Section 4 concludes the paper.

2   DiaGen and the DiaGen Designer

When using the former version of DIAGEN, i.e., before the DIAGEN designer has been provided, the editor developer – usually applying a regular text editor – had to create a textual specification for the syntax and semantics of a diagram language when he had to create an editor for this language (see Figure 1.) Additional program code which had to be written manually had to be supplied, too. This was necessary for the visual representation of diagram components on

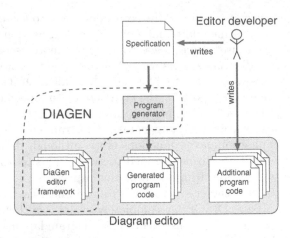

Fig. 1. Previous way of generating diagram editors with DiaGen

the screen, and for processing objects of the problem domain, e.g., for semantic processing when the editor is used as a component in another software system. The specification was then translated into Java classes[1] by the program generator. The diagram editor then consisted of these generated classes together with the manually supplied ones and the DiaGen editor framework, as a collection of Java classes, which provides the generic functionality needed for editing and analyzing diagrams.

The benefit of a plain textual specification was its conciseness. However, the developer was not guided by a tool when creating such a specification. Moreover, the developer was not able to specify graphical components (e.g., circles and arrows when specifying the language of all trees) in the textual specification. Instead, graphical components had to be programmed manually. And finally, after editing the textual specification, the developer had to run the DiaGen generator that checked the correctness of the specification and created Java code from it if it was correct. But if it contained errors, the developer had to improve the specification in the text editor again where the developer had to find the errors in the textual specification by means of line numbers that had been listed by the generator. This led to a cumbersome process of alternatively editing and generating.

These shortcomings have given rise to develop the DiaGen designer. Instead of supplying a textual specification together with additional Java classes that had to be programmed by hand, the DiaGen designer allows specifying all aspects of the diagram language's visual components. The user interacts now solely with the DiaGen designer (see Figure 2.) He no longer has to supply additional Java classes. The actual specification is managed as an XML document (instead of the previous, "human readable" textual specification language used by the DiaGen

---

[1] DiaGen is completely implemented in Java and is free software, available from http://www2-data.informatik.unibw-muenchen.de/DiaGen/.

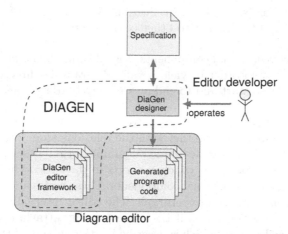

Fig. 2. Generating diagram editors with the DIAGEN specification tool

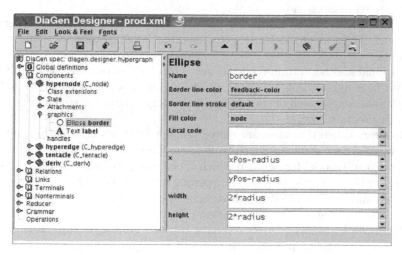

Fig. 3. A screenshot of the DIAGEN designer showing the graphics specification of a visual component

generator as in Figure 1) that is translated into Java classes by the DIAGEN designer's generator. The DIAGEN editor framework completes the generated editor.

Figure 3 shows a screenshot of the DIAGEN designer. A specification consists of the following main sections.

- Global definitions: This section contains specifications of global constants like colors or used text fonts.
- Visual components of the diagram languages: Directed graphs as a visual language, e.g., have circles and arrows as visual components; circles visualize nodes whereas edges are represented by arrows. Visual components have

attachment areas (see later in this Section), i.e., the parts of the components
that are allowed to connect to other components (e.g., start and end of an
arrow).

- Relations between visual components: Visual components have to be related
  in a certain way in order to yield a valid diagram. In directed graphs, e.g.,
  each arrow has to start and to end at a circle's border. These relations have
  to be specified in this section.
  DIAGEN editors (i.e., editors that have been generated by DIAGEN) represent visual components together with the relations among them by a hypergraph model: The most general and yet simple formal description of a visual
  component is a hyperedge[2] which connects to the nodes which represent the
  attachment areas of the visual components. These nodes and hyperedges first
  make up an unconnected hypergraph. The editor internally connects nodes
  by additional relation edges if the corresponding attachment areas are related in a specified way, which is described in this section of the specification.
  The result is the hypergraph model (HGM) of the diagram. More details are
  presented in Section 3.
- Links: These are additional hyperedge types of the internal hypergraph
  model. They are used for, e.g., connecting labeling text to visual components.
- Terminal and nonterminal edge types: The diagram language is mainly specified by a hypergraph grammar which makes use of such terminal and nonterminal edges. Each of these edges may carry attributes which are specified
  in this section, too.
- Reducer: When diagrams are edited using a DIAGEN editor, they are internally analyzed in a compiler-like fashion. They are first translated into the
  hypergraph model (see above) which is then lexically analyzed by a so-called
  reducer that creates the reduced hypergraph model (rHGM) [8]. The reducer's
  functionality is specified in this section by specifying its reduction rules that
  are essentially graph transformation rules.
- Hypergraph grammar: After reducing the internal hypergraph, it is syntactically analyzed, i.e., parsed with respect to a hypergraph grammar which has
  to be specified in this section. Each grammar production (for a screenshot
  see Figure 4) consists of the actual production together with its attribute
  evaluation rules (shown in Figure 4), attached constraints that are used for
  automatic layout, and additional (programmed) conditions that must hold
  if this production is used. Details on grammar productions are given in Section 3.
- Operations for structured editing: Each DIAGEN editor supports free editing,
  i.e., it can be used like a usual drawing tool. Diagram analysis distinguishes
  correct diagrams from invalid ones and provides visual feedback to the editor user. However, free editing is often to fussy; structured editing provides

---

[2] Hypergraphs consist of nodes as well as hyperedges and are similar to directed
graphs. Whereas edges of directed graphs connect to two nodes, hyperedges connect to an arbitrary number – which depends on the type of the hyperedge – of
nodes.

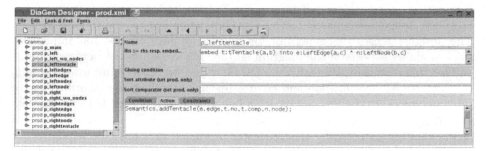

Fig. 4. A screenshot of the DIAGEN designer showing a hypergraph grammar production

easier ways for editing diagrams. Structured editing is specified by editing operations that are specified in this final section.

Compared to the previous approach to generate diagram editors with DIAGEN (cf. Figure 1), specification and generation of visual components was the most novel aspect. As each DIAGEN editor is a free editor, the specification of a visual component has to consist of the following sections (cf. Figure 3).

- Class extensions: Each component is represented by a Java class which is completely generated by the DIAGEN designer. However, if it might turn out that the generated class has to provide additional functionality, e.g., if the generated diagram editor is used as part of a larger system and the generated class has to implement external Java interfaces, arbitrary Java code can be added to this section. Contents of this section are then just copied to the generated class.
- State: The current state of a component on the screen is determined by its parameters for layout (e.g., position on the screen, size, etc.) and other member attributes of the generated Java class. These are specified in this section.
- Attachment areas: Each visual component has a specific number of areas where a component can be related to other components. In directed graphs, e.g., arrows have their end points as attachment areas (attachment points, actually) whereas circles have their border-lines as attachment areas.
- Handles: They are provided by every drawing tool for modifying the position or size of a visual component. A handle is generally shown as a small gray rectangle on the screen. When it is dragged with the mouse, its visual component is modified accordingly. An arrow usually has two handles, one at each of its end points. This section contains specifications of such handles. For each handle it has to be specified where it is positioned and what happens to the component's parameters when the handle is moved.
- Graphics: This section specifies a sequence of graphics primitives (e.g., lines, ovals, text) which have to be drawn on the screen in order to visualize the component. The parameters of these primitives are specified as expressions on the component's parameters.

Generation of visual components is the only completely new feature of the DIAGEN designer compared to the DIAGEN generator as shown in Figure 1. The designer reuses this generator for generating Java classes that implement the remaining aspects of the diagram editor. The designer just acts as a front-end for this generator. Basically, the DIAGEN designer collects information from the XML-based specification and creates textual specifications according to the previous, textual DIAGEN specification language which is then fed into the generator. However, this process is completely transparent to the user. Specification errors that are detected by the generator are listed in a message window. The DIAGEN designer automatically navigates to the corresponding editing dialog when the user selects such an error message.

Reusing the generator had simplified the process of realizing the DIAGEN designer. However, reduction rules, hypergraph grammar productions, and structured editing operations – although they are inherently visual constructs – still have to be specified textually. In order to overcome those limitations, we have specified and generated visual editors for them and included these editors into the DIAGEN designer. VISUALDIAGEN has been the result of this process.

3   VisualDiaGen – A extension of the DiaGen Designer

The DIAGEN designer has not been written "from scratch". It is actually based on a generic XML editor which offers a customizable graphical user interface for creating, editing, and visualizing XML documents [9].

The GUI of such an XML editor always comes with the usual appearance of a tree-view of the tree structure of the edited XML document on the left and an editing dialog for the selected XML node of the tree on the right. These dialogs can be customized rather freely as the screenshots of the DIAGEN designer (Figures 3 and 4) show. This customization is described in an XML document that is referred to by the document type definition (DTD) of the edited XML file. A customization tool, that is also an instance of this generic XML editor, allows convenient creating and editing such customization XML files [9]. XML editors, therefore, are quickly customized and easily adapted to changes to the XML-based language, e.g., the specification language.

One of the design criteria of the generic XML editor was to use graphical editors which have been generated with DIAGEN as dialogs, too. Each DIAGEN editor uses an XML format as a file format. Integrating DIAGEN editors, therefore, is easy:

- The customization file of the generic XML assigns the DIAGEN editor (actually the name of the generated Java class) to a specific XML element. A new instance of this DIAGEN editor is then opened as a sub-dialog within the XML editor whenever its XML element gets selected in the XML editor.
- When a DIAGEN editor instance is closed, e.g., when its XML element gets deselected, the editor's diagram is saved to an XML stream in the DIAGEN specific format. This XML stream is saved as a sub-tree of the XML element of the DIAGEN editor.

– If the DIAGEN editor is instantiated for a new XML editor, i.e., if the diagram is empty, the DIAGEN editor opens with an empty diagram. However, if the XML element contains a sub-tree representing the XML representation of a previously edited diagram, this XML stream is loaded into the DIAGEN editor instance and the previous diagram is restored.

– If the whole XML file being edited by the customized XML editor is saved, all sub-trees representing diagrams that have been edited by DIAGEN editors are saved, too, and they are automatically restored when the customized XML editor loads the XML file again.

As the DIAGEN designer is an customized extension of the generic XML editor, integration of DIAGEN editors into the DIAGEN designer is straight forward. However, in order to gain the visual DIAGEN designer that allows visual editing graph grammar productions etc. and creating Java code from them, requires two additional features:

First, visual editors for graph grammar productions etc. are no stand-alone editors. They rather have to take into account information from other parts of the specification. E.g., such productions are allowed using edge types which have been specified before. Furthermore, their arity and type (i.e., being terminal or nonterminal) has to be considered.

This flow of information from the XML editor to its DIAGEN editors is made possible by transferring the whole XML document that is just being edited as a context information to each DIAGEN editor as soon as it gets instantiated. The DIAGEN editor, therefore, has access to the complete specification information. If the diagram has become inconsistent with the rest of the specification, e.g., because the arity of a hyperedge type has changed, but the diagram has no yet been adjusted, diagram analysis by the diagram editor can find these inconsistencies and communicate this to the user in the usual way provided by DIAGEN editors.

Second, DIAGEN editors can translate a diagram into a Java data structure (i.e., its semantics) in a syntax-directed way. This translation process is performed by attribute evaluation based on the diagram's derivation structure [7, 8]. When VISUALDIAGEN uses such a DIAGEN editor, its semantic output has to be fed "back" into VISUALDIAGEN. In general, a specification will consist of more than a single diagram. Generating Java code from this specification requires that VISUALDIAGEN has immediate access to the diagrams' semantics. Starting the corresponding DIAGEN editor and translating each of the diagrams to its semantics "on the fly" is not feasible.

This problem has been solved by additionally saving a diagram's semantics to the XML document. An XML element that is edited by a DIAGEN editor, therefore, not only saves the diagram in the specific DIAGEN XML format as a subtree of the XML element, but also the diagram's semantics which have been created by the diagram editor's diagram analysis. VISUALDIAGEN, when generating Java code from the XML specification, is then able to access the diagram's meaning by just using the information in these "semantics sub-trees".

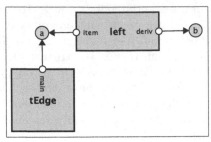

Fig. 5. Visual representation of a sample hypergraph consisting of two nodes and two hyperedges

After these considerations, creating VISUALDIAGEN was straight forward. First, details of the visual languages for reduction rules and grammar productions had to be drawn up.[3] As those languages are based on hypergraphs, a visual language for hypergraphs had to be designed. Figure 5 shows the chosen visual representation:[4] Nodes are represented by circles that optionally contain names (e., a). Hyperedges are represented by rectangles which are connected to their visited nodes by arrows. In the literature (e.g., [4]), the set of all nodes that are visited by hyperedges is arranged in an ordered sequence which allows distinguishing the different "tentacles" of an hyperedge. In DIAGEN where hyperedges represent graphical components, tentacles correspond to attachment points or areas of those components. Giving names to such tentacles, therefore, helps the user. Figure 6 shows how such names are visualized for the different hyperedges: The arrows start at small white circles at the rectangles' borders which carry the corresponding tentacle name (e.g., main.) When selecting the hyperedge, those circles can be freely positioned on the rectangle's border. Different fill colors of the rectangles indicate different kinds of edges (i.e., terminal vs. nonterminal edges). Each rectangle carries a label (e.g., tEdge) that is its edge type. This type has to correspond with a declared hyperedge type of the specification if such a hypergraph diagram is used in VISUALDIAGEN. The corresponding declaration then determines the set of tentacles of this edge.

As the next step towards VISUALDIAGEN, diagram editors for the designed visual languages had to be specified using the DIAGEN designer. This has required suitable representation of hypergraph diagrams by hypergraph models (HGMs). Figure 6 shows the HGM of the sample diagram in Figure 5: Nodes, hyperedges, and the small arrows (tentacles) are represented by hyperedge types hypernode, hyperedge, and tentacle. Hypernode edges visit a single node since circles (i.e., hypernodes) have a single attachment area, and arrows (i.e., tentacles) have two attachment areas. However, hyperedge edges are special: The HGM has

---

[3] Syntax-directed editing operations are not yet supported by visual editors as the graph transformation language will change in the near future [5].

[4] The graphics of Figures 5, 6, 7, and 8 have actually been produced by the print functionality of the corresponding DIAGEN editors.

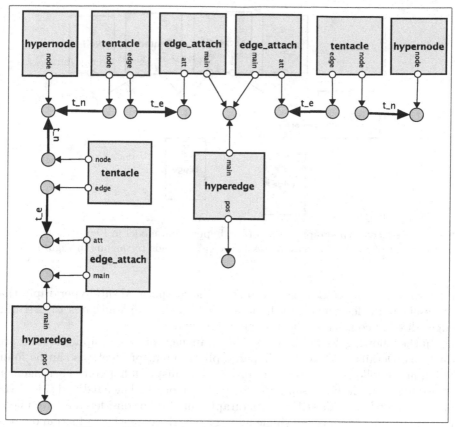

Fig. 6. Hypergraph model of the diagram in Figure 5

to represent each attachment area of the rectangle that visualizes the hyperedge. However, the number of its tentacles depends on the actual specification. As a consequence, we cannot declare just a single hyperedge type with a fixed number of tentacles for all hyperedges. Instead, each hyperedge component with, say, n tentacles is represented by $n+1$ hyperedges in the HGM: 1 hyperedge edge and n edge_attach edges, each representing one of component's n tentacles. The hyperedge edge visits two nodes: the pos one represents the rectangle, the main one acts as a connection point for edge_attach edges. Moreover, Figure 6 shows the HGM's relation edges as fat arrows. They are used to connect tentacle components to hyperedges (edge type t_e) resp. to nodes (edge type t_n).

This sample shows that HGMs grow large even for small diagrams. In order to improve efficiency of syntax analysis, HGMs are transformed into a reduced hypergraph model (rHGM). The translation process has to be specified in the reducer section of the DIAGEN designer and, hence, VISUALDIAGEN. Due to lack of space, the specification of the reducer is omitted here. However, Figure 7 shows the rHGM of the HGM in Figure 6: Hyperedge and hypernode edges are trans-

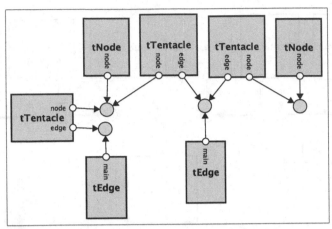

Fig. 7. Reduced hypergraph model of the hypergraph model in Figure 6. The spatial arrangement of the hyperedges in this figure corresponds to the one in Figure 6

lated to tEdge and tNode edges, resp., and the complicated sub-hypergraphs that represent tentacles connecting hyperedges to nodes are translates to tTentacle edges directly connecting the corresponding nodes.

In the following, we focus on the visual language of hypergraph grammar productions. Details of DIAGEN's hypergraph grammar productions can be found in [8]; in the following, we consider productions just as a horizontal arrangement of two hypergraphs being separated by the symbol :=. The left-hand side (LHS) and the right-hand side (RHS) hypergraph can also be considered as two unconnected sub-hypergraphs of a single hypergraph. Hence, each production diagram is a hypergraph diagram with an additional := symbol which distinguishes its LHS from its RHS.

Figure 8 shows the productions of the hypergraph grammar that describes its own language of hypergraph grammar productions. This grammar is based on rHGMs with the following structure: A distinguished node is visited by a tDerive edge that represents the := symbol. All the other nodes are connected to this node either by a left or right edge which allows distinguishing the LHS from the RHS of the production. LHS and RHS are represented by the nonterminal hyperedges Left resp. Right Please note the "stacked" rectangles in productions P_4 and P_6. Those stacks represent sets of the represented hyperedges, i.e., LeftEdge and RightEdge, resp. Please note, too, that hyperedge labels have now the form $x : y$ where the identifier x can be freely chosen, whereas y is the edge's type as it is defined elsewhere in the DIAGEN designer specification which also determines the set of tentacles of this edge. The edge identifier x, e.g., can be used in the specification of actions to be taken during attribute evaluation (cf. Figure 9.) Finally, the small number in the top-left corner of each rectangle is used to order the set of hyperedges of a hypergraph. This ordering is used as a hint to the hypergraph parser when it constructs the parsing plan.

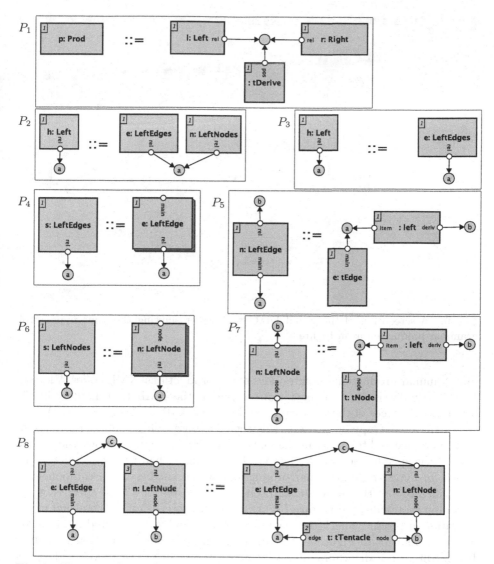

Fig. 8. Hypergraph grammar productions of the visual language of hypergraph grammar productions. Productions $P_9, \ldots, P_{15}$ are omitted. They are similar to $P_2, \ldots, P_8$, resp., with each occurrence of the word *left* replaced by *right*

The result of the editor's diagram analysis by attribute evaluation (omitted here) is a textual representation of the corresponding grammar production in exactly the same syntax as it is used by the DiaGen designer which is actually the same as it is used by the DiaGen generator (see Figure 1). The DiaGen designer's code generator, therefore, had to be adjusted in minor aspects, only. Whereas the DiaGen designer finds the textual representation of reduction rules

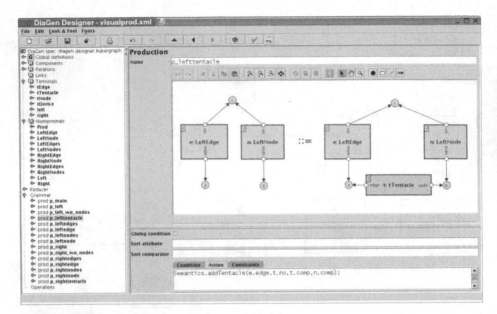

Fig. 9. A screenshot of the visual DIAGEN designer showing the same hypergraph grammar production as in Figure 4

and grammar productions as sub-nodes of the productions' XML nodes, VISUAL-DIAGEN finds the same information deeper in the XML tree, namely in the semantics subtrees of the rules' and productions' XML nodes (see above.)

In the bootstrapping process, this set of hypergraph grammar productions had to be specified textually in the DIAGEN designer. Figure 4 shows the screenshot of the specification of P_8. The DIAGEN designer had then created a visual editor for hypergraph grammar productions which has become a component of VISUALDIAGEN. Hence, VISUALDIAGEN now allows for visual specifying hypergraph grammar productions. Figure 9 shows a screenshot of the visual representation of production P_8, i.e., the same, but visually represented information as the screenshot in Figure 4. The hypergraph grammar production P_8 is now being edited in a diagram editor that is a component of the DIAGEN designer window. The screenshot shows also that some textual information is still needed, even when using visual editors, e.g., the textual specification of the action to be taken during attribute evaluation. This information is provided below the visual editor.

Figures 4 and 9 show that visually editing hypergraph grammar productions enhances readability of such productions. However, they show a shortcoming, too: Visual specifications require substantially more space on the screen and more time for editing them. As a consequence, the user should have the choice whether he prefers the textual or the visual representation. As VISUALDIAGEN always produces the corresponding textual notation, it is basically possible to switch from visual notation to textual notation. However, this feature is not yet supported by VISUALDIAGEN. The other direction, switching from textual to visual representation is less trivial and will be considered in further work.

Translating diagrams into their textual notation and then using the "old" code generator had the benefit of reusing code and, therefore, rapidly creating VISUALDIAGEN. However, it has the shortcoming that errors are detected by the generator with respect to the textual notation. The current implementation is not yet able to visualize errors in the visual notation. It just shows the erroneous textual one, and the user has to translate it back to the visual one.

4   Conclusion

The paper has described VISUALDIAGEN, a tool for visually specifying and generating visual editors. VISUALDIAGEN has been built on top of the existing DIAGEN tool that has required the user to write visual language specifications in a text-based representation. Actually, VISUALDIAGEN has been created by a bootstrapping process: The first DIAGEN version has required plain textual specification, and a compiler-like generator has generated Java code from it. In the next step, a specification tool with a graphical user interface (the DIAGEN designer) had been created. Specifications have been still textual in principle, however in a tree-like structure that has been supported by the designer. The designer has also integrated the generator such that specification errors could be visualized directly in the designer. In the final step, this designer has been used to specify and generate visual editors for reduction rules and hypergraph grammar productions. Those generated diagram editors have then been integrated into the DIAGEN designer, yielding VISUALDIAGEN. The user of VISUALDIAGEN now uses these diagram editors for visually editing those specification constructs.

Visual notations do not have benefits only, but some shortcomings, too, as this paper has shown. Visual notations require much more screen space and time for creating them. Further work, therefore, will investigate how the user can freely switch between visual and textual notation of reduction rules and hypergraph grammar productions.

The major benefit of the "old" textual specification was its conciseness. Such specifications could be easily printed which is no longer possible neither for the DIAGEN designer (primarily because of graphics specification of visual components that is not supported by the textual specification) nor for its visual counterpart VISUALDIAGEN. Further work will investigate methods for translating their XML-based specifications into easily obtainable and printable documents.

References

[1] Roswitha Bardohl. GENGED: A generic graphical editor for visual languages based on algebraic graph grammars. In *Proc. 1998 IEEE Symp. on Visual Languages, Halifax, Canada*, pages 48–55, 1998.   398, 399

[2] Sitt Sen Chok and Kim Marriott. Automatic construction of intelligent diagram editors. In *Proc. 11th Annual Symposium on User Interface Software and Technology, UIST'98*, November 1998.   399

[3] G. Costagliola, A. De Lucia, S. Orefice, and G. Tortora. A framework of syntactic models for the implementation of visual languages. In *Proc. 1997 IEEE Symp. on Visual Languages, Capri, Italy*, pages 58–65. 399

[4] Frank Drewes, Annegret Habel, and Hans-Jörg Kreowski. Hyperedge replacement graph grammars. In G. Rozenberg, editor, *Handbook of Graph Grammars and Computing by Graph Transformation. Vol. I: Foundations*, chapter 2, pages 95–162. World Scientific, Singapore, 1997. 406

[5] Frank Drewes, Berthold Hoffmann, and Mark Minas. Constructing shapely nested graph transformations. In *Proc. AGT'2002 (APPLIGRAPH Workshop on Applied Graph Transformation), Satellite Event to ETAPS 2002, Grenoble, France, 2002*, pages 107–118, 2002. 406

[6] Georg Frey and Mark Minas. Editing, visualizing, and implementing signal interpreted petri nets. In *Proc. 7. Workshop Algorithmen und Werkzeuge für Petrinetze (AWPN'2000)*, number TR 2/2000 in Fachberichte Informatik, pages 57–62. Universität Koblenz-Landau, 2000. 398

[7] Mark Minas. Creating semantic representations of diagrams. In Manfred Nagl, Andy Schürr, and Manfred Münch, editors, *Int'l Workshop on Applications of Graph Transformations with Industrial Relevance (AGTIVE'99), Selected Papers*, number 1779 in Lecture Notes in Computer Science, pages 209–224. Springer, 2000. 405

[8] Mark Minas. Concepts and realization of a diagram editor generator based on hypergraph transformation. *Science of Computer Programming*, 44:157–180, 2002. 398, 402, 405, 408

[9] Mark Minas. XML-based specification of diagram editors. In *Proc. Unigra'03*, Warsaw, Poland, 2003. Appears in ENTCS. 399, 404

[10] Mark Minas and Berthold Hoffmann. Specifying and implementing visual process modeling languages with DIAGEN. *Electronic Notes in Theoretical Computer Science*, 44(4), 2001. 398

[11] D.-Q. Zhang and K. Zhang. VisPro: A visual language generation toolset. In *Proc. 1998 IEEE Symp. on Visual Languages, Halifax, Canada*, pages 195–201, 1998. 398, 399

GenGED – A Visual Definition Tool for Visual Modeling Environments*

Roswitha Bardohl, Claudia Ermel, and Ingo Weinhold

Technische Universität Berlin
{rosi,lieske,bonefish}@cs.tu-berlin.de

Abstract. In this paper, we give a brief overview on GENGED that allows for the visual definition and generation of visual modeling environments. Depending on the underlying visual modeling language, different components are suitable in a visual modeling environment. GENGED supports the definition and generation of editors, parsers, and simulators.

1 Introduction

Today the use of visual modeling and specification techniques is indispensable in software system specification and development, so are corresponding visual modeling environments. As the development of specific visual modeling environments is expensive, generators have gained importance, especially in the field of rapid prototyping. In this contribution, we briefly present the current state of GENGED, a generator for visual modeling environments comprising editors, parsers, and, if suitable, simulators. The main aspect is the GENGED development environment for visually specifying visual languages (VLs) and corresponding environments.

The current GENGED version extends broadly the concepts and environment proposed in [1]. The basic structures given by an alphabet and grammars (represented by algebraic graph grammars and graphical constraint satisfaction problems) are used as before. These structures form the basis for the specification of syntax and behavior of visual models (diagrams over

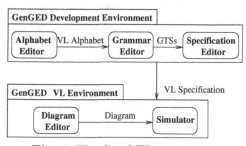

Fig. 1. The GENGED components

a specific VL, such as statecharts, automata, Petri nets, etc.).

The components of the GENGED environment are illustrated in Fig. 1. The alphabet editor supports the definition of symbol types and how they are linked,

---

* This research is partly funded by the German Research Council (DFG; Eh 65/8-3), and by the EU research training network SegraVis.

which results in a type system for all instances. Based on an alphabet, the grammar editor allows for defining distinguished kinds of graph grammars (GGs), namely for syntax-directed editing, parsing, and simulation.

The specification editor is to be used for extending the GGs by some control flow information, as e.g. an order for rule application. The resulting VL specification configures the aimed VL environment comprising at least a diagram editor and, if specified, a simulator. For the storage of the structures (VL alphabet, GGs, VL specification) we use XML w.r.t. [10].

In all GENGED components, we distinguish the abstract syntax and the concrete syntax (containing the layout) of VLs. The abstract syntax level of diagrams defines an attributed graph which is managed by the integrated graph transformation system AGG [11, 3]. The concrete syntax, represented by graphical expressions, is given by vertex attributes, and graphical constraints for the correct arrangement of symbol graphics. The constraint satisfaction problems are solved by the integrated graphical constraint solver PARCON [5].

2   The GenGED Development Environment

Alphabet Editor. The alphabet editor allows for defining the VL alphabet, i.e., the language vocabulary. Such an alphabet consists of visual symbols like state and transition symbols occurring in statecharts, and a definition how such symbols can be connected. According to these tasks, the alphabet editor comes up with a symbol editor and a connection editor, respectively.

The symbol editor (Fig. 2 (a)) works similar to well-known graphical editors like xfig, however, the grouping of primitive graphics (in order to define one symbol graphic) has to be done by graphical constraints. Moreover, according to the abstract syntax, each symbol type (which is represented by a vertex type) needs a unique symbol name. These symbol names are used in the connection editor (Fig. 2 (b)) for defining a connection which is represented by an edge. At the concrete syntax level, graphical constraints have to be defined in order to connect the corresponding symbol graphics (cf. [1]).

The resulting VL alphabet is represented by a type graph where the vertex types are given by the symbol names and the edge types are defined by the

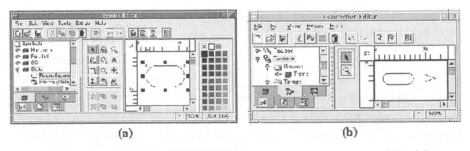

(a)   (b)

Fig. 2. The GENGED symbol editor (a), and the connection editor (b)

connections between two symbols, respectively. Considering, e.g., a state and a transition symbol, two connections are necessary: one describes the begin and the other one the end of the transition symbol. At the abstract syntax level, we obtain a type graph with edges Transition $\xrightarrow{begin}$ State and Transition $\xrightarrow{end}$ State. Concerning the concrete syntax level, symbol graphics define the graphical attributes of vertices, and all graphical constraints are collected to a constraint satisfaction problem: For both connections in our small state-transition example we would use border constraints expressing that a transition arrow begins and ends at the border of a state rectangle. Such high-level constraints are decomposed into low-level constraints (equations over position and size variables) in order to be solved by the graphical constraint solver PARCON [5]. GENGED environment.

Grammar Editor. Based on a VL alphabet, the grammar editor allows for defining distinguished GGs, namely for syntax-directed diagram editing as well as for parsing and simulating a diagram. A GG consists of a (start or stop) diagram and a set of rules, each one comprises a left- and a right-hand side, and optionally a set of negative application conditions (NACs), as well as attribute conditions.

The grammar editor works in a syntax-directed way: Based on a given VL alphabet so-called alphabet rules are generated automatically which function as basic edit commands. Fig. 3 shows a screenshot of the grammar editor, where the alphabet rule InsertState supporting the insertion of a state symbol is illustrated in the top. The left-hand side of this rule is empty, in its right-hand side a state symbol is generated. In the bottom of Fig. 3 we defined a new syntax-directed editing rule S-InsState for inserting a sub-state symbol. On the left-hand side we require already the existence of a state symbol which is preserved by the rule (indicated by number 1). On the right-hand side we inserted a further state symbol by applying the alphabet rule to the right-hand side of the syntax-directed editing rule. The structure view on the left of the grammar editor holds the names of all defined rules as well as start and/or stop diagrams. A suitable dialog supports the selection of rules and the export of GGs.

Specification Editor. Based on the definitions of the VL alphabet and the different GGs, the specification editor allows to establish a VL specification. Syntax-directed editing can be defined by loading the VL alphabet and a syntax-GG. A parse-GG may be extended by the definition of a layering function and a critical pair analysis in order to optimize the parsing process (cf. Fig. 4 (a)). The parse-GG together with these extensions result in a parse specification.

A simulation specification consists of a simulation-GG and a set of simulation steps, which specify how the rules of a simulation-GG have to be applied (Fig. 4 (b)). A simulation step describes an atomic step to be executed during the simulation process. The core of a simulation step is a simulation expression, a kind of iterative program executed when the step is being applied in the simulator of the VL environment.

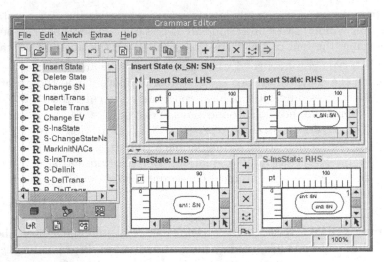

Fig. 3. The GENGED grammar editor

3   A Generated Environment

A VL environment generated by GENGED contains at least an editor, and
optionally, a parser, and a simulator if the underlying VL allows for the modeling
of behavior.

Diagram Editor. If syntax-directed editing is specified, in its current state,
the diagram editor shown in Fig. 3 supports diagram editing in a way similar
to the grammar editor: The structure view on the left holds all the names of
rules that can be applied for manipulating a diagram in the lower part of the
editor. As supported by the grammar editor, a rule can be selected such that it
is visualized in the upper part of the diagram editor. A rule can be applied after
defining mappings from the symbols in the rule's left-hand side to symbols in the
diagram. Please note that we are going to improve the user interface such that

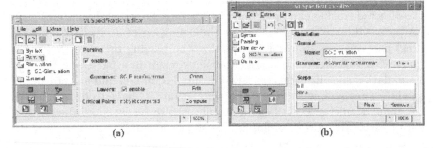

Fig. 4. The GENGED specification editor for parsing (a), and for simulation (b)

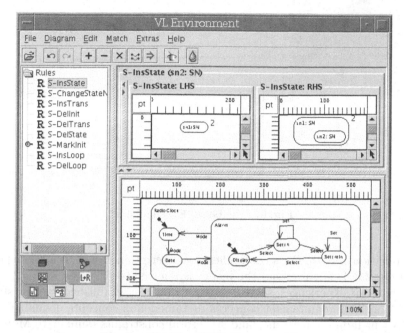

Fig. 5. The GENGED diagram editor

editing rules are put onto buttons and hence they are not visible to the user. However, a parse-GG must be available if free-hand editing should be possible in a diagram editor or if the syntax-GG does not build up the correct VL.

Simulator. A diagram can be simulated w.r.t. a given simulation specification. The simulator depicted in Fig. 3 can be activated from the diagram editor. It directly shows the diagram, and the names of specified simulation steps in the structure view. Selecting a name, the corresponding step is activated.

In Fig. 3 we can see an advanced simulation of a statechart modeling a radio clock. All the active states are marked by a red color (visible in a color print only). Choosing a further Step effects the dialog asking the user to define a parameter event, whereupon the red makring is moved to the respective successor state(s).

4   Related Work and Conclusion

In this contribution we have briefly sketched the current GENGED version for the visual definition and generation of visual modeling environments comprising editors, parsers, and simulators. The case studies we have regarded cover several kinds of (graph-like) UML diagrams, Petri nets, automata, (box-like) Nassi-Shneiderman diagrams, and an architectural specification language. Thereby, we considered pure syntax-directed editing, free-hand editing with parsing, and

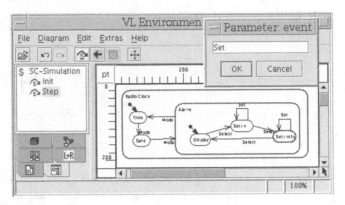

Fig. 6. The GENGED simulator

a combination of both. For a comprehensive description as well as for a download of the GENGED environment the reader is referred to [4].

Due to our experiences, several requirements are established we just started to work on. One requirement concerns metamodeling concepts definable for an alphabet including inheritance and structural constraints (similar to MOF [9]). Another requirement concerns the extension of (current flat) constraint satisfaction problems (towards hierarchical ones) in order to improve the performance. This is still conceptual work, whereas the animation concepts (cf. [2]) are being implemented at present. The animation concepts cover the visual definition of application-specific layout for a visual model like that of a radio clock specified by a statechart (cf. Fig. 3). In the case of a radio clock, e.g., a generated animation component would show a concrete clock.

In the literature one can find many concepts for generating VL environments, however, one can find few implementations only that are related to our work. Beside commercial tools like MetaEdit [7], we have to mention the meta CASE tool JKogge [6] which consists of three logical elements, namely a base system, components (plug-ins) and documents. For the latter one, distributed graphs are used as underlying structure for modelling documents on distinguished servers. In contrast to such meta CASE tools, at the moment, in GENGED the definition and generation of editors, parsers, and simulators/animators is in the foreground where all tasks are based on one formalism: graph transformation.

Closely related to GENGED is the DIAGEN system [8] since it supports the definition of structured and free editing, and simulation. The underlying internal structure is defined by hypergraphs. Although meanwhile also DIAGEN supports the visual specification of visual languages, a support like the automatically generated alphabet rules in GENGED is not available. These alphabet rules (cf. Grammar Editor) assist the language designer for defining the grammars.

In general, the definition and generation of VL environments can advantageously be used for industrial applications since it allows for rapid prototyping.

Moreover, the XML-based storage facilities in DiaGen as well as in GenGED allow for further processing of diagrams by external tools like model checkers.

References

[1] R. Bardohl, M. Niemann, and M. Schwarze. GenGEd – A Development Environment for Visual Languages. In M. Nagl, A. Schürr, and Münch, editors, *Int. Workshop on Application of Graph Transformations with Industrial Relevance (AGTIVE'99)*, number 1779 in LNCS, pages 233–240. Springer, 2000. 413, 414

[2] C. Ermel and R. Bardohl. Scenario Animation for Visual Behavior Models: A Generic Approach. *Journal on Software and System Modeling (SoSyM)*, 2003. to appear. 418

[3] C. Ermel, M. Rudolf, and G. Taentzer. The AGG-Approach: Language and Tool Environment. In H. Ehrig, G. Engels, H.-J. Kreowski, and G. Rozenberg, editors, *Handbook of Graph Grammars and Computing by Graph Transformation, volume 2: Applications, Languages and Tools*, pages 551–603. World Scientific, 1999. 414

[4] GenGED, 2003. http://tfs.cs.tu-berlin.de/genged. 418

[5] P. Griebel. *Paralleles Lösen von grafischen Constraints*. PhD thesis, University of Paderborn, Germany, February 1996. 414, 415

[6] JKOGGE, 2003. http://www.uni-koblenz.de/~ist/jkogge.html. 418

[7] MetaEdit, 2003. **http://www.dstc.edu.au/Products/metaSuite/ MetaEdit.html.** 418

[8] M. Minas. Bootstrapping Visual Components of the DiaGen Specification Tool with DiaGen. In J. Pfaltz and M. Nagl, editors, *Proc. Application of Graph Transformations with Industrial Relevance (AGTIVE'03)*, pages 391–405, Charlottesville, USA, 2003. http://www2.informatik.uni-erlangen.de/DiaGen/. 418

[9] OMG. Meta Object Facility (MOF) Specification, Version 1.4. Technical report, Object Management Group, April 2002. http://www.omg.org. 418

[10] G. Taentzer. Towards Common Exchange Formats for Graphs and Graph Transformation Systems. In *Proc. Uniform Approaches to Graphical Process Specification Techniques (UNIGRA'01), Electronic Notes of Computer Science (ENTCS)*, volume 44. Elsevier, 2001. 414

[11] G. Taentzer. AGG: A Graph Transformation Environment for Modeling and Validation of Software. In J. Pfaltz and M. Nagl, editors, *Proc. Application of Graph Transformations with Industrial Relevance (AGTIVE'03)*, pages 435–443, Charlottesville, USA, 2003. 414

CHASID – A Graph-Based Authoring Support System[*]

Felix H. Gatzemeier

RWTH Aachen University
Department of Computer Science III
fxg@i3.informatik.rwth-aachen.de

1 Introduction and Overview

CHASID experiments with ways to support authors by giving them an extension to their authoring applications that allows them to build a model of the content structure of their document and checks that.[1] The aim is to give the author confidence about not having some 'simple' error in the document. Like a spelling checker in another dimension, CHASID spots some errors (comparable to non-words found by a spelling checker), but cannot be expected to find everything a somewhat competent human reader would detect (like a spelling checker not detecting grammatical errors).

This demonstration begins with the author loading a simple XML document into CHASID, continues with him editing it there, introducing a structural problem that is likely to be overlooked in conventional authoring. CHASID, however, detects it and produces a warning. The demonstration closes with ways in which this condition can be fixed. This document generally follows this sequence of steps, after presenting with a brief overview of core concepts.

2 Core Concepts of CHASID

The CHASID document model as seen by the author consists of a conventional outline, which is an ordered hierarchy of divisions (hence called the document hierarchy) on the one hand and content structure on the other. The latter is a graph form closely related to conceptual graphs [4].

These two structures are connected by Import and Export relations. A division is said to export a topic (an element of the content structure) if reading this division should provide at least a working understanding of said topic. In the other direction, if a division imports a topic, at least some understanding of the topic is required to fully understand the division.[2]

From this view alone, the author can derive some information usually not available in text processing systems. Concepts that are only imported but never

---

[*] This work is funded by Deutsche Forschungsgemeinschaft grant NA 134/8-1.
[1] See [2] for an overview and [1] for a recent comparison with related work.
[2] For an example of a document graph, you might already want to look at figure 3.

J.L. Pfaltz, M. Nagl, and B. Böhlen (Eds.): AGTIVE 2003, LNCS 3062, pp. 420–426, 2004.

exported are assumed to be known to the reader. Conversely, concepts exported but never imported should contribute to the core meaning of the document. But there are more delicate patterns: if a topic is exported in the middle of a document and imported in the conclusion, it is very likely to be central to the document's meaning, too.

For the content of the document, we put the graph-based core in a context of standard and widely-used applications: ToolBook for Multimedia applications and Emacs psgml DocBook are the current incrementally online integrated content editing applications (to varying degrees of incompleteness).

Further outside integration is provided by imports and exports from and to XML documents. We concentrate here on an import from documents according to the PiG DTD (import and export relationships in a very simple DTD[3]). A noteworthy export is that to an extension of XML DocBook [5] with the content structure stored in a concept map element. That is used by a WWW server servlet to answer requests on the document, providing the parts that explain certain topics and all the ones they pre-require, minus the ones already known.

3   The Example Document

As an example document, we use an article on the correlation of E-mail interchange and perceived interaction. It is constructed according to the classical Introduction, Methods, Results, Discussion structure. The structure modelled by CHASID goes beyond that and helps identify some further problems.

We start out with the modelling available on the basis of XML (figure 1). The PiG DTD used for this document concentrates on the hierarchical structure and the way topics are exported and imported. The media content itself is not contained here.

This excerpt asserts that the subdivision "Interaction Questionnaire" exports (that is, explains) the topic of 'Interviews', while the "Introduction" imports (that is, expects to be known) the topic of 'Correlation'.

In the XML source, the hierarchical structure of the elements, which corresponds here to the hierarchical structure of the document, gives an impression of the document hierarchy. The content of the divisions are just imports and exports that are linked by names (ID/IDREF attributes in XML), making it hard to tell what is imported from where. This is, however, the level of cross-referencing support currently available in authoring systems.

Still within the XML world, there is some alleviation: a style sheet reformats the document in a more inspection-friendly format. For each division, exports are listed, imports and announcements are rendered in tables. All references may be followed are hyperlinks to their targets. Figure 2 shows a snippet of that derived file in a web browser.

---

[3] The DTD is a local development especially to capture the structure of [3], also known as the "Programming-in-the-large book", in German: "Programmieren im Großen". This has been abbreviated to "PiG".

```
<Buch Titel="EMi_And_pCI_correlate">
  <Div Titel="Introduction">
    <i ref="Questionnaire"/> <i ref="Correlation"/>
    <i ref="E-mail"/>        <e id="Context"/>
  </Div>
  <Div Titel="Methods">
    <e id="Comparative_Study"/>
    <Div Titel="Tools">
      <Div Titel="Interaction_Questionnaire">
        <e id="Interviews"/>
      </Div>
    </Div>
```

Fig. 1. Excerpt of the PiG model of the export/import structure

3.2. *ArchiveGrab*

Imports:

Concept	Imported from
EMailArchiveGrab	2.1.2. mailscan

Exports:

- *CapturedInteraction*

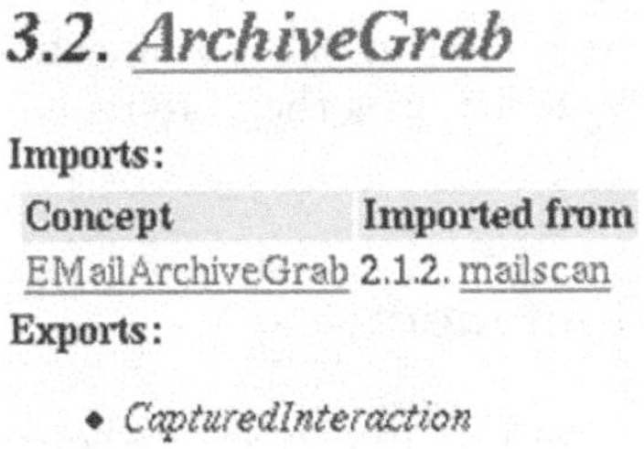

Fig. 2. Snippet of the HTML file derived from the export/import structure

This is helpful in analysis in showing at a glance what parts of the document a specific part relies upon, and how. In the snippet, the division "Archive Grab" depends upon the division "mailscan" by its use of the topic 'E-mail Archive Grab'. This help, however, is notconstructive authoring support: even this simple visualisation is batch-detached from the document by principle.

4   Loading the Document and Examining at the Graph

First, the above XML document is loaded into CHASID and the resulting graph is layouted. Figure 3 shows the graph describing the entire structure of the article.

The graph preserves the information content of the XML file: nested divisions are now division nodes in a tree layout (on the left side of the graph), the exported and imported topics are now nodes to the right of that tree, and the export and import relations are now edges.[4] The topics in this figure are not interconnected since such information is not contained in the PiG DTD. [1] contains some documents with relations.

This graphical display provides a better overview and a more direct feedback. Nodes and edges may be rearranged to accommodate for the current focus

---

[4] The import and export edges have a supporting icon in the middle since they may be referred to in functionality of CHASID not discussed here.

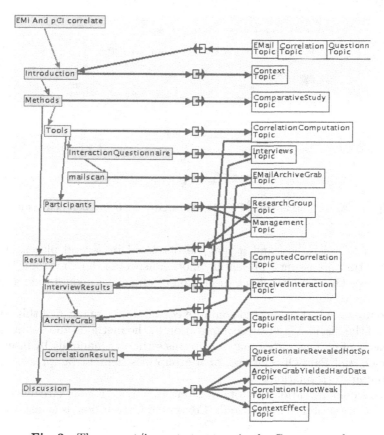

Fig. 3. The export/import structure in the CHASID graph

of work. Incremental layout algorithms, user-defined constraints and alignment operations support this.

The direction of import/export arrows indicates the information flow. So, an author sees by tendency whether an import is a forward or backward reference, or how central a topic is by the number of outgoing imports.

In this example, the first few divisions are expository explaining context that is not later specifically referred to. The "Methods" section, then, exports a number of topics (beginning with 'Correlation Computation') that are later imported in "Results".

Errors in the model are checked on demand so the author does not get confused during editing that produces incomplete intermediate stages.

5   Capturing Additional Structure

As an example for further modelling offered in CHASID, we look at the interplay of Aspect-Of and Part-Of relations. A topic may have several parts (modelled

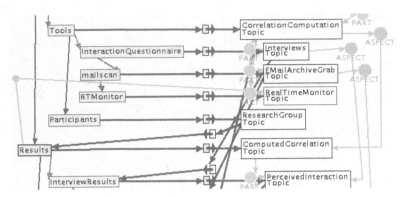

Fig. 4. A warning about a Part-Of hierarchy not being handled completely

with Part-Of relations), and it may be discussed in several places. A special variant of this discussing is presenting another aspect of the same topic (modelled with Aspect-Of relations). Each discussion of an aspect should then address all parts of the topic.

In our example, the ways to compute correlations are parts of this computation, and the discussion of this computation in the methods and results sections are aspects of it. The author may set up this structure manually by inserting the Aspect-Of and Part-Of relations, or a group of schemata (as presented in [1]) may contain such relations.

In this demonstration the topics 'Interviews' and 'E-mail Archive Grab' are parts of 'Correlation Computation', 'Computed Correlation' is an aspect of 'Correlation Computation', with parts 'Perceived Interaction' and 'Captured Interaction' which are aspects of their respective counterparts within 'Correlation Computation'. With these changes, the model is still correct.

6   Making an Error and Spotting It

The added structure serves to identify errors in the structure. We now make an error and have CHASID spot it.

We assume that the author wants to describe another correlation computation, real-time traffic monitoring. He inserts this into the methods section with its own division explaining the workings. Since this is another part of 'Correlation Computation', the author also adds the Part-Of relation.

The analysis, when activated, now detects an Aspect-Of/Part-Of anomaly. This is indicated by a small dot (as shown in figure 4) connected to 'Monitoring', the new part that is not discussed, and 'Computed Correlation', the aspect discussion that lacks that part. Expanding the dot yields a more detailed message. Further help is available in the context help system.

There are several ways to cope with this diagnostic: suppress the warning (prevent it from reappearing), ignore it (just remove its representation), or fix the condition.

production TSIEA_AI_Add_NoTopic * =

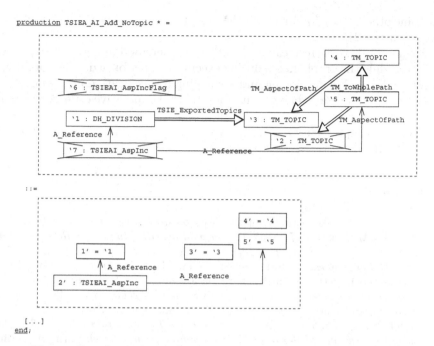

[...]
end;

Fig. 5. Sample analysis production: detecting an incomplete aspect

In the case of this warning, CHASID can even help in fixing the warning. Since this is just a matter of adding a discussion of the new part to the other aspect, there is a production taking care of that.

Such fixing would lead to rather mechanical documents. There does not have to be, however, a separate division exporting the result of real-time traffic monitoring. If this computation is just implemented, but has not been fully applied, it can just as well be handled within the 'Computed Correlation' division. Removing the monitoring computation result division and having the 'Computed Correlation' export its chief concept still yields a valid document, because there is some point in the discussion of 'Computed Correlation' where the result aspect of real-time traffic monitoring is exported.

7 Graph Transformations in Document Analysis

For the document analysis, *-qualified PROGRES productions are used so that for every match of their left-hand sides, the transformation to the right-hand side is performed. Figure 5 shows the production responsible for (one variant of) incomplete aspects.

The left-hand side can be read as follows: look for a division ('1) which exports a topic ('3) that is just an aspect of another topic ('4), which in turn has some other part ('5), for which there is at least one aspect ('6) for which no aspect exists ('2).[5]

Such a complex pattern can frequently be understood in only one sequence of matches, while the graph match during execution is performed –conceptually– in parallel. This makes developing and debugging a specification in the interesting parts more chanllenging than writing conventional imperative code. A graph pattern is a rather compact notation for such conditions. The problems inherent to this improved expressivity are set off by the integrated checks of the PROGRES editor in our case.

References

[1] F. Gatzemeier. Authoring support based on user-serviceable graph transformation. In *Applications of Graph Transformations with Industrial Relevance*, 2003. This volume. 420, 422, 424

[2] Felix H. Gatzemeier. Patterns, Schemata, and Types — Author Support through Formalized Experience. In Bernhard Ganter and Guy W. Mineau, editors, *Proc. International Conference on Conceptual Structures 2000*, volume 1867 of *Lecture Notes in Artificial Intelligence*, pages 27–40. Springer, 2000. 420

[3] Manfred Nagl. *Softwaretechnik: Methodisches Programmieren im Großen (Software Engineering: Methodological Programming in the Large)*. Springer, Berlin, 1 edition, 1990. 421

[4] John F. Sowa. *Conceptual Structures: Information Processing in Mind and Machine*. The Systems Programming Series. Addison-Wesley, Reading, MA, USA, 1984. 420

[5] Norman Walsh and Leonard Muellner. *DocBook: The Definitive Guide*. O'Reilly, October 1999. 421

---

[5] The remaining nodes, '6 and '7 serve to provide user configurability and idempotence of analysis, both of which are not addressed here.

Interorganizational Management of Development Processes*

Markus Heller[1] and Dirk Jäger[2]

[1] RWTH Aachen University
Department of Computer Science III
Ahornstrasse 55, 52074 Aachen, GERMANY
heller@cs.rwth-aachen.de
[2] Bayer Business Services GmbH
51368 Leverkusen, GERMANY
dirk.jaeger@bayerbbs.com

Abstract. The AHEAD system supports the management of development processes for complex end products in engineering disciplines. AHEAD is based on nearly ten years of ongoing research on development processes in different engineering disciplines and the underlying concepts have been applied to the above application domains.

Today, development processes tend to be distributed across organization boundaries. As the organizations which execute a shared process, usually have different goals and interests, a balance between all interests has to be achieved in a supporting concept.

This paper illustrates by a short sample scenario, how the AHEAD system supports distributed development processes thereby taking the interests of all cooperation partners into account.

1 Introduction

Development processes for complex end products in engineering disciplines are difficult to plan and manage in advance due to their highly creative character. It is very likely, that some product specifications are modified or some planned activities are reorganized during development. Therefore, development processes also have a very dynamic character.

The process management system AHEAD supports the management of development processes in an organization, e.g. a company or a department of a company. The AHEAD system has been developed within a long-term running Collaborative Research Center IMPROVE (Information Technology Support for Collaborative and Distributed Design Processes in Chemical Engineering)[4]. A net-based approach is used to model development activities and their relationships, resources and products, like technical documents or plans. A detailed description of AHEAD can be found in [5].

---

* Financial support is given by Deutsche Forschungsgemeinschaft, Collaborative Research Center 476.

J.L. Pfaltz, M. Nagl, and B. Böhlen (Eds.): AGTIVE 2003, LNCS 3062, pp. 427–433, 2004.
© Springer-Verlag Berlin Heidelberg 2004

Today, the development of an end product often is not accomplished by one organization alone. Instead, cooperating organizations execute a common development process together.

A common pattern for cooperation of organizations is a delegation relationship, where selected fragments of the overall development process are delegated from one contractor organization to another subcontractor organization. The subcontractor executes the delegated process and returns a specified product to the contractor. Both parties usually have different interests. Contractor organizations often try to gain full control over the delegated process fragments, while subcontractor organizations are interested in hiding the internal details of the delegated parts of the process.

The AHEAD system provides a management tool to manage distributed development processes. The proposed concept, called delegation of process fragments, takes the divergent interests of contractor and subcontractor organizations into account.

This paper describes a demonstration scenario for the support of distributed development processes in the AHEAD system. For sake of brevity, the underlying concepts and the internal realization of the concept are omitted here, they can be found in [1], [3] and [2].

2 Demonstration Scenario

In this section, the main features of the AHEAD system for the support of distributed development processes are illustrated. Figure 1 shows a schematic view on a delegation relationship between a contractor and a subcontractor organization. It is assumed that in both organizations AHEAD is used for the process management. In the demonstration session a detailed concrete sample scenario will be used that covers the steps described below.

The upper part of the figure shows a task net that is maintained in the management system of a contractor organization. The top-level task T is decomposed into four subtasks T1-T4. Furthermore, the task T3 has been refined by the tasks T3.1-T3.3. The tasks are connected by control flow edges, as indicated.

2.1 Export and Import of Delegated Process Fragments

The contractor selects a fragment of the process for delegation, in the example tasks T2 and T3 (shown in the dark grey box). The refining subnet of T3 is added to the delegated process fragment automatically.

Then the selected fragment is exported to a file. Copies of the delegated tasks remain in the contractor's management system. The file also includes contextual information about the delegated part of the process.

Next, the subcontractor imports the file and a corresponding task net is set up in his management system. In the lower part of Figure 1 the imported task net is shown in the grey box (the surrounding context information of the delegated process fragment is omitted).

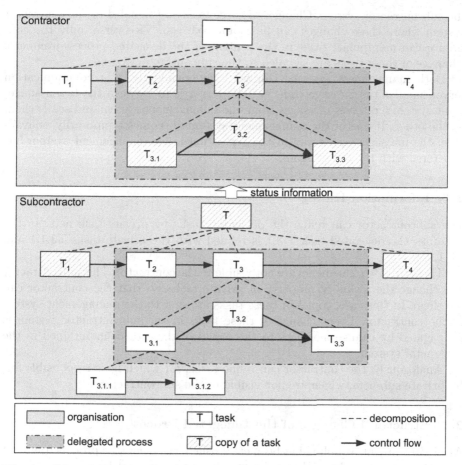

Fig. 1. Schematic view of a delegation relationship between two coupled systems (after [3])

After inspection of the imported task net, the subcontractor accepts the order of the contractor to execute the transferred process and to deliver the specified results to the contractor. In this way, the delegation of process fragments is interpreted as a contract between contractor and subcontractor. In contrast to the delegation of a single task, entire task nets can be delegated. Thereby, the subcontractor has to agree to the specified structure of the delegated process (milestone activities). Violations of this treaty may happen, but need not have legal consequences.

2.2   Run-Time Coupling

Both management systems are coupled at run-time and exchange m essages to inform each other about changes. All delegated tasks can only be manipulated in the subcontractor system. Changes of these tasks are signaled to the contractor system where these changes can be monitored, too. Vice versa, only the contractor can manipulate tasks in the context of the delegated process fragment. Changes of these tasks are signaled to the subcontractor system.

Both management systems can work independently, if a communication server handles the message transfer. The server is in between the two management systems, receives messages from one management system and sends them to the other. If one of the management systems becomes temporarily unavailable, the messages are stored in the server until this management system has registered with the server again.

2.3   Information Hiding

The subcontractor can refine the delegated tasks by private task nets. In the example, the delegated task T.3.1 is refined with the private tasks T3.1.1 and T3.1.2.

These private refinements are not visible to the contractor. The subcontractor can change the visibility property of private tasks, so that the contractor can see them. In this case, copies of these tasks are sent to the management system of the contractor. Every change of these tasks in the subcontractor system is propagated by change messages to the copies of these tasks maintained in the contractor system.

Analogously, the contractor can refine tasks, for which he is responsible for, by private sub-nets, which are not visible to the subcontractor.

2.4   Structural Changes of the Delegated Process

After the run-time coupling has been established, the delegated process fragment can be structurally changed in order to change the contract between contractor and subcontractor. For example, addititional tasks can be added to the delegated task net or new control flow relationships between delegated tasks can be inserted.

Both organizations have to communicate with each other about an intended change of the delegated task net, and a consensus has to be reached. This phase of negotiation is not supported in the AHEAD system. A formal change protocol is enforced, where the contractor carries out all changes, while the subcontractor can accept or refuse these changes (Figure 2).

The subcontractor starts the change procedure to allow changes of the delegated process fragment by executing the command **Allow Changes** within his instance of the AHEAD system . The delegated process fragment changes its state from Accepted to Change (according to the state transition diagram in Figure 2.a).

In the change phase only the contractor can change the delegated process fragment structurally. All changes are indicated in both AHEAD systems at the same time, while the contractor changes the task net. These changed parts of

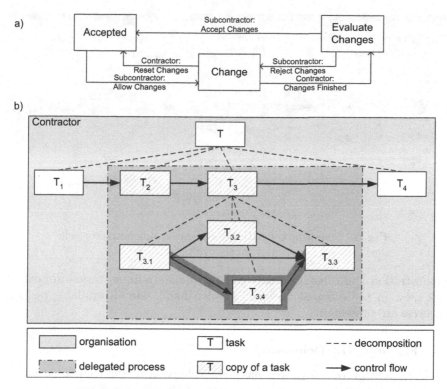

Fig. 2. Structural changes of the delegated process fragment

the task net are displayed by colored markings to indicate that they are just intented changes, since these changes are proposed by the contractor and are not yet accepted by the subcontractor. Figure 2.b shows an example, where a new task T3.4 is added to the delegated process fragment as a sub task of T3, together with two new control flow relationships between T3.1, T3.3, and T3.4. After all intended changes have been done, the contractor executes the command Changes Finished. The cancellation of all changes is possible by executing the command Reset Changes.

In an evaluation phase the subcontractor inspects the changed process fragment and decides, if he accepts these changes of the contract or not. To simplify the change protocol, all intended changes can only be accepted or rejected as a whole. In the example, the subcontractor accepts the changes and executes the command Accept Changes. The delegated process fragment changes its state from Change to Accepted. All intended changes are then executed in both AHEAD systems and the delegated task net is updated. Now the changed process fragment is visible in both systems.

A screenshot of the AHEAD system on the subcontractor side is shown in Figure 3. In the graphical view on the right-hand side the task net is presented to the subcontractor. All tasks of the delegated process fragment presented in the

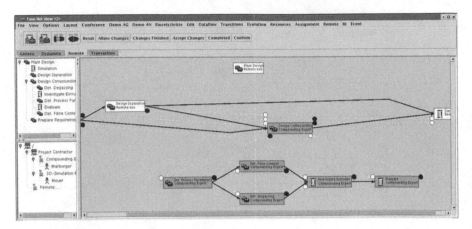

Fig. 3. Screenshot of AHEAD system on subcontractor side

demonstration (task `Design Compounding` and its refining tasks) are coloured dark grey. In the different views on the left-hand side all available tasks and resources are presented.

2.5 Finishing the Delegation

The delegated process fragment has been delegated to the contractor in order to receive a specific product as the result of the delegated process activities. After the subcontractor has produced the desired result, it is sent back to the contractor organization. The contractor evaluates the product and accepts the result by executing a special command. In this case, the delegation relationship is finished and the necessary coupling of the two management systems is terminated. Otherwise the contractor rejects the produced result by executing a special command.

3   Summary

The AHEAD system supports the management of development processes that are distributed across multiple organizations[3]. The cooperating organizations work together in a delegate relationship and are able to execute a shared process. In his instance of AHEAD, the contractor exports delegated parts of a development process into a file and a subcontractor imports this file into his management system. Next, the management systems of the cooperating partners are coupled at run-time and exchange messages to inform each other about changes relevant for the delegation relationship. The use of a communication server in between both management systems ensures that they can operate independently. Contractor and subcontractor can refine tasks with private task nets, which are not visible to the partner. If wanted, these private tasks can be made visible for the other organization.

Structural changes of the delegated process after enactment are possible. A fixed change protocol enforces that only the contractor can change the delegated process in its structure, and the subcontractor can only reject these changes.

The proposed delegation concept for supporting distributed development processes has been applied in software and chemical engineering within the IMPROVE project[4]. In this demonstration session, the coupling of two instances of the AHEAD system will be demonstrated.

References

[1] Simon Becker, Dirk Jäger, Ansgar Schleicher, and Bernhard Westfechtel. A delegation-based model for distributed software process management. In Vincenzo Ambriola, editor, *Proceedings 8th European Workshop on Software Process Technology (EWSPT 2001)*, LNCS 2077, pages 130–144, Witten, June 2001. Springer. 428

[2] Markus Heller and Dirk Jäger. Graph-based tools for distributed cooperation in dynamic development processes. In Manfred Nagl and John Pfaltz, editors, *Proc. AGTIVE 2003*, LNCS. Springer, 2004. In this volume. 428

[3] Dirk Jäger. *Unterstützung übergreifender Kooperation in komplexen Entwicklungsprozessen*, volume 34 of *Aachener Beiträge zur Informatik*. Wissenschaftsverlag Mainz, Aachen, 2003. 428, 429, 432

[4] Manfred Nagl and Bernhard Westfechtel, editors. *Integration von Entwicklungssystemen in Ingenieuranwendungen*. Springer, Heidelberg, 1998. 427, 433

[5] Bernhard Westfechtel. *Models and Tools for Managing Development Processes*. LNCS 1646. Springer, Heidelberg, 1999. 427

Conceptual Design Tools for Civil Engineering*

Bodo Kraft

RWTH Aachen University
Department of Computer Science III
Ahornstrasse 55, 52074 Aachen, Germany
kraft@i3.informatik.rwth-aachen.de

Abstract. This paper gives a brief overview of the tools we have developed to support conceptual design in civil engineering. Based on the UPGRADE framework, two applications, one for the knowledge engineer and another for architects allow to store domain specific knowledge and to use this knowledge during conceptual design. Consistency analyses check the design against the defined knowledge and inform the architect if rules are violated.

1 Introduction

Conceptual design in civil engineering is a vague and creative phase at the beginning of the construction process. Whereas in later phases only a small part of the building is the matter of elaboration, conceptual design considers the whole building, its usage, and functionality. Classical CAD systems give the architect less support in this early design, as they do not model conceptual design information, such as areas or accessibility between them. As a consequence, architects currently still draw conceptual sketches by hand without any support by a computer program. In a second step, they manually transfer the conceptual design using a CAD system into the constructive design. The drawback of this development process is not the creative and artistic way of designing, but the informal way of information storage and the lack of consistency analyses [3].

In this paper we introduce two graph-based tools to support conceptual design. Knowledge about a specific type of buildings can be inserted by a knowledge engineer, using the so called Domain Model Graph Editor. This editor allows to formally define rules and restrictions about a specific type of buildings, in this paper an office block. Using the Design Graph Editor, architects profit from this knowledge while designing an abstract sketch of an actual building. The Design Graph Editor imports the defined knowledge and supports the architects by consistency checks, design errors are found in this early phase.

We use the graph based tools developed at our department to create new applications. Starting from an initial version, an UPGRADE prototype [2] can be extended with new functionality and adapted to the needs of an application domain. As a result, the PROGRES specification [5] can be executed in a problem oriented visual tool.

* Work supported by Deutsche Forschungsgemeinschaft (NA 134/9-1)

J.L. Pfaltz, M. Nagl, and B. Böhlen (Eds.): AGTIVE 2003, LNCS 3062, pp. 434–439, 2004.
© Springer-Verlag Berlin Heidelberg 2004

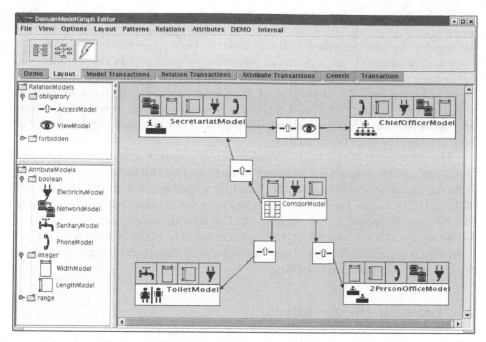

Fig. 1. Domain Model Graph Editor, a tool for the knowledge engineer

In this paper we concentrate on the provided functionality of the developed tools and how they give support. The underlying PROGRES specification and discussions of related work can be found in [4].

2 Tools for the Knowledge Engineer

Fig. 1 depicts a screenshot of the Domain Model Graph Editor, the tool used by the knowledge engineer. It is used to define domain specific knowledge, like room types, relations between rooms and room attributes. We call this knowledge basic models. Based on these models, the knowledge engineer defines a specific rule base, valid for one building type. The domain model graph in the screenshot represents knowledge about an office building, it can be used for each building project of this building type.

The editor is divided into two parts: On the left side two tree views represent relation and attribute definitions. The main part of the application shows the domain model graph, the data structure which represents specific domain knowledge. The first step of knowledge input is the definition of relations and attributes. In the upper tree view in Fig. 1, two obligatory relations, access to demand accessibility, and view to denote visibility between two areas, have already been defined. Integer attributes like length or width are used to demand size restrictions. Boolean attributes like network are used to demand or forbid an area to have a certain equipment installed, e.g. network sockets.

In the next step the knowledge engineer creates areas, in Fig. 1 e. g. the secretariat's room, the chief officer's room or the corridor. All room types that are needed for the described building have to be represented by an area. For each area, a minimal and a maximal number of occurrences in the building can be defined, to express that certain rooms have to be existent, or that others have to be unique. The definition of knowledge is refined through attributes, e. g. network or electricity, and relations. To demand e. g. two areas to have an access relation, an edge-node-edge construct is established between e. g. the corridor and the secretariat's office.

The knowledge stored in the domain model graph describes rules on a type level, and not for an actual building. Therefore the obligatory access relation between the 2PersonOffice and the Corridor expresses that in a building each 2PersonOffice has to have an access relation to a Corridor. Forbidden relations in the domain model graph express that a relation must not be established between two areas of a building. In the same way, the attributes do not describe restrictions for an actual area, they describe restrictions valid for all areas of the specified type in an actual building.

3   Tools for the Architect

The Design Graph Editor is the tool used by architects supporting them during the conceptual design phase. This tool allows the architect to concentrate on the coarse organization and functionality of the building without being forced to think about exact dimensions or material definition. If there exists a domain model graph as a knowledge base for a specific type of buildings, the architect directly profits from the specified rules.

The underlying building type of the domain model graph shown in Fig. 1 is an office building, it has been linked to the Design Graph Editor in Fig. 2. The predefined basic models are represented by the trees views on the left side of the screenshot. Using them, the architect sketches a building by selecting a model from the tree and inserting it into the sketch. In our idea conceptual design is based on rooms. Thus the architect initially inserts the rooms of the future building in the design graph. He then creates attributes and relations to further define the organization of the building.

The design graph, depicted on the right side of Fig. 2 represents a part of an office building, which has already been sketched by an architect. The corridor in the middle of the graph has direct access to all areas except to the chief officer's room, which is only accessible from the secretariat's room. Attributes define the existence of equipment in corresponding area, in Fig. 2, the chief officer's room has electricity, network access and a phone to be installed, furthermore some length and width restrictions.

The layout of the nodes does not represent the arrangement of the rooms in the future building, it is optimized to provide a good readability. Even if this sketch does not contain any geometric information about the sketch, the organization of the building becomes clear.

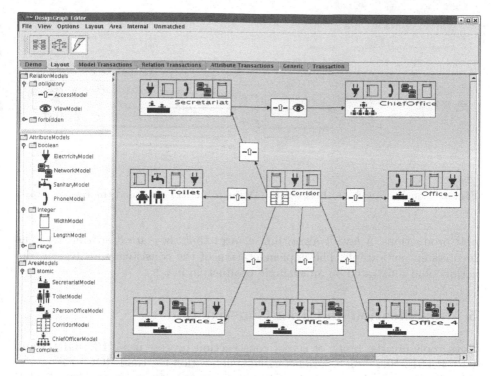

Fig. 2. Design Graph Editor, architect's tool to sketch buildings

4   Consistency Analyses

Fig. 3 depicts both tools, the D om ain M odelG raph Editor and the D esign G raph
Editor next to each other. The knowledge definition on type level is depicted on
the left side, the conceptual design on the level of instances on the right side.
With the aid of consistency analyses the actual design is checked against the
knowledge. These checks enclose the correct usage of the defined attributes and
relations, as well further as internal consistency checks.

An example of a consistency analysis is depicted in Fig. 3. As defined in
the domain model graph each 2PersonOffice should have telephone, network,
and electricity attributes, and an access relation to the Corridor. To check the
consistency, each 2PersonOffice in the design graph is examined, if it has (a)
access to the Corridor, (b) all the demanded attributes, and (c) none of the
forbidden ones. As shown in Fig. 3, one of the 2PersonOffices does not have
a network access. This inconsistency is found and marked by an error message.
The architect can react to this inconsistency in different ways. He is free to stay
in an inconsistent state, he can fix the error in the design graph, or he can change
the knowledge definition in the domain model graph.

The consistency checks are realized by graph tests comparing the domain
model graph against the design graph. Notifications are created by transactions

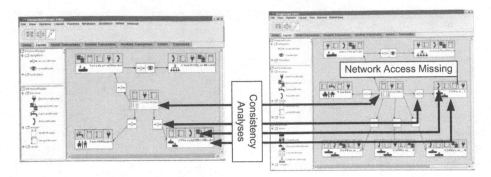

Fig. 3. Consistency checks between the Domain Model Graph and the Design Graph

and productions, a layout algorithm ensures that they are drawn next to the inconsistent sub graph. The implementation of the consistency analyses is described and illustrated on an example production in [4].

5   Specification and Tool Construction

Both tools are the result of a tool construction process using PROGRES and UPGRADE. Based on a parameterized specification [4], the PROGRES system generates C-code. This code is then compiled, together with the UPGRADE framework, to an UPGRADE prototype. In this initial version, the prototype already provides the functionality to execute productions and transactions, and to visualize the graph. Moreover, some basic layout algorithms, filter definitions, and node representations can be used. As UPGRADE prototypes are planned to become extended, further adoptions can easily be done [1]. The tools depicted in Fig. 1 and Fig. 2, are both extended UPGRADE prototypes. Unparsers written in Java change the representation of sub graphs e. g. into a table or a tree structure. New layout algorithms ensure that attributes are arranged next to the corresponding area, and that the areas themselves are clearly arranged. New node representations provide e. g. that an icon illustrating the attribute is drawn. A further extension is an import/export interface with an automatically generated HTML based documentation. Our goal is to give the user of our tools an abstract and more clear view on graph structure, adopted to the need of the application domain. Even if the screenshots do not look like graph editors, the underlying data structure is still a graph.

In a usual PROGRES specification, the domain knowledge, here area types, attribute types and relation types are fixed in the schema and in the operational part of the specification. Node types for access or view, for each relation type, and for each room type would exist. The disadvantage of the traditional specification method is the difficulty to extend or change the knowledge. If the knowledge engineer wants to use a new room type, e. g. Tea Kitchen, the specification has to be changed, new code has to be generated, and the UPGRADE prototype has be restarted. We want the knowledge engineer to elaborate the knowledge, e. g. which area types are necessary. Using the traditional specification method, the tool construction would have to be repeated several times. Moreover, a graph technology expert would have to assist the knowledge engineer, helping him to change the specification and to rebuild the tools. Therefore we introduced a new parameterized specification method [4] which allows storing the specific domain knowledge in the host graph.

We use the described prototypes as an experimentation platform, to elaborate adequate data structures and functionality for conceptual design support. In the long run our goal is to extend commercial CAD systems with functionality to semantically check the architect's sketch against formally defined knowledge.

References

[1] Thomas Haase, Oliver Meyer, Boris Böhlen, and Felix Gatzemeier. A domain specific architecture tool: Rapid prototyping with graph grammars. this volume, 2003. 438

[2] Dirk Jäger. UPGRADE - A framework for graph-based visual applications. In Mandred Nagl, Andy Schürr, and Manfred Münch, editors, *Proceedings Workshop on Applications of Graph Transformation with Industrial Relevance*, volume 1779 of *LNCS*, pages 427–432, Kerkrade, The Netherlands, September 2000. Springer, Berlin. 434

[3] Bodo Kraft, Oliver Meyer, and Manfred Nagl. Graph technology support for conceptual design in civil engineering. In M. Schnellenbach-Held and Heiko Denk, editors, *Proceedings of the 9th International EG-ICE Workshop*, pages 1–35, VDI Düsseldorf, Germany, 2002. 434

[4] Bodo Kraft and Manfred Nagl. Parameterized specification of conceptual design tools in civil engineering. this volume, 2003. 435, 438, 439

[5] Andy Schürr, Andreas J. Winter, and Albert Zündorf. PROGRES: Language and Environment. In Hartmut Ehrig, Gregor Engels, Hans-Jörg Kreowski, and Grzegorz Rozenberg, editors, *Handbook on Graph Grammars and Computing by Graph Transformation: Applications, Languages, and Tools*, volume 2, pages 487–550. World Scientific, Singapore, 1999. 434

E-CARES* – Telecommunication Re- and Reverse Engineering Tools

André Marburger and Bernhard Westfechtel

RWTH Aachen University, Department of Computer Science III
Ahornstrasse 55, 52074 Aachen, Germany
{marand,westfechtel}@cs.rwth-aachen.de
http://www-i3.informatik.rwth-aachen.de

Abstract. The E-CARES project addresses the reengineering of large
and complex telecommunication systems. Within this project, graph-
based reengineering tools are being developed which support not only
the understanding of the static structure of the software system under
study. In addition, they support the analysis and visualization of its
dynamic behavior. The E-CARES prototype is based on a programmed
graph rewriting system from which the underlying application logic is
generated. Furthermore, it makes use of a configurable framework for
building the user interface. In this demo, we show by example how the
different tools within the prototype work and how the analysis results
are represented to the user.

1 Introduction

The E-CARES research cooperation between Ericsson Eurolab Deutschland
GmbH (EED) and Department of Computer Science III, RWTH Aachen, has
been established to improve the reengineering of complex legacy telecommunication systems. It aims at developing methods, concepts, and tools to support
the processes of understanding and restructuring this special class of embedded systems. The subject of study is Ericsson's Mobile-service Switching Center
(MSC) for GSM-networks called AXE10. The AXE10 software system comprises
approximately 10 million lines of code spread over circa 1,000 executable units.

Maintenance of long-lived, large, and complex telecommunication systems
is a challenging task. In the first place, maintenance requires understanding of
the actual system. While design documents are available at Ericsson which do
support understanding, these are informal descriptions which cannot be guaranteed to reflect the actual state of implementation. Therefore, reverse engineering
and reengineering tools are urgently needed which make the design of the actual
system available on-line and which support planning and performing changes
to the system. Interesting enough, tools especially designed for the domain of
telecommunication systems can hardly be found and existing tools are often not

---

* E-CARES is an acronym for **E**ricsson **C**ommunication **AR**chitecture for **E**mbedded
  Systems [4]. The project is funded by Ericsson Eurolab Deutschland GmbH

J.L. Pfaltz, M. Nagl, and B. Böhlen (Eds.): AGTIVE 2003, LNCS 3062, pp. 440–445, 2004.
© Springer-Verlag Berlin Heidelberg 2004

suitable for various reasons — lack of language support, behavioral analysis, or domain appropriate visualization of results to name just some.

In E-CARES, a prototypical reengineering tool is being developed which addresses the needs of the telecommunication experts at Ericsson. Currently, it assumes that the systems under study are written in PLEX, a proprietary programming language that is extensively used at Ericsson. However, we intend to support other programming languages — e.g., C — as well, so that the prototype may handle multi-language systems. So far, tool support covers only reverse engineering, i.e., the first phase of reengineering. While structural analysis is covered as well, we put strong emphasis on behavioral analysis since the structure alone is not very expressive in the case of a telecommunication system [6]. These systems are process-centered rather than data-centered like legacy business applications. Therefore, tool support focuses particularly on understanding the behavior by visualizing and animating traces, constructing state diagrams, etc.

2   Conceptual Architecture

Internally, telecommunication systems are represented by various kinds of graphs. The structure of these graphs and the effects of graph operations are formally defined in PROGRES [8, 9], a specification language which is based on programmed graph transformations. From the specification, code is generated which constitutes the core part of the application logic. In addition, the E-CARES prototype includes various parsers and scripts to process textual information, e.g., source code. At the user interface, E-CARES offers different kinds of textual, graphical, and tree views which are implemented with the help of UPGRADE [1, 3], a framework for building graph-based applications. The basic architecture of the E-CARES prototype is outlined in Figure 1. The solid parts indicate the current state of realization, the dashed parts refer to further extensions.

Below, it is crucial to distinguish between the following kinds of analysis: Structural analysis refers to the static system structure, while behavioral analysis is concerned with its dynamic behavior. Thus, the attributes "structural" and "behavioral" denote the outputs of analysis. In contrast, static analysis denotes any analysis which can be performed on the source code, while dynamic analysis requires information from program execution. Thus, "static" and "dynamic" refer to the inputs of analysis. In particular, behavior can be analyzed both statically and dynamically.

We obtained three sources of information for the static analysis of the structure of a PLEX[1] system. The first one is the source code of the system. It is considered to be the core information as well as the most reliable one. Through code analysis (parsing) a number of structure documents is generated from the

---

[1] PLEX (Programming Language for EXchanges) is a proprietary programming language developed at Ericsson. The language is extensively used to implement software for telecommunication systems, e.g., switching systems like the AXE10.

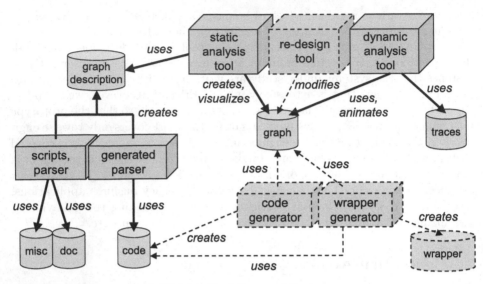

Fig. 1. Prototype architecture

source code, one for each block[2]. These structure documents form a kind of textual graph description. The second and the third source of information are miscellaneous documents (e.g., product hierarchy description) and the system documentation. As far as the information from these sources is computer processable, we use parsers and scripts to extract additional information, which is stored in structure documents, as well.

The static analysis tool processes the graph descriptions for individual blocks and creates corresponding subgraphs of the structure graph representing the overall application. The subgraphs are connected by performing global analyses in order to bind signal send statements to signal entry points. Moreover, the subgraphs for each block are reduced by performing simplifying graph transformations [4]. The static analysis tool also creates views of the system at different levels of abstraction. In addition to structure, static analysis is concerned with behavior (e.g., extraction of state machines or of potential link chains from the source code).

3   E-CARES Prototype

In Figure 2, some sample screen-shots of different static views are presented. In the lower left, the main window of the reengineering tool is shown. It consists of a project browser tree view on the left and a graph view showing subsystems and inter-subsystem communication. The larger window in the background shows

---

[2] Software systems implemented in PLEX are composed of a number of blocks, where blocks can be compared to program files or class files in other programming languages.

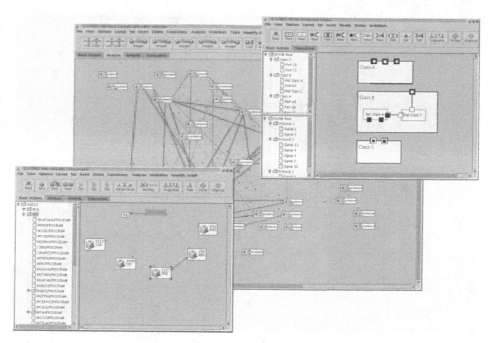

Fig. 2.   Screen-shots of some static views

a block view, which focuses on the block communication within one subsystem. On the right-hand side, there is the ROOM architecture window. It consists of two tree views – class browser and protocol browser, respectively – and a graph view showing class models.

There are two possibilities to obtain dynamic information (see Figure 1): using an emulator or querying a running AXE10. In both cases, the result is a list of events plus additional information in a temporal order. Such a list constitutes a trace which is fed into the dynamic analysis tool. Interleaved with trace simulation, dynamic analysis creates a graph of interconnected block instances that is connected to the static structure graph. This helps telecommunication experts to identify components of a system that take part in a certain traffic case. At the user interface, traces are visualized by collaboration and sequence diagrams.

Figure 3 shows different screen-shots of the dynamic analysis tool. The right-hand side shows two different states of the trace analysis tool. Both consist of a collaboration diagram view on top of a textual logging view. Additionally, the lower one has got a job queue viewer attached to the bottom. On the left-hand side, a sequence diagram is shown, which represents the same state of a trace as the collaboration diagram views, but in a different shape.

Both the static and the dynamic analysis tool calculate metrics which were designed to improve the understanding of both structure and behavior. These metrics are visualized, e.g., in the underlying structure graph.

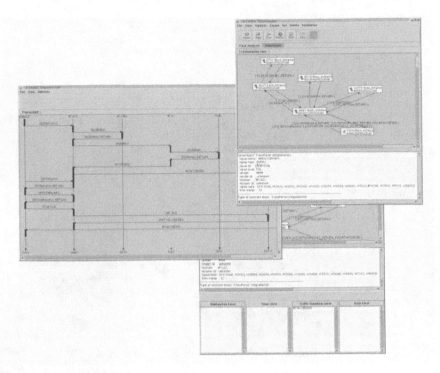

Fig. 3. Screen-shots of the dynamic analysis tool

The dashed parts of Figure 1 represent planned extensions of the current prototype. The re-design tool, which is under construction, is used to map structure graph elements to elements of the ROOM modeling language [10]. The open modular architecture of our prototype allows to easily add mappings to SDL [2] or UML [7] later. However, each mapping results in an architecture graph that can be used to perform architectural changes to the AXE10 system. The code generator will generate PLEX code according to changes in the structure graph and/or the architecture graph. The wrapper generator will enable reuse of existing parts of the AXE10 system written in PLEX in a future switching system that is written in a different programming language, e.g., C/C++.

To reduce the effort of implementing the E-CARES prototype, we make extensive use of generators and reusable frameworks [5]. Scanners and parsers are generated with the help of JLex and jay, respectively. Graph algorithms are written in PROGRES [8, 9], a specification language based on programmed graph transformations. As already mentioned, code is generated from the specification which constitutes the application logic of the E-CARES prototype. The user interface is implemented with the help of UPGRADE [3, 1], a framework for building interactive tools for visual languages.

This prototype demonstration is organized as a walk-through. Each part of the prototype is introduced by means of different kinds of examples. Non-visual parts of the prototype like parsers etc. are not explained in detail. Instead, we emphasize the interactive process of analyzing a complex legacy telecommunication system from a user perspective.

References

[1] Boris Böhlen, Dirk Jäger, Ansgar Schleicher, and Bernhard Westfechtel. UP-GRADE: Building interactive tools for visual languages. In Nagib Callaos, Luis Hernandez-Encinas, and Fahri Yetim, editors, *Proceedings of the 6th World Multiconference on Systemics, Cybernetics, and Informatics (SCI 2002)*, volume I (Information Systems Development I), pages 17–22. International Institute of Informatics and Systemics, July 2002. 441, 444

[2] Jan Ellsberger, Dieter Hogrefe, and Amardeo Sarma. *SDL - Formal Object-oriented Language for Communicating Systems*. Prentice Hall, 1997. 444

[3] Dirk Jäger. Generating tools from graph-based specifications. *Information Software and Technology*, 42(2):129–140, January 2000. 441, 444

[4] André Marburger and Dominikus Herzberg. E-CARES research project: Understanding complex legacy telecommunication systems. In *Proceedings of the 5th European Conference on Software Maintenance and Reengineering*, pages 139–147. IEEE Computer Society Press, 2001. 440, 442

[5] André Marburger and Bernhard Westfechtel. Graph-based reengineering of telecommunication systems. In Andrea Corradini, Hartmut Ehrig, Hans-Jörg Kreowski, and Grzegorz Rozenberg, editors, *Proceedings international conference on graph transformations ICGT 2002*, LNCS 2505, pages 270–285. Springer: Heidelberg, Germany, October 2002. 444

[6] André Marburger and Bernhard Westfechtel. Tools for understanding the behavior of telecommunication systems. In *Proceedings 25th International Conference on Software Engineering ICSE 2003*, pages 430–441. IEEE Computer Society Press:, May 2003. 441

[7] James Rumbaugh, Ivar Jacobson, and Grady Booch. *The Unified Modelling Language Reference Manual*. Addison Wesley: Reading MA, USA, 1998. 444

[8] Andy Schürr, Andreas Winter, and Albert Zündorf. Graph grammar engineering with PROGRES. In Wilhelm Schäfer and Pere Botella, editors, *Proceedings of the European Software Engineering Conference (ESEC '95)*, LNCS 989, pages 219–234. Springer, September 1995. 441, 444

[9] Andy Schürr, Andreas Winter, and Albert Zündorf. The PROGRES approach: Language and environment. In Hartmut Ehrig, Gregor Engels, Hans-Jörg Kreowski, and Grzegorz Rozenberg, editors, *Handbook on Graph Grammars and Computing by Graph Transformation: Applications, Languages, and Tools*, volume 2, pages 487–550. World Scientific, 1999. 441, 444

[10] Bran Selic, Garth Gullekson, and Paul T. Ward. *Real-Time Object-Oriented Modeling*. John Wiley & Sons, Inc., 1994. 444

AGG: A Graph Transformation Environment for Modeling and Validation of Software*

Gabriele Taentzer

Technische Universität Berlin, Germany
gabi@cs.tu-berlin.de

Abstract. AGG is a general development environment for algebraic graph transformation systems which follows the interpretative approach. Its special power comes from a very flexible attribution concept. AGG graphs are allowed to be attributed by any kind of Java objects. Graph transformations can be equipped with arbitrary computations on these Java objects described by a Java expression. The AGG environment consists of a graphical user interface comprising several visual editors, an interpreter, and a set of validation tools. The interpreter allows the stepwise transformation of graphs as well as rule applications as long as possible. AGG supports several kinds of validations which comprise graph parsing, consistency checking of graphs and conflict detection in concurrent transformations by critical pair analysis of graph rules. Applications of AGG include graph and rule-based modeling of software, validation of system properties by assigning a graph transformation based semantics to some system model, graph transformation based evolution of software, and the definition of visual languages based on graph grammars.

1 Introduction

Graphs play an important role in many areas of computer science and they are especially helpful in analysis and design of software applications. Prominent representatives for graphical notations are entity relationship diagrams, control flows, message sequence charts, Petri nets, automata, state charts and any kind of diagram used in object oriented modeling languages as UML. Graphs are also used for software visualizations, to represent the abstract syntax of visual notations, to reason about routing in computer networks, etc. Altogether graphs represent such a general structure that they occur nearly anywhere in computer science.

Graph transformation defines the rule-based manipulation of graphs. Since graphs can be used for the description of very different aspects of software, also graph transformation can fulfill very different tasks. E.g. graphs can conveniently be used to describe complex data and object structures. In this case, graph transformation defines the dynamic evolution of these structures.

* Research partly supported by the German Research Council (DFG), and the EU Research Training Network SegraVis.

J.L. Pfaltz, M. Nagl, and B. Böhlen (Eds.): AGTIVE 2003, LNCS 3062, pp. 446–453, 2004.
© Springer-Verlag Berlin Heidelberg 2004

Graphs have also the possibility to carry attributes. Graph transformation is then equipped with further computations on attributes. Since graph transformation can be applied on very different levels of abstraction, it can be non-attributed, attributed by simple computations or by complex processes, depending on the abstraction level. AGG graphs may be attributed by Java objects which can be instances of Java classes from libraries like JDK as well as user-defined classes.

The graphical user interface provides a visual layout of AGG graphs similar to UML object diagrams. Several editors are provided to support the visual editing of graphs, rules and graph grammars. Additionally, there is a visual interpreter, the graph transformation machine, running in several modes. If another than the standard layout is preferred, it is possible to just use the underlying graph transformation machine and to implement a new layout component or, moreover, a new graphical interface for the intended application.

AGG has a formal foundation based on the algebraic approach to graph transformation [1, 2]. Since the theoretical concepts are implemented as directly as possible – not leaving out necessary efficiency considerations – AGG offers clear concepts and a sound behavior concerning the graph transformation part. Clearly, Java semantics is not covered by this formal foundation.

Due to its formal foundation, AGG offers validation support like graph parsing, consistency checking of graphs and graph transformation systems as well as conflict detection of graph transformation rules. Graph parsing can be advantageously used to e.g. analyze the syntax of visual notations. Modeling a dynamic system structure as graph there is generally the desire to formulate invariants on the class of evolving graphs. AGG offers the possibility to formulate consistency conditions which can be tested on single graphs, but which can also be shown for a whole graph transformation system. If a consistency condition holds for a graph transformation system, all derived graphs satisfy this condition. The conflict detection of graph rules is useful to check dependencies between different actions. It is based on a critical pair analysis for graph rules, a technique which has been developed for term rewriting used in functional programming. Since graph rules can be applied in any order, conflict detection helps to understand the possible interactions of different rule applications.

2   Graph Transformation Concepts

Graph transformation based applications are described by AGG graph grammars. They consist of a type graph defining the class of graphs used in the following, a start graph initializing the system, and a set of rules describing the actions which can be performed. The start graph as well as the rule graphs may be attributed by Java objects and expressions. The objects can be instances of Java classes from libraries like JDK as well as user-defined classes. These classes belong to the application as well. Moreover, rules may be equipped by negative application conditions and attribute conditions. The type graph defines all possible node and edge types, their possible interconnections and all attribute types.

Please note that the type graph can contain additional application conditions for rules, since it offers the possibility to set multiplicity constraints for arc types. Type graphs have first be introduced in [6].

The way how graph rules are applied realizes directly the algebraic approach to graph transformation as presented in [1, 2]. The formal basis for graph grammars with negative application conditions (NACs) was introduced in [8].

Besides manipulating the nodes and arcs of a graph, a graph rule may also perform computations on the objects' attributes. During rule application, expressions are evaluated with respect to the variable instantiation induced by the actual match. The attribution of nodes and arcs by Java objects and expressions follows the ideas of attributed graph grammars as stated in [11] and further in [14] to a large extent. The main difference here is the usage of Java classes and expressions instead of algebraic specifications and terms. The combination of attributed graph transformation with negative application conditions has been worked out comprehensibly in [14]. The AGG features follow these concepts very closely.

Graph transformation can be performed in two different modes using AGG. The first mode to apply a rule is called Debug mode. Here, one selected rule will be applied exactly once to the current host graph. The matching morphism may be (partially) defined by the user. Defining the match completely "by hand" may be tedious work. Therefore, AGG supports the automatic completion of partial matches. If there are several choices for completion, one of them is chosen arbitrarily. All possible completions can be computed and shown one after the other in the graph editor. After having defined the match, the rule will be applied to the host graph once. The result is shown in the graph editor that is, the host graph is now transformed according to the rule and the match. Thereafter, the host graph can immediately be edited.

The second mode to realize graph transformation is called Interpretation mode. This is a more sophisticated mode, applying not only one rule at a time but a whole sequence of rules. The rule to be applied and its match are non-deterministically chosen. Starting the interpretation, all rules are applied as often as possible, until no more match for any rule can be found. Please note that in general the result graph is not unique, since the application of one rule may avoid the application of another rule.

The basic concepts of AGG are presented comprehensively in [7]. In the following, the presentation of AGG concentrates on its new features which cover mainly validation possibilities.

3   Validation Support

Besides editing and interpretation facilities, AGG also offers support for model validation. Since the main AGG concepts rely on a formal approach to graph transformation, i.e. the algebraic approach, validation techniques developed formally for this approach, can be directly implemented in AGG. In the following, three main techniques are presented.

3.1   Graph Parsing

The AGG graph parser is able to check if a given graph belongs to a certain graph language determined by a graph grammar. In formal language theory, this problem is known as the membership problem. Here, the membership problem is lifted to graphs. Three different parsing algorithms are offered by AGG, all based on back tracking, i.e. the parser is building up a derivation tree of possible reductions of the host graph. Leaf graphs are graphs where no rule can be applied anymore. If a leaf graph is isomorphic to the stop graph, the parsing process finishes successfully. Since simple back tracking has exponential time complexity, the simple back tracking parser is accompanied by two further parser variants exploiting critical pair analysis for rules.

Critical pair analysis can be used to make parsing of graphs more efficient: decisions between conflicting rule applications are delayed as far as possible. This means to apply non-conflicting rules first and to reduce the graph as much as possible. Afterwards, the conflicting rules are applied, first in uncritical situations and when this is not possible anymore, in critical ones. In general, this optimization reduces the derivation tree constructed, but does not change the worst case complexity.

A parsing process might not terminate, therefore so-called layering conditions are introduced. Using layers for rules and graph types such that each rule deletes at least one graph object, which is of the same or a lower layer, creates graph objects of a greater layer only, and has negative application conditions with graph objects of the current or lower layers only, it has been shown in [5] that a parsing process based on such a layered graph transformation always terminates.

3.2   Critical Pair Analysis

Critical pair analysis is known from term rewriting and usually used to check if a rewriting system is confluent. Critical pair analysis has been generalized to graph rewriting. Critical pairs formalize the idea of a minimal example of a conflicting situation. From the set of all critical pairs we can extract the objects and links which cause conflicts or dependencies.

A critical pair is a pair of transformations both starting at a common graph G such that both transformations are in conflict, and graph G is minimal according to the rules applied. The set of critical pairs represents precisely all potential conflicts, i.e. there exists a critical pair like above if, and only if, one rule may disable the other one. There are three reasons why rule applications can be conflicting: The first two are related to the graph structure while the last one concerns the graph attributes.

1. One rule application deletes a graph object which is in the match of another rule application.
2. One rule application generates graph objects in a way that a graph structure would occur which is prohibited by a negative application condition of another rule application.

3. One rule application changes attributes being in the match of another rule application.

Please note that using a type graph with multiplicity constraints usually leads to considerably less critical pairs.

3.3   Consistency Checking

Consistency conditions describe basic properties of graphs as e.g. the existence of certain elements, independent of a particular rule. To prove that a consistency condition is satisfied by a certain graph grammar, i.e. is an invariant condition, a transformation of consistency conditions into post application conditions can be performed for each rule [10]. A so-constructed rule is applicable to a consistent graph if and only if the derived graph is consistent, too. A graph grammar is consistent if the start graph satisfies the consistency conditions and the rules preserve this property.

In AGG, consistency conditions are defined on the basis of graphical consistency constraints which consist each of a premise graph and a conclusion graph such that the premise graph can be embedded into the conclusion graph. A graphical consistency constraint is satisfied by a graph G if for each graph pattern equivalent to the premise also an extension equivalent to the conclusion can be found. Based on graphical consistency conditions a propositional logics of formulae has been defined, to be used to formulate more complex consistency conditions.

4   The Tool Environment

Figure 1 shows the main graphical user interface of the AGG system. To the left, a tree view with all graph grammars loaded is shown. The current graph grammar is highlighted. A selected graph, rule or condition is shown in its corresponding graphical editor on the right. The upper editor is for rules and conditions showing the left and the right-hand sides or the premise and conclusion, respectively. The lower editor is for graphs. The attribution of graph objects is done in a special attribute editor that pops up when a graph object is selected for attribution. In Fig. 1 a grammar for a sample requirement specification on shopping is shown. This grammar is presented in detail in [9] where conflict detection of such a requirement specification is analyzed by critical pair analysis. Here, we extend the specification by two consistency conditions to show how they can look like. The graph editor on the lower right shows the type graph of this sample application. It contains typical object types around shopping. On top of the type graph, an atomic consistency condition "UniqueGoodPropriety" is depicted which is part of consistency condition "Proprieties" being a logical formula shown in a separate formula editor on the lower left. Condition "Proprieties" is defined as conjunction of atomic conditions "UniqueGoodPropriety" and "RackPropriety". Condition "UniqueGoodPropriety" has two conclusions,

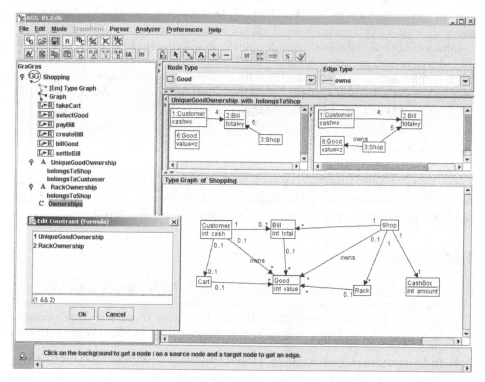

Fig. 1. Screen dump of AGG

i.e. this condition contains two sub-conditions (with the same premise) disjunctively connected. The one presented expresses that each good belongs to a shop. Otherwise, it has to belong to a customer.

Graphs and graph grammars can be stored as XML documents [15], especially AGG supports exchange in GXL [16], the quasi standard format for graphs.

5   Conclusion

This paper gives a rough overview on the graph transformation environment AGG. It consists of visual editors for graphs, rules and graph grammars as well as a visual interpreter for algebraic graph transformation. Moreover, standard validation techniques for graph grammars are supported. Applications of AGG may be of a large variety because of its very flexible attribution concept relying on Java objects and expressions. E.g. the application of AGG for visual language parsing [5] implemented in GenGEd [3], conflict detection in functional requirement specifications [9], consistency checking of OCL constraints in UML models [4] have been considered. AGG is not the only tool environment which is based on graph transformation. In this context we also have to mention PROGRES [13] and a variety of tools which apply graph transformation in a certain context, e.g. Fujaba [17], DiaGen [18], GenGED [3] and many more. But AGG is the only one which consequently implements the theoretical results available for algebraic graph transformation to support their validation. The development group of AGG at the Technical University of Berlin will continue implementing concepts and results concerning validation and structuring of graph transformation systems, already worked out formally. AGG is available at: http://tfs.cs.tu-berlin.de/agg.

Acknowledgement

Large parts of AGG have been developed by Olga Runge. She is carefully maintaining AGG, caring about all technical and documentation issues and puts a lot of efforts on integrating students' work smoothly into the whole project.

References

[1] A. Corradini, U. Montanari, F. Rossi, H. Ehrig, R. Heckel, and M. Löwe. Algebraic approaches to graph transformation part I: Basic concepts and double pushout approach. In G. Rozenberg, editor, *Handbook of Graph Grammars and Computing by Graph transformation, Volume 1: Foundations*, pages 163–246. World Scientific, 1997. 447, 448

[2] H. Ehrig, R. Heckel, M. Korff, M. Löwe, L. Ribeiro, A. Wagner, and A. Corradini. Algebraic approaches to graph transformation II: Single pushout approach and comparison with double pushout approach. In G. Rozenberg, editor, *The Handbook of Graph Grammars and Computing by Graph Transformations, Volume 1: Foundations*, pages 247–312. World Scientific, 1996. 447, 448

[3] R. Bardohl. A Visual Environment for Visual Languages. *Science of Computer Programming (SCP)*, 44(2):181–203, 2002. 451, 452

[4] P. Bottoni, M. Koch, F. Parisi-Presicce, and G. Taentzer. Consistency Checking and Visualization of OCL Constraints. In A. Evans and S. Kent, editors, *UML 2000 - The Unified Modeling Language*, volume 1939 of *LNCS*. Springer, 2000. 451

[5] P. Bottoni, A. Schürr, and G. Taentzer. Efficient Parsing of Visual Languages based on Critical Pair Analysis and Contextual Layered Graph Transformation. In *Proc. IEEE Symposium on Visual Languages*, September 2000. Long version available as technical report SI-2000-06, University of Rome. 449, 451

[6] A. Corradini, U. Montanari, and F. Rossi. Graph Processes. *Special Issue of Fundamenta Informaticae*, 26(3,4):241–266, 1996. 448

[7] C. Ermel, M. Rudolf, and G. Taentzer. The AGG-Approach: Language and Tool Environment. In H. Ehrig, G. Engels, H.-J. Kreowski, and G. Rozenberg, editors, *Handbook of Graph Grammars and Computing by Graph Transformation, volume 2: Applications, Languages and Tools*, pages 551–603. World Scientific, 1999. 448

[8] A. Habel, R. Heckel, and G. Taentzer. Graph Grammars with Negative Application Conditions. *Special issue of Fundamenta Informaticae*, 26(3,4), 1996. 448

[9] J.H. Hausmann, R. Heckel, and G. Taentzer. Detection of Conflicting Functional Requirements in a Use Case-Driven Approach. In *Proc. of Int. Conference on Software Engineering 2002*, Orlando, USA, 2002. To appear. 450, 451

[10] R. Heckel and A. Wagner. Ensuring Consistency of Conditional Graph Grammars – A constructive Approach. *Proc. of SEGRAGRA'95 "Graph Rewriting and Computation", Electronic Notes of TCS*, 2, 1995.
http://www.elsevier.nl/locate/entcs/volume2.html. 450

[11] M. Löwe, M. Korff, and A. Wagner. An Algebraic Framework for the Transformation of Attributed Graphs. In M.R. Sleep, M.J. Plasmeijer, and M.C. van Eekelen, editors, *Term Graph Rewriting: Theory and Practice*, chapter 14, pages 185–199. John Wiley & Sons Ltd, 1993. 448

[12] G. Rozenberg, editor. *Handbook of Graph Grammars and Computing by Graph Transformations, Volume 1: Foundations*. World Scientific, 1997.

[13] A. Schürr, A. Winter, and A. Zündorf. The PROGRES-approach: Language and environment. In H. Ehrig, G. Engels, J.-J. Kreowski, and G. Rozenberg, editors, *Handbook of Graph Grammars and Computing by Graph Transformation, Volume 2: Applications, Languages and Tools*. World Scientific, 1999. 451

[14] G. Taentzer, I. Fischer, M Koch, and V. Volle. Visual Design of Distributed Systems by Graph Transformation. In H. Ehrig, H.-J. Kreowski, U. Montanari, and G. Rozenberg, editors, *Handbook of Graph Grammars and Computing by Graph Transformation, Volume 3: Concurrency, Parallelism, and Distribution*, pages 269–340. World Scientific, 1999. 448

[15] *The Extensible Markup Language (XML)* http://www.w3.org/XML/, 2003. 451

[16] *GXL* http://www.gupro.de/GXL, 2003. 451

[17] *Fujaba Project Group*, 2003. Available at http://www.fujaba.de. 452

[18] M. Minas. Concepts and realization of a diagram editor generator based on hypergraph transformation. *Science of Computer Programming*, 44(3):157 – 180, 2002. 452

Process Evolution Support
in the AHEAD System[*]

Markus Heller[1], Ansgar Schleicher[2], and Bernhard Westfechtel[1]

[1] RWTH Aachen University, Department of Computer Science III
Ahornstraße 55, 52074 Aachen, Germany
{heller,bernhard}@i3.informatik.rwth-aachen.de
[2] DSA Daten- und Systemtechnik GmbH
Pascalstraße 28, 52076 Aachen, Germany
ansgar.schleicher@dsa.de

Abstract. Development processes are inherently difficult to manage. Tools for managing development processes have to cope with continuous process evolution. The management system AHEAD is based on long-term experience gathered in different disciplines (software, mechanical, or chemical engineering). AHEAD provides an integrated set of tools for evolving both process definitions and their instances. This paper describes a demonstration of the AHEAD system which shows the benefits of process evolution support from the user's point of view.

1 Introduction

D evelopm ent processes in different disciplines such as software, mechanical, or chemical engineering share many features. Unfortunately, one of these common features is that they are hard to manage. Development processes are highly creative and therefore can be planned only to a limited extent. In this paper, we present the comprehensive evolution support [7] offered by A H E A D [4], an A daptable and H uman-Centered E nvironment for the MA nagement of D evelopment Processes. The tool demonstration given here complements the conceptual paper in this volume [3]. AHEAD is based on nearly 10 years of work on development processes in different engineering disciplines. So far, we have applied the concepts underlying the AHEAD system in software engineering, mechanical engineering, and chemical engineering. In contrast to the conceptual paper, our tool demonstration will refer to the chemical engineering domain.

2 Process Evolution

For managing evolving development processes, AHEAD offers dynam ic task nets [2]. Within dynamic task nets, tasks describe units of work to be done.

---

[*] The work presented in this paper was carried out in the Collaborative Research Center IMPROVE which is supported by the Deutsche Forschungsgemeinschaft.

J.L. Pfaltz, M. Nagl, and B. Böhlen (Eds.): AGTIVE 2003, LNCS 3062, pp. 454–460, 2004.

Tasks are organized hierarchically, i.e., a task may be decomposed into a set of subtasks. Control flow relationships define the order of subtask enactment. Data flow relationships connect input and output parameters of tasks and allow to exchange documents between them. Feedback relationships model cycles within the process which may stem from planned iterations or occurring exceptional situations. Software processes evolve continuously and it is usually impossible to completely plan a process before its enaction. Therefore, dynamic task nets are designed to be editable and enactable in an interleaved fashion.

With respect to dynamic task nets, we have to distinguish between process definitions and process instances. A process instance represents a specific development process being executed. In contrast, a process definition describes a whole class of processes. It describes the structures of processes of this class, and the constraints that have to be met. Process instances refer to definitions; e.g., the task `ImplementFoo` might be an instance of the task type `Implement`.

Process evolution is supported by the AHEAD system in the following ways:

- Instance-level evolution. Planning and enactment of dynamic task nets may be interleaved seamlessly.
- Definition-level evolution. Process knowledge at the definition level may evolve as well. Evolution is supported by version control at the granularity of packages (modular units of process definitions).
- Bottom-up evolution. By executing process instances, experience is acquired which gives rise to new process definitions. An inference algorithm supports the semi-automatic creation of a process definition from a set of process instances.
- Top-down evolution. A revised process definition may be applied even to running process instances by propagating the changes from the definition to the instance level.
- Selective consistency control. By default, process instances are required to be consistent with their process definition. However, the process manager may allow for deviations resulting in inconsistencies. These deviations are reported to the process manager who may decide to reinforce consistency later on (e.g., when an improved process definition is available).

3   AHEAD System

Figure 1 displays the architecture of AHEAD. The tools provided for different kinds of users are shown on the right-hand side. "Process modeler", "process manager", and "developer" denote roles rather than persons: A single person may play multiple logical roles, and a single role may be played by multiple persons. The left-hand side, which will not be discussed further here (see [3]), shows internal components of the AHEAD system which are not visible at the user interface. Furthermore, the horizontal line separates definition and instance level.

The process modeler uses a commercial CASE tool — Rational Rose — to create and modify process definitions in the UML [1]. Rational Rose is adapted

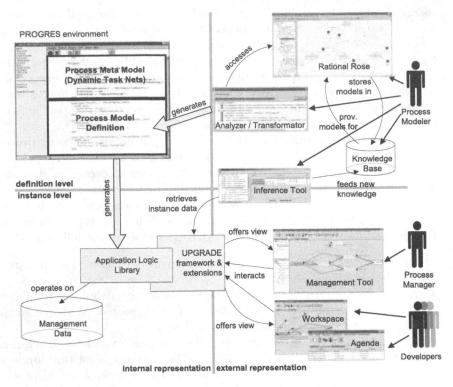

Fig. 1. Architecture of the AHEAD system

with the help of stereotypes which link the UML diagrams to the process meta model (dynamic task nets). The process definition introduces domain-specific types and constraints for operating on the instance level. Definition-level evolution is supported through version control for model packages.

On the instance level, the management tool assists a process manager in planning, analyzing, monitoring, and controlling a development process. The management tool support instance-level evolution (through dynamically evolving task nets), consistency control (with respect to the process definition generated from the UML model), and top-down evolution (migration of a process instance to a new version of the process definition). The management tool is coupled with tools provided to developers (or designers) which are used to display agendas of assigned tasks and to operate on these tasks in workspaces from which domain-specific tools may be activated in order to work on design documents.

Finally, the inference tool closes the loop by assisting in the inference of process definitions from process instances. The inference tool analyzes process instances and proposes definitions of task and relationship types. These definitions are stored in a knowledge base which may be loaded into Rational Rose. In this way, bottom-up evolution is supported (for a more detailed description, see [7].

4 Demonstration

AHEAD is being developed in the context of IMPROVE [5], a long-term research project which is concerned with models and tools for design processes in chem - ical engineering. Within the IMPROVE project, a reference scenario is being studied referring to the early phases (conceptual design and basic engineering) of designing a plant for producing Polyamide6 [6]. To a large extent, the requirements for process evolution support were derived from this scenario, even though we also studied processes in other domains (e.g., software engineering). In fact, the reference scenario constitutes a fairly challenging benchmark against which process evolution capabilities of management systems can be evaluated. Process knowledge is incomplete, instable, and rather fuzzy, feedback occurs frequently in the design process, multiple design variants have to be considered, etc.

Based on our work on the reference scenario, we have prepared a demo session which is sketched briefly below. First, the process modeler creates a process definition for design processes in chemical engineering. The process definition describes design processes at the type level. Figure 2 shows a class diagram which defines a part of the overall design process. The described part consists of one task for designing a set of flowsheet alternatives, a set of simulation tasks (one for each alternative), and one evaluation task for selecting the best alternative.

The design process is planned and executed according to this process definition. On the top level, the design process is essentially decomposed according to the structure of the chemical process, which consists of three steps (reaction, separation, and compounding). We will focus on the design of the separation, which requires input from the reaction design. The process manager insert tasks for elaborating flowsheet alternatives and for performing a final evaluation of

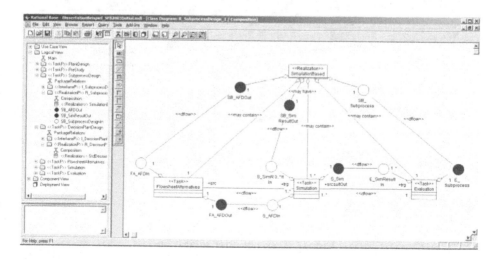

Fig. 2. Adapted UML class diagram for defining a chemical engineering design process

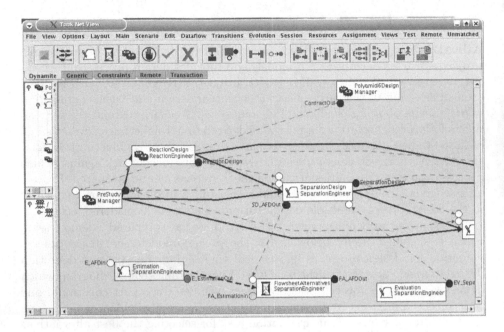

Fig. 3. Highlighting of inconsistencies in the management tool

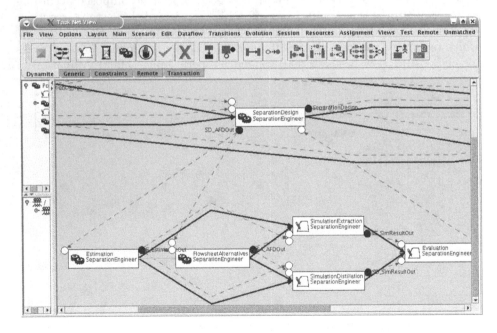

Fig. 4. Task net after migration

the alternatives. So far, the task net is still incomplete because the simulation tasks have not been instantiated yet.

Now, the process manager detects the need for performing a task which has not been anticipated in the process definition. To insert this task (a task for estimating the inputs to a certain component of the chemical plant in order to accelerate the overall design process by concurrent engineering), consistency enforcement is switched off. The task net is extended, resulting in inconsistencies which are displayed by colored markings (Figure 3).

The process definition is extended accordingly by introducing a new package version (not shown in a figure). In this way, traceability on the definition level is maintained. The corresponding class diagram differs from the previous version in the definition of a task class for estimation tasks.

The modified process definition is propagated to the process instance. In this way, running process instances may be migrated to improved definitions. In general, migration has to be performed in a semi-automatic way (i.e., some steps may be automated, but some have to be performed interactively by the process manager). Figure 4 shows a state of the task net which is nearly consistent with the improved definition. The control flow between the estimation task and the design task is still inconsistent: Although it has been defined as a sequential control flow, source and target are active simultaneously. This inconsistency will be removed when the estimation task is terminated. After that, consistency enforcement may be switched on again. This final step closes the process evolution roundtrip.

5   Conclusion

We have presented a management system supporting comprehensive evolution support for dynamic development processes. AHEAD provides round-trip process evolution, combining instance-level evolution, definition-level evolution, bottom-up and top-down evolution as well as toleration of inconsistencies into a coherent process evolution framework. We have applied our system to an industrially relevant reference scenario. In the future, we hope to transfer the implementation into industrial practice, resulting in feedback from practical use.

References

[1] Grady Booch, James Rumbaugh, and Ivar Jacobson. *The Unified Modeling Language User Guide*. Addison Wesley, Reading, Massachusetts, 1998. 455

[2] Peter Heimann, Carl-Arndt Krapp, Bernhard Westfechtel, and Gregor Joeris. Graph-based software process management. *Int. Journal of Software Engineering and Knowledge Engineering*, 7(4):431–455, December 1997. 454

[3] Markus Heller, Ansgar Schleicher, and Bernhard Westfechtel. Graph-based specification of a management system for evolving development processes. In Manfred Nagl and John Pfaltz, editors, *Proc. AGTIVE 2003*, LNCS. Springer, 2004. In this volume. 454, 455

[4] Dirk Jäger, Ansgar Schleicher, and Bernhard Westfechtel. AHEAD: A graph-based system for modeling and managing development processes. In Manfred Nagl, Andy Schürr, and Manfred Münch, editors, *Proc. AGTIVE '99*, LNCS 1779, pages 325–339. Springer, September 1999. 454

[5] Manfred Nagl and Wolfgang Marquardt. SFB-476 IMPROVE: Informatische Unterstützung übergreifender Entwicklungsprozesse in der Verfahrenstechnik. In Matthias Jarke, Klaus Pasedach, and Klaus Pohl, editors, *Proc. Informatik '97*, Informatik aktuell, pages 143–154, Aachen, September 1997. Springer-Verlag. 457

[6] Manfred Nagl, Bernhard Westfechtel, and Ralph Schneider. Tool support for the management of design processes in chemical engineering. *Computers & Chemical Engineering*, 27(2):175–197, February 2003. 457

[7] Ansgar Schleicher. *Management of Development Processes — An Evolutionary Approach*. Deutscher Universitäts-Verlag, Wiesbaden, Germany, 2002. 454, 456

Fire3: Architecture Refinement
for A-posteriori Integration*

Thomas Haase, Oliver Meyer, Boris Böhlen, and Felix Gatzemeier

RWTH Aachen University, Department of Computer Science III
Ahornstraße 55, 52074 Aachen, Germany
{thaase,omeyer,boris,fxg}@i3.informatik.rwth-aachen.de

Abstract. The Friendly Integration Refinement Environment (Fire3) is
an architecture design tool that supports a-posteriori application integration. It covers multiple refinement stages of the architectural design.
An initial coarse-grained view on an integration scenario is refined to
a logical architecture that mentions components and strategies to integrate them. That, in turn, is refined to a concrete architecture that
distributes these components on various processes. The internal application logic of the Fire3 prototype is based on a formal specification by a
programmed graph rewriting system. Its user interface is built with the
help of a framework for generating graph-based interactive tools.

1 Introduction

Creating an architecture design tool is no new idea. Especially, our department
has studied architecture design languages and accompanying tools for multiple
decades now [9, 10, 8]. Most of these tools (our own and those from others) claim
to be usable for every and any context of software development. The operations
offered to the user are thus common to all domains and do not give specific
support.

With the Friendly Integration Refinement Environment (Fire3) we developed
architecture design tool which is specific to the domain of a-posteriori application integration. It covers multiple refinement stages of design. The initial very
coarse-grained view on the integration scenario is refined to a logical architecture
that mentions components, e. g. wrappers to homogenize various data models,
and strategies, e. g. middleware-techniques such as COM or CORBA, to be used
for integration. The logical architecture, in turn, is refined to a concrete architecture that distributes these components on various processes regarding specific
distribution mechanisms.

For each of the refinement stages, there exist specific operations that support
the user in defining the results on that stage or refining them to get to the next
stage. These operations are formally specified as graph transformations using
the PROGRES language and programming environment [13, 14]. Additional

---

* Financial support is given by Deutsche Forschungsgemeinschaft, Collaborative Research Centre (CRC) 476 IMPROVE.

J.L. Pfaltz, M. Nagl, and B. Böhlen (Eds.): AGTIVE 2003, LNCS 3062, pp. 461–467, 2004.
© Springer-Verlag Berlin Heidelberg 2004

help texts explain the usage of these operations. Different views, implemented with UPGRADE [3, 6], a framework for building graph-based interactive tools, allow the software engineer to focus on the context specific to the refinement stage he is working with.

The demo shows how the refinement takes place, which decisions are made in each stage, and how the user is supported. It complements the paper [5], which explains how the demo prototype, depicted here, has been developed using graph rewriting systems. This paper focuses on the external side and describes the result of the development described in [5] as an example for the application of graph transformations.

2   Tool Support for Architecture Refinement

Knowledge about integration architectures has found its way into the tool in multiple ways:

Transformations Changes in the architecture often affect multiple and different parts simultaneously. When, for example, addressing a single component of an integrated application, particular wrappers are needed to access it. Therefore, they get demonstrated in all steps described in section 3.

Stereotypes The tool uses a variant of UML [4] to display its architecture diagrams. As the tool further classifies types of classes, packages, and components UML-stereotypes were introduced. They convey this more complex model to the software engineer immediately. All diagrams contain such stereotypes. They are mainly introduced in the steps described in sections 3.1 and 3.2.

Analyses Using particular types of components from a specific application domain, the use and arrangement of the components is restricted. While specific transformations prohibit erroneous conditions in the first place, analyses point to problem spots and incomplete specifications. The use of analyses is described in section 3.2 and in section 3.3.

Help Texts To inform the software engineer about the options he may choose from, and to give him arguments for his decisions, help texts offer the declarative information he needs. They help to alleviate analyzed problems and allow to directly activate the necessary repair actions. Help texts are described in section 3.2.

Illustration As the resulting architecture can get very complex even for a simple integration scenario[1], typical uses of the architecture can be illustrated. Section 3.4 mentions an animated collaboration diagram illustrating the interaction of the various components.

---

[1] In this demo only two applications get integrated.

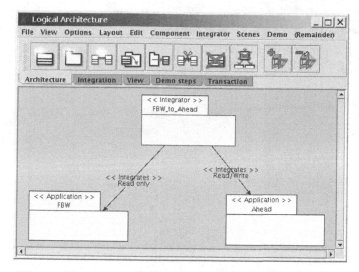

Fig. 1. Initial, coarse-grained architecture of the integrator

3 Demonstration

The section presents an example where the tool os applied. It demonstrates not only the different means of user support described above, but also the usefulness of domain specific architecture refinement implemented by it.

The scenario for our demonstration is taken from chemical engineering. In the early phases of developing a chemical plant the chemical process is designed and its corresponding plant is described on a conceptual level. This is done by an engineer with the help of a process flow diagram editor (PFD editor). Furthermore, the overall development process is organized using an administration system which assists the project manager in planning, executing, and monitoring the development process. In our scenario we use the AHEAD system [7, 12]. Activities to be carried out during the development process, their state and their dependencies are presented to the project manager by a task net.

Parts of this task net can be derived automatically from the process flow diagram which, in our scenario, is done by a third tool, the Integrator [1, 2]. It reads the data of the process flow diagram, determines the affected tasks in the task net, and adjusts it. Therefore, the Integrator has to be integrated with both applications. The integration is implemented in an a-posteriori manner, because adapting the source code of the tools to be integrated to our requirements is either not possible or causes to much effort.

3.1 Initial Integration Scenario

In a first step the initial, coarse-grained integration scenario gets created (see fig. 1). Here the **stereotypes** ≪Integrator≫ and ≪Application≫ help to define

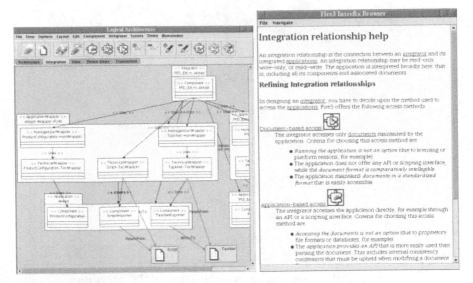

Fig. 2. Logical architecture of the integrator and help texts

what to model in this early stage. The `Integrates` relation is annotated with access requirements, showing the targeted integration direction. Here information from the PFD editor is transferred to the process configuration within AHEAD. The integrator therefore reads from the PFD editor and writes into AHEAD. To integrate with the data already present in AHEAD, read access to AHEAD is also required.

This scenario is built up using tool operations which allow to create typed, labelled packages and to relate them.

3.2   Logical Architecture

To arrive at the logical architecture (see fig. 2) the `Integrates` relation between Integrator and Application needs to be refined. The integrator can either access the application directly via an API (Application Programming Interfaces), access only documents maintained by the application, or access the documents as well as the API. **Help texts** provide the software engineer with the information needed for that decision.

Here the API is used to integrate with both applications a-posteriori. A tool wrapper may be created for each application if that refinement is chosen. The engineer is supported by **specific transformations**: The wrapper package is needed only in the context of application integration.

Then the application access itself is refined to specific application components. The user, therefore, selects the wrapper package and defines the Task Net as an application component he needs to access. As a result of a further **specific transformation**, not only the component but also the required homogenizer and technical wrappers are created.

The Task Net component provides no read access through its API. The logical architecture is, therefore, already influenced by the technical constraints. This is typical for a-posteriori integration scenarios. A common solution is to access the component's data by parsing an exported file in the technical wrapper component. A specific command transforms access to a component: The Task Net component is transformed into a Task Net Exporter that writes into a file that, in turn, gets read from the technical wrapper.

With the Task Net Exporter as the only component in the wrapper, the specified read/write access (see section 3.1) cannot be provided. This is noticed by the analyses running in the background. The affected component is shown in light red. An associated help text recommends to gain write access through another component. In the example a Script component allows to execute arbitrary scripts that also can change the Task Net. Again a specific transformation creates the substructure that reflects component access through a file input.

For the Script component no homogenizer wrapper is needed. With a generic architecture transformation it is removed from the architecture. The additional Product Configuration component and a monolithic PFD component complete the logical architecture.

3.3 Concrete Architecture

The next step for the engineer is to develop the concrete architecture (see fig. 3). An isomorphic, initial version is created automatically. All components are placed in processes, a process package. Process internal and external procedure calls are created accordingly.

Stereotypes in the concrete architecture are processes, wrapper implementation components, and middleware support components (like ≪COM_Interface≫, ≪CorbaStub≫, and ≪CorbaSkel≫). MethodInvocation and InterProcessCall, as well as, Corba_IIOP and COM_Call are specific relations used in this stage of development.

To reduce the overall number of processes, the software developer decides to remove the Ahead-Wrapper process. He, therefore, moves all components into the integrator process. Method calls between the components are transformed from interprocess calls to simple method invocations and vice versa where needed.

Analyses show that the architecture is still incomplete. The software engineer has still to decide for every interprocess call which middleware is used to realize it. Again, a light red warning is used to indicate the incompleteness.

A **specific transformation** converts the interprocess call to a COM or CORBA call. Interfaces for COM or Stub and Skeleton for CORBA are created and connected automatically.

3.4 Specific Distribution Mechanisms and Collaboration

As Fire3 was developed in the context of IMPROVE [11], a long-term research project, it also includes specific knowledge about the integration scenario of this

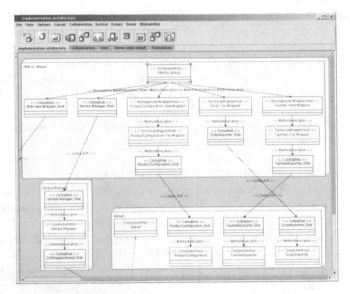

Fig. 3. Concrete architecture of the integrator

project. For example, the IMPROVE specific Service Manager is a component for distributing and controlling the execution of operating system processes. To make a process location independent it must be started from the Service Manager. Other components using that process then query the Service Manager for the current location of the started process.

In the demo, the Service Manager controls the execution of Ahead. Declaring Ahead as location independent automatically creates the Service Manager and a coarse-grained wrapper for Ahead. The integrator gets access to the Service Manager as it is a caller of Ahead (see fig. 3).

To illustrate as a result of all above steps the rather complex network of components and their interprocess calls, a typical usage scenario is shown in the form of an animated collaboration diagram. The prototype offers commands to define a collaboration sequence. Yet in the demo one was already prepared. The **Illustration** shows how the integrator calls the Task Net Exporter with help of the Service Manager.

References

[1] Simon Becker, Thomas Haase, Bernhard Westfechtel, and Jens Wilhelms. Integration Tools Supporting Cooperative Development Processes in Chemical Engineering. In *Proc. of the 6th Biennial World Conf. on Integrated Design and Process Technology (IDPT-2002)*, Pasadena, Ca., 2002. 10 pages. 463

[2] Simon Becker and Bernhard Westfechtel. Incremental integration tools for chemical engineering: An industrial application of triple graph grammars. In Hans L. Bodlaender, editor, *Proc. Workshop WG 2003*, volume 2880 of *LNCS*, pages 46–57. Springer, 2003. 463

[3] Boris Böhlen, Dirk Jäger, Ansgar Schleicher, and Bernhard Westfechtel. UP-GRADE: Building Interactive Tools for Visual Languages. In Nagib Callaos, Luis Hernandez-Encinas, and Fahri Yetim, editors, *Proc. 6th World Multiconf. on Systemics, Cybernetics, and Informatics (SCI 2002)*, volume I (Information Systems Development I), pages 17–22, Orlando, Fa., 2002. 462

[4] Grady Booch, James Rumbaugh, and Ivar Jacobson. *The Unified Modeling Language User Guide*. Addison-Wesley, Reading, Ma., 1999. 462

[5] Thomas Haase, Oliver Meyer, Boris Böhlen, and Felix Gatzemeier. Fire3 – architecture refinement for a-posteriori integration. this volume, 2003. 462

[6] Dirk Jäger. Generating tools from graph-based specifications. *Information and Software Technology*, 42:129–139, 2000. 462

[7] Dirk Jäger, Ansgar Schleicher, and Bernhard Westfechtel. AHEAD: A graph-based system for modeling and managing development processes. In Manfred Nagl, Andy Schürr, and Manfred Münch, editors, *Proc. Workshop AGTIVE'99*, volume 1779 of *LNCS*, pages 325–339. Springer, 2000. 463

[8] Peter Klein. *Architecture Modeling of Distributed and Concurrent Software Systems*. Wissenschaftsverlag Mainz, Aachen, Germany, 2001. PhD thesis. 461

[9] Claus Lewerentz. *Interaktives Entwerfen großer Programmsysteme: Konzepte und Werkzeuge*, volume 194 of *Informatik-Fachberichte*. Springer, 1988. Doctoral dissertation. 461

[10] Manfred Nagl, editor. *Building Tightly Integrated Software Development Environments: The IPSEN Approach*, volume 1170 of *LNCS*. Springer, 1996. 461

[11] Manfred Nagl and Bernhard Westfechtel, editors. *Integration von Entwicklungssystemen in Ingenieuranwendungen*. Springer, 1999. 465

[12] Ansgar Schleicher. *Management of Development Processes: An Evolutionary Approach*. Deutscher Universitätsverlag, Wiesbaden, Germany, 2002. PhD thesis. 463

[13] Andreas Schürr. *Operationelles Spezifizieren mit programmierten Graphersetzungssystemen*. Deutscher Universitätsverlag, Wiesbaden, Germany, 1991. PhD thesis, in German. 461

[14] Andreas Schürr, Andreas Joachim Winter, and Albert Zündorf. The PROGRES Approach: Language and Environment. In H. Ehrig, G. Engels, H.-J. Kreowski, and G. Rozenberg, editors, *Handbook of Graph Grammars and Computing by Graph Transformation*, volume 2, pages 487–550. World Scientific, Singapore, 1999. 461

A Demo of OptimixJ

Uwe Aßmann and Johan Lövdahl

Research Center for Integrational Software Engineering (RISE)
Programming Environments Lab (PELAB), Linköpings Universitet, Sweden*
{uweas,jolov}@ida.liu.se

Abstract. OptimixJ is a graph rewrite tool that generates Java code from rewrite specifications. Java classes are treated as graph schemas, enabling OptimixJ to extend legacy Java applications through code weaving in a simple way. The demo shows how OptimixJ has been used to implement graph rewriting for RDF/XML documents in the context of the Semantic Web.

1 Introduction to OptimixJ

OptimixJ [2] lets the user specify graph transformations in a declarative, Datalog style language . From these transformation specifications (also called rewrite system s) OptimixJ generates Java code that navigates/modifies Java objects. OptimixJ specifications consists of a data m odel (i.e. the type graph) and one or several graph rewrite m odules.

Rather than having a specialized language for describing the data model OptimixJ uses ordinary Java classes. Each Java class defines a possible node type and the fields in the class define the possible outgoing edges. Hence OptimixJ data model specifications allow directed graphs where the nodes have one or several types and where the edges are binary and labelled. Furthermore the range and domain of edges are typed.

One advantage with using Java as the data model language is that it becomes easy to integrate OptimixJ rewrite specifications with e.g. legacy Java implementations.

A graph rewrite module consists of one or several rewrite system s which specify the actual graph transformations to be carried out. The rewrite specification language has a higly declarative style, it is based on relational graph rewriting and supports both edge addition/deletion as well as node deletion/addition.

OptimixJ generates one Java method for each rewrite system in the rewrite module, the method will have the same name and parameter list as given in the rewrite system. If the name of the module containing the rewrite system

---

* Work partially supported by European Community under the IST programme - Future and Emerging Technologies, contract IST-1999-14191-EASYCOMP. The authors are solely responsible for the content of this paper. It does not represent the opinion of the European Community, and the European Community is not responsible for any use that might be made of data appearing herein.

J.L. Pfaltz, M. Nagl, and B. Böhlen (Eds.): AGTIVE 2003, LNCS 3062, pp. 468–472, 2004.

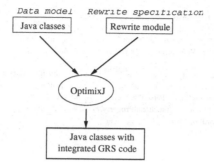

Fig. 1. OptimixJ uses the data definition parts of Java (classes,fields and method signatures) as its language for data models. Rewrite specifications are typechecked against the data model before generating Java code

matches the name of a Java class the generated method will be weaved in to that Java class, if no matching class is found a new class with static methods will be generated. See figure 2 for an example of a rewrite module.

During code generation OptimixJ also type-checks the rewrite rules: predicates in the rules are looked up in the data model and the domain/range types is checked. OptimixJ also analyzes the inheritance structure of the data model so relations applicable to a class are also applicable to its sub-classes.

To learn more about OptimixJ see [1, 2].

The next section describes how the OptimixJ tool can be used in the context of the Semantic Web.

```
public class Node {                    MODULE Node
    public Node[] basicEdge;           EARS transitiveClosure(nodes:Node[]) {
    public Node[] transitiveEdge;          RANGE node <= nodes; RULES
}                                          transitiveEdge(node, n) :- basicEdge(node, n);
                                           transitiveEdge(node, n) :- basicEdge(node, s),
                                                                      transitiveEdge(s, n);

a)                                     }    b)
```

```
public class Node {
    public Node[] basicEdge;
    public Node[] transitiveEdge;
    public void transitiveClosure(Node[] nodes) {
        /* The generated transformation code */
    }
}    c)
```

Fig. 2. A tiny example of an OptimixJ specification. a) shows the data model specification, it defines two relations (Java fields) *basicEdge* and *transitiveEdge* that can be used in transformation specifications. b) shows how the transitive closure can be computed given the relations in the data model. c) shows the result after code generation and weaving into the class

```
<Class ID="Person"/>
<Class ID="Male">                        <Male ID="John">
   <subClassOf resource="Person"/>         <spouse resource="Mary"/>
   <disjointWith resource="Female"/>     </Male>
</Class>
<ObjectProperty ID="spouse">            <Female ID="Mary">
   <domain resource="Person"/>             <spouse resource="John"/>
   <range resource="Person"/>           </Female>
   <type resource="functionalProperty"/>
</ObjectProperty>
a)                                       b)
```

Fig. 3. a) shows a sample OWL ontology defining some classes and a property. b) shows a fragment of an RDF document defining some instances of the ontology

2   OptimixJ and the Semantic Web

The focus of this demonstration is to show how graph rewriting and the OptimixJ tool can be used as an executable representation of Semantic Web ontologies.

First we will give a brief description of the Semantic Web languages.

2.1   A Brief Description of the Semantic Web

The purpose of the Semantic Web [6] is to express knowledge about things, called resources, (a resource can be a webpage, person or just anything that has a Uniform Resource Locator, URI) in a way that enables automated agents to reason about these resources, i.e. infer knowledge from the given facts.

The idea is that one creates ontologies that models a domain of interest. There are a few different ontology languages (RDFS, DAML+OIL, OWL) providing somewhat different language constructs, we will discuss the Web Ontology Language OWL [6] defined by the web consortium.

The basic modelling primitives of OWL are classes and properties, classes can inherit from other classes and have properties. But OWL also supports more expressive constructs that allows the modeller to express things like:

- characteristics of properties (e.g. a property can be symmetric or transitive).
- property inheritance (hasFather is a subproperty of hasParent).
- relations between classes (e.g. disjointness).

See figure 3 for an example ontology.

The classes and properties defined in an ontology can then be used to create instances (see figure 3b), the instances are represented in a format called RDF (Resource Description Framework) [5]. In terms of graphs RDF is an XML format for directed, edge-labelled graphs where the nodes can have one or several types. As we pointed out in the previous section this is exactly the kind of graphs OptimixJ is designed to work on, in the next section we show how this can be used in order to execute inference rules and constraints defined in OWL.

2.2   Compiling OWL Ontologies to Java and OptimixJ

In order to do something interesting with the ontologies and instances we need some kind of inference engine. OWL is based on the Description Logic formalism for which reasoning engines exist, these systems are very fast but it is difficult to extend the system to handle more expressive rules.

Another approach that is used is to translate the ontologies and instances to Frame Logic [3], this allows for the addition of more expressive rules. The problem with this approach is that the Frame Logic engine is quite slow (it is running on top of Prolog).

Another potential problem with these approaches is integration with the application programs that wants the result of the inferences, it can be difficult to communicate between the two parts.

To overcome the problems described above (additional expressivity and interfacing to inference engine) we are working on generating Java code and OptimixJ rules from ontologies defined in a subset of OWL. This will give both a more powerful language for expressing rules/constraints as well as seamless integration with Java applications.

As we can see from the description of OWL in section 2.1 there are some similarities between the OWL modelling language and Java. The OWL class and property constructs maps to Java's classes/interfaces and fields/methods, OWL inheritance can also be translated to Java inheritance/implementation. Hence part of the OWL language is used for describing the graph schema and it can be mapped to Java classes that are used to define the data model for the OptimixJ rewrite specifications (as discussed in section 1).

But there are other OWL modelling primitives that are less straightforward to map to Java. For example if the OWL property spouse has been declared as symmetric we need to generate some Java code to make sure that this rule is satisfied. Interestingly it turns out that the OptimixJ rewrite language to a large extent is sufficient for implementing the inference rules and constraints expressible by OWL. We can also add more expressive constraints to the Java code by writing OptimixJ specifications by hand and merging them with the rest of the code.

The Semantic Web Ontology Development Environment (SWEDE) [4] toolset is able to translate a subset of OWL to Java class definitions and OptimixJ rewrite modules, figure 4 shows the process of translating OWL to Java.

Each OWL class is mapped to one Java class and one interface, the interface is used for encoding the inheritance structure (OWL allows multiple inheritance). Each OWL property defined for a class is mapped to a Java field in the corresponding Java class. Once the OptimixJ has generated code from the rewrite specifications and merged the code with the Java classes we can use them to create Java objects from RDF instance documents.

The RDFReader parses the RDF document and instantiates a new Java object according to the types of the RDF instance, the reader also inserts property values into the Java object. For example, the RDF document in figure 3b would create two new Java objects: one of the type Female with the spouse field point-

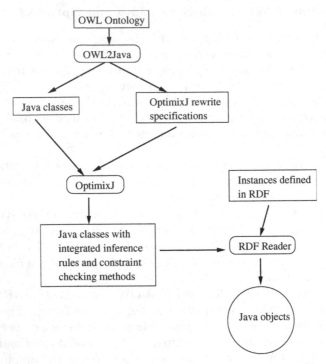

Fig. 4. Translating OWL ontologies to Java. Once the Java code generation is finished we can read RDF instances with a program that creates new Java objects

ing to the second object, the second object would have the type M a̲le and the spouse field would point to the first object.

The SWEDE tools can also translate XML DTD's to OWL ontologies, which enables us to generate Java classes and OptimixJ rewritings for XML trees.

References

[1] Uwe Aßmann. *OPTIMIX, A Tool for Rewriting and Optimizing Programs.* Chapman-Hall, 1999. 469
[2] Uwe Aßmann. OPTIMIX website. http://optimix.sourceforge.net, September 2003. 468, 469
[3] Flora. Flora2 website. http://flora2.sourceforge.net, March 2003. 471
[4] Johan Lövdahl. An editing environment for the semantic web. Master's thesis, Linköpings Universitet, January 2002. 471
[5] W3C. RDF specification. http://www.w3.org/TR/REC-rdf-syntax/, September 2003. 470
[6] W3C. The Semantic Web Activity. http://www.w3.org/2001/sw, September 2003. 470

Visual Specification of Visual Editors with VisualDiaGen

Mark Minas

Institute for Software Technology, Department of Computer Science
University of the Federal Armed Forces, Munich, 85577 Neubiberg, Germany
minas@acm.org

Abstract. VISUALDIAGEN is a tool for specifying a visual language and generating a graphical editor from such a specification. The specification is mainly based on graph transformation and graph grammars. This paper demonstrates VISUALDIAGEN and its features by briefly describing how the tool can be used to generate a simplified version of an application which is used in education courses for electrical engineers.

1 Introduction

Programming graphical editors for visual languages like UML class diagrams, Petri nets, Statecharts etc. is hard work. Therefore, various generators for diagram editors have been proposed and developed that create diagram editors from specifications. DIAGEN [2] is one of such tools which has been described in detail in [4]. Continued work on DIAGEN, in particular on VISUALDIAGEN, is also presented in another paper of this volume [3].

This paper briefly outlines a system demonstration of VISUALDIAGEN which gives a step-by-step description how DiaGen is used to generate a simplified version of a visual programming environment (VPE) for Programmable Logic Controllers (PLCs). The VPE is used as component of a commercial programming environment for PLCs and also in education courses for electrical engineers at the University of Kaiserslautern [6]. The visual programming language (SIPN, Signal Interpreted Petri Net) is based on a special type of Petri nets. This paper focuses on the process of generating a plain Petri net editor which – after visually editing a Petri net – provides the net's structure, i.e., the set of its places and transitions as well as they are interconnected. Fig. 1 shows an example. The VPE extends such editors by hierarchical Petri nets, code generation etc. [5].

2 VisualDiaGen

VISUALDIAGEN comes with a visual specification tool that is described in more detail in another paper of this volume [3]. VISUALDIAGEN allows for specifying all aspects of a visual language, e.g., the set of visual components and how they are drawn on the screen, and the language structure. Fig. 2 shows a screenshot of VISUALDIAGEN while editing a reduction rule for the Petri net editor

J.L. Pfaltz, M. Nagl, and B. Böhlen (Eds.): AGTIVE 2003, LNCS 3062, pp. 473–478, 2004.
© Springer-Verlag Berlin Heidelberg 2004

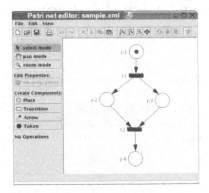

Places: [p1*,p2,p3,p4]
Transitions:
  t1:
    pre: [p1*]
    post: [p2, p3]
  t2:
    pre: [p2, p3]
    post: [p4]

Fig. 1. Screenshot of a Petri net editor and a textual description of the depicted net

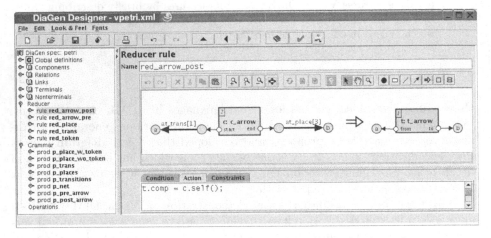

Fig. 2. Screenshot of VISUALDIAGEN with a reduction rule

(see below.) VISUALDIAGEN translates such a specification to Java code that, together with the DIAGEN framework, implements an editor for the visual language. Generated editors can be easily plugged into other systems as the VPE for SIPNs has shown.

The following sections briefly survey the main concepts of VISUALDIAGEN and the Petri editor which has been specified and generated with this tool.

3   Hypergraphs and Grammars

DIAGEN editors are based on hypergraphs as internal diagram models and hypergraph grammars as a means for syntax specification. These concepts are briefly outlined.

Hypergraphs are generalizations of directed graphs: they have a set of labeled hyperedges instead of edges. Each hyperedge has a fixed number of labeled tentacles which is determined by the hyperedge's label. Tentacles connect the hyperedge with its visited nodes. Nodes are represented (cf. Fig. 4) by filled circles, directed edges by arrows, and hyperedges by boxes containing the hyperedge label. Small circles at the border of a box together with tiny arrows starting at those circles are used to represent tentacles connecting the hyperedge with visited nodes. Tentacle labels are depicted by the small circles' names.

Hypergraph grammars are similar to string grammars. Each hypergraph grammar consists of two sets of term inal and nonterm inal hyperedge labels and a starting hypergraph which contains nonterminally labeled hyperedges only. Syntax is described by a set of productions of the form $L ::= R$ with L (left-hand side, LHS) and R (right-hand side, RHS) being hypergraphs. A production $L ::= R$ is applied to a (host) hypergraph H by finding L as a subgraph of H and replacing this match by R obtaining hypergraph H'. We say, H' is derived from H (written $H \rightarrow H'$) in one step. The grammar's language is then defined by the set of terminally labeled hypergraphs which can be derived from the starting hypergraph in a finite number of steps.

There are different types of hypergraph grammars which impose restrictions on a production's LHS and RHS. C ontext-free hypergraph grammars are the simplest ones: each LHS has to consist of a single nonterminally labeled hyperedge together with the appropriate number of nodes. Application of such a production removes the LHS hyperedge and replaces it by the RHS. Matching node labels of LHS and RHS determine how the RHS has to fit in after removing the LHS hyperedge. $P_1 \ldots P_6$ of Fig. 5 are context-free ones. Context-free hypergraph grammars w ith em beddings are more expressive than context-free ones. They additionally allow em bedding productions which consist of the same LHS and RHS, but with an additional ("embedded") hyperedge on the RHS, i.e., this hyperedge is embedded into the context provided by the LHS when applying such a production (P_7 of Fig. 5). Parsing algorithms and a more detailed description of both grammar types can be found in [4].

VISUALDIAGEN uses hypergraphs as diagram representations and hypergraph grammars for specifying syntactically correct diagrams. The following section describes how these concepts are used by the Petri net editor which has been generated with VISUALDIAGEN.

4   Petri Net Editing

The Petri net editor mainly consists of a free hand diagram editor which translates drawings into a hypergraph model, creates its syntactic structure and thus checks its syntactic correctness with respect to the Petri net syntax. As a result of this process, the editor has to provide visual feedback to the editor user if the drawing contains errors. The editor performs this task in a sequence of four steps after each editing operation: the scanning, the reduction, the parsing, and the attribute evaluation step. These steps are illustrated for the Petri net in Fig. 1.

476   Mark Minas

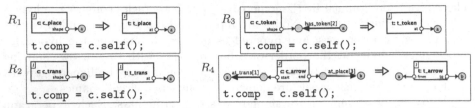

Fig. 3. Reduction rules for the Petri net editor. Rule R_5, which is similar to R_4 with reverse arrow direction, has been omitted. The rules' graphical representation has been created by VISUALDIAGEN and its printing functionality

Scanning Step. Diagram components (e.g., places, transitions, tokens, and arrows) have attachm ent areas, i.e., the parts of the components that are allowed to connect to other components (e.g., start and end of an arrow). The most general and yet simple formal description of such a component is a hyperedge which connects to the nodes which represent the attachm ent areas of the diagram components. These nodes and hyperedges first make up an unconnected hypergraph. The scanner connects nodes by additional edges if the corresponding attachment areas are related in a specified way, which is described in the specification. The result of this scanning step is the hypergraph m odel (HGM) of the diagram.

Reduction Step. HGMs tend to be quite large even for small diagrams. In order to allow for efficient parsing, a reduced hypergraph m odel (rHGM) is created from the HGM first. The reducer is specified by some transformations that identify those sub-hypergraphs of the HGM which carry the information of the diagram and build the HGM accordingly. This step is similar to the lexical analysis step of traditional compilers. Fig. 3 shows the reduction rules for the Petri net language whereas Fig. 4 shows the rHGM for the Petri net of Fig. 1. Unary edges of type t_place, t_trans, and t_token represent places, transitions, and tokens, resp. Binary edges of type t_arrow represent arrows.

Parsing Step. The syntax of the hypergraph models of the diagram language– and thus the syntax of the language–is defined by a hypergraph grammar. Fig. 5 shows a context-free hypergraph grammar with embeddings.[1] The starting hypergraph of the grammar consists of a Net hyperedge which does not visit any node.

Similar to compilers for (textual) programming languages, a hypergraph parser which is built-in into each DIAGEN editor is used for creating the syntactic structure of the HGM of the diagram, i.e., for finding a derivation sequence

---

[1] Please note that productions P_2 and P_3 are so-called *set productions* which are indicated by the "stacked" edges on the RHS. This more readable notation can be regarded as a shorthand for recursive productions. However, this is actually a special kind of production which allows for more efficient parsing.

from the starting hypergraph to the rHGM. The parser is capable of identifying syntax errors which are then visualized to the editor user.

Attribute Evaluation Step. The final step of the translation process creates the semantic representation of the diagram by some kind of syntax-directed translation based on a attribute grammar as it is also used in compilers for (tex-

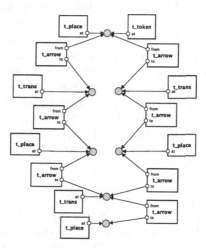

Fig. 4. Reduced hypergraph model of the Petri net of Fig. 1

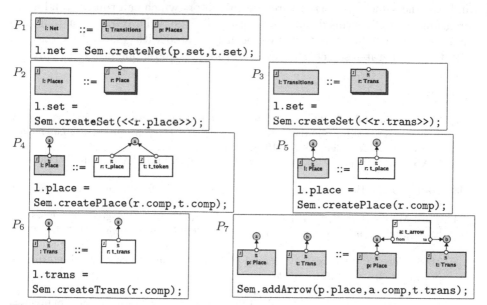

Fig. 5. Hypergraph grammar productions of the visual language of petri nets. Production P_8 which is similar to P_7 has been is omitted

tual) programming languages: terminal and nonterminal hyperedges are augmented by attributes, and hypergraph grammar productions by evaluation rules. Initial attribute values are defined by assignment rules being attached to reduction rules. Fig. 3 shows such assignments. Attributes are specified by $e.a$ where e is the name of the edge (i.e., the string in front of a colon in a hyperedge label), and a is the attribute label. Each terminal hyperedge has an attribute whose value is a reference to the corresponding visual component as a Java object. Method `self()` provides such a reference. Evaluation rules of the Petri net grammar productions (cf. Fig. 5) are implemented by methods of the Java class Sem. Please note the special syntax `<<...>>` in P_2 and P_3. As these productions are set productions, `<<r.place>>` resp. `<<r.trans>>` describe arrays of all place resp. trans attributes of the corresponding Place resp. Trans edges.

The implementation of the attribute evaluation methods and their required data structures is omitted here. Their task is generating an object structure that represents the analyzed Petri net. This object structure is referenced by the net attribute of the resulting Net edge of the parsing process. Fig. 1 shows the textual representation of this value for the depicted Petri net.

5   Conclusions

This paper has briefly demonstrated how VISUALDIAGEN is used for specifying and generating a Petri net editor. As a tool for specification and generation of visual editors, VISUALDIAGEN is most closely related to GENGED [1] which is also a Java-based, visual tool for specifying visual editors. However, there are also some major differences. While VISUALDIAGEN is able to generate free-hand and – at the same time – structured visual editors which can run as stand-alone programs as well as inside of other systems (VISUALDIAGEN as a tool contains several visual editors that have been generated with VISUALDIAGEN [3]), GENGED generates structured visual editors as stand-alone programs only.

References

[1] Roswitha Bardohl. GENGED: A generic graphical editor for visual languages based on algebraic graph grammars. In *Proc. 1998 IEEE Symp. on Visual Languages, Halifax, Canada*, pages 48–55, 1998. 478
[2] DIAGEN homepage. http://www2-data.informatik.unibw-muenchen.de/DiaGen/. 473
[3] Mark Minas. Bootstrapping visual components of the diagen specification tool with diagen. In this volume. 473, 478
[4] Mark Minas. Concepts and realization of a diagram editor generator based on hypergraph transformation. *Science of Computer Programming*, 44:157–180, 2002. 473, 475
[5] Mark Minas and Georg Frey. Visual PLC-programming using signal interpreted petri nets. In *Proceedings of the American Control Conference 2002 (ACC2002), Anchorage, Alaska*, pages 5019–5024, May 2002. 473
[6] http://www.eit.uni-kl.de/litz/ENGLISH/software/SIPNEditor.htm. (September 2003). 473

The GROOVE Simulator:
A Tool for State Space Generation

Arend Rensink

University of Twente
P.O.Box 217, 7500 AE Enschede, The Netherlands
rensink@cs.utwente.nl

1 Introduction

The tool described here is the first part of a tool set called GROOVE (GRaph-based Object-Oriented VErification) for software model checking of object-oriented systems. The special feature of GROOVE, which sets it apart from other model checking approaches, is that it is based on graph transformations. It uses graphs to represent state snapshots; transitions arise from the application of graph production rules. This yields so-called Graph Transition Systems (GTS's) as computational models.

The simulator does a small part of the job of a model checker: it attempts to generate the full state space of a given graph grammar. This entails recursively computing and applying all enabled graph production rules at each state. Each newly generated state is compared to all known states up to isomorphism; matching states are merged, in the way proposed in [4]. No provisions are currently made for detecting or modelling infinite state spaces. Alternatively, one may choose to simulate productions manually.

This paper describes two examples: Sect. 2 shows the behaviour of a circular buffer and Sect. 3 the concurrent invocation of a list append method. In both cases the behaviour is defined by a graph grammar, but to provide some intuition, Fig. 1 approximately describes the behaviour, using Java code. We conclude in Sect. 4 with a summary of tool design, implementation and planned future extensions.

2 Circular Buffer Operations

We assume the principles of circular buffers to be known. Their representation as graphs is relatively straightforward (see also Fig. 1). The buffer has a set of cells connected by next-edges. One of the cells is designated first and one last. Insertion will occur at last (provided this cell is empty) and retrieval at first (provided this is filled). A value contained in a cell is modelled by a val-labelled edge to an unlabelled node. The cell is empty if and only if there is no outgoing val-edge. (In the Java code of Fig. 1 this corresponds to a null value of the val attribute.)

J.L. Pfaltz, M. Nagl, and B. Böhlen (Eds.): AGTIVE 2003, LNCS 3062, pp. 479–485, 2004.
© Springer-Verlag Berlin Heidelberg 2004

```
public class Buffer {                  class Node {
    private class Cell {                   private Node next;
        Cell next; Object val;             private int val;
    }
    private Cell first, last;              public void append(int x) {
                                               if (this.val == x) {
    // Precondition: last.val == null             // Rule "stop"
    // Executed atomically                         return;
    public void put(Object val) {              } else if (this.next == null) {
        last.val = val;                            // Rule "append"
        last = last.next;                          Node aux = new Node();
    }                                              aux.val = x;
                                                   this.next = aux;
    // Precondition: first.val != null             return;
    // Executed atomically                     } else {
    public Object get() {                          // Rule "next"
        Object result = first.val;                 this.next.append(x);
        first.val = null;                          // Rule "return"
        first = first.next;                        return;
        return result;                         }
    }                                      }
}                                      }
```

Fig. 1. Approximate Java descriptions of the examples

Fig. 2 shows the simulator tool after loading the relevant graph grammar. The GUI of the simulator has two panels: a directory of the available rules with their matches in the current graph, and the current graph itself — in this case the initial graph, modelling a three-cell empty circular buffer. The latter panel can also display the currently selected rule and the resulting GTS (insofar generated), instead of the current graph. The example grammar has two rules: **get** for the retrieval of an element from the buffer and **put** for insertion. As usual, each rule prescribes when it applies to a given graph and what the effect of its application is. There are four types of nodes and edges:

- Thin black solid nodes and edges, which we call readers: they are required to be in a graph in order for the rule to apply, and are unaffected by rule application;
- Thin blue double-bordered nodes and dashed edges, which we call erasers: they are required in order for the rule to apply, and are deleted by rule application;
- Fat green solid nodes and edges, which we call creators: they are not required to be in the graph, and are created by rule application.
- Fat red double-bordered nodes and dashed edges, which we call embargoes: they are forbidden to occur in a graph in order for the rule to apply.

More precisely, a rule application is based on a matching, which is a mapping of the readers and erasers of the rule to corresponding elements of the graph that cannot be extended with any of the embargoes. For instance, as Fig. 2 shows,

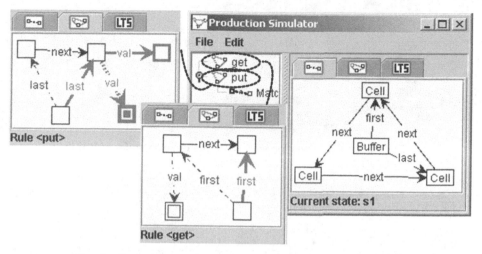

Fig. 2. Rules and initial graph of the circular buffer example

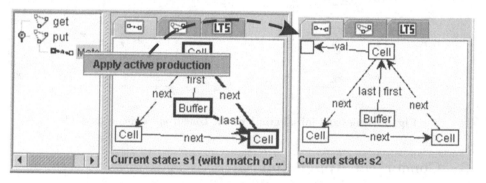

Fig. 3. Application of the put rule to the initial graph

put has a single match in the initial graph whereas get has none. The application deletes the elements matching the erasers and adds elements matching the creators. Fig. 3 shows how one may select and apply a matching in the simulator (the graph is fat where the matching applies). Besides "walking through" the rule applications in this fashion, the simulator tool also can (attempt to) recursively compute all possible applications. This gives rise to a GTS, with the set of all graphs generated by the grammar as states, connected by rule applications. Although the GTS is generally infinite, there are many cases in which it is not — in fact, for grammars that model computing systems, an infinite state space is arguably an error indication. For the circular buffer, the state space is quite small, consisting a single state for each possible number of filled cells: see Fig. 4.

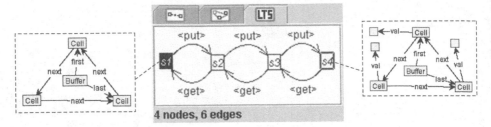

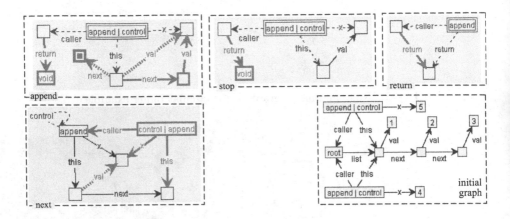

Fig. 4. Graph transition system of the circular buffer example

Fig. 5. Rules and initial graph of the concurrent append example

3   Concurrent Appending

The second example is a list append method. In this case we do not assume the
method to be atomic; instead we model it as a recursive invocation. The example shows many features typical for the dynamics of object-oriented programs.
Running methods are modelled by nodes, with local variables as outgoing edges,
including a this-labelled edge pointing to the object executing the method.
Each (recursive) method invocation results in a fresh node, with a caller edge
to the invoking method. Upon return, the method node is deleted, while creating a return edge from its caller to a return value — in this example always
void. It follows that the traditional call stack is replaced by a chain of method
nodes. The top of the stack is identified by a control label or an outgoing return
edge. The particular method in this example only appends a value that is not
already in the list, as suggested by the code in Fig. 1. The corresponding graph
grammar is shown in Fig. 5. The initial graph contains two concurrently enabled
invocations; we expect the simulator to tell us whether these invocations may
interfere. Due to the ensuing race condition, the system has two legal outcomes:
either the value 4 is appended before 5, or vice versa. For instance, Fig. 6 shows

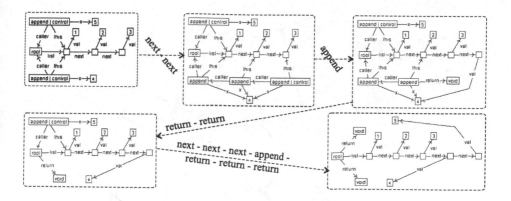

Fig. 6. Trace of graph transformations for the **append** function

(part of) the sequence of states in which the "bottom" **append** proceeds first, so that 4 "wins" from 5.

The full GTS, comprising 57 states and 92 transitions, is shown in Fig. 7. The race condition can be readily recognised, as can the fact that the final states are precisely the two legal ones. We conclude that the **append** method as modelled by this graph grammar is non-interfering.

4   Design and Implementation

We finish by discussing a number of issues in the design and implementation of the GROOVE tool set in general, and the simulator in particular.

Theoretical Background. There is a long history of research in graph transformation; see [5] for an overview. GROOVE follows the algebraic approach; we hope to reap the benefit of the resulting algebraic properties in future extensions (see below). GROOVE uses non-attributed, edge-labelled graphs without parallel edges (the node labels shown in the examples above are actually labels of self-edges) and implements the single-pushout approach with negative application conditions (see [3, 2]); however, the design of the tool is modular with respect to this choice, and support for the double-pushout approach can be added with a single additional class.

Design. The tool is designed for extensibility. The internal and visual representation of graphs are completely separated, and interfaces are heavily used, for instance to abstract from graph and morphism implementations. The most performance-critical parts of the simulator are: finding rule matchings, and checking graph isomorphism. The first problem is, in general, NP-complete; the second is in NP (its precise complexity is unknown). Fortunately, the graphs we

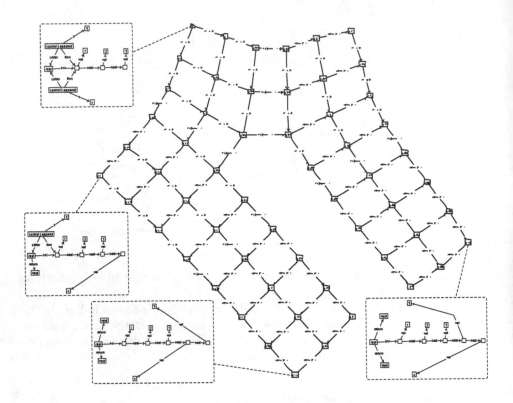

Fig. 7. Graph transition system generated from the **append** function

are dealing here tend to be "almost deterministic" (we will not make this precise) so that the complexity is manageable. The GROOVE tool uses strategies to ensure modularity for the algorithms used.

Implementation. The tool is implemented in Java, and developed under Eclipse. It currently consists of (approximately) 70 classes in 8 packages, comprising 25,000 lines of code. The implementation makes use of several existing open-source components: **jgraph** for graph visualisation, **xerces** and **castor** for handling XML, and **junit** for unit testing. The implementation is a research prototype; we are yet far from being able to tackle realistic problems. Currently the largest example we have simulated is approximately 20,000 states; this takes in the order of half an hour.

Interchange Formats. The modularity of GROOVE also extends to the serialisation and storage of graphs and graph grammars. Currently the tool uses GXL (see [8]), but in an ad hoc fashion: production rules are first encoded as graphs by tagging the erasers, creators and embargoes, and then saved individually; thus, a grammar is stored as a set of files. In the future we plan to migrate to GTXL (see [7]) as primary format.

Extensions. The simulator can be improved in many ways. Some extensions planned for the (near) future are: support for the double-pushout approach and graph types, partial order reduction using production rule independence; and especially, shape graph-like abstractions (cf. [6]), for the (approximate) representation of infinite state spaces. Furthermore, we plan to extend the GROOVE tool set beyond the simulator functionality. This includes a front-end to compile code to graph grammars (a prototype of a Java byte code translator has been developed) and a back-end to model check the resulting GTS's. In [1] we have taken a first step toward developing a logic to reason about the dynamic aspects of GTS's. See [4] for a global overview of the programme.

The GROOVE simulator tool and some sample graph grammars (the above examples among them) can be downloaded from:
`http://www.cs.utwente.nl/~groove`.

References

[1] Dino Distefano, Arend Rensink, and Joost-Pieter Katoen. Model checking birth and death. In R.A. Baeza-Yates, U. Montanari, and N. Santoro, editors, *Foundations of Information Technology in the Era of Network and Mobile Computing*, pages 435–447. Kluwer, 2002. 485

[2] Annegret Habel, Reiko Heckel, and Gebriele Taentzer. Graph grammars with negative application conditions. *Fundamenta Informaticae*, 26(3/4):287–313, 1996. 483

[3] M. Löwe. Algebraic approach to single-pushout graph transformation. *TCS*, 109(1–2):181–224, 1993. 483

[4] Arend Rensink. Towards model checking graph grammars. In Leuschel, Gruner, and Lo Presti, editors, *Proceedings of the 3rd Workshop on Automated Verification of Critical Systems*, Technical Report DSSE–TR–2003-2, pages 150–160. University of Southampton, 2003. 479, 485

[5] Grzegorz Rozenberg, editor. *Handbook of Graph Grammars and Computing by Graph Transformation*, volume I: Foundations. World Scientific, 1997. 483

[6] Mooly Sagiv, Thomas Reps, and Reinhard Wilhelm. Solving shape-analysis problems in languages with destructive updating. *ACM ToPLaS*, 20(1):1–50, January 1998. 485

[7] Gabriele Taentzer. Towards common exchange formats for graphs and graph transformation systems. In Ehrig, Ermel, and Padberg, editors, *UNIGRA*, volume 44 of *ENTCS*. Elsevier, 2001. 485

[8] Andreas Winter, Bernt Kullbach, and Volker Riediger. An overview of the GXL graph exchange language. In Diehl, editor, *Software Visualization*, volume 2269 of *LNCS*, pages 324–336. Springer, 2002. 484

AGTIVE'03: Summary from the Outside In

Arend Rensink

University of Twente
P.O.Box 217, 7500 AE Enschede, The Netherlands
rensink@cs.utwente.nl

1 The Perspective

Occasionally it happens that, while researching a particular subject, one encounters a previously unfamiliar field that turns out to be surprisingly appropriate and applicable. This is what occurred to me when, in investigating techniques for capturing the semantics of object-oriented programs, I came across graph transformations.

For someone coming from outside a field it is of the first importance to get an overview and to gain awareness of the issues. For this purpose there are several instruments at one's disposal:

Literature Survey. Traditionally, an excellent technique to get acquainted with a field is to read textbooks and handbooks, and after that, technical articles. On the other hand, without any guidance the abundance of material can make it difficult to separate the chaff from the wheat.

Personal Contacts. Having direct access to an expert in the field and asking his or her considered opinion is a much quicker way to obtain an impression of the important and the irrelevant, the recent and the abandoned. Of course, the trick is to find such an expert and get him to hold still for you.

Own Research. At the risk of repeating existing research, one may choose to dive into the subject and try to work out some theory or application, before having that complete overview. I have found that this is a good way to get more personally acquainted with some of the technical issues.

Conference Visits. Many of the desirables described above can be combined by joining a meeting of the experts, and so trying to make both their and their work's acquaintance. Optimally, the meeting should be visited by many of the main players, and yet the visitor count should not exceed that undefinable threshold where participants stop speaking to "outsiders" and only cluster in old, familiar circles.

It was with the last of these in mind that I went to Virginia, to visit a workshop which purposed not only to be about those graph transformations which looked so promising to me, but to actually show industrially relevant applications of the same, many of them tool-supported.

J.L. Pfaltz, M. Nagl, and B. Böhlen (Eds.): AGTIVE 2003, LNCS 3062, pp. 486–488, 2004.

2   The Programme

The workshop programme was thematically structured, and contained "full" and "short" presentations, of 30 and 15 minutes, respectively, as well as 45 minute invited presentations. The result was a large and (to me) surprising diversity of graph-based models in different application areas, ranging from natural language to architecture and from multimedia standardisation to biology. The interest in several of the contributions was heightened by the fact that they were backed up by a tool. The tools were shown at two demo sessions, separate from the presentations themselves — to my opinion unfortunately, since the demos provided definite added value that would have been even more worthwhile when combined with the talks.

Given my perspective, I was particularly interested in knowing where and how the many theoretical advances in graph transformations have been applied in practice. To my regret, this sometimes tended to be obscured, the focus lying more on the application domain than the connection with the underlying theory. This observation has made me wonder: is it perhaps the case that even in this field, ancient to computer science standards, the link between theory and application is weak, and there is, despite the best of intentions, a split between theoreticians and practitioners? Or is the situation rather that the connections are there, but not visible to a non-expert like myself?

Four days was hardly enough to digest the information of the many presentations, let alone to discuss research content with all those present. We found relief in an excursion, which was suffused with the very different subject of the life and deeds of Mr. Thomas Jefferson, in which graphs have played no role.

3   The People

Quite apart from considerations about the programme, a workshop stands or falls with the participants. Indeed, as stated above, meeting the researchers in the flesh has been my primary motivation in visiting the workshop. On this count I must say that I can hardly recall enjoying a more open and pleasant atmosphere at any meeting. As I had hoped, representatives from many of the most well-known research groups were there, and they were quite willing to discuss issues which, to them, must have been old hat. In this, it probably helps that the core community is small enough so that everyone knows one another. Nor have I seen signs of the inter-tribal wars that plague some other disciplines: maybe the graph transformation area is mature enough to have outgrown such tendencies — or maybe the wars were fought out of my earshot.

It also helps if you speak some German. This observation, although called "that $%&# nationality issue" by one of the organisers, was so inescapable that leaving it out of this summary would be to omit part of the reality of the event. Many of the main results in the field are due originally to researchers in a small region of the world (including also Italy, from which for some reason the delegation was much smaller); it is really unfortunate, and perhaps a source of concern to the community, that at least at this event the turnout was rather dominated by this same region.

4   The Conclusion

But I do not want to end this summary on a cautionary note, and I do not want to inflate a side issue out of proportion. Let me finish instead by looking back on the event holistically. I believe that in the coming years the workshop will continue stand out as a high point in my memory, due to a combination of content and ambience. I am grateful for this opportunity to thank the organisers once more for what was, for me, a very fruitful meeting that not only met but exceeded my expectations.

AGTIVE'03: Summary from the Theoretical Point of View

Gabriele Taentzer

Technische Universität Berlin, Germany
gabi@cs.tu-berlin.de

Attendees of the workshop, especially those coming from the very applied side, might ask themselves why it is interesting to consider the workshop from the theoretical perspective. A workshop which is clearly dedicated to applications of graph transformation concepts! Let us have a short look back to research on graph transformation about ten years ago. In the beginning of the nineties, the research community on graph transformation worked mainly on theoretical problems. Nowadays, graph transformation is applied in very different fields and the amount of practical work increased considerably. Compare e.g. the contributions to the 1. International Conference on Graph Transformation last year in Barcelona [1]. So the question arises: What is the significance of theoretical work on graph transformation nowadays?

Now we concentrate our attention on this workshop: A large variety of applications of graph transformation have been shown, coming from universities and from industry, and a lot of graph transformation-based tools have been demonstrated. Does this mean that the theory on graph transformation is widely applied now? Yes and no, a lot of applications are developed directly by those research groups which worked more theoretical before. Here, the established approaches to graph transformation are applied together with their theoretical results. But there are also a lot of applications by practitioners who developed their own graph transformation approaches. Those approaches are more or less ad-hoc defined, mostly come up with little or no theory and are seldom related to the already existing approaches. We may ask ourselves about the reasons of this development. One answer is certainly that, in general, the theory-driven approaches to graph transformation are not easy to understand. Furthermore, newcomers might be confused being confronted with a number of graph transformation approaches and do not find the right one for their needs.

So what can theoreticians draw out of this workshop? One task, a big task, is certainly to clarify the picture of graph transformation approaches. First of all, work has to be done to interrelate graph transformation concepts more closely. This is a purely theoretical work on one hand, on the other hand the commonalities and specialties of the approaches have to be elaborated such that a practitioner gets help to choose the right approach for the intended application. This approach comparison should not only comprise graph and transformation concepts but also theoretical results achieved on top of the basic transformation concepts. Then, dependent on the kind of application, new theoretical results might be needed.

J.L. Pfaltz, M. Nagl, and B. Böhlen (Eds.): AGTIVE 2003, LNCS 3062, pp. 489–490, 2004.

Considering the concrete contributions to this workshop, especially the invited talks raise interesting theoretical questions:

– Hawley Rising of the Sony US Research Lab, San Jose, USA presented the MPEG-7 format where graph transformation is used to describe semantic descriptions. Four different graph transformation approaches are taken to describe certain graph operations, like e.g. graph blending, which might be rather complex. This interesting application raises a number of questions: Does each of the transformation approaches implement the graph operation it is meant for? How do these approaches interact? Is it possible to integrate all these approaches into one and how would it look like?

– Gabor Karsai of the Vanderbilt University, Nashville, USA, presented the application of graph transformation in OMG's concept of Model-Driven Architectures. He showed the use of a new graph transformation approach for model transformation. For reasoning about this kind of model transformation it is worthwhile to compare this new approach with existing ones. A formalization by an approach with enough theory behind is necessary to argue for the correctness of model transformations. A functional behaviour and consistency are the main two properties of correct model transformations.

Besides the invited talks also other presentations were interesting from the theoretical prespective. For example, there is the idea to apply model checking to graph transformation systems, mentioned by Fernando Dotti in the context of specifiying and analyzing fault behaviours by graph transformation and discussed by Ralph Depke applying graph transformation to agent-oriented modelling. Arend Rensink presented his tool GROOVE for which he plans a support for model checking of graph transformation systems. Of course, there were further contributions raising interesting theoretical questions. The selection above is a very subjective one, following my personal interests.

References

[1] A. Corradini, H. Ehrig, H-J Kreowski, and G. Rozenberg, editors. *Proc. 1st Int. Conference on Graph Transformation (ICGT'02)*. Springer LNCS 2505, 2002. 489

AGTIVE'03: Summary from the Viewpoint of Graph Transformation Specifications

Mark Minas

Institute for Software Technology, Department of Computer Science
University of the Federal Armed Forces, Munich, 85577 Neubiberg, Germany
minas@acm.org

Graphs are a natural means for representing interrelationships on an abstract level. This is one of the reasons for the success of UML that is used for abstract representations, i.e., specifications, of object-oriented programs. And when graphs are well-suited for specifications, graph transformations are particularly appropriate for specifications of dynamic changes. It is, therefore, not astonishing that graph transformation specifications was one of the major applications of graph transformations at this workshop on Applications of Graph Transformations with Industrial Relevance. As the term industrial relevance suggests development of large program systems, one might have expected that specification of program dynamics has dominated the workshop in the same way as UML currently dominates the field of program specification. However, this was not the case. The following briefly summarizes the aspect of graph transformation specifications of the workshop from two different view points: We first look at the areas where graph transformation specifications are currently used. And this shows that graph transformation specifications is not just on program specification. And second, we look at the purposes what graph transformation specifications are used for.

When looking at the proceedings and after watching the talks, we can actually identify four major areas where graph transformation specifications are currently used: Of course, software specifications is on of these areas. And specification of object-oriented programs as well as object-oriented program systems are topics of this area, but also specification of software systems in general and, very specifically, specification of data types.

Language specification is another obvious area which can be seen as a descendant of specification of textual languages with Chomsky grammars. Remarkably, graph transformation specifications are now used also in the field of linguistics which has been the origin of the text-based Chomsky grammars. However, graph transformation systems are also applied to specification of visual languages resp. graphical systems that support processing of visual languages. This is in particular one of the traditional applications of graph transformation specifications.

The other two areas where graph transformation systems are used for specification are visualization & design and processes. The former covers approaches for geometric modeling and conceptual modeling in architecture. The latter comprises physical processes where it is used for process modeling as well as for

J.L. Pfaltz, M. Nagl, and B. Böhlen (Eds.): AGTIVE 2003, LNCS 3062, pp. 491–492, 2004.
© Springer-Verlag Berlin Heidelberg 2004

specifying fault models, but also organizational processes where graph transformation systems are used for process management specifications.

When looking at purposes what graph transformation specifications are used for, we can identify applications where such specifications are used as a formal basis for extending existing systems by new features. Papers on knowledge-based design and model transformation that fall into this field have been presented at the workshop. But graph transformation specifications are also used to build completely new systems. Code generation and rapid prototyping are the main applications that have been reported on the workshop. Moreover, such specifications are used for code refactoring, consistency analysis, code verification and test generation, for reasoning and simulation as the talks given at this workshop have shown.

Specifications, when they shall be used in applications of industrial relevance, must be accompanied by tools which are based on or use such specifications. Much progress has been made when we compare work presented at this workshop with work presented at its predecessor, AGTIVE in 1999 at the Monastery Rolduc, Kerkrade, The Netherlands. 17 papers or demonstrations of this workshop have been based on or related to graph transformation specifications. Eleven of them have already working tools; some of them have also been presented in the tool demonstration sessions. Five papers have presented plans for building such tools in the future. And one single paper was using an existing tool.

Two conclusions can be drawn from these observations. First, most specification techniques based on graph transformation have reached a maturity level that allows to build tools based on them, and these tools – if they are already existing – are used in applications as the system demonstrations have shown. However, the second observation is that most specification techniques require their own tools; they are not based on other, existing systems. Maybe that is an inherent property of specification techniques in general and graph transformation specification techniques specifically, but it also shows that most tools are not that frequently used by other groups as one might expect or wish. Perhaps this might have changed by the time when the successor of this AGTIVE workshop will take place.

AGTIVE'03: Summary from a Tool Builder's Viewpoint

Bernhard Westfechtel

Computer Science III
RWTH Aachen
D-52056 Aachen, Germany

1 The Past

Graph grammars and graph rewriting systems are not at all a new invention. They may be traced back to the fundamental work of John Pfaltz and Hans-Jürgen Schneider, who introduced them independently of each other around 1969. Since, the field has evolved into an established discipline, whose results have been documented in numerous books, journal articles, and proceedings of international workshops and conferences.

However, for a long time the focus has been set on the evolution of the theory underlying graph transformations. Several theoretical frameworks have been established, but practical applications have been neglected for a long time. In particular, this statement holds for the development of tools for graph transformations. Development of tools was considered difficult or even impossible. For example, pioneering work on integrated software development environments based on graph transformations in the 1980's faced serious efficiency problems. For many years, NP completeness of the subgraph search problem was perceived by many researchers as a killing argument against implementation efforts based on graph transformations.

Nevertheless, this attitude has changed considerably over the years. More and more, researchers tend to accept that NP completeness does not exclude feasible implementations because in many practical settings the problem at hand may be solved efficiently. As a simple example, consider pre-selected anchor nodes which reduce the global subgraph search problem to a local one (e.g., user-selected nodes in graphical editors). This insight, combined with significant gains in hardware speed and memory space, has had a catalyzing effect on the development of tools for graph transformations.

2 The Present

Currently, more and more graph transformation tools are being developed, covering a wide range of applications such as bioinformatics, process management, object-oriented modeling, architectural design, reengineering, distributed systems, etc. This is illustrated by the tools demonstrated

J.L. Pfaltz, M. Nagl, and B. Böhlen (Eds.): AGTIVE 2003, LNCS 3062, pp. 493–495, 2004.

at this workshop: GenGED, DiaGen, and AGG, which support the execution of graph transformations and the generation of visual environments, GROOVE (software model checking of object-oriented systems), OptimixJ (graph rewrite rules for Java), Esther (an operating shell with dynamic type checking), Fujaba (a CASE tool combining graph transformations with the UML), and several tools built with the help of the PROGRES environment (a meta-case environment for specifying and generating tools for visual languages): AHEAD (a management system for dynamic and interorganizational development processes), CHASID (a graph-based authoring support system), a conceptual design tool for civil engineering, FIRE3 (a tool for the refinement of logical software architectures), and ECARES (reverse engineering of telecommunication systems).

All of the tools mentioned above are research prototypes. This observation may be generalized: Graph transformation based tools may have gained industrial relevance to some extent, but they have hardly made their way into commercial products. On the other hand, the landscape of tools has changed significantly over the last years. Many of the tools mentioned above started as fairly small tools and have evolved into full-fledged environments with completely implemented functionality (graphical editing, both in syntax-directed and in free-hand mode, automatic and incremental layout, multiple views, code generation, seamless integration into Java environments, etc.). Thus, the term "research prototype" might be misleading since this is often associated to a quickly hacked small demonstrator which basically implements the functions required on the demo path. This characterization does not apply e.g. to PROGRES and Fujaba, which have been developed for quite a number of years and both comprise about 1 million lines of source code. And other tools such as DiaGen, AGG, or GenGED have been extended and improved considerably; there is a fairly large delta compared to the state at the first AGTIVE workshop in 1999. On the other hand, the term "research prototype" is still justified with respect to stability, documentation, and efficiency, which do not meet the requirements for commercial tools.

3   The Future

In the next stage, graph transformation based tools will hopefully make their way into commercial products. The research prototypes which are currently available constitute much more than just demonstrators. A lot of knowledge has been acquired concerning the implementation of graph-based tools, e.g. with respect to building parsers, unparsers, graphical editors, with respect to implementing subgraph replacement, etc. Furthermore, graphical languages are gaining more and importance. For example, in the area of software engineering tool support focused on tools for textual programs for a long time. Currently, the focus shifts away from textual programming languages to graphical modeling languages (e.g., UML). Graph transformations are ideally suited for defining such modeling languages and for building tools supporting these languages. Multi-media constitutes another application area where graphs can be used for modeling non-hierarchical structures of mutually linked multi-media documents and operations on these structures may be described by graph transformations. Thus, there are "hot" application domains for graph transformations. This may help in promoting the application of graph transformations significantly.

Best Presentation and Demonstration Awards

Bernhard Westfechtel

Computer Science III
RWTH Aachen, D-52056 Aachen, Germany

As on AGTIVE 1999, the Program Committee decided to decorate the best paper presentation and the best tool demonstration with respective awards. The winners were selected by the workshop participants' votes. At the end of the workshop, each winner received a certificate which testifies the quality of his presentation and motivates him to give high-quality presentations on future events as well.

Colin Smith received the award for the *best paper presentation*. He reported on joint work with Przemyslaw Prusinkiewicz and Faramarz Samavati, both University of Calgary. The talk dealt with the generation of polygon meshes for the geometric modeling of surfaces. Colin presented a novel approach to the specification of polygon mesh algorithms using graph rotation systems. His talk conveyed the underlying idea very clearly, and it was appreciated very much in particular because of the beautiful animations illustrating the generation of polygon meshes.

There was a strong competition for the *best tool demonstration*. Here, two winners shared the award. **Arend Rensink**, University of Twente, presented GROOVE (*GR*aph-based *O*bject-*O*riented *VE*rification), a tool set for software model checking of object-oriented systems. The demonstration was devoted to a part of the tool set, namely the GROOVE simulator. Unlike other tools for model checking, GROOVE represents states as graphs and performs state transitions with graph transformations. The GROOVE tool was well received by the workshop participants because it offers an attractive visualization of the state generation process.

Mark Minas, University of the Federal Armed Forces, Munich, gained as many votes as Arend Rensink for his presentation of the DiaGen system for generating tools for visual languages. Mark, who has been working on DiaGen for quite a number of years, presented a mature and comprehensive system which is based on hypergraph grammars. Free-hand editing of diagrams is a particularly distinguishing feature of tools generated with the help of Diagen. The award should encourage Mark to continue his work on DiaGen — and to give fascinating demos on future workshops and conferences as well.

Category	Winner	Title
Best paper presentation	Colin Smith	Local specification of surface subdivision algorithms
Best tool demonstration	Mark Minas	Visual specification of visual editors with DiaGen
	Arend Rensink	The GROOVE simulator: A tool for state space generation

J.L. Pfaltz, M. Nagl, and B. Böhlen (Eds.): AGTIVE 2003, LNCS 3062, p. 496, 2004.
© Springer-Verlag Berlin Heidelberg 2004

Author Index

Books on Graph Transformation

V. Claus, H. Ehrig, G. Rozenberg (Eds.): *Graph Grammars and Their Applications to Computer Science and Biology*, Proc. Intl. Workshop, Bad Honnef, Germany, 1978, Lecture Notes in Computer Science Vol. 73, Berlin: Springer-Verlag (1979), ISBN 0-387-09525-X

M. Nagl: *Graph Grammars: Theory, Applications, and Implementation (in German)*, Braunschweig: Vieweg Verlag (1979), ISBN: 3-528-03338-X

H. Ehrig, M. Nagl, G. Rozenberg (Eds.): *Graph Grammars and Their Applications to Computer Science*, Proc. 2nd Intl. Workshop, Osnabrück, Germany, 1982, Lecture Notes in Computer Science Vol. 153, Berlin: Springer-Verlag (1983), ISBN 0-387-12310-5

H. Ehrig, M. Nagl, G. Rozenberg, A. Rosenfeld (Eds.): *Graph Grammars and Their Applications to Computer Science*, Proc. 3rd Intl. Workshop, Warrenton, USA, 1986, Lecture Notes in Computer Science Vol. 291, Berlin: Springer-Verlag (1987), ISBN 0-387-18711-5

H. Göttler: *Graph Grammars in Software Engineering (in German)*, IFB 178, Berlin: Springer-Verlag (1988), ISBN: 3-540-50243-2

A. Schürr: *Operational Specification by Programmed Graph Transformation Systems (in German)*, Deutscher Universitäts-Verlag, Wiesbaden (1991), ISBN: 3-8244-2021-X

H. Ehrig, H.-J. Kreowski, G. Rozenberg (Eds.): *Graph Grammars and Their Applications to Computer Science*, Proc. 4th Intl. Workshop, Bremen, Germany, 1990, Lecture Notes in Computer Science Vol. 532, Berlin: Springer-Verlag (1991), ISBN 0-387-54478-X

A. Habel: *Hyperedge Replacement: Grammars and Languages*, Lecture Notes in Computer Science Vol. 643, Berlin: Springer-Verlag (1992), ISBN: 3-540-56005-X

H. J. Schneider, H. Ehrig (Eds.): *Graph Transformations in Computer Science*, Proc. Intl. Workshop Dagstuhl Castle, Germany, 1993, Lecture Notes in Computer Science Vol. 776, Berlin: Springer-Verlag (1994), ISBN 0-387-57787-4

A. Zündorf: *Programmed Graph Transformation Systems (in German)*, Deutscher Universitäts-Verlag, Wiesbaden (1995), ISBN: 3-8244-2075-X

J. Cuny, H. Ehrig, G. Engels, G. Rozenberg (Eds.): *Graph Grammars and Their Applications to Computer Science*, Proc. 5th Intl. Workshop, Williamsburgh, USA, 1994, Lecture Notes in Computer Science Vol. 1073, Berlin: Springer-Verlag (1996), ISBN 3-540-61228-9

M. Nagl (Ed.): *Building Tightly Integrated Software Development Environments: The IPSEN Approach*, Lecture Notes in Computer Science Vol. 1170, Berlin: Springer-Verlag (1996), ISBN 3-540-61985-2

G. Rozenberg: *Handbook of Graph Grammars and Computing by Graph Transformation: Foundations*, Vol. 1, Singapore: World Scientific (1997), ISBN 981-02-2884-8

B. Westfechtel: *Models and Tools for Managing Development Processes*, Lecture Notes in Computer Science Vol. 1646, Berlin: Springer-Verlag (1999), ISBN: 3-540-66756-3

H. Ehrig, G. Engels, H.-J. Kreowski, G. Rozenberg (Eds.): *Handbook of Graph Grammars and Computing by Graph Transformation: Applications, Languages, and Tools*, Vol. 2, Singapore: World Scientific (1999), ISBN 981-02-4020-1

H. Ehrig, H.-J. Kreowski, U. Montanari, G. Rozenberg (Eds.): *Handbook of Graph Grammars and Computing by Graph Transformation: Concurrency, Parallelism, and Distribution*, Vol. 3, Singapore: World Scientific (1999), ISBN 981-02-4021-X

H. Ehrig, G. Engels, H.-J. Kreowski, G. Rozenberg (Eds.): *Theory and Application of Graph Transformations*, Proc. 6th Intl. Workshop TAGT'98, Paderborn, Germany, 1998, Lecture Notes in Computer Science Vol. 1764, Berlin: Springer-Verlag (2000), ISBN 3-540-67203-6

M. Nagl, A. Schürr, M. Münch (Eds.): *Applications of Graph Transformation with Industrial Relevance*, Proc. Intl. Workshop AGTIVE'99, Castle Rolduc, The Netherlands, 1999, Lecture Notes in Computer Science Vol. 1779, Berlin: Springer-Verlag (2000), ISBN 3-540-67658-9

A. Corradini, R. Heckel (Eds.): *Proc. ICALP 2000 Workshop on Graph Transformation and Visual Modeling Techniques*, Geneva, Switzerland, 2000, Carleton Scientific

H. Ehrig, G. Taentzer (Eds.): *Proc. Joint APPLIGRAPH/GETGRATS Workshop on Graph Transformation Systems GraTra'2000*, Berlin, Germany, 2000, http://tfs.cs.tu-berlin.de/gratra2000/

H. Ehrig, C. Ermel, J. Padberg (Eds.): *UNIGRA 2001: Uniform Approaches to Graphical Process Specification Techniques*, Genova, Italy, 2001, Electronic. Notes in Theoretical Computer Science Vol. 44.4, Elsevier Science Publishers

L. Baresi, M. Pezzé, G. Taentzer (Eds.): *Proc. ICALP 2001 Workshop on Graph Transformation and Visual Modeling Techniques*, Heraklion, Greece, 2001, Electronic Notes in Theoretical Computer Science Vol. 50.3, Elsevier Science Publishers

W. Brauer, H. Ehrig, J. Karhumäki, A. Salomaa (Eds.): *Formal and Natural Computing: Essays Dedicated to Grzegorz Rozenberg*, Lecture Notes in Computer Science Vol. 2300, Berlin: Springer-Verlag (2002), ISSN: 0302-9743

A. Corradini, H. Ehrig, H.-J. Kreowski, G. Rozenberg (Eds.): *Proc. First International Conference on Graph Transformation, ICGT 2002*, Barcelona, Spain, 2002, Lecture Notes in Computer Science Vol. 2505, Berlin: Springer-Verlag (2002), ISSN: 0302-9743

D. Plump (Ed.): *Proc. Intl. Workshop on Term Graph Rewriting (Termgraph'02)* Barcelona, Spain, 2002, Electronic Notes in Theoretical Computer Science Vol. 72.1, Elsevier Science Publishers

T. Mens, A. Schürr, G. Taentzer (Eds.): *ICGT'02 Workshop on Graph-Based Tools (GraBaTs 2002)*, Barcelona, Spain, 2002, Electronic Notes in Theoretical Computer Science Vol. (72.2), Elsevier Science Publishers

P. Bottoni, M. Minas (Eds.): *Proc. Int. Workshop on Graph Transformation and Visual Modeling Techniques GT-VMT'02*, Barcelona, Spain, 2002, Electronic Notes in Theoretical Computer Science Vol. 72.3, Elsevier Science Publishers

R. Heckel, T. Mens, M. Wermelinger (Eds.): *Proc. ICGT 2002 Workshop on Software Evolution through Transformations*, Barcelona, Spain, 2002, Electronic Notes in Theoretical Computer Science Vol. 72.4, Elsevier Science Publishers

H. Ehrig , R. Bardohl (Eds.): *Workshop on Uniform Approaches to Graphical Specification Techniques UniGra 2003 at ETAPS'03*, Warsaw, Poland, 2003, Electronic Notes in Theoretical Computer Science Vol. 82.7, Elsevier Science Publishers

J. L. Pfaltz, M. Nagl, B. Böhlen (Eds.): *Applications of Graph Transformation with Industrial Relevance*, Proc. 2nd Intl. Workshop AGTIVE'03, Charlottesville, USA, 2003, this volume, Berlin: Springer-Verlag (2004)

There are more books on Graph Transformation either describing specific approaches, applications, or systems built by Graph Transformation methodologies.

Lecture Notes in Computer Science

For information about Vols. 1–2970

please contact your bookseller or Springer-Verlag

Vol. 3022: T. Pajdla, J. Matas (Eds.), Computer Vision - ECCV 2004. XXVIII, 621 pages. 2004.

Vol. 3021: T. Pajdla, J. Matas (Eds.), Computer Vision - ECCV 2004. XXVIII, 633 pages. 2004.

Vol. 3019: R. Wyrzykowski, J. Dongarra, M. Paprzycki, J. Wasniewski (Eds.), Parallel Processing and Applied Mathematics. XIX, 1174 pages. 2004.

Vol. 3016: C. Lengauer, D. Batory, C. Consel, M. Odersky (Eds.), Domain-Specific Program Generation. XII, 325 pages. 2004.

Vol. 3015: C. Barakat, I. Pratt (Eds.), Passive and Active Network Measurement. XI, 300 pages. 2004.

Vol. 3014: F. van der Linden (Ed.), Software Product-Family Engineering. IX, 486 pages. 2004.

Vol. 3012: K. Kurumatani, S.-H. Chen, A. Ohuchi (Eds.), Multi-Agnets for Mass User Support. X, 217 pages. 2004. (Subseries LNAI).

Vol. 3011: J.-C. Régin, M. Rueher (Eds.), Integration of AI and OR Techniques in Constraint Programming for Combinatorial Optimization Problems. XI, 415 pages. 2004.

Vol. 3010: K.R. Apt, F. Fages, F. Rossi, P. Szeredi, J. Váncza (Eds.), Recent Advances in Constraints. VIII, 285 pages. 2004. (Subseries LNAI).

Vol. 3009: F. Bomarius, H. Iida (Eds.), Product Focused Software Process Improvement. XIV, 584 pages. 2004.

Vol. 3008: S. Heuel, Uncertain Projective Geometry. XVII, 205 pages. 2004.

Vol. 3007: J.X. Yu, X. Lin, H. Lu, Y. Zhang (Eds.), Advanced Web Technologies and Applications. XXII, 936 pages. 2004.

Vol. 3006: M. Matsui, R. Zuccherato (Eds.), Selected Areas in Cryptography. XI, 361 pages. 2004.

Vol. 3005: G.R. Raidl, S. Cagnoni, J. Branke, D.W. Corne, R. Drechsler, Y. Jin, C.G. Johnson, P. Machado, E. Marchiori, F. Rothlauf, G.D. Smith, G. Squillero (Eds.), Applications of Evolutionary Computing. XVII, 562 pages. 2004.

Vol. 3004: J. Gottlieb, G.R. Raidl (Eds.), Evolutionary Computation in Combinatorial Optimization. X, 241 pages. 2004.

Vol. 3003: M. Keijzer, U.-M. O'Reilly, S.M. Lucas, E. Costa, T. Soule (Eds.), Genetic Programming. XI, 410 pages. 2004.

Vol. 3002: D.L. Hicks (Ed.), Metainformatics. X, 213 pages. 2004.

Vol. 3001: A. Ferscha, F. Mattern (Eds.), Pervasive Computing. XVII, 358 pages. 2004.

Vol. 2999: E.A. Boiten, J. Derrick, G. Smith (Eds.), Integrated Formal Methods. XI, 541 pages. 2004.

Vol. 2998: Y. Kameyama, P.J. Stuckey (Eds.), Functional and Logic Programming. X, 307 pages. 2004.

Vol. 2997: S. McDonald, J. Tait (Eds.), Advances in Information Retrieval. XIII, 427 pages. 2004.

Vol. 2996: V. Diekert, M. Habib (Eds.), STACS 2004. XVI, 658 pages. 2004.

Vol. 2995: C. Jensen, S. Poslad, T. Dimitrakos (Eds.), Trust Management. XIII, 377 pages. 2004.

Vol. 2994: E. Rahm (Ed.), Data Integration in the Life Sciences. X, 221 pages. 2004. (Subseries LNBI).

Vol. 2993: R. Alur, G.J. Pappas (Eds.), Hybrid Systems: Computation and Control. XII, 674 pages. 2004.

Vol. 2992: E. Bertino, S. Christodoulakis, D. Plexousakis, V. Christophides, M. Koubarakis, K. Böhm, E. Ferrari (Eds.), Advances in Database Technology - EDBT 2004. XVIII, 877 pages. 2004.

Vol. 2991: R. Alt, A. Frommer, R.B. Kearfott, W. Luther (Eds.), Numerical Software with Result Verification. X, 315 pages. 2004.

Vol. 2990: J. Leite, A. Omicini, L. Sterling, P. Torroni (Eds.), Declarative Agent Languages and Techniques. XII, 281 pages. 2004. (Subseries LNAI).

Vol. 2989: S. Graf, L. Mounier (Eds.), Model Checking Software. X, 309 pages. 2004.

Vol. 2988: K. Jensen, A. Podelski (Eds.), Tools and Algorithms for the Construction and Analysis of Systems. XIV, 608 pages. 2004.

Vol. 2987: I. Walukiewicz (Ed.), Foundations of Software Science and Computation Structures. XIII, 529 pages. 2004.

Vol. 2986: D. Schmidt (Ed.), Programming Languages and Systems. XII, 417 pages. 2004.

Vol. 2985: E. Duesterwald (Ed.), Compiler Construction. X, 313 pages. 2004.

Vol. 2984: M. Wermelinger, T. Margaria-Steffen (Eds.), Fundamental Approaches to Software Engineering. XII, 389 pages. 2004.

Vol. 2983: S. Istrail, M.S. Waterman, A. Clark (Eds.), Computational Methods for SNPs and Haplotype Inference. IX, 153 pages. 2004. (Subseries LNBI).

Vol. 2982: N. Wakamiya, M. Solarski, J. Sterbenz (Eds.), Active Networks. XI, 308 pages. 2004.

Vol. 2981: C. Müller-Schloer, T. Ungerer, B. Bauer (Eds.), Organic and Pervasive Computing – ARCS 2004. XI, 339 pages. 2004.

Vol. 2980: A. Blackwell, K. Marriott, A. Shimojima (Eds.), Diagrammatic Representation and Inference. XV, 448 pages. 2004. (Subseries LNAI).

Vol. 2979: I. Stoica, Stateless Core: A Scalable Approach for Quality of Service in the Internet. XVI, 219 pages. 2004.

Vol. 2978: R. Groz, R.M. Hierons (Eds.), Testing of Communicating Systems. XII, 225 pages. 2004.

Vol. 2977: G. Di Marzo Serugendo, A. Karageorgos, O.F. Rana, F. Zambonelli (Eds.), Engineering Self-Organising Systems. X, 299 pages. 2004. (Subseries LNAI).

Vol. 2976: M. Farach-Colton (Ed.), LATIN 2004: Theoretical Informatics. XV, 626 pages. 2004.

Vol. 2973: Y. Lee, J. Li, K.-Y. Whang, D. Lee (Eds.), Database Systems for Advanced Applications. XXIV, 925 pages. 2004.

Vol. 2972: R. Monroy, G. Arroyo-Figueroa, L.E. Sucar, H. Sossa (Eds.), MICAI 2004: Advances in Artificial Intelligence. XVII, 923 pages. 2004. (Subseries LNAI).

Vol. 2971: J.I. Lim, D.H. Lee (Eds.), Information Security and Cryptology -ICISC 2003. XI, 458 pages. 2004.